New Windows to the Universe presents a timely and stimulating exposition of six major areas in astrophysics: the sun and solar like stars, stellar structure and evolution, structure and evolution of galaxies, active galaxies, active galaxies and cosmology, interstellar and intergalactic medium, astronomical instrumentation. Over three hundred scientists from all over the world met in Tenerife to give their perspective on their area of expertise.

New windows to the universe

XIth European Regional Astronomy Meeting of the International Astronomical Union

Vol I

This book is dedicated to the Instituto de Astrofísica de Canarias

New windows to the universe

XIth European Regional Astronomy Meeting of the International Astronomical Union
Vol I

Edited by

F. SANCHEZ

Instituto de Astrofisica de Canarias, La Laguna, Tenerife

M. VAZQUEZ

Instituto de Astrofisica de Canarias, La Laguna, Tenerife

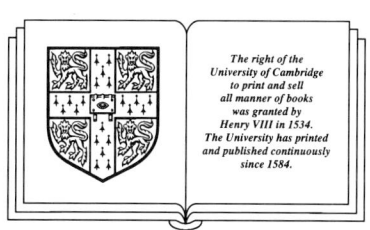

CAMBRIDGE UNIVERSITY PRESS

Cambridge

New York   Port Chester

Melbourne   Sydney

Published by the Press Syndicate of the University of Cambridge
The Pitt Building, Trumpington Street, Cambridge CB2 1RP
32 East 57th Street, New York, NY 10022, USA
10 Stamford Road, Oakleigh, Melbourne 3166, Australia

© Cambridge University Press 1990

First published 1990

Printed in Great Britain at the University Press, Cambridge

British Library cataloguing in publication data available

Library of Congress cataloguing in publication data available

ISBN 0 521 38429 X hardback

CONTENTS

Preface

The Sun and solar-like stars

Scientific Organiser: Prof. E. H. Schröter
Local Organizer: Dr. J. I. Garcia de la Rosa

Variability of the quiet photosphere of the Sun in the context of the activity cycle R. Muller	1
Superficial activity as a trace of the internal mechanisms of the solar cycle J. I. Garcia de la Rosa	27
Chromospheric phenomena in late-type stars P. Ulmschneider	45
Solar radio corona M. Pick and G. Trottet	65
The solar wind and the winds from cool stars R. Hammer	77
Eruptive phenomena on the Sun Z. Svetska	99
Radiative transfer problems in the Solar and Sun-like atmospheres J. Trujillo Bueno	119
Evolution of solar and stellar rotation S. Catalano	161
Spectroscopy of solar-type stars P. E. Nissen	179
Solar analogs seen at high spectral resolution and very high S/N ratios G. Cayrel de Strobel	195
Surface structures and flares in solar-like stars B. H. Foing, S. Char, S. Jankov and E. Houdebine	213
Quiescent X-ray emission of cool stars C. J. Schrijver	233
Duplicity of solar-like stars in the solar neighbourhood A. Duquennoy and M. Mayor	253

Stellar structure and evolution

Scientific Organizer: Prof. E. Schatzman
Local Organizer: Dr. R. Rebolo

Activity of young low-mass stars C. Bertout	269
Turbulent shear flow and rotation J. P. Zahn	291
Surface abundances of light elements in stars R. Rebolo	301

On the ages of galactic globular clusters F. Fusi Pecci and C. Cacciari	335
Cosmochronology: an introductory overview M. Arnould and K. Takahashi	355
Rotation, age and lithium E. Schatzman	375
Asteroseismology as a probe of stellar structure and evolution T. Roca Cortes and J. A. Belmonte	377
White dwarfs J. Isern, E. Garcia-Berro, M. Hernanz and R. Mochkovitch	391
The supernova 1987A W. Hillebrandt and P. Höflich	401
Supernova statistics H. E. Jorgensen	423

Astronomical instrumentation

Scientific Organizers: Prof. B. Fort and J. P. Picat
Local Organizer: Dr. P. Alvarez

ESO site evaluation for the VLT M. Sarazin	435
The LEST project O. Engvold	451
The instrumentation plan for the VLT of the ESO S. D'Odorico	463
CCDs for the 1990s P. Jorden	465
New trends in ground-based astronomy: Fast real time processing needed J. P. Picat	483
The ISO instrumentation and expected performances Th de Graauw	497
Interferometry with large optical telescopes P. Lena	507
New developments, concepts and dreams: Summary talk of the Instrumentation Meeting I. Appenzeller	515

General Lectures

Ground-based European Astronomical projects A. Boksenberg	521
ESA Astronomical projects M. C. E. Huber	535
Prospects of the development of ground-based anol space astronomy in the USSR A. A. Boyarchuk	551

PREFACE

These proceedings contain the invited review papers and general lectures delivered at the "XI European Regional Astronomy Meeting" of the IAU, which was hosted by the Instituto de Astrofisica de Canarias, and held at the University of La Laguna from 3 to 9 July 1989. Shorter contributed papers are being published separately in a dedicated volume of Astrophysics and Space Science. The broad title "New Windows to the Universe" was chosen intentionally, not only because it may encompass any research which is being carried out in Europe, but because astronomical investigations are continuously opening new windows in our knowledge of the Universe and because the newly established observatories in the Canaries are precisely new windows to the Universe.

A large number of European astronomers already know the Canary Islands having been here on an observing run at the Teide Observatory (Tenerife) or at the Roque de Los Muchachos Observatory (La Palma), or through participation in other astronomical conferences and events. Many have also spent some time working at the Instituto de Astrofisica, which is also on the campus of La Laguna University, and others have met us during this meeting. At any event, we all hope that you have had the opportunity to see the three astronomical centres of the Canary Islands, three points of a triangle for the study of the Universe. They are interconnected and each is designed to support each of the others while constituting together a powerful international astronomical facility. The Canary Islands are already becoming the natural European Northern Observatory, as is proclaimed by the important group of European Telescopes already installed here, and by a law approved by the Spanish Parliament to protect the Canarian skies which makes us an espccially preserved "astronomical reserve".

Besides our acknowledgement to the International Astronomical Union and the European Physics Society, we also express our gratitude for the cooperation of the University of La Laguna, which made available the rooms of the Faculties of Economics and Law; the Spanish "Comision Interministerial de Ciencia y Tecnologia" for providing the necessary funds for invited speakers; the town councils of Puerto de La Cruz, La Orotava and La Laguna, and the Cabildos (Island representative assemblies) for their generous contribution to the social events.

The vast majority of IAC staff were involved in some way or another in the preparation of the meeting. They are too numerous to mention here but their enthusiasm is gratefully acknowledged. The local organizing committee members, who devoted many hours to ensure the success of the event, deserve a special mention. Most of the work related to the organization of the meeting was carried out by Monica Murphy and Judith de Araoz. The Editors wish to congratulate them for their dedication.

The IAC was honoured to be appointed to organize the XI European Regional Astronomy Meeting, which was met as a challenge. All those individuals who made it possible would feel satisfied if the aims of this particular step of European Astronomy were achieved.

F. Sanchez
M. Vazquez
1990

NB. All the manuscripts contained here were supplied in camera-ready form by the authors, who are responsible for the layout and style.

THE SUN

AND

SOLAR-LIKE STARS

Scientific Organizer: E. H. Schröter
Local Organizer: J. I. García de la Rosa

VARIABILITY OF THE QUIET ATMOSPHERE OF THE SUN

IN THE CONTEXT OF THE ACTIVITY CYCLE

R. MULLER
Observatoire du Pic du Midi
65200 Bagnères de Bigorre - France

ABSTRACT

During the last few years, it has been found that the structure of the solar photosphere, outside active regions, varies over the 11-year activity cycle. The structure of the photosphere and of the convection zone varies as indicated by the observed variation of the granulation size, the shape of photospheric lines as well as the solar radius, luminosity and effective temperature ; however there is not a good agreement between these various results as yet. Outside active regions, the variation of the magnetic flux, which is an important parameter necessary to understand the solar variability, is not well known yet : while no variation is revealed by magnetograph observations, a variation in antiphase with the sunspot number is suggested by indirect indicators like NBPs, XBPs, spicules. The differential rotation as well as the meridional circulation are variable with a period of 11 years. All those variable phenomena have to be taken into account to understand the origin of the solar activity.

1 INTRODUCTION

The sun is an active star varying with a period of 11.2 years. One of the major problem of solar physics is to understand this cycle of activity. The solar active cycle has been studied, mainly through the properties and the evolution of active centers and of magnetic field related features outside active regions. The 11-year variation of the sunspot number, the equatorward migration of the latitude of appearance of the active centers, the poleward migration of magnetic flux and the polarity reversal are of particular importance.

However, there is more and more evidence that the properties of the quiet atmosphere outside active regions are also variable with an 11-year period. These variations have to be taken into account (and this is often the case now) to understand the interaction between the convection and the magnetic field of the sun, which is at the origin of the solar (and stellar) variability. This paper will be restricted to the photosphere.

It is divided into three sections describing respectively the variation of the structure of the photosphere (including small scale and global scale), the variation of the magnetic flux outside active regions, and the variation of the differential rotation and meridional circulation.

2 VARIATION OF THE PHOTOSPHERIC STRUCTURES

2.1 Variation of the profile of photospheric lines
A photospheric line can be characterised by the equivalent width (EW), the central depth (D) and the asymmetry (A). EW, D and A have been found to vary over the solar cycle.

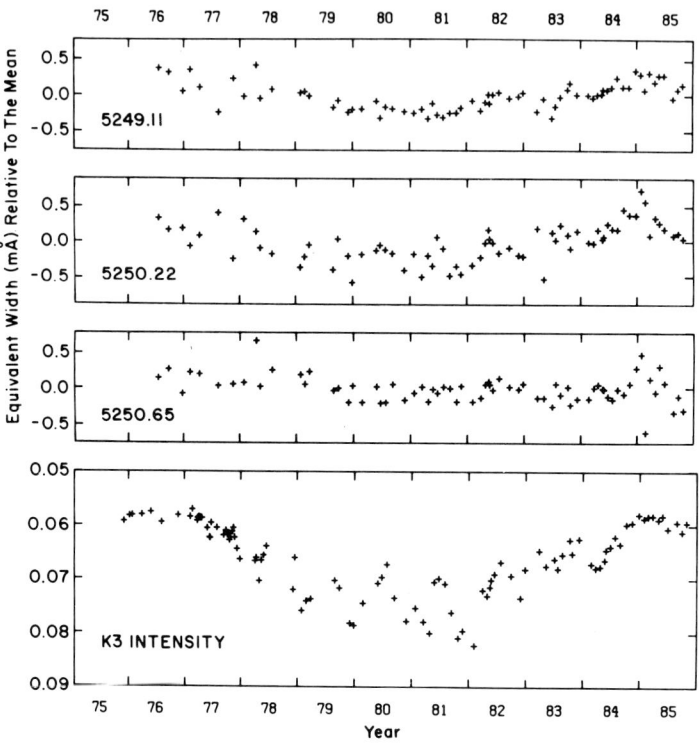

Figure 1. Variation of the Equivalent Width of three photospheric lines in the 5250 Å region and of the K3 Intensity over the solar cycle (Livingston and Wallace, 1987, Figure 4).

2.1.1 Equivalent Width and Central Depth.

Livingston and Wallace (1987) have measured the variation of both parameters, integrated over the visible solar hemisphere (i-e. including quiet sun and active regions), from 1976 to 1985 for a number of photospheric lines. They have found (Figure 1) that they vary like the K3-Index (the intensity of the absorption core of the line CaIIK 3933, see Figure 13), i.e. in phase with the activity cycle. However, from the comparison of the lines FeI 5250 and MnI 5394, Livingston and Wallace come to the conclusion that the source of variability is not confined to solar plage regions alone. The variation of the central depth of CI 5380 indicates a decrease in Teff of 0.7 K in the period 1979-1985, which is consistent with the solar constant variation derived from the solar irradiance measurements by ACRIM on board of SMM (Figure 2). However other photospheric lines do not show a variation above the noise level.

2.1.2 Asymmetry

Because of the penetration of convective motions into the photosphere, photospheric lines are not symmetric, the line bisector showing a characteristic C-shape. A change of asymmetry can be interpreted by a change of the convection in the upper solar atmosphere. A variation of the asymmetry has been reported by several authors. Livingston (1984) has observed a decrease of the curvature of the bisector as the activity cycle proceeds from minimum (1976) to maximum (1979-1982). Bruning and Labonte (1985) have found that the position of the mean line bisector (which is related to the shape of the line bisector) has not changed throughout the period from May 1982 to February 1983, while the magnetic flux has dropped during the same period. On the other hand accurate line asymmetry measurements by Cavallini, Ceppatelli and Righini (1986) show a decrease of bisector curvature during the period 1983-1984. Putting all these results together, it appears that the line asymmetry was continuously decreasing from 1976 to 1984, which indicates a reduce of the granule convection, according to Livingston (1984). This regular decrease does not follow the activity cycle as represented by the sunspot number which was very low in 1976, very high in 1979 and which has decreased between 1979 and 1984.

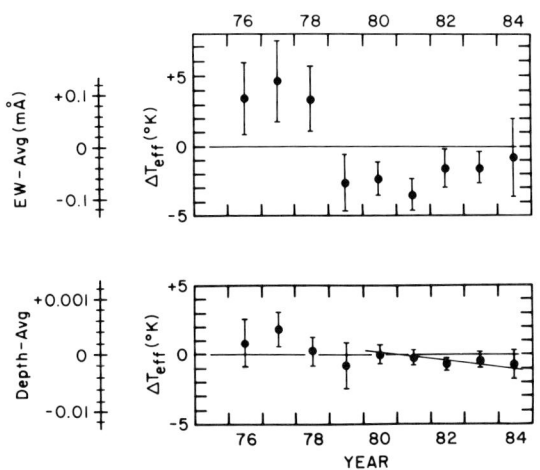

Figure 2. Variation in the annual mean equivalent width and central depth of CI 5380.3 interpreted as effective temperature. Error bars are 3σ Solid line is a linear least-squares fit to the ACRIM data.

2.2 Variation of solar granulation

The variation of the solar granulation is a direct indication of the variation of the convection in the upper convection zone ; it was first suggested by Macris in 1951. Several authors, all using the Pic du Midi collection of granulation photographs, reported a variation of the mean size of granules in phase with the solar cycle, the granules having been found smaller near the sunspot minimum. The Pic du Midi granulation collection starts in 1966. It is however not very homogeneous : from 1966 to 1971 the photographs were taken with a 38 cm refractor, then with a 50 cm one of very high optical quality ; the image quality has been improved in 1978, when some heating problems of the optical parts were solved. Thus, from 1978 on, a homogeneous set of photographs is available, closely approaching the resolution of the 50 cm refractor (0"25).

Macris and Rösch (1983), using an inhomogeneous set of photographs taken in the period 1966-1978, found a variation of the mean distance between granule centers as high as 20 % (Figure 3). Subsequently, Muller and Roudier reported a variation of the number of granules per surface unit (numerical density) of 20 %, corresponding to a 10 % variation of the mean size of granules (Figure 4). They used a more homogeneous set of granulation photographs than Macris and Rösch did. These results are published in Macris et al. (1984) and Roudier (1986).

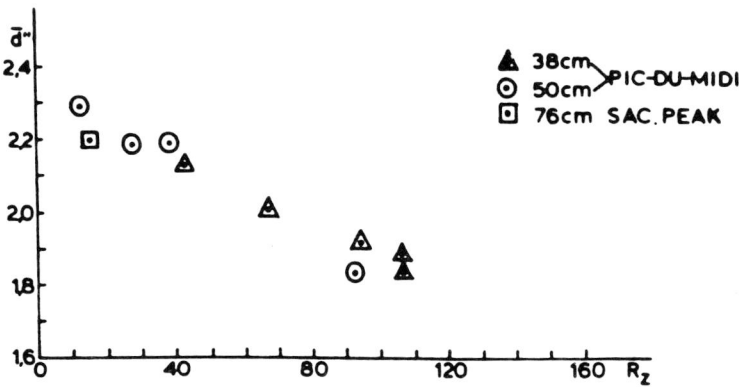

Figure 3. Mean distance between granule centers d", as a function of the Wolff number, R_z (Macris et al. 1984, Figure 4).

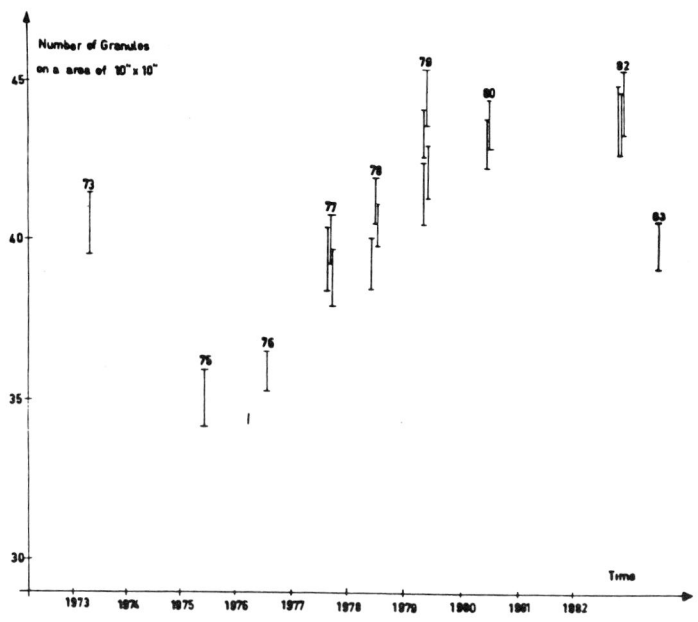

Figure 4. Number of granules per 10" x 10" area unit, over the duration of one solar cycle (Macris et al., 1984, Figure 7).

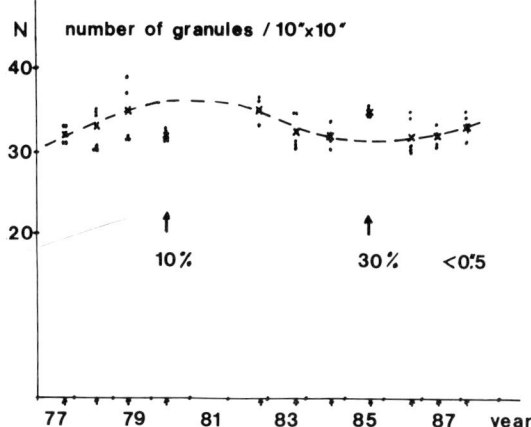

Figure 5. Variation of the number of granules (per surface unit of 10" x 10") at the center of the solar disc. Only the granules larger than 0"5 are taken into account. Each point represents a count within an area of 25" x 25" ; each cross the average value over 6 such areas. The arrows correspond to the granulation photographs of poorest and highest resolutions, where 10 % and 30 % of granules smaller than 0"5 were respectivily identified. (Muller, 1988, Figure 3).

A new measurement was made by Muller (1988b), using the best 1978-1988 homogeneous set of photographs. Only considering the granules larger than 0"5, he was able to confirm the variation of the number density of granules over the solar cycle; however he found a smaller amplitude of the variation : 10 % of the number density, 5 % of the mean size (Figure 5). This variation was confirmed by Jain (1989) who repeated independently the measurements made by Muller.

Thus several authors, using different sets of granulation photographs, have found a variation of the mean size of granules, in phase with the sunspot number. This gives us some confidence in the result, despite the fact that the visual indentification of granules is subjective and can influence the result.

Of course an objective analysis is desirable. For this purpose, autocorrelation functions of the granulation have been computed. Surprisingly they do not show any secondary maximum which we expected to be a measure of the mean intergranular distance. There is probably no secondary maximum because, when the resolution of the frames is increased, the granule fine structures are enhanced and the number of small granules is increased, both of which contribute to decrease the degree of correlation as well as the non-uniform distribution of granules at the surface of the sun (Muller, 1988a, 1988b, Muller, Roudier and Vigneau, 1989).

2.3 Variation of the 5 min oscillations

The possibility of observable changes in the frequencies of solar p-mode oscillations has been investigated by several authors. Woodard and Noyes (1985), Woodard (1987), Fossat et al. (1987), and Gelly, Fossat and Grec (1988) have analysed the SMM/ACRIM solar total irradiance and reported a slight but significant decrease (about 1 part in 10^4) of low spherical-harmonic degree near sunspot minimum (1984-1985) compared to solar maximum (1980). Analysing, in addition, South Pole data, Fossat and his co-workers

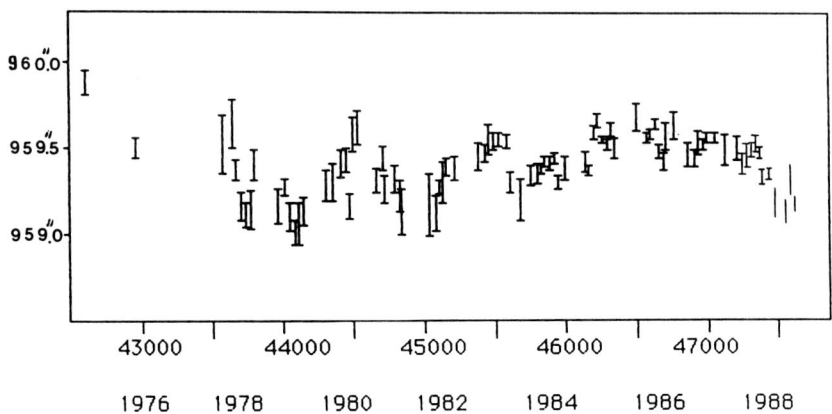

Figure 6. Semi-diameter measurements of the sun with the CERGA solar astrolabe (Laclare, Journet and Merlin, 1989).

have found that the oscillation frequency was stable between 1984 and 1986. However, some other authors have not found any clear variation in the p-mode frequencies (Pallé et al., 1986, from Doppler data obtained at Izana from 1977 through to 1984 ; Rhodes et al., 1988, from the comparison of narrow-band intensity data from the South Pole, 1981, and of Dopplergrams from Mt Wilson, 1985). More surprisingly, Isaak et al., (1987), employing some of the data analysed by Pallé et al. have found an increase between 1980 and 1984. The results thus seem to be strongly influenced by the technique of analysis.

This brief review clearly shows that we do not know yet for sure whether the frequencies of the polar p-mode oscillations vary over the solar cycle or not. More observations and carefull analysis are required.

2.4 Variation of the solar diameter and oblateness

2.4.1 Diameter
Analysing several independant sets of solar diameter measurements with meridian circles, Gilliland (1981) found a significant correlation with the sunspot cycle ; the half amplitude of the variation is 0"1.

Measurements of the solar diameter made with an astrolabe (Laclare, 1983 ; Delache et al., 1985 ; Laclare, Journet, Merlin, 1989) also clearly show a variation in phase with the solar cycle (Figure 6) : the diameter of the sun is 0"10 - 0"15 smaller at sunspot maximum. Such a decrease is not apparent in other measurements (Brown, 1987).

2.4.2 Oblateness
Dicke, Kuhn and Libbrecht (1986, 1987) have found different values for the solar oblateness measured in 1966, 1983, 1984, 1985 ; including the value obtained for 1973 by Hill and Stebbins (1975), they conclude that the oblateness amplitude could vary with a 11-year period (Figure 7), (which, according to the authors, might be due to a torsional oscillation of a rotating magnetic core).

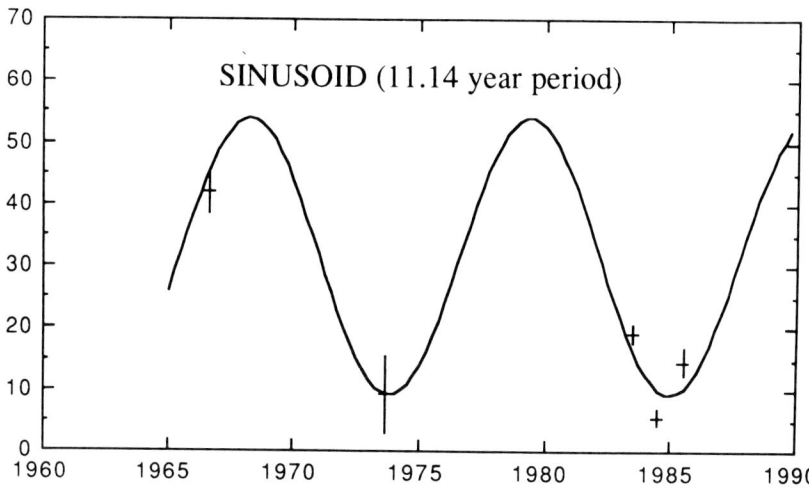

Figure 7. Weighted least-squares fit of simusoïdal curve of solar cycle period to the four plotted solar oblateness points of 1966, 1983, 1984 and 1985. The ordinate is the solar oblateness in arc milliseconds (Dicke, Kuhn and Libbrecht, 1987, Figure 1).

Figure 8. The solar temperature distributions between 1983 and 1987, derived from limb profile measurements after faculae and sunspots removal. The range from light to dark corresponds to a fractional intensity variation of about 2×10^{-3}. Starting in the upper left and moving down and then to the right column, the images correspond to the sun in the summers of 1983, 1984, 1985 and 1987. (Kuhn, Libbrecht and Dicke, 1988, Figure 2).

This interpretation has been criticized by Chapman and co-workers (i.e. Chapman and Klabunde, 1982) who argue that the oblateness signal can be modulated by the limb faculae, which are known to vary in phase with the activity cycle.

2.5 Variation of the limb profile

Spectral line shape and granule size are indicators of the upper convection zone and photosphere vertical structure. The profile of the extrem limb is also sensitive to variations in density, temperature, radius, luminosity, convective efficiency ; in turn, limb profile variation can be interpreted in terms of Teff variations (Petro, Foukal and Kurucz, 1985). Limb profiles are obtained at Kitt Peak with the Mc Math telescope since 1980. However, no variations have been detected in the period 1980-1982 (Petro et al. 1984). The observing program is still going on.

Recently, Kuhn, Libbrecht and Dicke (1988), reported a very exciting result (Figure 8) : the limb profile is latitude dependent and is different in 1983, in 1984, in 1985 and in 1987. They attribute these limb observations to a latitude variation of Teff, which is active cycle dependent. The variation of solar irradiance revealed by ACRIM/SMM can be explained by these observations.

2.6 Variation of the luminosity

2.6.1 Definitions
The irradiance S, is the total energy received on the earth from the sun. The solar Luminosity L, is the energy radiated by the sun. The Photosphere Luminosity L_{ph}, is the brightness of the quiet photosphere outside active regions and magnetic features; L_{ph} is related to Teff.

$$S = -P_s + P_f + P_N \pm L_{ph}$$

- P_s is the sunspot deficit, P_f and P_N are the emission from faculae and the network, respectively.

Solar luminosity variations are derived from Irradiance measurements with radiometers on satellites : ERB on board of Nimbus 7, since 1978, ACRIM on board of SMM, since 1980.

The solar irradiance exhibits short (days) and long (months to years) time scale variations.

2.6.2 Short time-scale variations
The solar irradiance varies over time-scale of a few days, showing dips of 0.1 - 0.2 %, associated with large active centers (great deficit for large sunspots) crossing the solar disk (Hickey et al., 1980 ; Willson et al., 1981). Sunspots can explain 70 % of S ; the contribution of faculae and network, has not been accurately evaluated yet, because of their low contrast and poorly defined areas. It is thus not possible as yet to derive a variation of the brightness of the photosphere outside active regions ; direct measurements have not been performed.

An important question is behind these measurements ; namely : is the energy flux blocked by sunspots re-radiated almost immediately over a time span comparable to an active region lifetime or less, in faculae, network or the quiet photosphere ? Or over a longer period such as the 11-year activity cycle (luminosity modulated by the sunspot cycle) ?

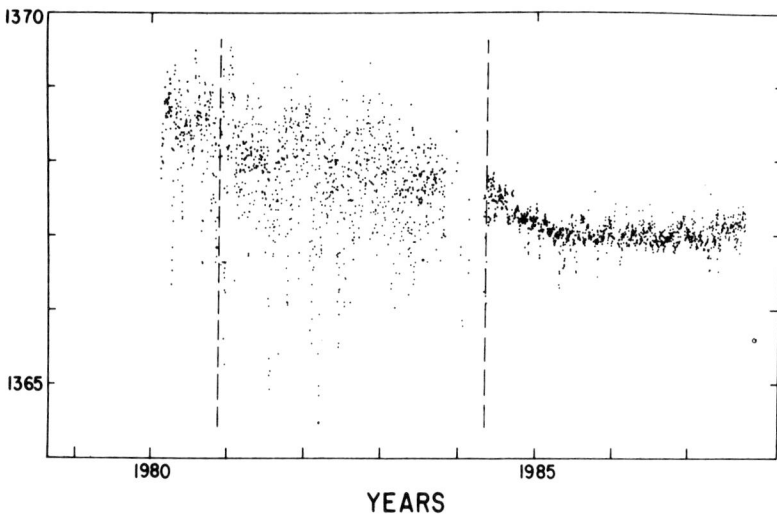

Figure 9. Time series from the ACRIM instrument on board the Solar Maximum Mission, showing the daily measures of total solar irradiance (the "solar constant") for the period 1980-86. From 1981 to 1984, in the interval between the vertical lines, the satellite had lost its fine-pointing capability. The decrease of scatter in 1984 is partly due to the higher sampling frequency that returned when the satellite was repaired. Thes "dips" due to large sunspot groups and the longer term trends (Willson et al. 1986) are both visible in the plot. (Hudson, 1988, Figure 2).

2.6.3 Long time-scale variations

The solar total irradiance measured by ERB and ACRIM shows a long term variation related to the solar cycle (Figure 9, from Hudson, 1988 ; Figure 10, from Foukal and Lean, 1988). The general *downtrend* observed in both independant measurements strongly suggests a decrease of the total solar irradiance with declining magnetic activity. A minimum value was reached when the sunspot number was also minimum.

Foukal and Lean (1988) analysed the residual irradiance $S-P_s$, which is the contribution of the faculae, the network and the non-magnetic photosphere, to the total irradiance variation. They found that $S-P_s$ decreases slowly (Figure 11, from Foukal and Lean, 1988). This slow decrease cannot be reproduced when only the contribution of faculae in active regions is taken into account : consequently, Foukal and Lean propose that the observed decrease of luminosity comes from the decrease of the network contribution during the descending phase of the activity cycle.

But there is another plausible explanation. In section III, we have seen, on the one hand that the number of NBPs, which form the network *increases* with decreasing activity (both near the equator and near the poles), and on the other hand that the calcium index remains constant.

This means that the contribution from the photospheric network is increasing or constant, but in no way decreasing. Then the S-Ps decrease can be explained by a decrease of the brightness of the non-magnetic atmosphere, that is to say of its effective temperature Teff. It can be noted that this is in agreement with the Teff decrease derived by Livingston and Wallace (1987), from observed line strength variation.

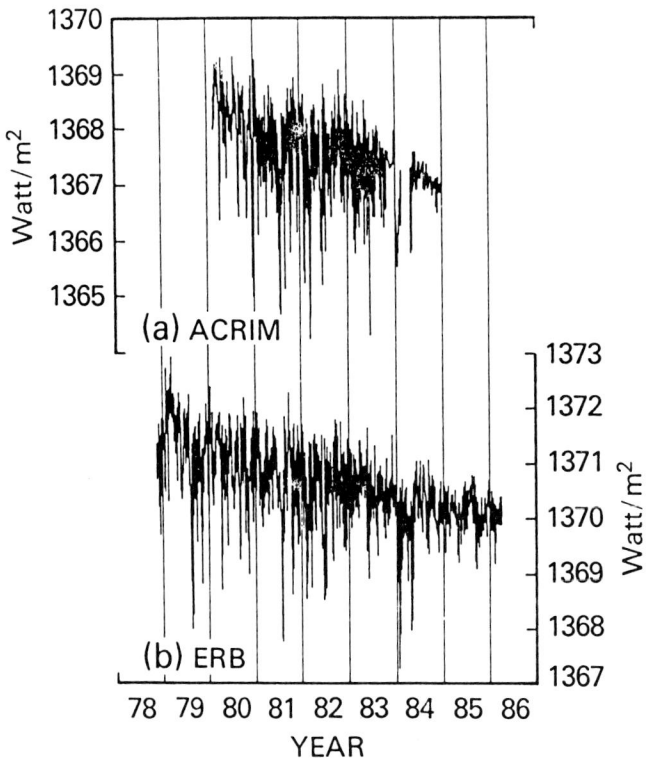

Figure 10. Daily irradiance measurements (a) for 1980-1984 from the ACRIM and (b) for 1978-1986 from the ERB. (Foukal and Lean, 1988, Figure 1).

2.6.4 Possible reasons for luminosity (and diameter) variations

The most attractive explanation is a modified efficiency of convection by the presence of many embedded magnetic flux tubes in the convection zone (Peckover and Weiss, 1978). Various mechanisms of interaction between convective and magnetic flux have been investigated. At the base of the convection zone, where the magnetic flux is believed to be generated, the superadiabatic temperature gradient is perturbed ; the transport rate of the energy flux is changed, resulting in a change of solar luminosity and radius (Spiegel and Weiss, 1980 ; Gilliland, 1982). The convective flux is inhibited in magnetic regions ; the blocked energy is diffused into the photosphere, modifying the temperature structure, the convective flux and the luminosity (Foukal, 1981). The observed variations can also be simply explained by magnetic buoyancy (Jensen, 1955 ; Thomas, 1979 ; Deaborn and Blake, 1982) : because the magnetic flux tubes are relatively empty, a variation of magnetic flux provokes a variation of volume and luminosity.

Luminosity variation may also be explained by temporal changes of the differential rotation in the interior of the sun, or by the presence of large-scale convective cells.

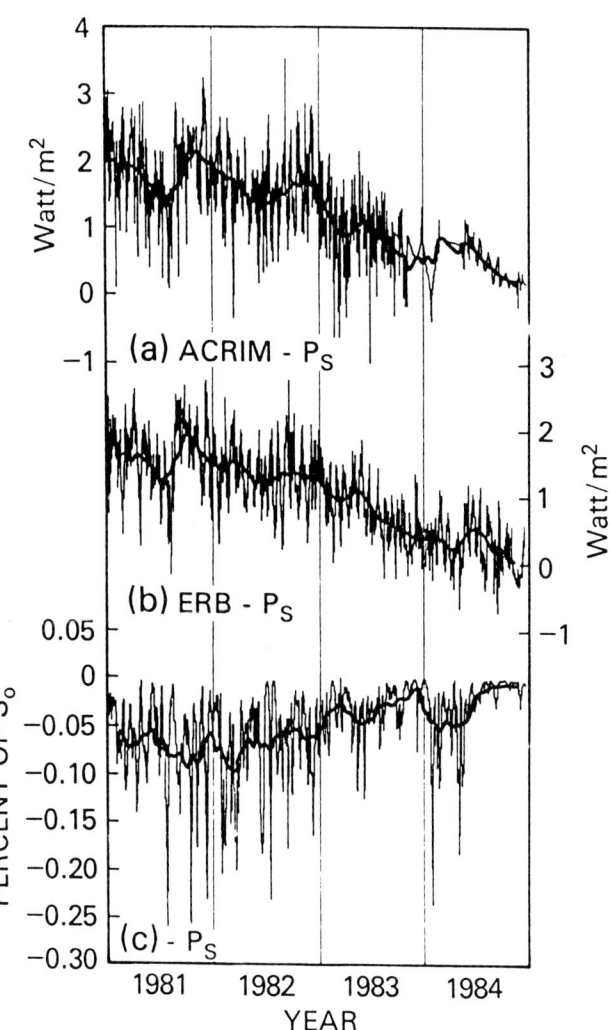

Figure 11. The residuals S-P_s for 1981-1984 obtained after substracting the sunspot blocking function, $-P_s$ from the ACRIM and ERB data are shown in (a) and (b). The function $-P_s$ is shown in (c). Smooth curves are 81 day running means (Foukal and Lean, 1988, Figure 4).

3 VARIATION OF THE MAGNETIC FLUX OUTSIDE ACTIVE REGIONS

The observed variability of solar features is the result of the interaction between the convection and a variable magnetic field. The 11-year variation of the magnetic flux which appears in active regions is well known. It is also important to know whether there is a magnetic flux variation outside active regions, in the quiet atmosphere. This field can be investigated either by direct magnetograph measurements or indirectly by some magnetic indicators.

3.1 Direct magnetograph measurements
Using the Mt Wilson magnetograph, La Bonte and Howard (1982) did not find any variation of the magnetic flux outside active regions (Figure 12a).

These results deserve a comment. The total magnetic flux of the sun, including active and quiet regions, has been found to vary by a factor no larger than 2 or 3 (Howard and La Bonte, 1981 ; Figure 12b). But the magnetic flux of active regions alone varies by a factor as large as 10 or so during the solar cycle, which implies that the magnetic flux outside active region should vary by a factor 3 to 5 in antiphase to the sunspot number. Some inconsistency thus appears between global and quiet sun magnetic flux measurements, which is not yet understood.

3.2 Magnetic Field Indicators
3.2.1 Calcium Index
The emission in the CaIIK3933 line is closely connected to the photospheric magnetic field in plages and in the network. The most widely used indexes are the K3 index (intensity of the line center) and K-1Å average intensity of the 1A window (Figure 13).

Both the K3 and K-1Å, integrated over the sun, vary in phase with the activity cycle (White and Livingston, 1981 ; Keil and Worden, 1984 ; Sivaraman et al., 1987). K3 increased by 30 % on average, from 1976 to 1980 ; the K-1Å increased by 18 % according to White and Livingston (Figure 14).

Concerning the quiet sun, no variation was detected by these authors (Figure 15) at the center of the disk. On the other hand, Skumanich et al. (1984) found it necessary to introduce an additional "active network" away from the equator, in order to explain the variation they have observed. This active network could be made with the remnant of plages, dispersed by supergranular and meridional motions. It can also be noted that Raghavan (1983) reported a decrease of the area covered by the quiet network when the solar activity increases. When these parameters are converted into magnetic fluxes, the results appear to be contradictory.

3.2.2 Variation of the number of NBPs
The Network Bright Points (NBPs) are tiny bright points which form the photospheric network. Their size is smaller than 0"5 and their lifetime is 20 min in average (ranging from 5 to 50 min, Muller, 1983) ; their brightness, as observed with a 50cm refractor under perfect seeing conditions, is 1.08 times the mean photosphere ; the true brightness may be as high as 1.3 - 1.5 (Muller and Keil, 1983). The important point is that they are associated to magnetic fields as strong as 1 to 2 kG.

It has been found (Figure 16) that the number of NBPs at the center of the disk, varies by a factor 5, in antiphase with the sunspot number (Muller and Roudier, 1984 ; Muller, 1988b). If we consider that the characteristic size of an NBP is 150 km and the magnetic field strength 1 500 G, each NBP carries a quantum of magnetic flux of 2.5×10^{17} Mx.

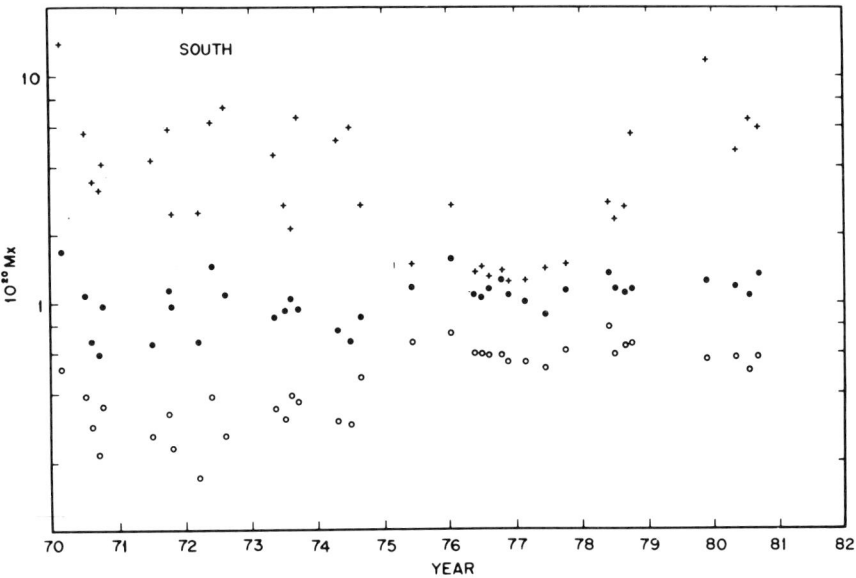

Figure 12a.

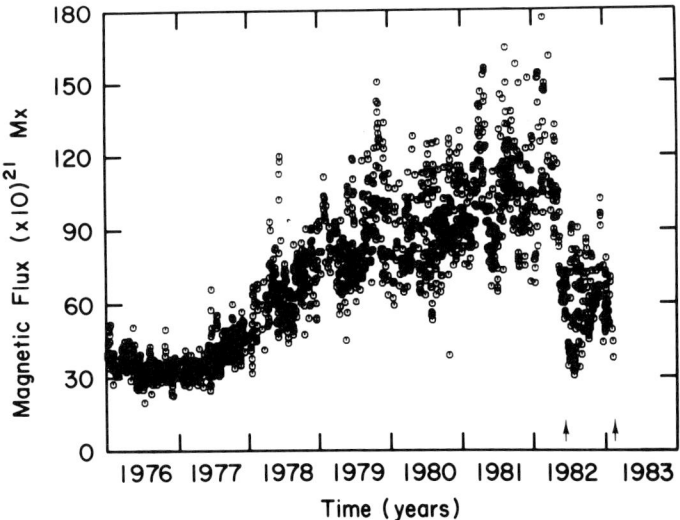

Figure 12b.

Figure 12a. Variations of total magnetic fluxes (sum of positive and negative fluxes in the longitudinal segment ±6°8) of the quiet sun in the northern hemisphere. Values for latitude zones centered on 1°7, 15°3, and 58°5 are plotted as filled circles, crosses and open circles respectively. Latitude zones have equal widths of 1/17 in sine latitude (Labonte and Howard, 1982, Figure 2).

Figure 12b. The total magnetic flux of the sun in Maxwells, including active regions and the quiet sun, as a function of time (Bruning and Labonte, 1985, Figure 4).

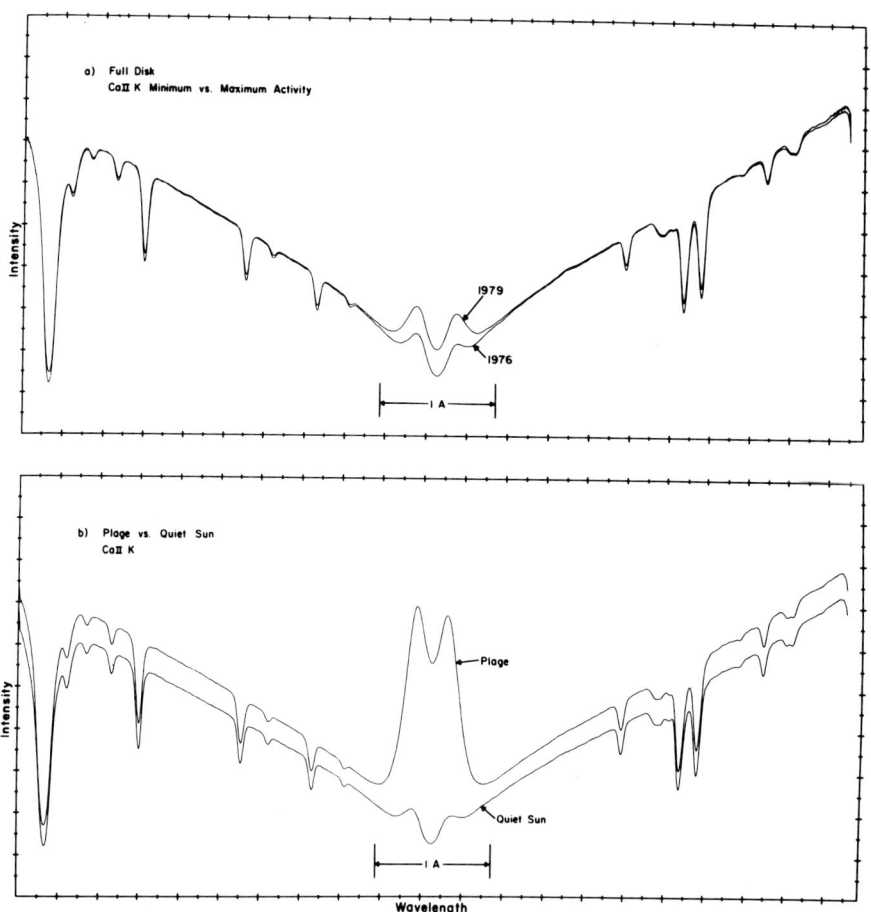

Figure 13. The variability of CaIIK line profiles : (a) full disk K line profiles at minimum and maximum, and (b) average profiles for an active region and a nearly quiet region at the same distance from the center of the solar disk (White and Livingston, 1981 ; Figure 1).

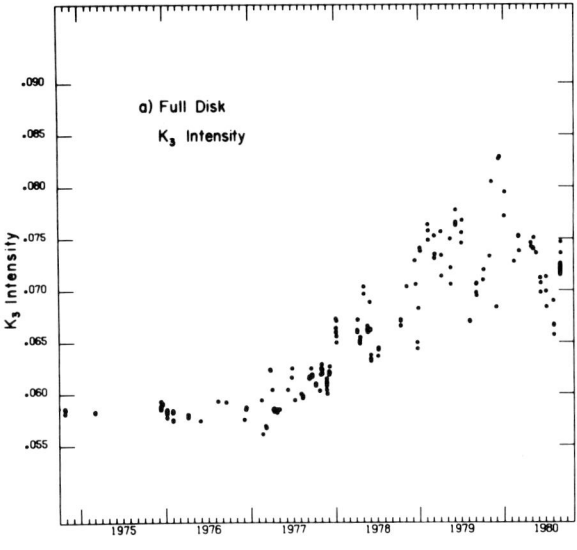

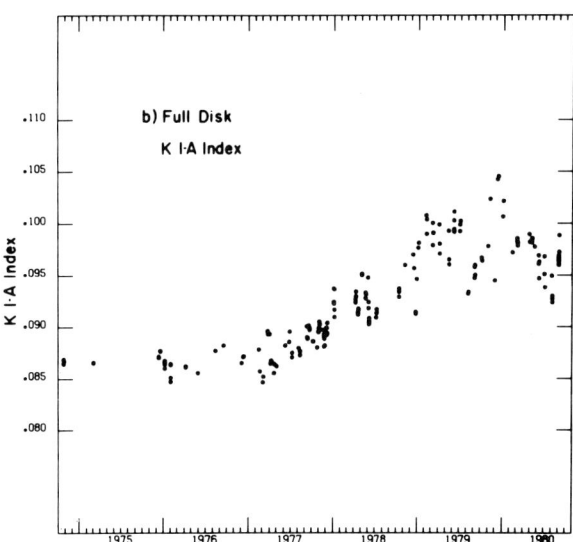

Figure 14. The cycle variation of line intensity parameters measured for the full solar disk : (a) variation of the central intensity of the K line, and (b) variation of the 1Å K index (White and Livingston, 1981 ; Figure 2).

16   The Sun and solar-like stars

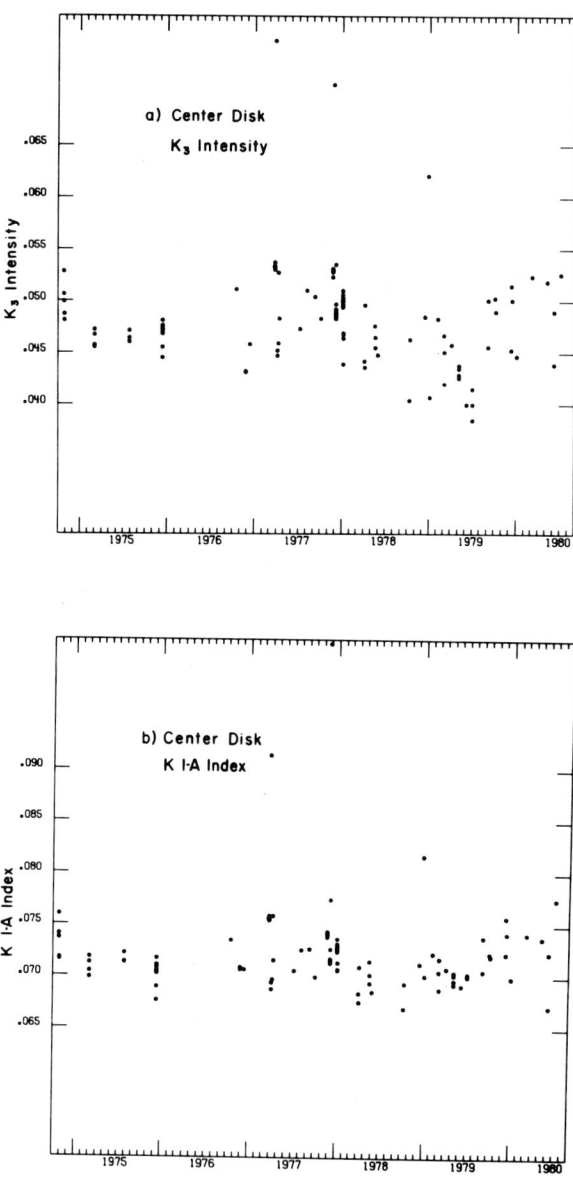

Figure 15. The cycle variation of line intensity parameters measured over a 1' x 3' area at the center of the solar disk : (a) variation of the central intensity (K3) of the K line, and (b) bariation of the K-1Å index. (White and Livingston, 1981 ; Figure 3).

The variation of the number of NBPs thus indicates the magnetic flux outside active regions is variable, at least near the disk center, the variation being in antiphase with the sunspot cycle.

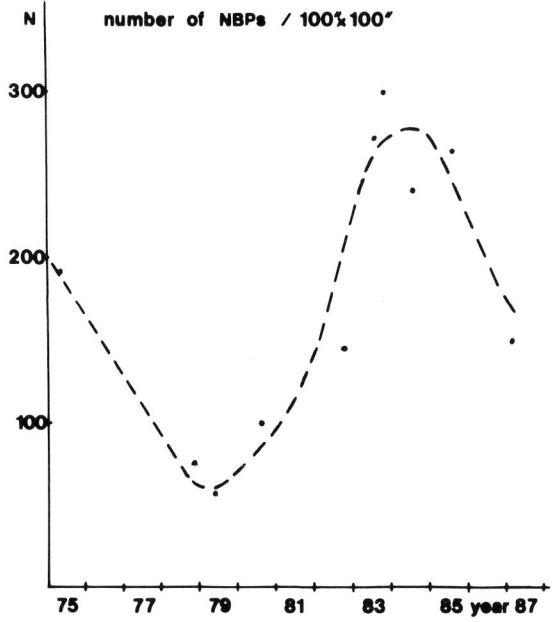

Figure 16. Variation of the number of Network Bright Points, per surface unit of 100" x 100". (Muller, 1988b, Figure 5).

3.2.3 Variation of the number of polar faculae

The number of polar faculae varies in antiphase with the sunspot cycle (Figure 17 from Sheeley, 1984) like the polar magnetic field strougth (Svalgaard et al., 1978), which is not surprising, owing to their magnetic origin. However a 1 to 2 year shift is observed compared to the sunspot cycle.

3.2.4 Variation of the number of spicules

Gulyaev (1975) reported a variation of the number of spicules in antiphase with the sunspot number. As it is believed that, in the quiet sun, spicules are related to individual flux tubes, this is another indirect indication that the magnetic flux outside active regions is variable.

3.2.5 Variation of the number of XBPs

X-Ray Bright Points (XBPs) are identified as small emission features in soft X-ray spectroheliograms (Vaiana et al., 1973). Their size is about 30", with a bright core not exceeding 5" - 10" ; their lifetime is less than 2 days. Temperatures as high as $1.3 - 1.7 \ 10^6 K$ and densities 2 to 4 times larger than the mean corona have been measured in XBPs (Golub et al. 1974). They are associated, at the photospheric level, with ephemeral active regions (which are considered as mini-active regions), that is with $10^{19} - 10^{20}$ Mx magnetic fluxes.

Davis (1983) found that the number of XBPs, integrated over the entire surface of the sun, varies in antiphase with the sunspot cycle (Figure 18), like NBPs and spicules. The amplitude of the variations is about 5.

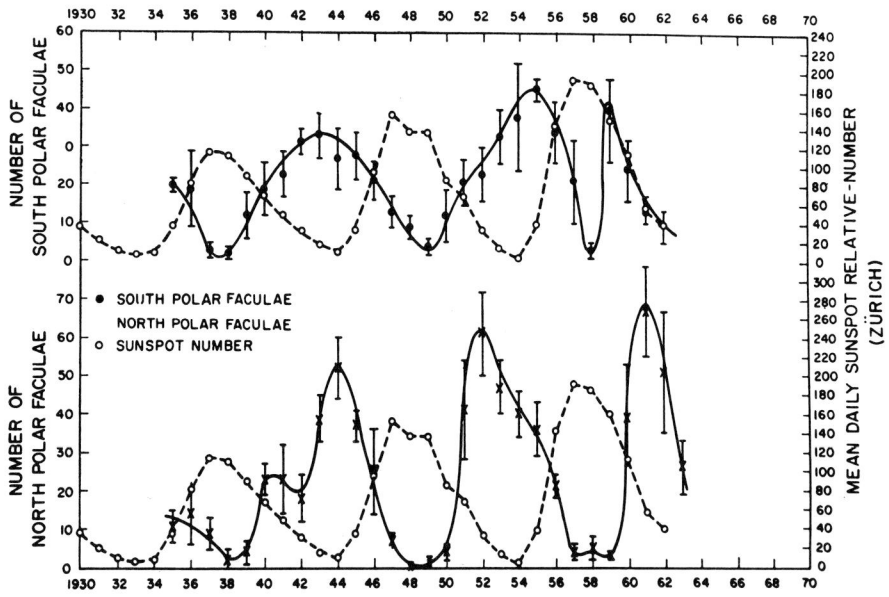

Figure 17. Comparison between the numbers of north and south polar faculae, respectively full lines, and the sunspot number for the whole solar disk (dashed lines). (Sheeley, 1964, Figure 2).

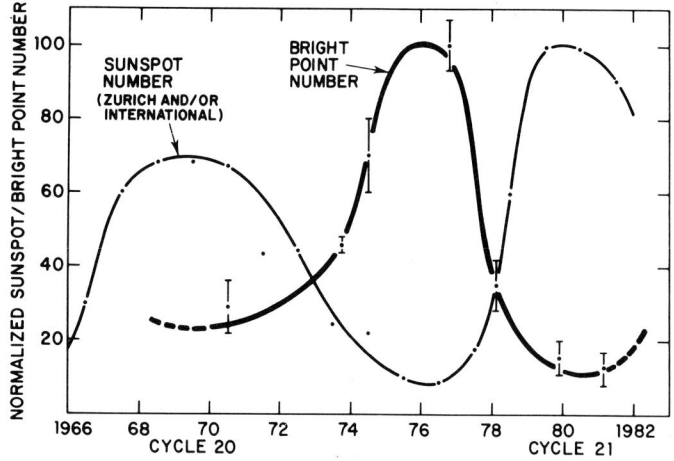

Figure 18. Variation of the number of X-Ray Bright Points over sunspot cycles 20 and 21 (Davis, 1983, Figure 2).

3.2.6 Variation of the magnetic flux outside active regions : conclusions

The variations of the magnetic flux presented above contradict each other. From K-Index and direct magnetographic measurements, no variation of the magnetic flux has been found outside active regions. On the contrary, a variation of the magnetic flux (which decreases with increasing activity) is suggested by the variation of the number of various magnetic features (NBPs, XBPs, spicules, polar faculae).In addition, the variation of the total magnetic flux of the sun requires, in order to be understood, a decreasing magnetic flux outside active regions, with increasing activity, while the variation of the integrated K-Index seems to require an increase of flux in active latitudes. The reason of these discrepencies is not yet understood. The variation of magnetic flux and magnetic features deserves to be measured for many more years.

4 DIFFERENTIAL ROTATION AND LARGE SCALE MOTIONS

The differential rotation $\Omega(\emptyset)$ (the equatorial atmosphere of the sun rotates faster than the polar atmosphere) is a result of the interaction between the convection and the magnetic field ; it plays an essential role in the dynamo process in stars. The differential rotation is commonly attributed to a meridional circulation, which is maintained by an anisotropic viscosity or by a latitude dependent convective energy transport. It turns out that the measurement of $\Omega(\emptyset)$ is a search for systematic large-scale and long-lived motions at the surface of the sun. For this reason, both phenomena are reviewed in the same section.

The solar rotation can be measured either by spectroscopy (Doppler maps of plasma motion) or by tracing the motion of solar features, usually of magnetic origin : sunspots, faculae, Ca^+ mottles, coronal holes, filaments. It can be noted that spectroscopic measurements refer to non-magnetic plasma, while tracer measurements refer to magnetic features. A very detailed review can be found in the paper by Schröter (1985). Each method has its own instrumental and solar noise limitations. An accuracy of 1 % is required for the detection of the variation of the solar rotation, which is not an easy task. As a general result, sunspots rotate faster than the surrounding plasma (14.55 - 14.35°/day, compared to 14.1°/day) ; among sunspots, the long-lived ones rotate faster than sunspots at their early stage of development. The fast sunspot rotation can be explained by their deep anchorage, reflecting the rotation of these layers and the radial differential rotation.

The studies of a cycle dependence of the solar differential rotation is based on long-term data sets. They include Greenwich, Mt Wilson and Kanzelhöhe plate collections (sunspots, facula), Ca^+K spectroheliograms and filtergrams from Meudon and Sac Peak, plasma rotation maps from Mt Wilson and Stanford. The variation, if it exists, is expected to be less than 20 ms^{-1} (1 %) ; only sunspot data seems reliable enough to reach such accuracy (Schröter, 1985). The only established fact regarding a cycle dependence of $\Omega(\emptyset)$ from sunspot seems to be a faster rotation of the equatorial belt around the sunspot minimum.

From Mt Wilson dopplergrams, a very important discovery has been made : the existence of torsional waves (La Bonte and Howard, 1982). The differential rotation has the mathematical form :

$$\Omega(\emptyset) = A + B\sin^2\emptyset + C\sin^4\emptyset$$

The parameters A, B and C are variable. The torsional waves are revealed by the time-variation (daily) of the residuals from the average (11-years) differential rotation (<10 ms^{-1}). Successive improvements in the data reduction have been made by Snodgrass

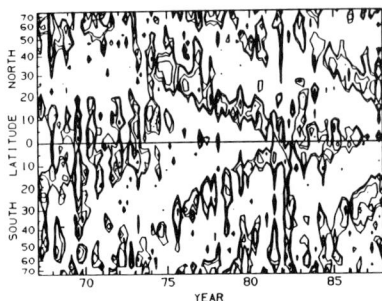 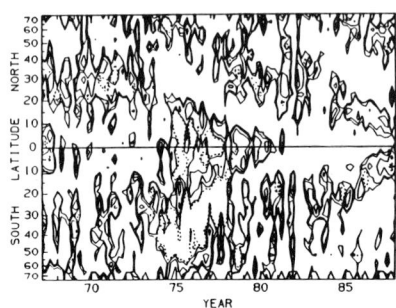

Figure 19. The zonal velocity pattern of accelerated and decelerated rotation. The summations have been taken over three solar rotations. Each plotted point represents the average over the full circumference at the indicated latitude. In order to make the structure of the result more readily apparent, the flows rotating faster (left figure) and slower (right figure) than average, have been separated. (Ultrich et al., 1988, Figure 16).

(1984, 1987) and by Ulrich et al. (1988). Torsional waves consist of alternative latitude zones of accelerated and decelerated rotations (Figure 19, from Ulrich et al., 1988). Two waves are present simultaneously in each hemisphere (Snodgrass, however, can find only one wave). Waves are born near the poles and migrate equatorward, reaching the equator in about 18 years. Successive waves build up at intervals of about 11 years, and thus overlap. The magnetic activity of the cycle tends to maximize in the regions where the torsional wave nearest the equator enhances the shear between accelerated and decelerated latitudes. The high latitude wave and shear coïncide with regions of Ephemeral Active Regions (ERs) and also, possibly, with polar faculae and $H\alpha$ filaments (Makarov and Sivaraman, 1989).

Torsional waves and meridional motions can be explained by a pattern of azimuthal convective rolls migrating equatorward (Snodgrass, 1987 ; Wilson, 1987a). The fluid in adjacent rolls rotates in opposite directions, alternatively accelerating (in rolls rotating poleward) and decelerating (in rolls rotating equatorward) the surface rotation. This yields to regions of maximum shears where activity tends to emerge (Figure 20, from Snodgrass, 1987). According to Wilson (1987a) the rolls brake up into giant cells as they progress equatorward. In a subsequent paper, Wilson (1987b) gives an explanation for the formation of convective rolls in polar regions and for the equatorward migration. Convection is more easily initiated along or near the axis of rotation, i-e near the poles, because elsewhere the effect of buoyancy is reduced by the angular momentum effect of the radial motions. The rolls are then carried along by the azimuthal magnetic toroid, which is a preferred region of downflow, because of thermal shadow which suppresses the buoyancy.

Meridional motions as derived from tracers yield, unfortunatelly, to quite a different description of the large-scale circulation at the surface of the sun. Ribes and co-workers (Ribes, Mein, Mangeney, 1985 ; Ribes, 1986 ; Ribes, 1989) analysed the proper motion of young sunspots ; they found that they alternately move poleward and equatorward at different phases of the solar cycle, inside belts of unipolar magnetic polarity (Figure 21, from Ribes, 1989) ; the belt boundaries were derived from the $H\alpha$ filaments proper motion. This behavior is explained by the formation of successive azimuthal rolls (every 2

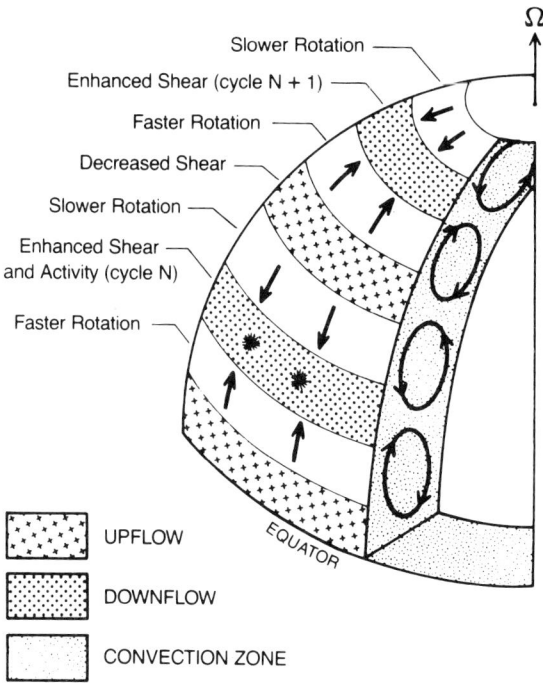

Figure 20. Schematic diagram of the azimuthal-roll model for the solar cycle, showing the rolls and associated surface motions at solar maximum for cycle N (sunspots in active zone) along with the emergence at high latitude of cycle N + 1. The bands on the surface indicate the various predicted large-scale motions produced by the rolls shown as ovals in the convection zone. Arrows on the surface depict meridional flow and arrows on the ovals show the directions in which the rolls are turning.

or 3 years) moving toward the poles ; the first roll appears at mid-latitude, 2 or 3 years after the maximum ; as the cycle proceeds, the latitude of the roll formation is shifted toward the equator. The Ribes's rolls can explain the butterfly sunspot diagram, the polarity reversal, as well as the existence of torsional waves.

Both spectroscopic and proper motion results seem to prove the existence of azimuthal rolls. But the location of formation and the direction of propagation are very controversed. More observations and data analysis are required.

5 CONCLUSION

In order to understand the solar cycle and the generation of magnetic flux, it is fundamental to have a good knowledge of the variation of the quiet atmosphere of the sun (including the photospheric, chromospheric, transition zone and coronal layers) and if possible, of the interior (through the surface oscillations). Active centers have to be considered as only a part of what we need to know ; it is the most spectacular part, not necessary the most useful. Unfortunately it is not yet possible to establish a realistic

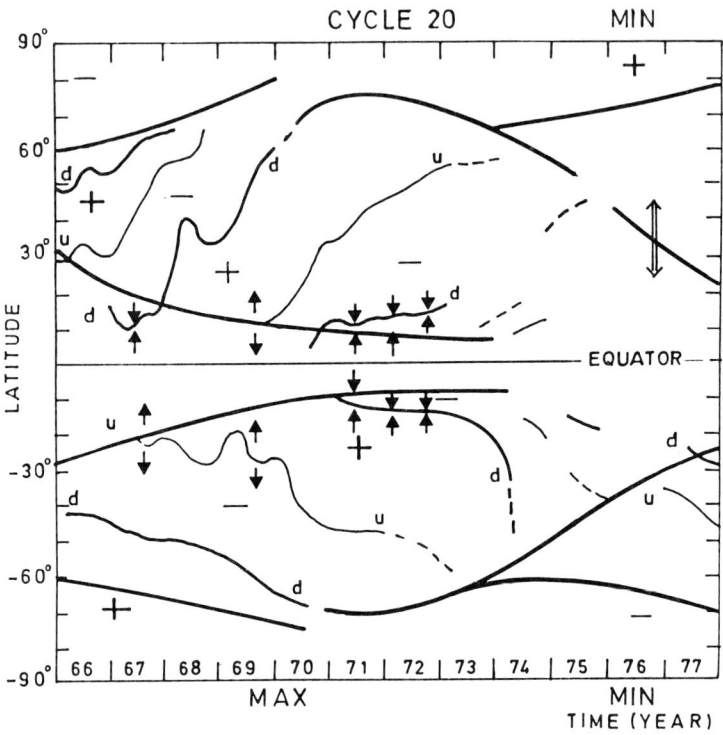

Figure 21. Azimuthal roll pattern detected on the spectroheliograms of Meudon Observatory, for solar cycle n°20. The Hα filament trajectory (thin and extra-thin lines) defines the borders of large-scale unipolar magnetic regions. Arrows indicate the direction of the meridional circulation traced by young sunspots. The "d" and "u" symbols refer to converging and diverging rolls respectively. The rotation of the rolls is related to their magnetic polarity. The solar activity, symbolized by the green coronal emission line at 5303 (thick lines), starts from the poles, shortly after the sunspot maximum and move equatorward through the solar cycle. There is a high-latitude branch of the coronal emission which outlines the coronal holes. The solar activity is confined to the extremities of the azimuthal roll pattern rather than being superimposed upon it. (From Ribes, 1989, Figure 4).

description of the variable quiet sun, because of the lack of converging results. The knowledge of global variations is not sufficient ; we also need to know latitude variations, as most parameters are probably latitude dependent because of the existence of azimuthal rolls and active latitudes ; for example latitude variations have been found for NBPs, limb profile and differential rotation.

The latitude and cycle variation of the magnetic flux outside active regions, the effective temperature (through limb profile and luminosity measurements), the temperature gradient (through the line shape and granulation analysis), the oscillations, oblateness, differential rotation and meridional motions are of particular importance.

ACKNOWLEDGMENTS.

I like to thank Y. Cabes for preparing the photographic reproductions of the figures and M.P. Sacristan for typing the paper.

REFERENCES.

Brown, Th. : 1987, private communication.

Bruning, D.H. and La Bonte, B. : 1985, Solar Phys. 97, 1.

Cavallini, F., Ceppatelli, G., and Righini, A. : 1986, Astron. Astrophys. 158, 275.

Chapman, G.A. and Klaburde, D.P. : 1982, Astrophys. J. 261? 387.

Davis, J.M. : 1983, Solar Phys. 88, 337.

Deaborn, D.S. and Blake, J.B. : 1982, Astrophys. J. 257, 896

Delache, Ph., Laclare, F., and Sadsaoud, H. : 1985, Nature 317, 416.

Dicke, R.H., Kuhn, J.R., and Libbrecht, K.G. : 1986, Astrophys.J. 311, 1025.

Dicke, R.H., Kuhn, J.R., and Libbrecht, K.G. : 1987, Astrophys.J. 318, 451.

Fossat, E., Gelly, B., Grec, G., and Pomerantz, M. : 1987, Astron. Astrophys. Letters, 177, L 47.

Foukal, P. : 1981 in Physics of Sunspots, L. Cram, and J. Thomas (eds.) (Sacramento Peak Obs. Publ.), p. 391.

Foukal, P. and Lean, J. :1988, Astrophys. J. 328, 347.

Gelly, B., Fossat, E., and Grec, G. : 1988, Astron. Astrophys. Letters, 200, L 29.

Gilliland, R. : 1981, Astrophys. J. 248, 1144.

Gilliland, R. : 1982, Astrophys. J. 257, 896.

Golub, L., Krieger, A.S., Silk, J.K., Timothy, A.F., and Vaiana, G.S. : 1974, Astrophys. J. 189, L 93.

Gulyaev : 1975

Hickey, J. Griffin, F., Jacobwitz, H., Stowe, L., Pellegrino, P., and Masckhoff, R. : 1980, Eos 61, 355.

Hill, H.A. and Stebbins, R.T. : 1975, Astrophys. J. 200, 471.

Howard, R. and La Bonte, J.L. : 1981, Solar Phys. 74, 131.

Hudson, H.S. : 1988, Ann. Rev. Astron. Astrophys. 26, 473.

Isaak, G.R., Jefferies, S.M., Mc Leod, C.P., New, R., and van der Raay, H.B. : 1987, in IAU Symposium 123, Advances in Astero-and Helioseismology, ed. J. Christensen - Dalsgaard and S. Frandsen (Dordrecht, Reidel).

Jensen, E. : 1955, Ann. Astrophys. 18, 127.

Keil, S.L. and Worden, S.P. : 1984, Astrophys. J. 276, 766.

Kuhn, J.R., Libbrecht, K.G. and Dicke, R.H. : 1988, Science 242, 908.

La Bonte, J.L. and Howard, R. : 1982, Solar Phys. 80, 15.

Laclare, F. : 1983, Astron. Astrophys. 125, 200.

Laclare, F., Journet, A., and Merlin, G. : 1989, Symp. UAI n°138, Solar Photosphere, J.O. Stenflo (ed.) in press.

Livingston, W. : 1984, in S. Keil (ed.), Small-Scale Dynamical Processes in Quiet Stellar Atmospheres, Sacramento Peak Observatory, Sunspot, New Mexico, p.265.

Livingston, W. and Wallace, L. :1987, Astrophys. J. 314, 808.

Macris, C.J., Muller, R., Rösch, J. and Roudier, Th. : 1984, in Small-Scale Processes in the Solar and Stellar Atmospheres, S. Keil ed., Sunspot, NM. USA.

Macris, C.J. and Rösch, J. : 1983, Compte Rend. Acad. Sci. Paris, 296, 265.

Makarov, V.I. and Sivaraman, K.R. : 1989, Solar Phys. 121

Muller, R. : 1983, Solar Phys. 85, 113.

Muller, R. : 1988a, in Solar and Stellar Granulation, NATO series, R. Rutten and G. Severino ed. p. 101.

Muller, R. : 1988b, Adv. Space Res. Vol.8, n°7, 159.

Muller, R. and Keil, S. : 1983, Solar Phys. 87, 243.

Muller, R. and Roudier, Th. : 1984, Solar Phys. 94, 38.

Muller, R., Roudier, Th. and Vigneau, J. : 1990, Solar Phys. in press.

Pallé, P.L., Perez, J.C., Regulo, C., Roca Cortes, T., Isaak, G.R., Mc Leod, C.P., and van der Raay, H.B. : 1986, Astron. Astrophys. 170, 114.

Peckover, R.S. and Weiss, N.O. : 1978, MNRAS 182, 189.

Petro, L.D., Foukal, P.V., and Kurucz, R.L. : 1985, Solar Phys. 98, 23.

Petro, L.D., Foukal, P.V., Rosen, W.A., Kurucz, R.L., and Pierce, A.K. : 1984, Astrophys. J. 283, 426.

Raghavan, N. : 1983, Solar Phys. 89, 35.

Rhodes, E.J., Woodard, M.F., Cacciani, A., Tomczyk, S., Korzennik, S.G., and Ulrich, R.K. : 1988, Astrophys. J. 326, 479.

Ribes, E. : 1986, Adv. Space Res. 6, 221.

Ribes, E. : 1989, Astrophys. J., in press.

Ribes, E., Mein, P., and Mangeney, A. : 1985, Nature 318, 170.

Roudier, Th. : 1986, Thesis, Université de Toulouse.

Schröter, E.H. : 1985, Solar Phys. 100, 141.

Sheeley, N.R. : 1964, Astrophys. J. 140, 731.

Skumanich, A., Lean, J.L., White, O.R., and Livingston, W.C. : 1984, Astrophys. J. 282, 776.

Snodgrass, H.B. : 1984, Solar Phys. 94, 13.

Snodgrass, H.B. : 1987, Astrophys. J. 316, L 91.

Spiegel, E.A. and Weiss, N.O. : 1980 : Nature 287, 616.

Svalgaard, L., Duvall, T.L. and Scherrer, P.H. : 1978, Solar Phys. 58, 225.

Thomas, J.H. : 1979, Nature 280, 663.

Ulrich, R.K., Bogden, L.W., Snodgrass, H.B., Padilla, S.P., Gilman, P., and Shieber, T. : 1987, Solar Phys. 117, 291.

Vaiana, G.S., Davis, J.M;, Giacconi, R., Krieger, A.S., Silk, J.K., Timothy, A.F. and Zombeck, M.V. : 1973, Astrophys. J. 185, L 47.

White, O.R. and Livingston, W.C. : 1981, Astrophys. J. 249, 798.

Willson, R., Gulkis, S., Janssen, M., Hudson, H., and Chapman, G.: 1981, Science, 211, 700.

Wilson, P.R. : 1987a, Solar Phys. 110, 59.

Wilson, P.R. : 1987b, Solar Phys. 117, 217.

Woodard, M.F. : 1987, Solar Phys. 114, 21.

Woodard, M.F. and Noyes, R.W. : 1985, Nature, 318, 449.

Superficial Activity as a Trace of the Internal Mechanisms of the Solar Cycle

J.I. GARCIA DE LA ROSA

Instituto de Astrofísica de Canarias, 38200-La Laguna, Tenerife, Spain.

Astrophysicists who study the Solar Cycle are particularly lucky because its most controversial phases take place right in front of their eyes. One of the reasons for the poor use of this advantage is probably the sophisticated mathematical approach which has dominated the study of the Cycle along the last 20 years. The apparent success of those parametrised approaches in describing the general aspects of the Cycle has diminished the role played by the observations. New observations have been generally incorporated to the models with just a tune of their parameters instead of a, sometimes required, re-consideration of their physical foundations.

We are approaching the 30th anniversary of Babcock's purely observational model of the Solar Cycle (Babcock, 1961). The occasion seems apt to check its health, after almost three decades of improved solar observations. Although the present review cannot accomplish this whole task it will attempt to give some hints in that direction.

Our discussion will be mainly devoted to the stages 3, 4 and 5 of Babcock's model.

1 THIRD STAGE OF BABCOCK'S MODEL

1.1 Introduction

Magnetic buoyancy will tend to lift concentrated flux ropes to the surface... When loops or stitches of the toroidal flux ropes rise to the surface, they emerge to form Bipolar Magnetic Regions. (Babcock, 1961)

The emergence of solar activity is the most obvious link between the observable surface and the internal source of the magnetic cycle. An approach to its study can be carried out from three different directions, related to the three coordinates of a spherical geometry.
(1) Depth of origin of the active regions
(2) Latitude distribution of the active regions.
(3) Longitude distribution of the active regions.

First of all, let us re-consider the activity features which rank the largest known units of emerging magnetic flux: **Complexes of Activity.** They were defined by Bumba and Howard (1965) as units of activity which grow rapidly by the birth of new large active regions. This causes an expansion of the complex both in latitude and longitude, eventually covering 150°. Gaizauskas et al. (1983) used the same name for a different feature: a cluster of active regions emerging within a small area. To avoid confusion, Castenmiller et al. (1986) propose to rename these: "Nests of Activity". In the following, we shall use the early definition by Bumba and Howard (1965).

1.2. Complexes of Activity

A detailed study of these large units of solar activity has been carried out by García de la Rosa and Carlos Reyes (1989) using the Solar-Geophysical Data Bulletins along the period 1970-1988. The conspicuous alignment of the active regions of the complexes, besides other regularities, led us to make the following reasonable assumption: **All the active regions which make the complex belong to the same sub-photospheric flux tube** (see Figure 1(a)).

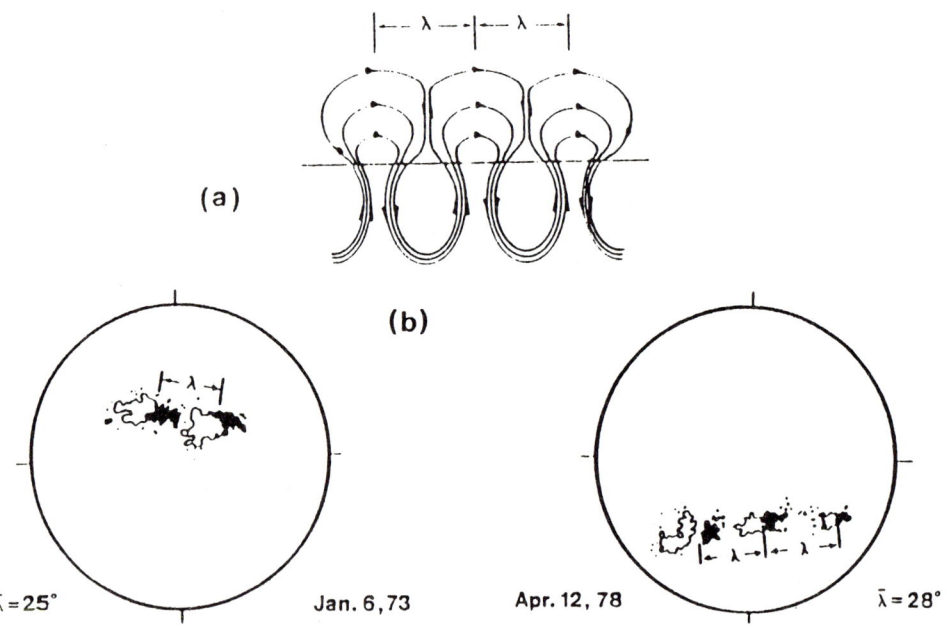

Fig. 1. (a) Assumed sub-photospheric connection between the active regions of a complex. (b) Two examples of complexes of activity taken from Mt. Wilson magnetograms.

The assumption suggests a few selective rules to distinguish genuine complexes from accidental clusterings:

(1) All member active regions must be roughly aligned.
(2) All members must show comparable magnetic flux.
(3) The complex must make a small angle with the solar equator ($\leq 30°$).

143 complexes were selected during the period studied. Figure 1(b) shows two different examples taken from Mt. Wilson magnetograms.

Complexes show the following general behaviour:
(1) The members of the complex emerge at regular spatial intervals.
(2) The emergence of the members of the complex is not simultaneous, but generally progresses in longitude from its center to both sides.
(3) When X-ray or EUV pictures are available, the active regions of the complex show interconnecting loops.
(4) When members of the complex age, they eventually share the same plage.
(5) The typical magnetic flux content of the members is: $10^{21} - 10^{22}$ Mx.

1.3 Depth of Origin of the Active Regions

Both theoretical and observational arguments suggest that active regions are launched from the base of the Convection Zone (see the review by Moreno Insertis, 1987). Observational arguments point out that the typical dimensions of the active regions are comparable to the length scales of the deep Convection Zone, rather than to those of the photosphere. Parker (1984a) calculated the equilibrium configuration of an arch-shaped flux tube and deduced the depth of the anchor points by measuring the distance between the p- and f-parts of the active region. This estimate relies very much on the sub-photospheric geometry of the arch and therefore we propose taking a different approach based on the regular intervals shown by the members of the complexes.

A sinusoidal perturbation can develop a kink instability and loops in a flux rope, provided that its wavelength is not too short to produce a strong restoring magnetic force against the buoyancy (Parker, 1979; Spruit and Van Ballegooijen, 1982; Moreno Insertis, 1986). The critical wavelength to develop the instability in the magnetic tube has been estimated with a varied degree of precision. The most basic approach, made by Parker (1979, chapter 6), studies the equilibrium of a magnetic loop embedded in both isothermal and polytropic atmospheres. He concludes that any flux tube anchored at points separated by more than $\lambda_{crit} = 2\pi\Lambda$ (Λ: Pressure Scale Height) cannot be held in equilibrium. This is a rough result that should be taken with caution.

Figure 2 shows the evolution of the average wavelength of our selected 143 complexes of activity. The wavelength clearly decreases along the cycle, indicating a decrease

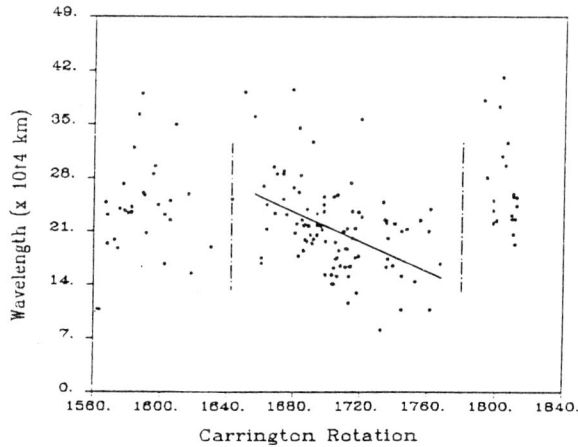

Fig. 2. Evolution along the cycle of the average separation of the members of complexes of activity. Least squares regression line for cycle 21 is given. Vertical lines show separation with contiguous cycles 20 and 22.

of the pressure scale height of the departure level. It can then be concluded that the departure level slowly moves upward along the cycle. A rough quantitative calculation using $\lambda_{obs} = \lambda_{crit}(= 2\pi\Lambda)$ and the model of the Convection Zone by Spruit (1977) suggests a depth of origin which varies from $2,8\ 10^5$ to $1,4\ 10^5$ km from the beginning to the end of the cycle. More sophisticated calculations of the critical wavelength would help to enhance the potential use of the method. The works by Spruit and Van Ballegooijen (1982) and by Parker (1984a) advance in this direction.

The discovery by Albregtsen and Maltby (1978) (see also Maltby et al., 1984) of an increase of the umbra/photosphere intensity of large sunspots along the Cycle has been equally interpreted by Yoshimura (1983) as a decrease in the depth of the roots of the flux ropes.

1.4 Latitude Distribution of the Active Regions

One of the most far reaching discoveries in Solar Physics during the present decade was the so called Torsional Oscillation (TO) (Howard and LaBonte, 1980). Its profound consequences on the study of the Solar Cycle are just starting to show up. As presented in Figure 3, the TO pattern is made by zones divided by parallels of latitude, that rotate alternately faster or slower (6 m/sec) than the average rotation rate for that latitude (2000 m/sec at the equator). In general, there are four such zones in each hemisphere, which migrate slowly toward the equator in 22 years. Furthermore, it was discovered that the shear boundary between fast and slow

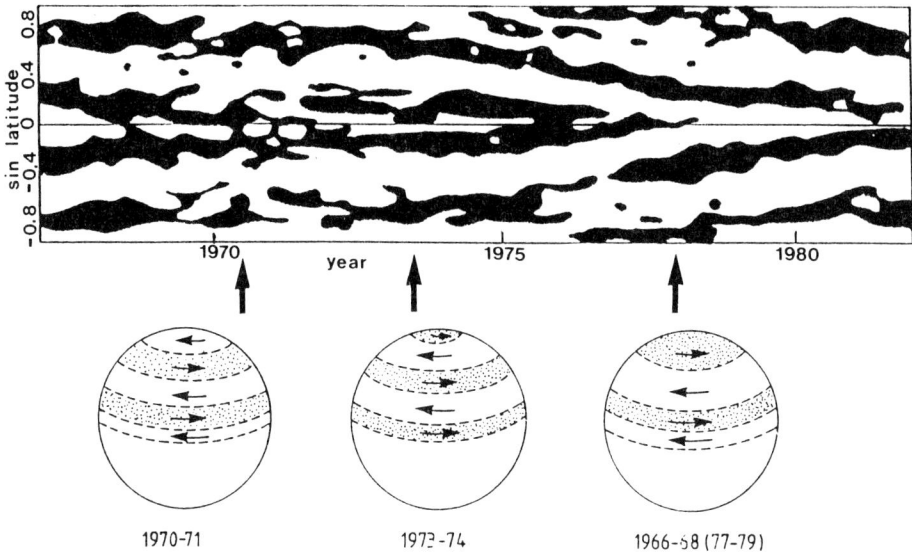

Fig. 3. Evolution of the Torsional Oscillation, adapted from Howard and LaBonte (1983). Dark zones represent faster than average latitudes. The diagrams below, due to Wilson (1987b), show instantaneous pictures of the Oscillation at several times.

zones is the locus (below 30°) for the emergence of active regions (see Figure 4).

So far there are three main suggestions concerning the origin of the TO (see the review by Wilson, 1987a):
(1) The TO is explained by Yoshimura (1981) and by Schussler (1981) in terms of Lorentz Forces caused by the sub-photospheric magnetic fields.
(2) LaBonte (see Wilson, 1987a) suggests that the TO is the surface manifestation of a deep-seated torsional wave which drives the dynamo action.
(3) Snodgrass and Wilson (1987) (see also Ribes et al. (1985)) suggest that the TO results from a large scale system of convective rolls stretched in longitude and migrating slowly toward the equator.

This third explanation is supported by several independent observations:
(1) Line-of-sight residuals of the Mount Wilson velocity data measured by Snodgrass (1987) show regions of upflow and downflow motions. Their position agrees with the interpretation of the TO as a result of Coriolis accelerations
(2) The position of latitudes of limb temperature excess and deficit, measured by Kuhn et al. (1985) and Kuhn et al. (1987), fits well with the upflow and downflow zones determined in (1).

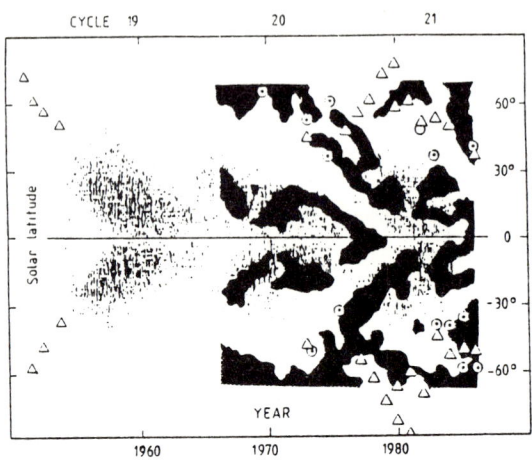

Fig. 4. Extended Solar Cycle shown by the correspondence between the shear zones (dark) with: (1) The Butterfly diagrams of sunspots; (2) Latitudes where reversed polarity Ephemeral Active Regions predominate (circles ⊙) and (3) The high latitude enhancements of the green line emission (triangles △) (adapted from Wilson et al. (1988)).

The low latitude connection between the torsional shear and the Butterfly Diagram, has induced observers to look for the high latitude counterpart of that connection. Both Ephemeral Active Regions (EARs) and Green-Line Emission (GLE) apparently show the expected behaviour.

Ephemeral Active Regions: Martin and Harvey (1979), after studying the magnetic polarities of EARs along three periods of cycle 20, concluded that at high latitudes, they preferentially showed reverse polarities to those dominating the current cycle. A similar behaviour was found by Wilson et al. (1988) during the declining phase of cycle 21 (see Figure 4).

Green-Line Emission: Early observers, as Trellis (1957), detected high latitude enhanced emission of the green coronal line several years before the beginning of the sunspot cycle. This emission migrates equatorward, reaching a latitude of 40° in coincidence with the appearance of the new-cycle spots. Recent studies by Altrock (1986) and by Wilson et al. (1988) find that the high latitude coronal bands start at $60° - 70°$, 2-3 years after solar minimum and migrate equatorward at a rate of $2,5°\ year^{-1}$, until they fuse with normal activity zones (see Figure 4).

Those observations have lead to a certain consensus in what has been called **The Extended Solar Cycle** (see the reviews by Wilson, 1987a, 1988). This extended cycle begins at high latitudes, near sunspot maximum and progresses toward the equator over a period of 18-22 years. In it, the classical sunspot cycle would only be the "tip of the iceberg". This extended cycle is quite different from the 22 year magnetic cycle, which simply consists of two 11-year sunspot cycles.

These ideas are quite recent and it would be unfair to ask them for complete results. However, they leave in the air the distressing suggestion that magnetic flux is migrating from poles to equator, in contradiction with the observations. Howard and LaBonte (1981) report that large scale magnetic fields on the Sun, especially the polar ones, originate only in the spot zones, with **no** other source. In Section 2.5 we shall return to this apparent contradiction.

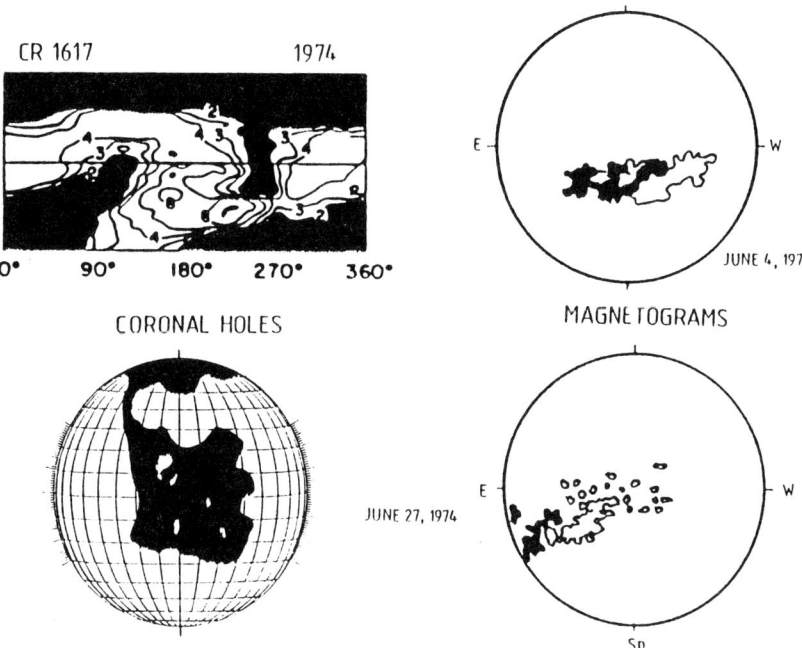

Fig. 5. Contribution of a complex of activity to the maintenance of two coronal holes. Above-left gives the position of the two persistent coronal holes (after Hundhausen et al. (1981)). Above-right presents the position of the complex of activity on June 4, 1974 in a magnetogram whose Central Meridian Longitude is $L_o = 192°$. Below-left shows the position of the Northern hole (after Underwood and Broussard (1977)) on June 27, during the next Carrington Rotation. Below-right shows the unbalanced p-flux of the Easternmost active region of the complex, which feeds the coronal hole.

1.5 Longitude Distribution of Active Regions

Active regions do not emerge randomly in longitude, but the existence of Magnetic Active Longitudes (MALs) indicates that a specific pattern exists in their emergence. The distribution of MALs determines the background field sector structure and the position of the persistent coronal holes.

Bohlin (1977) and Levine (1982) found that coronal holes, always located in Unipolar Magnetic Regions, are usually associated with a succession of nearby active regions showing local unbalance of magnetic flux. The reason for such a particular behaviour of some individual active regions is far from clear. Zwaan (1987) claims that "the process of formation and maintenance of coronal holes is not yet well understood", but at the same time, he suggests that the relationship between coronal holes and complexes of activity is the key to understand the fundamental process in solar magnetism.

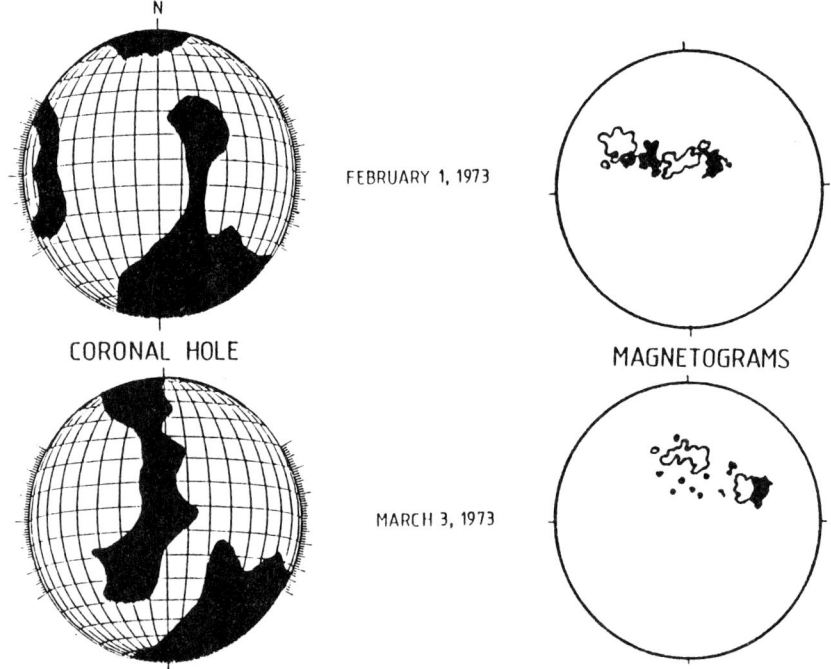

Fig. 6. Two coronal holes maintained by a complex of activity. Above-left presents the position of the coronal holes on February 1, 1973 during Carrington Rotation 1597 (after Underwood and Broussard (1977)). Above-right shows the position of the complex of activity during the same day. Below-left presents the position of the Northern hole in the next Carrington Rotation. Below-right is clearly evident how the f-flux of the Westernmost active region of the complex contributes to the coronal hole.

The concept of Complex of Activity discussed in Section 1.2 proves to be a positive step in the understanding of the mentioned relationship, as suggested, for instance, by the two examples presented in Figures 5 and 6. Our detailed study of the complexes suggests the following typical behaviour (see Figure 7(a)):

(1) The coronal holes of each hemisphere are fed by the f-polarity flux of the Easternmost active regions of the complexes. Their unbalance arises naturally from the mutual cancellation of the flux of the inner magnetic concentrations of the complex.

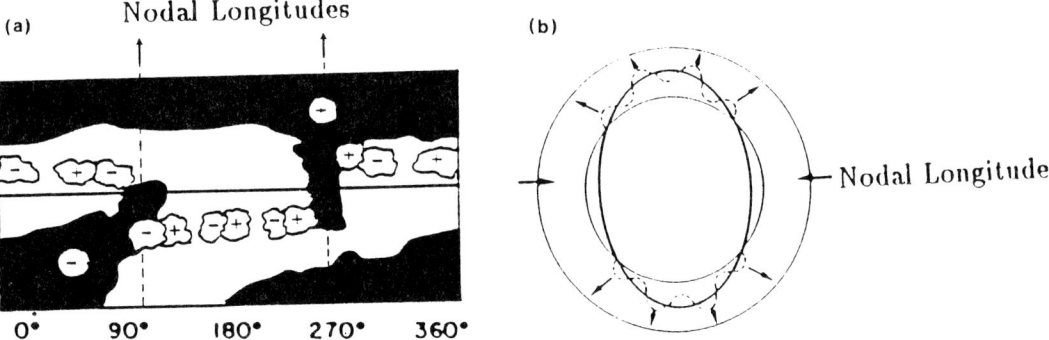

Fig. 7. (a) Scheme of the proposed contribution of complexes of activity to the maintenance of coronal holes. (b) Unstable m=2 mode of a toroidal magnetic field showing nodal longitudes at the longitude zones where the field becomes burried below the Convection Zone.

(2) The unbalanced p-polarity of the Westernmost active region of the complex contributes to the enlargement of the coronal hole of the other hemisphere, which temporarily crosses the equator.

(3) The location of coronal holes or Unipolar Active Regions, where no emergence takes place, act as nodal longitudes for the large scale magnetic emergence in the Sun.

A new question readily arises, **Why should such nodal longitudes exist ?**. At least two explanations have been suggested:

(1) Spruit and Van Ballegooijen (1982) study the instability of a toroidal flux system, concluding that modes with low azimuthal numbers (m=1,2..) are unstable at any field strength. Fig. 7(b) shows a m=2 mode with nodal longitudes separated by 180°, as it is usually observed.

(2) Moreno Insertis (1984) and McIntosh and Wilson (1985) study the action of giant convective cells on the toroidal flux system, concluding that the flux would be expelled at the upflow regions and dragged at the downflow ones.

2 FOURTH STAGE OF BABCOCK'S MODEL

2.1 Introduction

The respective p and f parts of Bipolar Magnetic Regions (BMRs) generally expand and draw apart, while the flux loops rooted in them are being pushed outward into the corona...It is often observed that the p-part of a BMR expands or shifts toward the equator while the f-part expands or migrates in the direction of the pole... (Babcock, 1961)

Leighton (1964) suggested that the random motions of magnetic elements produced by the supergranular flows, together with their inclination to the equator, cause an equatorward diffusion of p-flux and poleward diffusion of f-flux. He could simulate

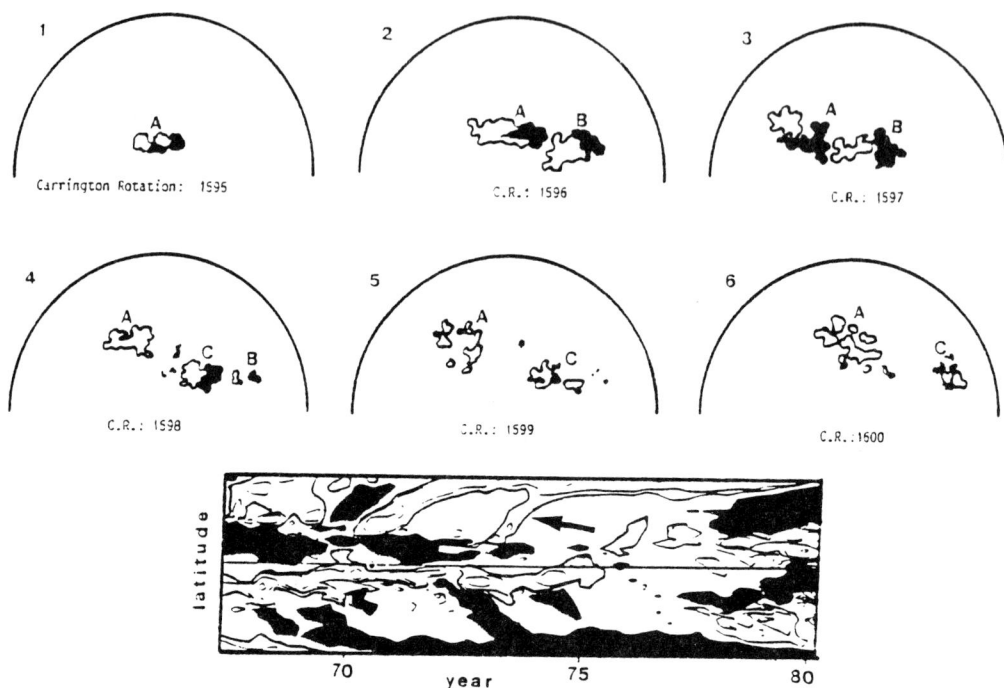

Fig. 8. Contribution of a complex of activity to the episodic poleward migration of f-flux. Magnetograms from 1 to 6 show the Northern hemisphere at a Central Meridian Longitude of $L_o \approx 130°$. The complex is made of three regions: A, B and C. After inner field cancellations the unbalanced f-flux of region A remains at high latitudes, contributing to the episode of poleward migration marked on the diagram of magnetic evolution.

the observed polar field reversals using a "diffusion constant"(D) of 770-1540 km^2/s. However, such a high value of D has never been observed and Mosher (1977) had to assume a poleward meridional flow to assist the diffusion. The attempts to measure meridional motions by Doppler shifts have been spoiled by the poor knowledge of the latitude variation of the limb shift (see the review by Schröter, 1985). Tracer motions, also affected by unknown hydromagnetic forces generally, although not always, show equatorward motions predominating at low latitudes and poleward motions at latitudes $\geq 20°$ (Tuominen and Virtanen, 1984; Howard and Gilman, 1986).

After the study on the evolution of large scale magnetic fields (see Figures 8 or 9), Howard and LaBonte (1981) conclude:
(1) The spot zone becomes the source of any large scale magnetic field.
(2) Polar fields are built by the episodic migration of f-polarity dominated fields.
(3) The migration does not occur by diffusion, but the fields are carried poleward by a meridional flow of average speed 10 m/s.
(4) Magnetic flux is being continuously removed from the photospheric level at a rate of almost 10^{22} Mx/day.

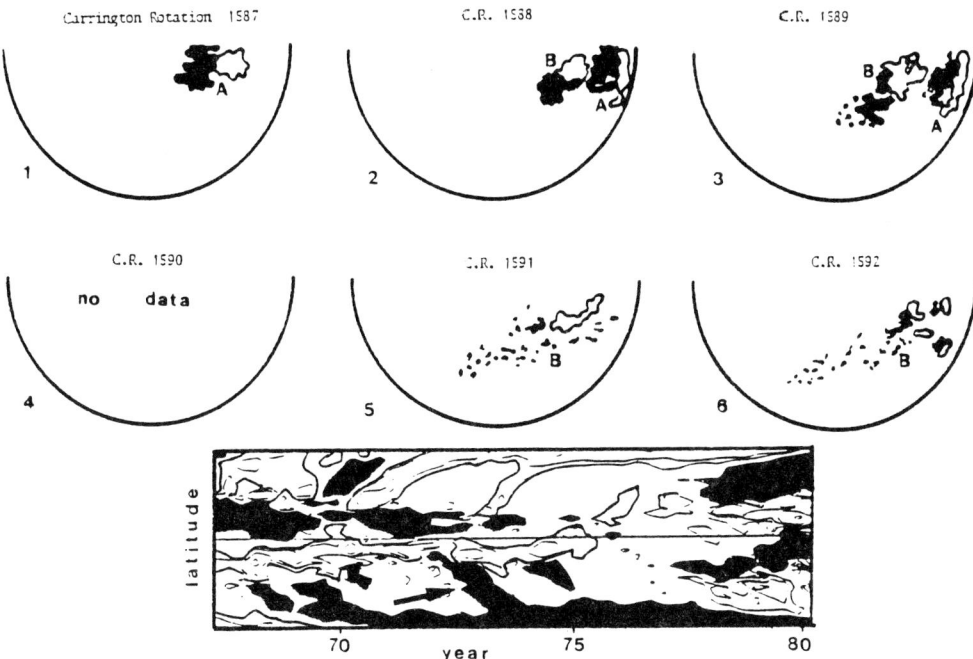

Fig. 9. Contribution of a complex of activity to the episodic poleward migration of f-flux. Magnetograms from 1 to 6 show the Southern hemisphere during consecutive rotations at a Central Meridian Longitude of $L_o \approx 285°$. The activity complex located at the Western side is made of regions A and B. After the cancellation of inner magnetic flux, the unbalanced f-flux of region B is left to contribute to the episodic poleward migration marked in the diagram below. This diagram, adapted from Howard and LaBonte (1981), shows the evolution of the large scale magnetic fields.

A further work by Topka et al.(1982) on the migration of Hα filaments, concludes that, although there is a constant poleward flow of balanced bipolar magnetic flux, what seems to be episodic is the availability of unbalanced magnetic concentrations.

Several questions arise around the fourth stage of Babcock's Model. We shall deal with some of them in the following sections.

2.2 Why the f-magnetic fields should elect to move poleward while the leading fields do not ? (Parker, 1987).

The best place to look for an answer is in the history of the magnetic precursors of the episodes of f-polarity migrations. Two examples are presented in Figures 8 and 9. The following conclusion is drawn from our study:

The episodes of unbalanced f-flux which migrates poleward are related with the emergence of complexes of activity. A single active region with a typical 10° separation between p- and f-extremes, only spans a small 2° interval of latitude, whereas the longest complexes show their p- and f-extremes separated by latitude intervals of 10° − 20°. In those cases, the unbalanced extremes of the complex can be independently acted by the high and low latitude meridional motions (see Figure 10).

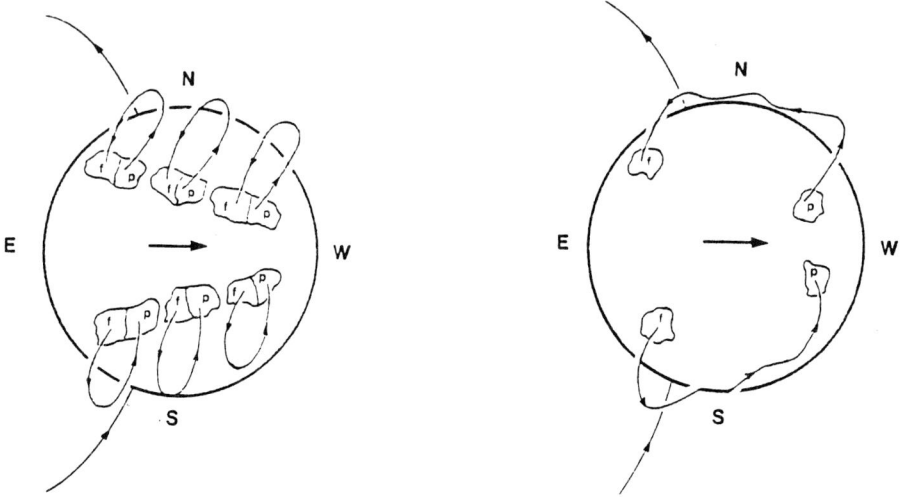

Fig. 10. Poleward migration of the f-part of a complex of activity. After the cancellation of the magnetic flux of the inner active regions, the p- and f-extremes are left locally unbalanced and with a large latitude interval. The high latitude f-flux is likely to be dragged poleward by the meridional flows.

2.3 How is the magnetic flux removed from the photospheric level ?

Diffusion only spreads magnetic flux over larger areas, but does not remove it from the photospheric level. Flux needs to be pulled (upward or downward) out of the photosphere (Zwaan, 1978). On the quiet Sun, Martin (1988) reports three observed patterns of magnetic field disappearance:

(1) **Cancellation** (Livi et al.,1985): mutual disappearance of closely spaced magnetic flux of opposite polarity, involving network, intranetwork or EAR elements.
(2) **Fading** in situ of single polarity fragments (Wilson and Simon, 1983).
(3) **Fragmentation** of the field before fading.

On active regions, Zirin (1985), Martin et al. (1985) and García de la Rosa et al. (1989) have observed similar cancellations to those of the quiet Sun, but at a higher rate which, according to Martin et al.(1985), can amount to 10^{19} Mx/hour.

Due to the close packing of active regions inside complexes of activity, the reconnection should proceed with great efficiency at the high atmospheric levels, as proposed by Parker (1984b)(see Figure 11(a)). After the reconnection has been completed a

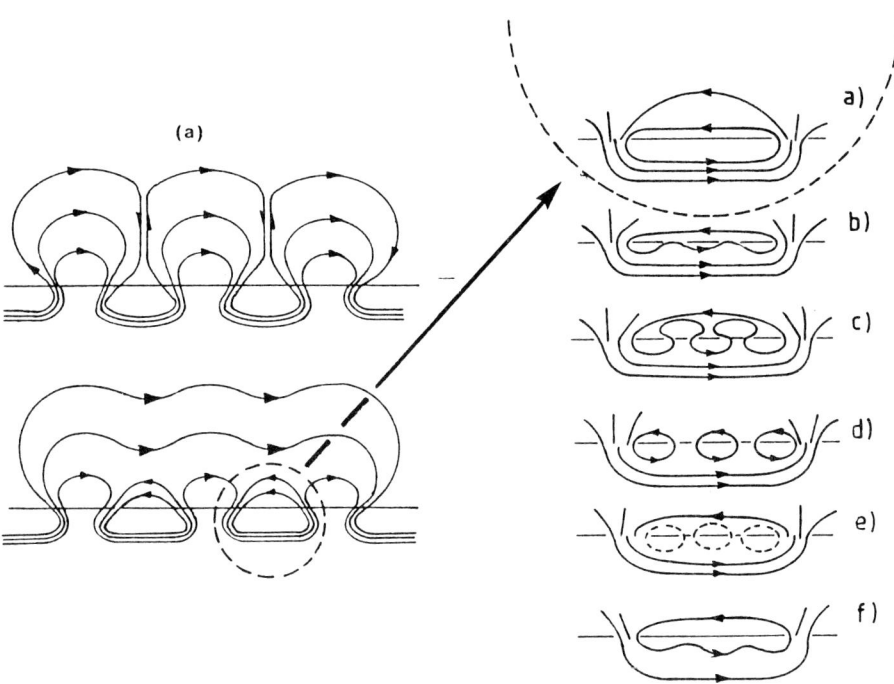

Fig. 11. (a) On the left-hand side, the process of reconnection on closely spaced active regions is roughly represented (adapted from Parker (1984b)). (b) On the right-hand side, the further process of U-loop escape is depicted according to Spruit et al. (1987).

long loop interconnects both extreme poles of the complex and several submerged U-loops are left between them. A possible mechanism for the escape of U-loops has been described by Spruit et al. (1987) (see Figure 11(b)). From our point of view, the intermediate active regions of complexes of activity are the best candidate places to observe the processes of massive flux removal from the photosphere.

2.4. Is the reconnection really taking place at the coronal level ?

Babcock (1961) stated: *In the high corona the full-blown flux loops above old BMRs continue to expand toward each other... The result is a neutralization of long segments of the standing legs of the flux loops, with severing and reconnection of the ends... The low latitude end of each such section of rope merges with the corresponding end of a flux rope on the opposite side of the equator.*

A decade later this view was verified by the EUV and soft X-ray observations made from the Skylab. During its operation period, 100 interconnecting loops were observed, 20 of them crossing the equator (Chase et al. , 1976). Several workers analysing the Skylab data (Sheeley et al. (1975); Svestka et al. (1977); Svestka and Howard (1981) etc.) suggest that the origin of loops is related to reconnections of magnetic field lines extending from the active regions.

2.5 Does the concept of Extended Cycle imply a magnetic flux migration from poles to equator ?

As already mentioned in Section 1.4, there is an apparent contradiction between the following two observations:
(1) The mean latitude of the EARs with reversed polarity (and also coronal emission) drifts equatorward, following the migration of the shear zone of the Torsional Oscillation.
(2) The high latitude and polar magnetic flux is originated at the sunspot zones.

Both observations can be reconciled if we accept that EARs are not small active regions but **a surface phenomenon perhaps resulting from convection interacting with subphotospheric fields**, as proposed by Harvey (1984) and indirectly suggested by other authors:
(1) García de la Rosa (1983), after a study on the emergence of active regions, concluded that the small active regions ($\Phi_B \leq 5\ 10^{21}$ Mx), and perhaps the EARs, result from the recycling of decayed flux of larger active regions.
(2) Tang et al. (1984) show that the EARs outnumbered the small active regions in three orders of magnitude, during a certain period, concluding that EARs are not the tail end of active regions.

The combination of an equatorward migrating shear zone with a poleward migration of decayed magnetic flux, could possibly explain the observed pattern of high latitude EARs.

This statement needs to be checked, but Bumba and Růžičková-Topolová (1969) have already presented a few cases of coronal green emissions related to tongues of f-polarity extending to poles from the sunspot belts.

3 FIFTH STAGE OF BABCOCK'S MODEL

The reversed dipolar field is the residual of the foregoing process... The analogues of stages 2, 3 and 4 take place to complete the whole 22-year magnetic cycle

A vivid proof, that the polar magnetic field at sunspot minimum governs the amplitude of the next cycle, is given by the remarkable success of the Long-Term Prediction methods which are based on the observation of the polar magnetic fields.

Different methods have been used to evaluate the polar field at sunspot minimum:
(1) Direct measurement of the field and its bending angle observed during eclipses (Schatten and Sofia, 1987).
(2) The degree of flattening of the heliospheric current sheet (Schatten and Sofia, 1987).
(3) The interplanetary magnetic field (Brown, 1981).
(4) Some indexes of geomagnetic activity, linked to the solar fields through the wind originated at the polar coronal holes (Ohl and Ohl, 1979; Kane, 1987).

As a reliability test of these prediction methods for the present cycle 22, Schatten (1988) gives an amplitude of $R_m = 170 \pm 25$ peaking near October, 1989 ± 7 months.

4 CONCLUSIONS

Despite the almost three decades of new solar observations which have elapsed since the release of the Babcock model of the Solar Cycle, our opinion is that, not only has it survived, but it enjoys remarkable good health. New observations have come just to refine the model, but none have deeply questioned its foundations. Certainly many questions remain unanswered. We have proposed to re-consider the concept of **complexes of activity**, defined by Bumba and Howard (1965), as a possible answer to some of those questions:
(1) Regularities observed in the complexes of activity suggest that their active regions belong to the same sub-photospheric magnetic flux tube.
(2) The regular intervals between active regions of the complexes can teach us about the departure level of the magnetic loops. Our own study suggests that this level rises along the cycle.
(3) Complexes of activity contribute to the maintenance of persistent coronal holes and large scale magnetic structures. The cancellation of the magnetic flux of the inner active regions, leaves unbalanced unipolar magnetic regions, separated by long intervals of latitude and longitude.
(4) The unbalanced f-flux of the Easternmost extreme of the complex shows a large latitude difference with the p-extreme. This can explain the observed preference for the poleward migration of the f-flux.
(5) Due to their compactness, the inner active regions of complexes are possibly the best places to observe massive reconnection and removal of magnetic flux from the photosphere.

REFERENCES

Albregtsen,F., Maltby,P.: 1978, *Nature* **274**, 41
Altrock,R.C.: 1986, *Publ. Astron. Soc. Pacif.* **98**, 1100. Abstract.

Babcock,H.W.: 1961, *Astrophys. J.* **133**, 572
Bohlin,J.B.: 1977, in *Coronal Holes and High Speed Wind Streams*, ed. J.B. Zirker, Colorado Associated University Press, p. 27
Brown,G.M.: 1981, *Solar Phys.* **74**, 125
Bumba,V., Howard,R.: 1965, *Astrophys. J.* **141**, 1502
Bumba,V., Růžičková-Topolová,B.: 1967, *Solar Phys.* **1**, 216
Castenmiller,M.J.M., Zwaan,C., Van der Zalm,E.B.J.: 1986, *Solar Phys.* **105**, 237
Chase,R.C., Krieger,A.S., Svestka,Z., Vaiana,G.S.: 1976, *Space Res.* **16**, 917
Gaizauskas,V., Harvey,K.L., Harvey,J.W., Zwaan,C.: 1983, *Astrophys. J.* **265**, 1056
García de la Rosa,J.I.: 1983, *Solar Phys.* **89**, 51
García de la Rosa,J.I., Aballe,M.A., Collados,M.: 1989, *Solar Phys.* in press
García de la Rosa,J.I., Carlos Reyes,R.: 1989, in preparation
Harvey,K.L.: 1984, in *The Hydromagnetics of the Sun*, ESA SP-220, p. 231
Howard,R., Gilman,P.A.: 1986, *Astrophys. J.* **307**, 389
Howard,R., LaBonte,B.J.: 1980, *Astrophys. J.* **239**, L33
Howard,R., LaBonte,B.J.: 1981, *Solar Phys.* **74**, 131
Howard,R., LaBonte,B.J.: 1983, in *Solar and Stellar Magnetic Fields: Origins and Coronal Effects*, IAU *Symp.* **102**, cd. J.O. Stenflo, Reidel, Dordrecht, p. 101
Hundhausen,A.J., Hansen,R.T., Hansen,S.F.: 1981, *J. Geophys. Res.* **86**, 2079
Kane,R.P.: 1987, *Solar Phys.* **108**, 415
Kuhn,J.R., Libbrecht,K.G., Dicke,R.H.: 1985, *Astrophys. J.* **290**, 758
Kuhn,J.R., Libbrecht,K.G., Dicke,R.H.: 1987, *Nature* **328**, 326
Leighton,R.B.: 1964, *Astrophys. J.* **140**, 1547
Levine,R.H.: 1982, *Solar Phys.* **79**, 203
Livi,S.H.B., Martin,S.F., Wang,J.: 1985, *Australian J. Phys.* **38**, 855
Maltby,P., Barth,S.B., Lilje,P.B., Vikanes,F.W.: 1984 in *The Hydromagnetics of the Sun*, ESA SP-220, p. 233
Martin,S.F.: 1988, *Solar Phys.* **117**, 243
Martin,S.F., Harvey,K.L.: 1979, *Solar Phys.* **64**, 93
Martin,S.F., Livi,S.H.B., Wang,J.: 1985, *Australian J. Phys.* **38**, 929
McIntosh,P.S., Wilson,P.R.: 1985, *Solar Phys.* **97**, 59
Moreno Insertis,F.: 1984, PhD Thesis. University of Munich
Moreno Insertis,F.: 1986, *Astron. Astrophys.* **166**, 291
Moreno Insertis,F.: 1987, in *The Role of Fine-Scale Magnetic Fields on the Structure of the Solar Atmosphere* eds. E.-H. Schroter, M. Vazquez and A.A. Wyller, Cambridge University Press, p. 167
Mosher,J.M.: 1977, PhD Thesis. California Institute of Technology
Ohl,A.I., Ohl,G.I.: 1979, in *Solar-Terrestrial Predictions Proceedings* ed. R.F. Donnelly, NOAA, Boulder, **Vol II**, 258
Parker,E.N.: 1979, *Cosmical Magnetic Fields*, Oxford University Press
Parker,E.N.: 1984a, *Astrophys. J.* **280**, 423

Parker,E.N.: 1984b, *Astrophys. J.* **281**, 839
Parker,E.N.: 1987, *Astrophys. J.* **312**, 868
Ribes,E., Mein,P., Mangeney,A.: 1985, *Nature* **318**, 170
Schatten,K.H.: 1988, in *Seismology of the Sun and Sun-like Stars*, ESA SP-286, 3
Schatten,K.H., Sofia,S.: 1987, *Geophys. Res. Letters* **14**, 632
Schröter,E.H.: 1985, *Solar Phys.* **100**, 141
Schüssler,M.: 1981, *Astron. Astrophys.* **94**, L17
Sheeley,N.R., Bohlin,J.D., Brueckner,G.E., Purcell,J.D., Scherrer,V., Tousey,R 1975, *Solar Phys.* **40**, 103
Snodgrass,H.B.: 1987, *Astrophys. J. Letters* **316**, L91
Snodgrass,H.B., Wilson,P.R.: 1987, *Nature* **328**, 696
Spruit,H.C.: 1977, PhD Thesis. Utrecht University
Spruit,H.C., Title,A.M., Van Ballegooijen,A.A.: 1987, *Solar Phys.* **110**, 115
Spruit,H.C., Van Ballegooijen,A.A.: 1982, *Astron. Astrophys.* **106**, 58
Svestka,Z., Howard,R.: 1981, *Solar Phys.* **71**, 349
Svestka,Z., Krieger,A.S., Chase,R.C., Howard,R.: 1977, *Solar Phys.* **52**, 69
Tang,F., Howard,R., Adkins,J.M.: 1984, *Solar Phys.*, **91**, 75
Topka,K., Moore,R., LaBonte,B.J., Howard,R.: 1982, *Solar Phys.* **79**, 231
Trellis,M.: 1957, *Ann. D'Astrophys. Suppl.* No. 5
Tuominen,I., Virtanen,H.: 1984, *Astr. Nach.* **305**, 225
Underwood,J.H., Broussard,R.M.: 1977, *Atlas of Coronal Hole Boundaries fro Observations Prior to The Skylab Mission*, Aerospace Rep. No. ATR-77(7405)
Wilson,P.R.: 1987a, *Solar Phys.* **110**, 1
Wilson,P.R.: 1987b, *Solar Phys.* **110**, 59
Wilson,P.R.: 1988, *Solar Phys.* **117**, 205
Wilson,P.R., Altrock,R.C., Harvey,K.L., Martin,S.F., Snodgrass,H.B.: 1988, *Natu* **333**, 748
Wilson,P.R., Simon,G.: 1983, *Astrophys. J.* **273**, 805
Yoshimura,H.: 1981, *Astrophys. J.* **247**, 1102
Yoshimura,H.: 1983, *Solar Phys.* **87**, 251
Zirin,H.: 1985, *Astrophys. J.* **291**, 858
Zwaan,C.: 1978, *Solar Phys.* **60**, 213
Zwaan,C.: 1987, *Ann. Rev. Astron. Astrophys.* **25**, 83

Chromospheric Phenomena in Late-Type Stars

P. ULMSCHNEIDER

Institut für Theoretische Astrophysik,
Universität Heidelberg,
Im Neuenheimer Feld 561
D–6900 Heidelberg, Federal Republic of Germany

Summary: Chromospheres and coronae are hot shell-like layers around stars which are characterized by strong H_α, CaII, UV, IR and X-ray emission and by large spatial inhomogeneity. Chromospheres of late-type stars have a unique physics, where optically thick radiation fields with departures from thermodynamic equilibrium, strong magnetic fields, mechanical heating and mass flows dominate. In this short overview first the characteristic NLTE thermodynamics of chromospheres is discussed. Then the observational evidence for magnetic fields and their relation to the chromospheric emission activity of the sun and of stars (e.g. the magnetic field- radiative emission-, the rotation-activity-, the flux-flux- correlations, the two component theory, the upper and lower limits of chromospheric emission and the Wilson-Bappu effect) is summarized. A third unique feature of chromospheres and coronae is that they need persistent mechanical heating. Proposed heating mechanisms and in particular recent advances in acoustic heating are outlined. Finally the role of chromospheres and coronae in the generation of eruptive flows is discussed.

1. Introduction

Ground based and satellite observations have shown that probably all stars with the possible exception of A-stars have shells or regions in their outer atmosphere in which the temperature is much higher than the photospheric value. These chromospheric and coronal layers show a unique physics, distinctively different from the underlying photosphere and the overlying interstellar medium, which for late-type stars is characterized by four essential ingredients: 1. large departures from local thermodynamic equilibrium (NLTE) occur in the presence of optically deep radiation fields, 2. strong magnetic fields are present, 3. the energy balance is dominated by mechanical heating and 4. eruptive-type mass flows which can lead to stellar mass-loss are present. In the photosphere departures from LTE, magnetic fields (except in concentrated field areas like sunspots) and mechanical heating are not important, and one has only convective type flows. The interstellar medium has a NLTE thermodynamics with optically thin radiation fields, weak magnetic fields, no need of a persistent mechanical heating and exhibits different gas flows.

The physics of late-type chromospheres and coronae is also distinctively different from that of similar layers or regions in early-type stars which is dominated by intense radiation fields. The outer atmosphere of early-type stars is characterized by a NLTE thermodynamics, strong radiatively driven winds and by localized heating due to radiatively amplified acoustic shock waves. In Section 2 the NLTE thermodynamics of chromospheres and coronae is outlined, Section 3 deals with the solar and stellar magnetic fields and their relation with the chromospheric and coronal emission activity. Section 4 reviews the heating mechanisms and discusses why heating is necessary, while Section 5 describes mass flows generated in the chromospheres and coronae. The conclusions are presented in Section 6. Chromospheric phenomena have been reviewed recently by Linsky (1988), Hammer (1987) as well as Jordan and Linsky (1987), see also Linsky (1980) and Ulmschneider (1979).

2. Chromospheric NLTE thermodynamics

A characteristic property which is unique to chromospheres is encountered when one tries to construct empirical chromosphere models. Detailed models of the chromospheric temperature structure in the solar network and supergranulation cell interior are derived from UV continuum observations obtained from the Skylab space station (Vernazza et al. 1981). Assuming a run of temperature versus depth and enforcing hydrostatic equilibrium, an empirical model is constructed and the emergent intensity as function of wavelength is simulated. This run of temperature is subsequently varied such that the theoretical intensities optimally fit the observed intensities of the CI, SiI and HI continua. A characteristic property in this procedure is the strong departure from thermodynamic equilibrium (NLTE).

Assume two gas elements which have the same temperature: one deep in the photosphere and the other in the chromosphere (Fig. 1). Consider the energy levels and transition rates e.g. for the CI continuum. As for typical chromospheric temperatures the stimulated emission can be neglected for the UV continua, the radiative

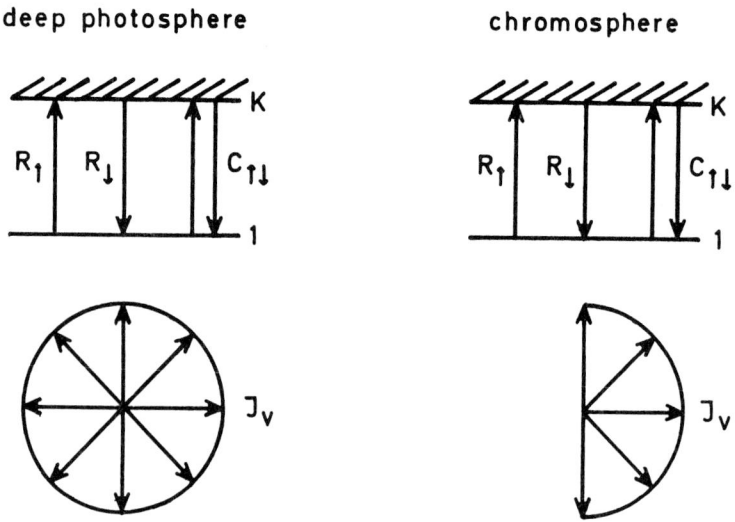

Fig. 1 Radiative and collisional transitions as well as the radiation field in chromospheric and deep photospheric gas elements.

recombination rate is given by $R_\downarrow = n_K(n_1^*/n_K^*) \int \alpha_1 \frac{4\pi}{h\nu} \frac{2h\nu^3}{c^2} \exp(-h\nu/kT) d\nu$ and the radiative absorption rate by $R_\uparrow = n_1 \int \alpha_1 \frac{4\pi}{h\nu} J_\nu d\nu$. Consider now the frequency integrals in $R_\downarrow$ and $R_\uparrow$. As we have the same temperature the frequency integrals for $R_\downarrow$ are the same, but for $R_\uparrow$ very different in the two gas elements. This is due to the fact that in the deep photosphere the mean intensity J_ν derives from an isotropic radiation field, while in the chromosphere only from the outwardly directed radiation field. With the collision rates $C_\uparrow = n_1 n_e \Omega_{1K}, C_\downarrow = n_K(n_1^*/n_K^*) n_e \Omega_{1K}$ one finds for the departure coefficient

$$b_1 = \frac{n_1}{n_1^*} \frac{n_K^*}{n_K} = \frac{\int \alpha_1 \frac{4\pi}{h\nu} \frac{2h\nu^3}{c^2} e^{-h\nu/kT} d\nu + n_e \Omega_{1K}}{\int \alpha_1 \frac{4\pi}{h\nu} J_\nu d\nu + n_e \Omega_{1K}}, \quad (1)$$

where starred quantities are values in local thermal equilibrium (LTE). To enforce a balance between the radiative rates valid for statistical equilibrium, the rate $R_\uparrow$ must increase by overpopulating $n_1 = b_1 n_1^*(n_K/n_K^*)$ with $b_1 \approx 20$. An added difficulty is that J_ν is not a local function of height but must be computed solving the transfer equation using the NLTE source function S_ν:

$$J_\nu = \Lambda\{S_\nu\} \quad , \quad S_\nu = \frac{2h\nu^3}{c^2} \frac{1}{b_1 e^{h\nu/kT} - 1} \quad . \quad (2)$$

The *characteristic chromospheric difficulty* thus is a twofold one, that one needs to iterate between Eqs. (1) and (2) and that J_ν depends in a complicated way on the entire atmospheric structure. In the *transition layer and the corona* the thin plasma approximation applies, that is, radiative absorptions and return collisions can be neglected and Eq. (1) becomes a simple local function $b_1 = \int \alpha_1 \frac{4\pi}{h\nu} \frac{2h\nu^3}{c^2} \exp(-h\nu/kT) d\nu/n_e \Omega_{1K}$ while in the *interstellar medium*, e.g. for planetary nebulae, the intensity in Eq. (1) is given by an optically thin value $J_\nu = W B_\nu$ where W is a dilution factor and $B_\nu(T_{eff})$ the Planck function of a very hot central star. Here with very low collision rates one then has $b_1 = \int \alpha_1 \frac{4\pi}{h\nu} \frac{2h\nu^3}{c^2} \exp(-h\nu/kT) d\nu / \int \alpha_1 \frac{4\pi}{h\nu} W B_\nu d\nu$, which again is a local function.

The described procedure was used to derive empirical chromosphere models for different solar surface regions from bright network elements to dark supergranulation cell interiors and plage areas (Vernazzza et al. 1981, Avrett 1985). In all models a rapid outward temperature increase was found with an almost discontinuous temperature rise in the transition region. Transition layer models have been constructed exploiting the thin plasma approximation discussed above and deriving emission measures from the transition layer line fluxes (Jordan and Brown 1981, Jordan et al. 1987). This way the chromospheric temperature structure can be followed to layers of increasing height in the lines L_α near $10^4 K$, He II 304 Å near $2 \cdot 10^4 K$, C IV 1335 Å near $10^5 K$ to Mg X 625 Å near $2 \cdot 10^6 K$, the temperature maximum of the corona.

For stars other than the sun a large number of empirical chromosphere models have been constructed by reproducing the Ca II K and Mg II h+k line observations (see Linsky 1980), similarly as has been described above for the solar UV continuum observations. These models have the same characteristic difficulty with the NLTE thermodynamics as the continuum models, with the added problem that the line shape and the energy balance depend sensitively on the treatment of coherence in the line-wings (PRD). These problems are even more severe in time-dependent chromospheric wave calculations, for a review see Ulmschneider and Muchmore (1986). It should be pointed out that the above procedures to infer chromospheric temperatures depend on the assumption of a smoothly varying atmosphere and could be different if a very wavy atmosphere is assumed in which the emission occurs predominantly at the hottest parts of the wave (see Kalkofen et al. 1984).

The solar models of Vernazza et al. can be used to infer the chromospheric radiation losses which give a clue to the amount of mechanical heating necessary to support the chromosphere. As is shown in Section 4, persistent mechanical heating is absolutely essential to maintain both chromospheres and coronae. Vernazza et al. on basis of the empirical models identified the H^- continuum, and the Ca II, Mg II lines as the main chromospheric emitters. They also find, that the H_α losses are essentially balanced by radiative heating in the Balmer continuum. Recently Anderson and Athay (1989) have improved the computation of the net radiative cooling rate, Φ_R, by including FeII line losses. For the average chromosphere, Fig. 2, adopted from their work, shows Φ_R ($erg\ cm^{-3}\ s^{-1}$) as function of height together with the mechanical flux F_M which is obtained by integrating Φ_R over height. It is seen that a mechanical flux of about $10^7\ erg\ cm^{-2}\ s^{-1}$ at 400 km height is needed and that this flux at greater height decreases proportional to the gas pressure p, roughly like $F_M \approx 2.5 \cdot 10^9\ m \approx 9.2 \cdot 10^4\ p$, where m is the mass column density.

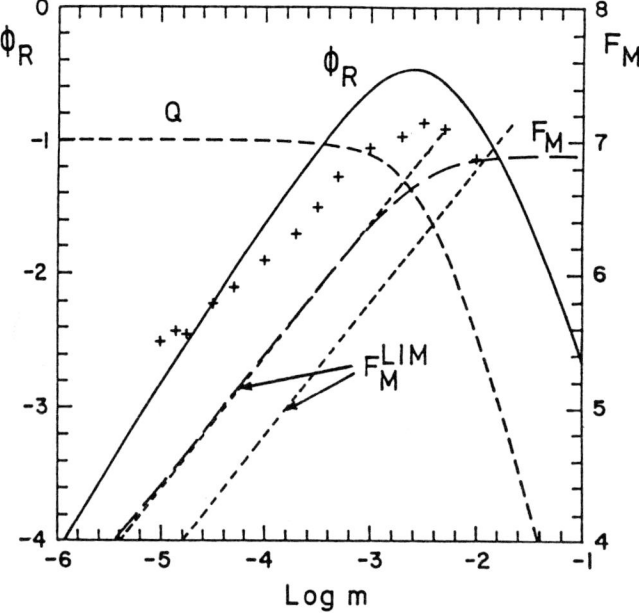

Fig. 2 Net radiative cooling rate Φ_R and mechanical Flux F_M after Anderson and Athay (1989), together with theoretical acoustic fluxes.

It is interesting that observations of short period (compared to the acoustic cut-off period P_A, see Eq. (7)) acoustic waves find energy fluxes of the same magnitude (Deubner et al. 1988) and that limiting strength acoustic shock waves have the same pressure dependence (see Section 4). Fig. 2 shows a comparison of the empirical F_M dependence with the theoretical limiting shock strength results of Eq. (8) for waves of period $P = 44, 22\ s$, where one has $F_M = 9.2 \cdot 10^4 p$, $2.4 \cdot 10^4 p$, respectively.

3. Magnetic fields and chromospheric activity

Since about 10 years it is known that the solar surface is a two component medium with nonmagnetic areas where hot granulation cells come to the surface, and intense magnetic flux tubes concentrated in the intergranular lanes where the cool material flows back into the solar interior (Stenflo 1978, Zwaan 1978). The magnetic flux tubes are situated mainly on the boundaries of the supergranulation cells and in plage regions. Whether there are also flux tubes inside the supergranulation cell is presently controversial.

From observation one finds that the chromospheric emission e.g. in the Ca II H+K line cores is strongly concentrated in the magnetic areas above the supergranulation boundaries and plage regions. By detailed comparison of a Ca II K spectroheliogram and a magnetogram of an active region complex, Schrijver et al. (1989) have newly quantified the relation between the Ca II flux and the magnetic flux. They find

$$\Delta F_{CaII} = 0.6\ \log <fB> + 4.8\ , \qquad (3)$$

where f is the filling factor, B the magnetic field strength in G, and ΔF_{CaII} is the Ca II excess flux in $erg\ cm^{-2}\ s^{-1}$. Here $<fB> = \Phi/A$ where Φ is the measured average magnetic flux and A the area. The excess flux is computed by subtracting from the observed flux F_{CaII} the lower limit flux $F_{CaII\ LL}$ which consists of a non-magnetic photospheric radiative equilibrium contribution and a non-magnetic chromospheric basal flux contribution attributed to the heating by acoustic waves. For $F_{CaII\ LL}$ a stellar lower limit flux (see Fig. 3) was taken. This stellar lower limit is essentially identical to the flux $F_{CaII\ LL}$ observed from the supergranulation cell interior.

The X-ray and transition layer line emission, although more diffuse, also depends on the photospheric magnetic field. Similar as for the Ca II flux a correlation between the coronal X-ray emission flux and the magnetic flux has been inferred by Schrijver et al. (1985), Schrijver (1987a, 1989, Fig. 2) from a comparison of active regions of different size, quoted here after Vilhu (1987):

$$F_X = 4 \cdot 10^4 f\ B\ . \qquad (4)$$

Fig. 3 after Rutten (1987) shows the Ca II emission flux for dwarfs (dots) and giants (crosses) as function of colour. Note that all dwarfs of a given colour in a vertical slice in Fig. 3 are stars of the same kind, having the same effective temperature T_{eff} and gravity g. It is seen that these stars can have up to a factor of ten different Ca II emission fluxes which implies a corresponding difference in the magnetic flux coverage. Similar stars are thus shown to have a considerable *chromospheric emission variability*. Note, however, that towards small B-V this variability decreases markedly. Vaiana et al. (1981) show in their Fig. 5 a similar plot of X-ray emission flux versus spectral type for main sequence stars. Here the *X-ray emission variability* of three to four orders of magnitude for stars of the same T_{eff} and g is considerably larger.

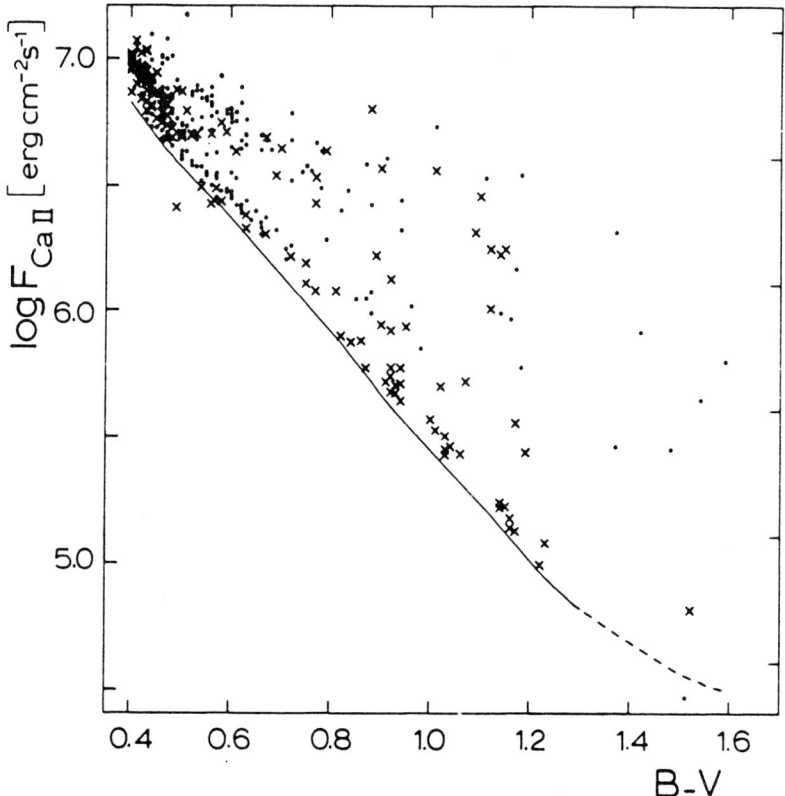

Fig. 3 Chromospheric emission fluxes F_{CaII} of stars versus colour after Rutten (1987). The drawn line is the lower limit flux $F_{CaII\ LL}$.

As shown in Fig. 4 after Rutten (1987) this chromospheric emission variability of late-type stars is due to the different rotation rates of similar kinds of stars. An X-ray flux-rotation relation of the same kind is shown in Fig. 5 of Pallavicini et al. (1981). These emission-rotation correlations are attributed to the dynamo mechanism. In the convection zones of late-type stars the dynamo mechanism leads to a greater magnetic flux generation the faster the stars rotate. In early- type stars there is no X-ray emission - rotation correlation but an X-ray emission - bolometric luminosity correlation as shown by Pallavicini et al. (1981, Figs. 2 and 3). Here the X-rays are generated in the hot regions behind strong acoustic shocks which have been amplified by the intense radiation field. For a recent review of the rotation-emission flux correlation and the evolution of rotation see Catalano (1989, this volume).

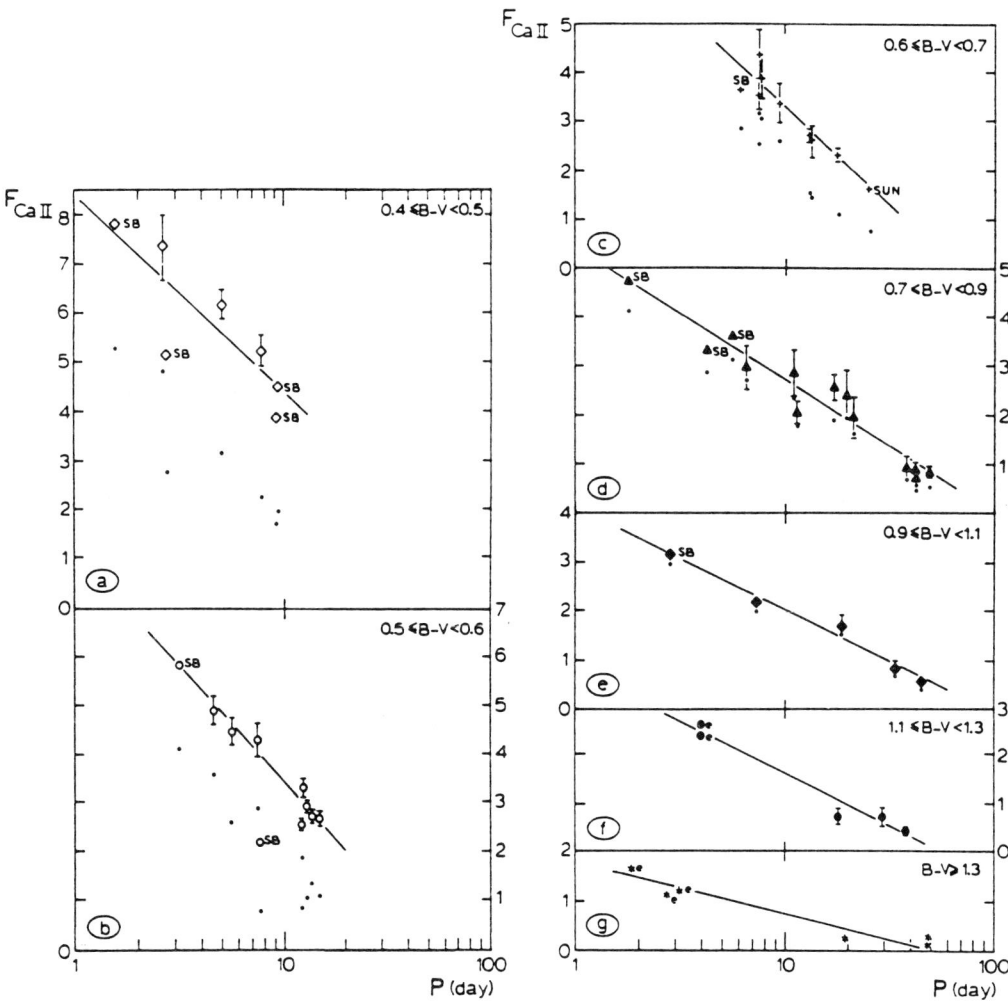

Fig. 4 Chromospheric emission fluxes F_{CaII} of stars versus rotation period for different colour intervals after Rutten (1986).

An interesting idea by Oranje and Zwaan (1985) and particularly by Schrijver (1987b) was to separate the chromospheric emission into two components: a *nonmagnetic component* which is independent of rotation and only depends on T_{eff} and possibly slightly on g, and a *magnetic field related component* which depends on rotation. For late-type stars the heating by the nonmagnetic component leads to a *basal chromospheric emission flux* which is tentatively identified as due to the heating by acoustic shock waves. This basal flux constitutes a low background emission observable in stars of very low rotation rate but for faster rotating stars usually is greatly exceeded by the more energetic magnetic heating component.

The question is how to disentangle the nonmagnetic from the magnetic component, both for the sun and for stars? Stepien and Ulmschneider (1989) as well as Hammer and Ulmschneider (1989, see also Hammer 1989, this volume) show that pure acoustic wave heating either produces very weak coronae or no coronae at all. It thus appears that the coronal X-ray emission can only be produced by the magnetic field related heating component and should serve as an excellent indicator for the magnetic chromospheric emission component. It is thus not surprising that subtracting from the measured emission flux at a given colour the lowest observed emission flux from stars at this colour, Schrijver (1983) found that the correlation between the observed X-ray flux and e.g. the Ca II emission flux is considerably improved. The improvement of the magnetic field - Ca II flux correlation by subtracting a lower limit flux $F_{CaII\ LL}$ has already been discussed above.

The idea, that in the heating of chromospheres two components, a basal nonmagnetic, probably acoustic component and a magnetic, rotation dependent component are at work, can explain several other observations. We have already seen that for vanishing rotation only the nonmagnetic component survives which leads to the basal chromospheric emission. For Ca II after adding the unavoidable photospheric background line core and wing emission, one then finds the observed lower limit of the Ca II emission. For Mg II where the photospheric background is very weak the basal chromospheric emission is directly observed as lower limit flux. In the C IV and X-ray emission, arising in the transition layer and the corona, respectively, the observed lower limits are detection limits as shown by Schrijver (1987b).

Consider in Fig. 3 the low variability of the F-stars (see also Fig. 1 of Walter and Schrijver 1987). The theoretical acoustic energy generation computations of Bohn (1984) shown in Fig. 5 exhibit a strong increase towards higher T_{eff}. Although this calculation has been criticized, the general result, that earlier type stars as well as giants have more acoustic energy generation is not affected. This is a consequence of the fact that the acoustic energy generation depends on a high power of the convective velocity u. In fairly efficient convection zones one has $\sigma T_{eff}^4 \approx \rho u^3$, that is, the total stellar flux is carried mainly by the convective flux. As the convective velocity of stars is greatest near the surface where the densities are lowest, the acoustic energy generation is large only in the surface layers. Consequently acoustic energy generation does not depend on the depth of the stellar convection zone. Despite the fact that F-stars have shallow convection zones, the fact that in these stars T_{eff} is largest, before the convection zones disappear towards earlier spectral type, ensures that F-stars have the largest amount of acoustic energy production. Going towards earlier type stars, one thus adds an increasingly stronger acoustic heating component to a given variable magnetic heating component, which results in a reduced variability of the total emission.

By the same token as seen in Fig. 5, the acoustic energy generation increases when going from dwarfs to giants, because the density in the atmospheres of giant stars is much smaller than in dwarfs, and thus requires larger convective velocities to transport the same total flux σT_{eff}^4. From this one would expect a higher basal flux limit for giants than for dwarfs. Actually observations show the opposite, that the dwarfs appear to have a slightly higher basal flux limit than the giants (Schrijver

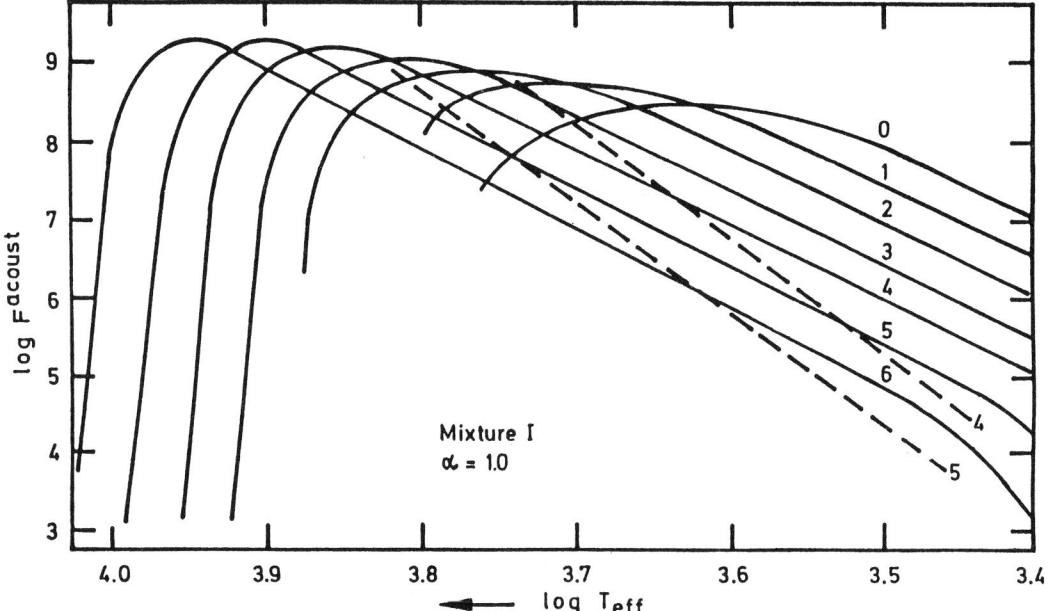

Fig. 5 Theoretical acoustic energy fluxes generated in stellar surface convection zones versus $\log T_{eff}$ with $\log g$ as parameter after Bohn (1984).

et al. 1989). Ulmschneider (1988, 1989) has shown that this can be explained by greater radiation damping of the acoustic waves in the giants and by the limiting shock strength behaviour of the acoustic waves, which leads to a lower limiting acoustic flux due to the lower gas pressure in giant star atmospheres.

Red giant stars, due to their low rotation rate resulting from angular momentum conservation during the large evolutionary increase in radius and from angular momentum loss by massive stellar winds, are another class of stars, where the two component chromospheric heating theory can be tested. Middelkoop (1982, Fig. 4a-c) showed that the chromospheric emission variability decreases very much toward late spectral type and there becomes a low basal emission. The same was found by Judge (1989) who studied late-type giants with peculiar chemical abundances. These are highly evolved stars in their red giant and AGB evolutionary state. Judge finds that these late M- and C-stars (the latter after correction for the Mg II circumstellar shell absorption) all fall on the same emission versus T_{eff} relation and do not show a chromospheric emission variability, consistent with the fact that in these stars only the nonmagnetic acoustic heating component seems to be present.

If one now considers the other extreme, the very quickly rotating stars, it is found that these stars approach an *upper limit of the chromospheric emission*. In Fig. 6 after Vilhu (1987) it is seen that this limit (dashed) is populated by different types

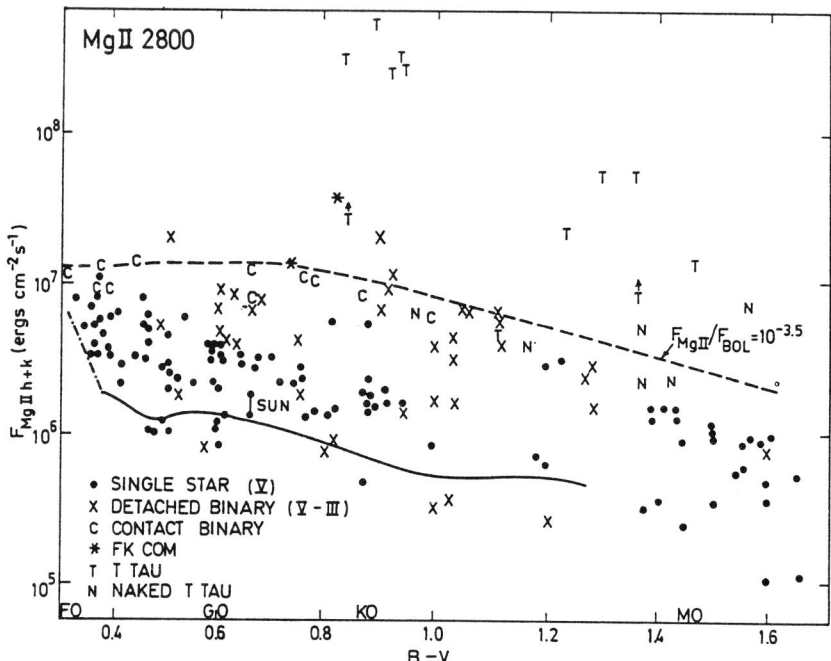

Fig. 6 Upper limit of the chromospheric emission flux F_{MgII} versus colour after Vilhu (1987).

of binaries where, the fast stellar rotation is apparently produced by tidal interaction between the binary stars. From the chromospheric emission - magnetic flux correlation it is clear that there must be a saturation, where the surface of the star is nearly completely filled by magnetic fields, that is, where the filling factor of the magnetic field is nearly one. This is supported by X-ray observation which also show a distinctive upper limit for rapidly rotating stars. Vilhu (1987, Fig. 5) showed that this X-ray flux agrees well with the solar relation found by Schrijver given above (Eq. 4) for a filling factor $f = 1$ and a field strength $B = 2000\ G$.

It is interesting that Fig. 6 shows stars with Mg II emission larger than the upper limit. These stars are T-Tau stars where the emission comes from the innermost part of an accretion disk around these young stars. Here the energy which balances the chromospheric radiation losses is derived from mass accretion and not from the stellar convection zone inside the star. Note that the naked T-Tau stars, which apparently have lost their accretion disk, fall again on the upper limit for rapidly rotating stars.

Another interesting chromospheric effect is the well known Wilson-Bappu effect. Wilson and Bappu (1957) discovered that there is a tight correlation between the width of the Ca II K line emission core and the absolute visual magnitude of the stars which extends over roughly 15 magnitudes and is nearly independent of the emission strength of the line. A similar Wilson-Bappu relation has been found for the MgII k line (Elgarøy 1988). Hammer (1987) has recently reviewed the discussion on the origin of the Wilson-Bappu effect and concludes that several processes could produce this effect and that therefore the effect appears unpredictive as far as the heating mechanisms are concerned.

4. Mechanical heating

Stellar chromospheres and coronae are layers with persistent mechanical heating. Let us first discuss why chromospheres and coronae constantly require mechanical heating. Consider a chromospheric gas element. Due to the solar wind this element will slowly move and also exchange energy with its surroundings. A powerful bookkeeping quantity for monitoring energy exchange is the entropy S [$erg\ g^{-1}\ K^{-1}$], which in chromospheres changes mainly due to mechanical heating and radiation. In transition layers and coronae, thermal conduction, viscous- and Joule heating as well as other mechanisms described below are at work. From the second law of thermodynamics the energy equation describing the heat addition to a chromospheric gas element can be written

$$\rho T \left(\frac{\partial S}{\partial t} + \mathbf{v} \cdot \nabla S \right) = \Phi_M - \Phi_R \quad , \tag{5}$$

where ρ is the density, T the temperature, $\mathbf{v}$ the flow speed, Φ_M [$erg\ cm^{-3}\ s^{-1}$] the mechanical heating rate and Φ_R the radiative cooling rate. For illustration let us assume grey radiation, then we have $\Phi_R = 4\pi\kappa(B - J)$, where κ is the absorption coefficient, B the frequency integrated Planck function and J the mean intensity. Other expressions for Φ_R are given by Ulmschneider and Muchmore (1986).

Since the chromosphere has not changed much over millions of years we can consider it as a steady state phenomenon and have $\partial S/\partial t = 0$. The entropy gradient can be approximated by $dS/dx = g/T$, valid for an isothermal atmosphere. For the heating due to the potential energy flux one then gets $\rho T \mathbf{v} \cdot \nabla S = \rho v g \approx 8 \cdot 10^{-7}\ erg\ cm^{-3}\ s^{-1}$. Here $\rho \approx 4 \cdot 10^{-11}\ g/cm^3$ and $v \approx 0.75\ cm/s$, were used for the middle chromosphere. Compared with the value $\Phi_R \approx 0.3$ found by Anderson and Athay (1989) (see Fig. 2), the left hand side of Eq. (5) can thus be safely neglected and there remains a balance between mechanical heating and radiative cooling.

Let us suppose for a moment that there is no mechanical heating. Then radiative equilibrium prevails, and one has in the grey case $B = J$. With $B = \sigma T^4/\pi$ and $J = \frac{1}{2}\sigma T_{eff}^4/\pi$, where the factor $1/2$ takes into account that there is only outgoing radiation, one finds $T^4 \approx \frac{1}{2}T_{eff}^4$ or $T \approx 0.8 T_{eff}$. This is the well known result that grey radiative equilibrium atmospheres have a boundary temperature with $T < T_{eff}$. From the empirical chromospheric and coronal temperature distributions of Section 1 with $T >> T_{eff}$ it is thus clear that *mechanical heating is absolutely essential* and that in Eq. (5) one can neglect the radiative heating term $4\pi\kappa J$, and has a balance of mechanical heating and radiative cooling, $\Phi_M = 4\pi\kappa B$. The heating moreover must be *constantly* applied, because if it were switched off, then the chromosphere would rapidly cool down to the boundary temperature at a time scale of the radiative damping time $t_{rad} = \rho c_v/(16\kappa\sigma T^3) \approx 200\ s$. Here c_v is the specific heat at constant volume. For the corona one finds $t_{rad} = \rho c_v T/(7 \cdot 10^{-17} T^{-1} n^2) \approx 4.3 \cdot 10^3 \ s \approx 1 \ h$ for the radiative damping alone, while thermal conduction decreases this time further. Here we have assumed the thin plasma approximation with $T \approx 2.5 \cdot 10^6 \ K$ and $n \approx 2 \cdot 10^9 \ cm^{-3}$.

The mechanisms which have been proposed to heat the chromosphere and corona have recently been summarized by Narain and Ulmschneider (1989). Tab. 1 gives a list of these mechanisms.

mechanism	dissipation
acoustic waves	shock dissipation
slow-mode mhd waves	shock dissipation
fast-mode mhd waves	Landau damping
Alfvén waves	mode-coupling
	resonant heating
	turbulent heating
	Landau damping
	phase mixing*
	resonant absorption*
current heating	Joule heating
	reconnection
	turbulent heating

Table 1 *Chromospheric and coronal heating mechanisms and their mode of dissipation. Dissipation modes labeled * depend on transverse Alfvén speed variations.*

We first discuss the acoustic heating mechanism. Here acoustic waves generated in the stellar surface convection zone run down the steep density gradient of the outer stellar atmosphere and, due to energy conservation, grow to large amplitude and form shocks. Acoustic wave heating is due to shock dissipation, while direct viscous and thermal conductive heating is unimportant. There are two effects which severely influence the behaviour of acoustic waves. First, the acoustic wave energy flux is strongly affected by *radiation damping*, when the wave propagates through the radiation damping zone, which in the sun extends to heights of about 200 km, but for other stars can be much more extended (Ulmschneider 1988). Second, it is a persistent result of time-dependent acoustic wave calculations that acoustic shock waves, once formed, tend to quickly reach *limiting shock strength*.

This behaviour, where the wave amplitude becomes essentially constant with height, and independent from the initial amplitude, results from the balance of shock dissipation which decreases the wave amplitude and amplitude growth which is caused by the steep density gradient (Ulmschneider 1970, Cuntz and Ulmschneider 1988, Ulmschneider 1989). In an isothermal atmosphere the limiting shock strength is given by (Ulmschneider 1970)

$$M_S^{Lim} = 1 + \frac{\gamma g}{4 c_S} P \quad, \tag{6}$$

where $\gamma = 5/3$ is the ratio of specific heats, $c_S(T_{eff})$ the sound speed and P the wave period. After the calculations of Bohn (1984) the acoustic wave spectrum generated by the turbulent motions in the stellar convection zone extends from the acoustic cut-off period P_A roughly 1.5 decades to shorter wave periods with a maximum near $P_A/10$. This maximum shifts towards $P_A/5$ and longerwards for late type dwarf stars. Thus acoustic waves typically have wave periods of

$$P = \frac{1}{10} P_A = \frac{1}{10} \frac{4\pi c_S}{\gamma g} \quad . \tag{7}$$

From Eqs. (6) and (7) one has $M_S^{Lim} = 1.3, 1.6$ for $P = P_A/10, P_A/5$, respectively. Note that this result is roughly valid for all late-type stars, independent of gravity and T_{eff}. Consequently the velocity-, temperature- and pressure amplitudes of limiting acoustic shock waves are also roughly the same for all late-type stars. With $v \approx 2c_S(M_S^{Lim} - 1)/(\gamma+1) \approx 0.23 c_S$ one finds

$$F_M^{Lim} = \frac{1}{3}\rho v^2 c_S \approx 2.4 \cdot 10^4 \, p \quad , \tag{8}$$

for $P = P_A/10$, or $F_M^{Lim} \approx 9.2 \cdot 10^4 \, p$ for $P = P_A/5$. These limiting acoustic wave fluxes, which are roughly valid for all late-type stars, have been plotted in Fig. 2.

Slow-mode mhd waves in homogeneous fields and longitudinal mhd tube waves in magnetic flux tubes, in situations where the Alfvén speed, $v_A > c_S$, are essentially acoustic waves propagating along the magnetic field (Herbold et al. 1985). They very likely are an important heating mechanism for the lower and middle chromosphere (see Ulmschneider 1986). In addition to the direct generation of these waves in the convection zone they are also generated in the chromosphere and corona from transverse or torsional Alfvén waves by mode-coupling. Like acoustic waves, they heat by shock dissipation. From the acoustic nature of these waves it is clear that they suffer radiation damping and show the limiting shock strength behaviour (Herbold et al. 1985), when they travel along magnetic flux tubes with slowly varying cross- section (e.g. tubes where the field has fully spread over the available space).

Fast-mode mhd waves due to the strongly increasing Alfvén speed in the upper chromosphere are thought to be strongly refracted away from the high chromosphere and thus do not form shocks. However, the generation by mode-coupling of these waves in the corona, their property to refract away from regions of high Alfvén speed and to dissipate by Landau damping could be important for the explanation of coronal loops with cool cores (Habbal et al. 1979).

Transverse and torsional Alfvén waves are pure shear waves and to first order do not show temperature or density fluctuations. Consequently they do not suffer from radiation damping. They are also not easily dissipated in the chromosphere by simple Joule heating, viscous- or ion-neutral collisional heating. As they are thought to be efficiently generated in the convection zone (Ulmschneider 1986), these waves should carry a considerable energy into the upper chromosphere, corona and the solar wind (Withbroe 1988). A surprising number of more complex, nonlinear processes, some of which derived from heating experiments in plasma fusion, have been proposed to dissipate Alfvén waves (Tab. 1). Yet most of these very promising dissipation mechanisms have presently not been investigated enough to give detailed predictions in specific chromospheric or coronal loop situations. Only order of magnitude estimates are presently available for most mechanisms showing their viability (see Narain and Ulmschneider 1989).

Mode-coupling is an efficient and very general method by which mhd waves in the complex coronal magnetic field geometry generate other types of mhd waves. Fig. 7 from Ulmschneider et al. (1989, see also Zähringer and Ulmschneider 1987) is an illustrative example. Here a thin, vertically directed solar magnetic flux tube has been excited at the bottom by purely transverse shaking. The restoring forces are due to the tension of the magnetic field and are directed towards the center of curvature of the magnetic field distorted by the wave (c.f. Fig. 7). The curvature forces can be divided into vertical and horizontal components of which the horizontal components drive the pure transverse Alfvén wave and the vertical components lead to compression and expansion of the gas in the flux tube, that is, generate a longitudinal tube wave. It is easy to visualize related situations where mode-coupling should occur: bent magnetic loops (Wentzel 1974) or magnetic loops meeting adjacent magnetic fields must lead to deformations of the gas columns or sheets and thus give rise to acoustic-type waves which subsequently are easily dissipated by shocks.

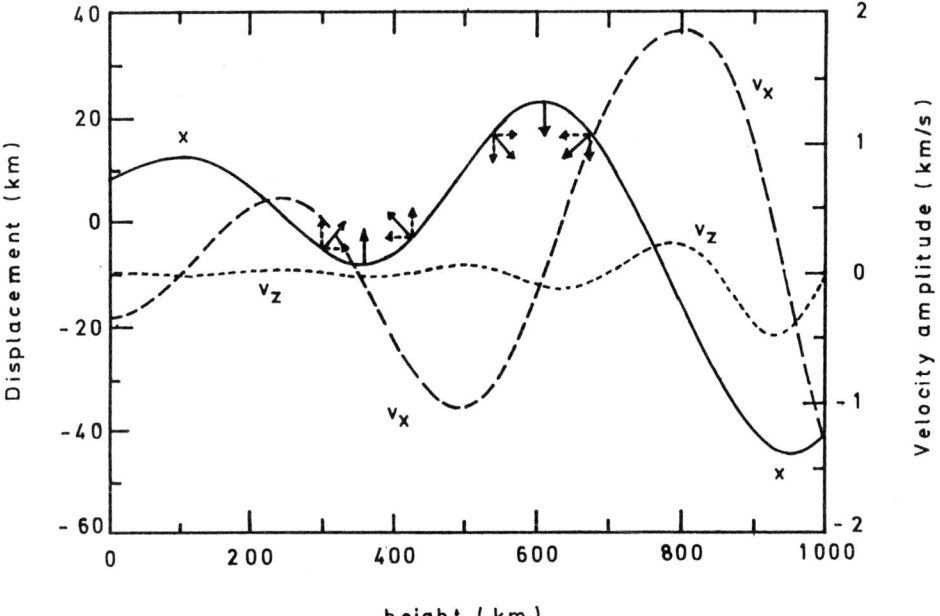

Fig. 7 *Longitudinal-transverse mhd tube wave initiated by purely transverse shaking at the bottom. The components of the curvature force vector are shown. x is the horizontal displacement, v_x and v_z are the horizontal and vertical velocity components.*

Resonant heating lives from the reflection of Alfvén waves at the foot points of a coronal loop which is very analogous to the situation in an electric resonant circuit (Ionson 1984). From a wide spectrum of Alfvén waves generated in the convection zone, similarly to an electric circuit, only those waves which are in a narrow band around the resonant frequency get absorbed.

Turbulent heating possibly is a very promising method for the dissipation of Alfvén wave energy but also for the magnetic energy stored at much slower time scales (see current heating below). Here the magnetic field perturbations in the closed or open coronal flux tubes are supposed to be dissipated by turbulence developing perpendicular to the tube axis. Dahlburg et al. (1988) have recently performed three-dimensional time- dependent numerical simulations in a vertical coronal gas column, which show this effect and the ensuing formation of current sheets where reconnection takes place and vortex structures where viscous dissipation occurs.

Phase-mixing and resonant absorption are processes which live from a transverse Alfvén speed gradient of the coronal flux tube. After coherent transverse shaking at the foot point of the tube, the transverse Alfvén speed variation leads to a variation of the phase speed (phase-mixing) in adjacent field lines resulting in the build-up of transverse magnetic field gradients which results in current sheets and reconnective heating (Heyvaerts and Priest 1983). Resonant absorption occurs, when at a certain magnetic field line the Alfvén speed matches the phase speed of the Alfvénic surface wave which travels in the loop boundary region where there is a transverse Alfvén speed variation. In a narrow layer around the resonant field line considerable energy dissipation occurs (Ionson 1978). Resonant absorption should not be confused with resonant heating discussed above, which is due the constructive interference from waves reflected at the photospheric foot points of a coronal loop. While resonant heating occurs only in loops, resonant absorption also occurs in open structures. In realistic situations loop resonances and resonant absorption occur simultaneously as has been shown by Grossmann and Smith (1988), who studied the resonant absorption of a photospheric spectrum of Alfvén waves in coronal loops.

Slow photospheric footpoint motions with timescales much longer than the loop transit timescales can store large amounts of magnetic energy in coronal loops which is thought to be dissipated by currents flowing along the magnetic field. In addition these motions lead to interweaving and winding magnetic field filaments (Van Ballegooijen 1986). In both cases the currents are unsteady and associated with reconnection and turbulent heating as discussed above. It is widely assumed that the current heating processes correspond to micro- and nanoflare events (see Section 5).

5. Mass flows

The fourth essential characteristic of chromospheres and coronae is that they are layers where mass flows and mass loss originate. A well known mass flow phenomenon on the sun are the *spicules* (Beckers 1972). Observed mainly on the solar limb, spicules are jets of cool plasma with temperatures of $T_{sp} \approx 1.5 \cdot 10^4 \ K$ protruding into the hot corona, which rise with velocities of $v_{sp} \approx 20 \ km/s$. The lifetime of spicules is about 1/2 hr. Athay and Holtzer (1982) estimated that with a density of $n_{sp} \approx 6 \cdot 10^{10}$ cm^{-3} and an area filling factor of $A_{sp} \approx 0.01$ a particle flux density of $F_{sp} = n_{sp} v_{sp} A_{sp} \approx 1.2 \cdot 10^{15}$ cm^{-2} s^{-1} of cool spicular material enters the corona. This is about a factor 100 more than the particle flux density of the solar wind of $F_{wind} = 1.4 \cdot 10^{13}$ cm^{-2} s^{-1}, which shows that most of the spicular mass must return to the sun. The question is how. As there is little doubt that the spicular flow is along magnetic loops, the question is whether the spicular flow goes up one leg of the loop and comes down the other leg, as is proposed by the so called siphon flow models, or whether it comes down the same leg as is supposed by the eruptive flow models. In any case the spicular material is rapidly heated and comes down as gas of transition layer temperature.

Systematic *downflows* with velocities of $v \approx 4 - 17$ km/s have been observed in UV lines by Doschek et al. (1976), Brueckner (1981), Gebbie et al. (1981) and Feldman et al. (1982) particularly in the transition layer line C IV at temperatures of $T \approx 10^5$ K. Athay and Holtzer (1982) assuming a downflow velocity of $v_{df} \approx 4$ km/s, a particle density $n_{df} \approx 6 \cdot 10^9$ cm^{-3} and an area filling factor of $A_{df} \approx 0.45$, find a particle flux density of $F_{df} = n_{df} v_{df} A_{df} \approx 1.1 \cdot 10^{15}$ cm^{-2} s^{-1} in the downflows which agrees well with the spicular mass flow. The observation of double neutral lines in C IV dopplergrams moreover shows that the eruption model appears to be the correct picture with siphon flows occurring only occasionally (Athay et al. 1983, Athay 1987).

Mass flows do not depend on collimating magnetic fields, but as a result of a stochastic production of acoustic energy, can also occur in nonmagnetic situations over certain areas of the stellar surface. Cuntz (1987) describes episodic mass loss events in late-type giant stars which are caused by the nonlinear interaction of a stochastic field of acoustic shock waves. Fig. 8 shows how a slightly faster shock of small amplitude from a spectrum of acoustic shock waves catches up with the shock propagating in front and thereby strengthens to a larger shock. The now even more quickly moving shock eats up more shocks in front and eventually by cannibalizing a large number of shocks grows to large amplitude which leads to episodic mass loss.

Two different types of flows called *turbulent events* and *high velocity jets* have been observed by Brueckner and Bartoe (1983) in the corona above the quiet sun. Turbulent events are confined to small areas (< 1500 km) and have average lifetimes of 40 s. The average energy of a turbulent event is $7 \cdot 10^{23}$ erg. With 753 events per sec one has a heating flux of $9 \cdot 10^3$ erg cm^{-2} s^{-1} for the whole sun in turbulent events. High-velocity jets have moving material with velocities (≈ 400 km/s) exceeding the sound speed (≈ 120 km/s) and are confined to areas < 3000 km. A single jet with an energy of $3 \cdot 10^{26}$ erg carries a mass of $3 \cdot 10^{11}$ g to an altitude of $4000 - 16000$ km and has a maximum lifetime of 80 s. With 24 jets per second one finds a heating flux of $1 \cdot 10^5$ erg cm^{-2} s^{-1} over the whole sun.

Parker (1983, 1986, 1988) suggests that the turbulent events and high velocity jets are only the tip of the iceberg of a large number of events which are associated with the heating of the corona by reconnective dissipation at many small current sheets

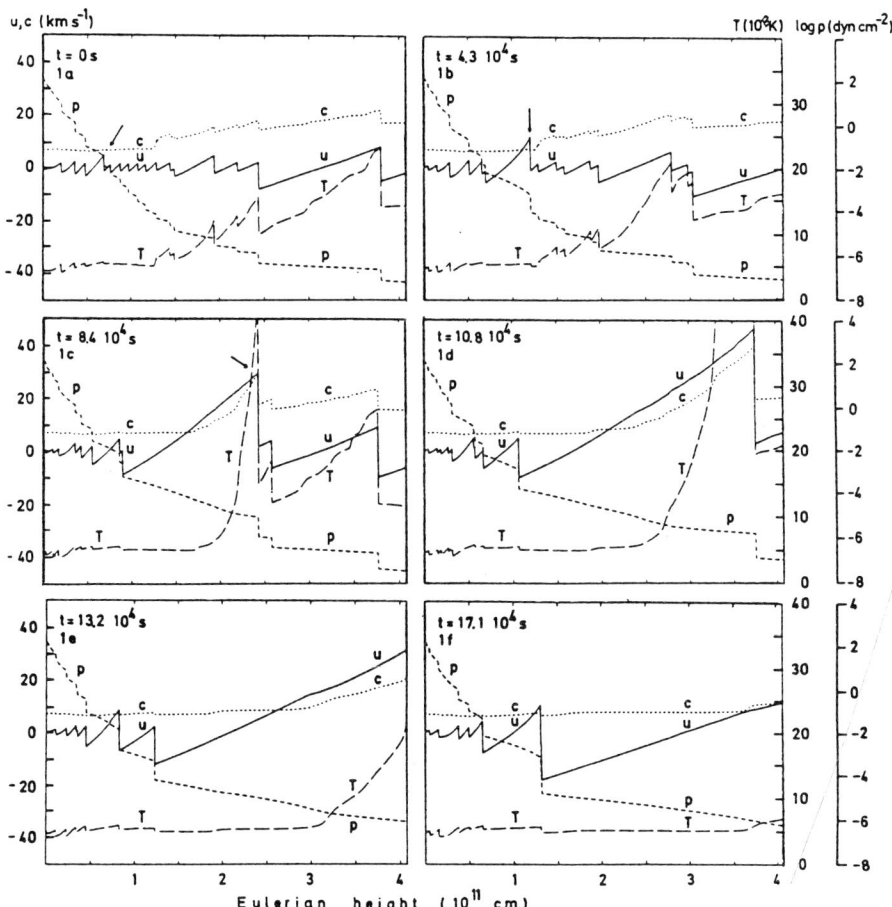

Fig. 8 Shock overtaking for a spectrum of acoustic shock waves in late-type giant star atmospheres after Cuntz (1987).

which are formed all the time as tangential discontinuities between interweaving and winding magnetic filaments. The individual reconnection events are called *nanoflares*. From SiIV and OIV emission line fluctuations and using an estimate of the heating flux of $1 \cdot 10^7$ erg cm^{-2} s^{-1} for active regions by Withbroe and Noyes (1977), Parker estimates that there are about $8 \cdot 10^6$ nanoflares per sec over the whole sun with powers of 10^{23} erg/s each and a lifetime of 20 s. The largest nanoflares reach the powers $10^{27} erg/s$ of microflares and produce the isolated turbulent events and high velocity jets observed by Brueckner and Bartoe (1983). Parker suggests that the observed X-ray corona is simply the superposition of a very large number of nanoflares.

In addition to the above discussed individual small scale or intermittent flows the chromospheres and coronae are the seat of systematic large scale persistent stellar wind flows which emerge from the open magnetic field configurations in coronal holes. These winds in solar-like stars are thermally driven, that is, the escape speed is close to the sound speed of the high-temperature coronae. For non-solar late-type giant stars, Alfvén wave driven winds are important. Even for the sun the momentum deposition due to Alfvén waves at distances greater than $15 R_\odot$ (Withbroe 1988) is necessary to explain the observed interplanetary solar wind speed. For a review of late-type stellar winds see Dupree (1986) and Hammer (1989, this volume).

6. Conclusions

We have seen that chromospheres and coronae are layers with a unique physics. The uniqueness is due to their basic energetic support by mechanical heating, to the dominant influence of the magnetic fields, to their special geometry as stellar boundary layers with an NLTE thermodynamics, and to their function as source of eruptive type flows. Unfortunately the details of the mechanical heating are presently not well understood. Acoustic waves appear to play a significant role in stars of low rotation rate and as weak general background. The chromospheres and especially the coronae of more rapidly rotating stars are dominated by magnetic type heating: by acoustic-like slow-mode and Alfvén waves as well as by unsteady currents. Similarly, the role of the acoustic and magnetic mechanisms in driving eruptive flows is poorly known. Here more detailed stellar and particularly high-resolution solar observations are needed, together with more detailed theoretical developments of the various proposed heating mechanisms.

References

Anderson, L.S., Athay, R.G..: 1989, Astrophys. J., in press
Athay, R.G.: 1987, Nature **327**, 685
Athay, R.G., Gurman, J.B., Henze, W.: 1983, Astrophys. J. **269**, 706
Athay, R.G., Holzer, T.E.: 1982, Astrophys. J. **255**, 743
Avrett, E.H.: 1985, in: Chromospheric Diagnostics and Modelling, B.W. Lites Ed., Natl. Solar Observatory, Sunspot NM, p.67
Beckers, J.M.: 1972, Ann. Rev. Astron. Astrophys. **10**, 73
Bohn, H.U.: 1984, Astron. Astrophys. **136**, 338
Brueckner, G.E.: 1981, in Solar Active Regions, F.Q. Orrall Ed., Colorado Ass. Univ. Press, Boulder CO, USA, p. 113
Brueckner, G.E., Bartoe, J.D.F.: 1983, Astrophys. J. **272**, 329
Cuntz, M.: 1987, Astron. Astrophys. **188**, L5
Cuntz, M., Ulmschneider, P.: 1988, Astron. Astrophys. **193**, 119
Dahlburg, R.B., Dahlburg, J.P., Mariska, J.T.: 1988, Astron. Astrophys. **198**, 300
Dahlburg, R.B., Dahlburg, J.P., Mariska, J.T.: 1988, Astron. Astrophys. **198**, 300
Deubner, F.-L., Reichling, M., Langhanki, R.: 1988, in Advances in Helio- and Asteroseismology, IAU Symp. **123**, J. Christensen-Dalsgaard, S. Frandsen Eds., p. 439
Doschek, G.A., Feldman, U., Bohlin, J.D.: 1976, Astrophys. J. **205**, L177
Dupree, A.K.: 1986, Ann. Rev. Astron. Astrophys. **24**, 377
Elgarøy, Ø.: Astron. Astrophys. **204**, 147
Feldman, U., Cohen, L., Doschek, G.A.: 1982, Astrophys. J. **255**, 325

Gebbie, K.B., et al.: 1981, Astrophys. J. **251**, L115
Grossmann, W., Smith, R.A.: 1988, Astrophys. J. **332**, 476
Habbal, S.R., Leer, E., Holzer, T.E.: 1979, Solar Phys. **64**, 287
Hammer, R.: 1987, in: Solar and Stellar Physics, E.H. Schröter, M. Schüssler Eds., Lecture Notes in Physics **292**, Springer, Berlin, Germany, p. 77
Hammer, R., Ulmschneider, P.: 1989, Astron. Astrophys., in press
Herbold, G., Ulmschneider, P., Spruit, H.C., Rosner, R.: 1985, Astron. Astrophys. **145**, 157
Heyvaerts, J., Priest, E.R.: 1983, Astron. Astrophys. **117**, 220
Ionson, J.A.: 1978, Astrophys. J. **226**, 650
Ionson, J.A.: 1984, Astrophys. J. **276**, 357
Jordan, C., Linsky, J.L.: 1987, in: Chromospheres and Transition Regions, Y. Kondo Ed., D. Reidel, Dordrecht, p.259
Judge, P.G.: 1989, in The Evolution of Peculiar Red Giant Stars, IAU Coll. **106**, in press
Kalkofen, W., Ulmschneider, P., Schmitz, F.: 1984, Astrophys. J. **287**, 952
Linsky, J.L.: 1980, Ann. Rev. Astron. Astrophys. **18**, 439
Linsky, J.L.: 1988, in: Multiwavelength Astrophysics, F. Cordova Ed., Cambridge Univ. Press, p. ?
Middelkoop, F.: 1982, Astron. Astrophys. **113**, 1
Narain, U., Ulmschneider, P.: 1989, Space Science Reviews, in press
Oranje, B.J., Zwaan, C.: 1985, Astron. Astrophys. **147**, 265
Pallavicini, R., Golub, L., Rosner, R., Vaiana, G.S., Ayres, T., Linsky J.L.: 1981, Astrophys. J. **248**, 279
Parker, E.N.: 1983, Astrophys. J. **264**, 642
Parker, E.N.: 1986, in Coronal and Prominence Plasmas, A.I. Poland Ed., NASA CP-2442, p. 9
Parker, E.N.: 1988, Astrophys. J. **330**, 474
Rutten, R.G.M.: 1986, Astron. Astrophys. **159**, 291
Rutten, R.G.M.: 1987, Astron. Astrophys. **177**, 131
Schrijver, C.J.: 1983, Astron. Astrophys. **127**, 289
Schrijver, C.J.: 1987a, Astron. Astrophys. **180**, 241
Schrijver, C.J.: 1987b, in Cool Stars, Stellar Systems and the Sun, J.L. Linsky, R.E. Stencel Eds., Lecture Notes in Physics **291**, Springer, Berlin, Germany, p. 135
Schrijver, C.J., Coté, J., Zwaan, C., Saar, S.H.: 1989, Astron. Astrophys. **337**, 964
Stenflo, J.O.: 1978, Rep. Prog. Phys. **75**, 3
Stepien, K., Ulmschneider, P.: 1989, Astron. Astrophys. **216**, 139
Ulmschneider, P.: 1970, Solar Phys. **12**, 403
Ulmschneider, P.: 1986, Adv. Space Res. **6**, No. 8, 39
Ulmschneider, P.: 1988, Astron. Astrophys. **197**, 223
Ulmschneider, P.: 1989, Astron. Astrophys. in press
Ulmschneider, P., Muchmore, D.: 1986, in Small Scale Magnetic Flux Concentrations in the Solar Photosphere, W. Deinzer, M. Knölker, H.H. Voigt Eds., Vandenhoeck and Ruprecht, Göttingen, Germany, p. 191
Ulmschneider, P., Zähringer, K., Musielak, Z.: 1989, Astron. Astrophys., in press
Vaiana, G.S., et al.: 1981, Astrophys. J. **245**, 163

Van Ballegooijen, A.A.: 1986, Astrophys. J. **311**, 1001
Vernazza, J.E., Avrett, E.H., Loeser, R.: 1981, Astrophys. J. Suppl. **45**, 635
Vilhu, O.: 1987, in Cool Stars, Stellar Systems and the Sun, J.L. Linsky, R.E. Stencel Eds., Lecture Notes in Physics **291**, Springer, Berlin, Germany, p. 110
Walter, F.M., Schrijver, C.J.: 1987, in Cool Stars, Stellar Systems and the Sun, J.L. Linsky, R.E. Stencel Eds., Lecture Notes in Physics **291**, Springer, Berlin, Germany, p. 262
Wentzel, D.G.: 1974, Solar Phys. **39**, 129
Wilson, O.C., Bappu, M.K.V.: 1957, Astrophys. J. **125**, 661
Withbroe, G.L.: 1988, Astrophys. J. **325**, 442
Withbroe, G.L., Noyes, R.W.: 1977, Ann. Rev. Astron. Astrophys. **15**, 363
Zähringer, K., Ulmschneider, P.: 1987, in The Role of Fine- Scale Magnetic Fields on the Structure of the Solar Atmosphere, E.H. Schröter, M. Vazquez, A.A. Wyller Eds., Cambridge Univ. Press, Cambridge, England, p. 243
Zwaan, C.: 1978, Solar. Phys. **60**, 213

SOLAR RADIO CORONA

M. PICK and G. TROTTET

URA 324, DASOP, Observatoire de Paris,
Section de Meudon
92195 MEUDON PRINCIPAL CEDEX, France

ABSTRACT. This review is restricted to the structure of the corona including three topics : the quiet corona, the active corona and the topology of flaring sites. The interest of imaging radio observations with high spatial and time resolutions is emphasized.

1. INTRODUCTION.

The solar corona is a medium which is highly structured because of the magnetic field. Such an atmosphere is far from being static and the different structures evolve on various time scales varying from hours to several days or tens of days. Radio observations are one of the most powerful way to study the corona because : (1) they are performed over a wide wavelength domain ranging from centimetric to decametric wavelengths which allows to probe the solar atmosphere from the photosphere to the outer corona ; (2) they are obtained both on the disk and on the limb ; (3) they provide continuous information over large periods of time with high temporal resolution; (4) the emission processes are known and the problem of radiative transfer is rather simple.

New possibilities have been opened over the past two decades for the study of the solar corona. In particular: (i) in the microwave domain, observations with spatial resolutions ranging from 1" to 10" have been obtained by using large synthesis arrays, built primarily for sideral radio astronomy ; (ii) larger radioheliographs as the Nançay, Clark-Lake, instruments provided two dimensional maps from decimetric-metric to decametric wavelengths.

This review will be resticted to the structure of both the quiet and active corona and to the topology of flaring sites.

2. THE QUIET CORONA.

2.1. Active Regions

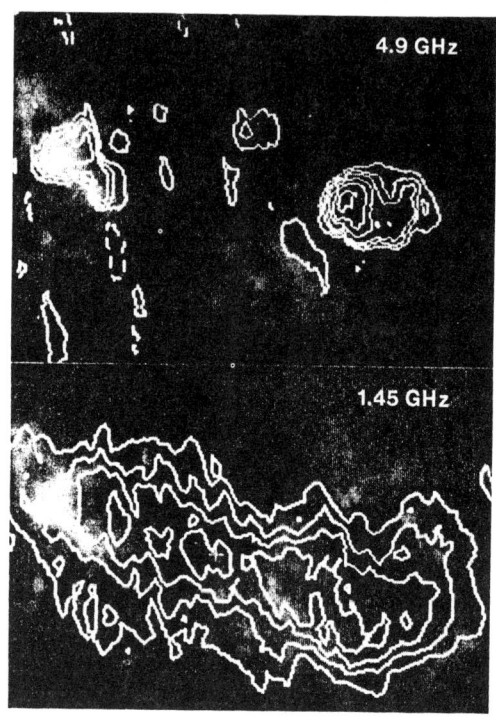

Figure 1 . (from Gary and Hurford 1987).
The 4.9 and 1.45 GHz VLA eclipse maps overlaid on magnetograms obtained at Kitt-Peak. Note the excellent correspondence with high field regions at 4.9 GHz and the poor correspondence at 1.45 GHz. The 4.9 GHz sources are due to gyroresonance emission and the 1.45 GHz source is due to free-free emission.

In the microwave domain, most of the information came from the Westerbork Synthesis Radio Telescope (WSRT) and the Very Large Array (VLA) and is obtained with high spatial resolution down to a few arc seconds. Among the various features which have been observed, active regions have been extensively studied. The overall picture is in agreement with a combination of free-free emission and gyroresonance emission. The brighest sources have brightness temperatures up to a few times 10^6 K, are strongly polarized and are associated with sunspots. This "sunspot" component is consistent with gyromagnetic emission which requires magnetic field between about 500G and 1000G. At shorter wavelengths, millimetric active regions present only low contrast features generated by bremsstrahlung : both the field strength and the electron temperature are too low to yield significant gyromagnetic emission. At decimetric wavelengths typically longer than 10cm the picture is different : observations show extended sources which overly the plage faculae. The emission is bremsstrahlung from electrons at coronal temperatures (figure 1). On the other hand, observations of an active region with the RATAN 600 at 5 wavelengths between 2 and 4cm have revealed that in addition to gyromagnetic sources associated to sunspots, sources with quite different spectra were found to be associated with active region filaments. As shown in figure 2, these sources have quite different spectra : the D component located above the filament has a high brightness temperature and a steep spectrum which requires either the presence of currents that enhance the magnetic field or non thermal radiation of suprathermal electrons.

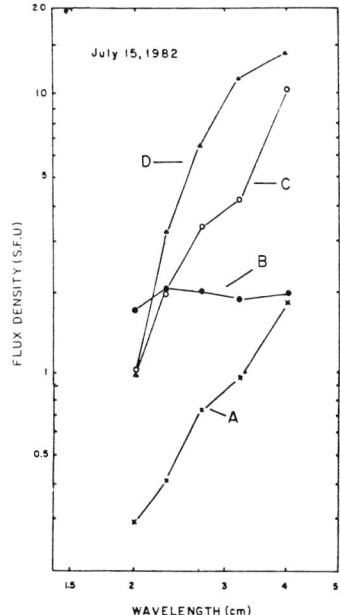

Figure 2. (Achmedov et al, 1986).
Radiation spectrum of four core sources. A and C are sources located above sunspot. B and D are sources located above a filament. Here the flux density in solar flux units (sfu) has been plotted at a function of wavelength.

Summarizing, microwave observations with high spatial resolution contain a great deal of information on the coronal structure associated with active regions provided they are obtained simultaneously at different wavelengths. Detailed calculations by Brosius and Holman 1989 show that the centimetric wavelength emission is particularly sensitive to the magnetic field and can be used, in principle, to determine the 3 dimensional structure of the sunspot magnetic field.

2.2. Microwave bright points

Coronal bright points were first identified on X-UV spectroheliograms and have now been observed on VLA maps at 6cm and 20cm (Fu et al., 1987 - Habbal et al, 1986). They appear at all latitudes on the sun and are associated with bipolar features. They may as well be considered as manifestations of the active corona : indeed they fluctuate on time scales of minutes and may be radiosignature of intermittent heating. Systematic investigations of their distribution over a solar cycle and determination of their physical parameters, may bring important clues for the understanding of coronal heating. Future joint observations will reveal if these radio bright points have the same origin as in one hand some of the hard X-ray microflares which have been detected with high sensitivity experiments (Lin et al, 1984) and on the other hand the "high energy jets" identified in UV (Bruekner and Bartoe, 1983).

2.3. Metric decametric observations

Metric and decametric observations of the quiet sun probe coronal structures from the chromosphere-corona transition region to the outer corona. In the absence of X-UV full sun images they constitue the only means to probe the solar atmosphere on the disk over this altitudinal range and complement optical observations on the limb. Two dimensional maps of the corona have been obtained at 327 MHz with the VLA (Lang et al. 1988) and at different frequencies between 408 MHz and 30 MHz with the Nançay and Clark Lake radioheliographs (e.g. The Radioheliograph Group 1989, Lantos et al 1987). In this wavelength range all observations indicate that none of the emitting sources is associated with active region. This indicates that the sunspot magnetic field does not reach metric levels. Figure (3) shows the typical corona as observed at 169 MHz. The emission pattern comprises depressions with brightness temperature of $\simeq 10^5$ K corresponding to coronal holes. In addition diffuse sources of $T_b \leq 10^6$ K are also observed. Some of these sources overly quiescent filaments and represent the radio counterpart of coronal streamers as was pointed out in a former study by Axisa et al. 1971. Other sources have no clear association with chromospheric Hα features and may arise from large scale coronal loops that, as shown by Skylab X-ray pictures cover a large fraction of the corona. The similarity between K corona and radio synoptic maps led Lantos et al. 1989 to show that the bright radio emission observed on both sides of the solar equator traces the general shape of the heliosheet in the middle corona.

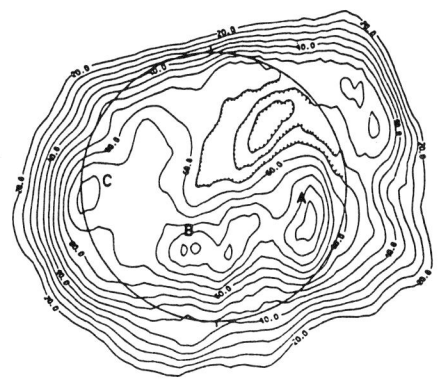

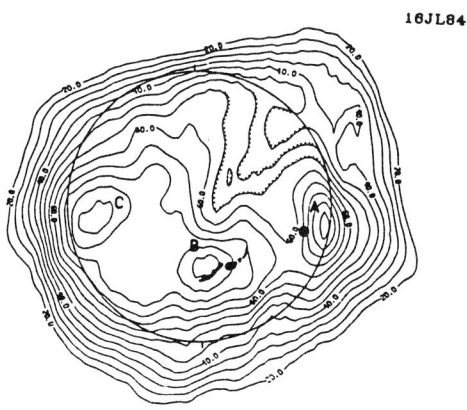

Figure 3. (from Lantos et al, 1989).
Maps of the sun at 169 MHz. Hatched contours are outlining the brightness depressions. The brightness temperature is labelled in unit 10.000 K. ("A" source is a noise storm center associated to active region indicated by a dot). The Hα filament related to emission B is indicated.

We emphasize that such systematic radio survey of the corona is a unique and necessary input for the future space missions devoted to the corona, the interplanetary space and the earth magnetosphere such as ULYSSES, SOHO, and Cluster.

3. THE ACTIVE CORONA

Suprathermal electrons spiralling along the magnetic field generate a variety of radio bursts which in turn can be looked as tracers of magnetic flux tubes. Among the various types of bursts which are produced we shall focuss on type III bursts and noise storm centers which are respectively indicators of open and closed magnetic fields regions.

3.1. Type III bursts

3.1.1. *Type III bursts tracers of open coronal structures.*
Type III bursts are due to electrons of a few tens of keV travelling from the low to the high corona along the magnetic field lines and producing at each level radiowaves at the local plasma frequency or the second harmonic. Imaging observations of the type III radio sources have revealed that most of the type III sources can be resolved into multiple components excited quasi simultaneously. This has been interpreted in terms of electron beams propagating along a wide range of diverging magnetic field lines (Raoult and Pick, 1979). Direct evidence of these structures has been obtained with the SMM C/P instrument : they are small, discrete and short lived. Their typical life time is one or two days (figure 4) Multifrequency radio observations show that they are often strongly inclined from the radial direction. This was also observed by Gergely et al. at decametric wavelengths. Such observations are of fundamental interest for the understanding of the electron propagation and injection into the interplanetary medium.

3.1.2. *Type III bursts and the coronal magnetic field.*
The magnetic field strength can in principle be obtained from the spatially resolved observations of type III burst polarization. The degree of circular polarization is proportional to the radio f_b/f_p, when f_b is the gyrofrequency and f_p the plasma frequency. At low frequencies, below 100 MHz, no systematic variation of the degree of polarization with frequency was observed (Suzuki and Dulk, 1985). Higher in frequency recent observations obtained in the range 164-435 MHz have revealed that the degree of polarisation in general increases with frequency and is not uniform inside the radiosource. The magnetic field values deduced from the

theory are rather high, corresponding typically to 50 and 13 Gauss respectively at 435 and 164 MHz level (Mercier, 1990). This variation of the magnetic field with the altitudes is consistent with the existence of a strong divergence between the site of the electron injection in the low corona and the higher located meter sources.

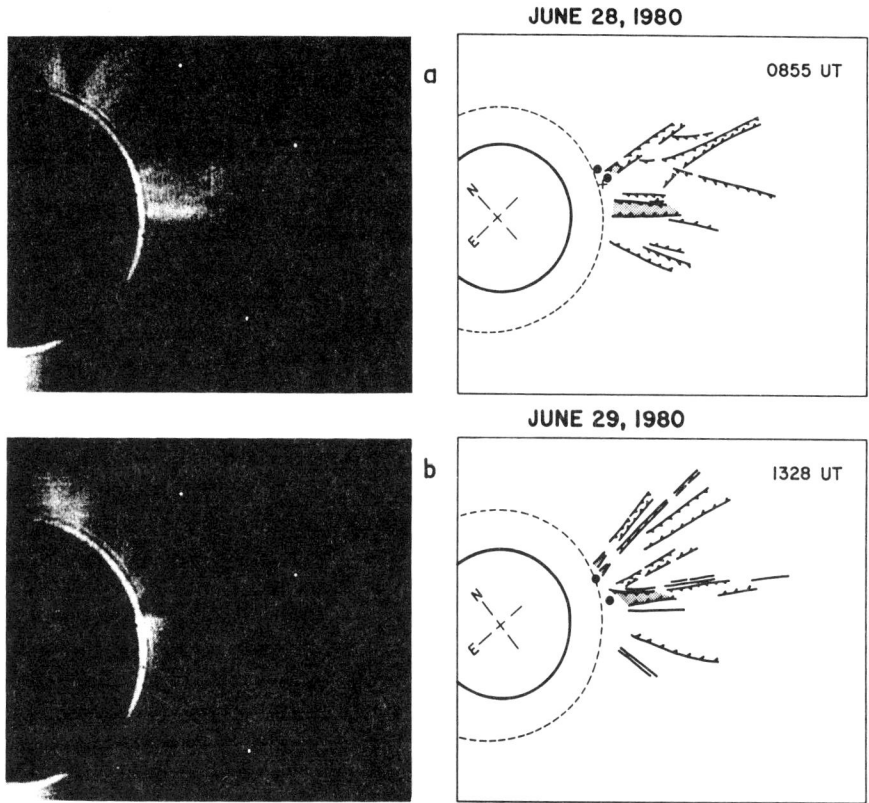

Figure 4. (from Trottet et al, 1982).
The coronal structure overlying the west limb on 28 June a, 29 June b. For clarity, the most important coronal features have been drawn on the right hand side of the figures where type III elementary source positions are shown by dots. Note the narrow dense ray overlying the radio sources.

3.1.3. *Type III bursts and coronal inhomogeneities.*
If the corona is highly inhomogeneous, the type III emission at a given frequency will come from a range of altitude. All the former models have assumed that a given frequency, the source geometry is a quasiplanar surface. Recently a model has been developed for the type III emission in a fibrous corona. It was shown that the density variations $(\delta \ln N)_{rms}$ can be derived from the

observations. Then radio emission may become a useful way to obtain density diagnostics in a region where direct measurements are presently not reachable (Roelof et Pick, 1990).

3.2. Noise storms : tracers of closed structures.

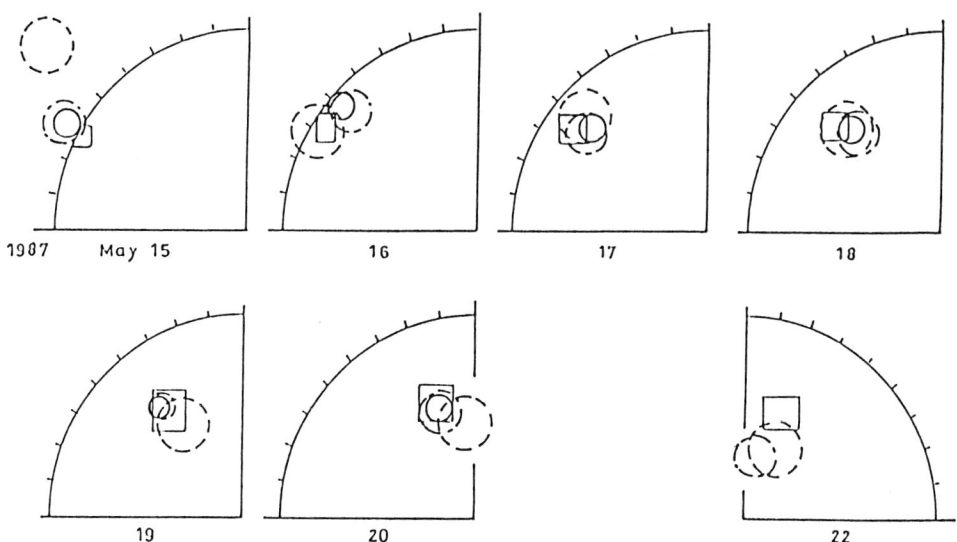

Figure 5. (from Mercier et al, 1988).
Noise storm sources observed with the Nançay Radioheliograph during seven days (from May 15, to May 22 1987). The photospheric active region is indicated by rectangles (Solar Geophysical Data). The circles indicate the period of the highest spatial harmonic at each observing frequency. Their diametre increases with decreasing frequency (408, 327, 236,6 and 164 MHz).

Noise storms are the most common observed activity at metric wavelengths. They are produced by suprathermal electrons spiralling along coronal magnetic arches. Multifrequency imaging observations have been obtained with the Nançay Radioheliograph (Mercier and al 1988). The observed emission pattern suggests that noise storms are emitted from one of the legs of large scale coronal loops. Figure 5 also shows that for a given active region the emission pattern changes from day to day indicating restructuration of the magnetic field or injection of electrons in different structures. Further theoretical work and further observations are needed to fully understand the origin of noise storm. Nevertheless this above schematic discussion show that a detailed study of noise storm is a potential interest to access the dynamic of large coronal loops on time scale of one day.

4. RADIO EMISSION AND THE TOPOLOGY OF THE ACCELERATION SITE

Combined radio imaging obvservations and hard X-ray spectral and imaging data have brought important new results on the magtnetic field topology giving rise to particle acceleration in the different phases of solar flares and on the physical conditions in the acceleration site itself. Microwave intensity increases and changes of the source structure in both intensity and circular polarization have been observed prior to flares. These observations have been often interpreted as the emergence of magnetic structures and their interaction with preexisting ones (Priest et al, 1986, Kundu, 1986). Hard X-ray emission has been also observed prior to the flash phase in energetic events (Benz et al. 1983). During this period, metric/decametric type III bursts are also often observed (see figure 6). At the onset of the flash phase which is characterized by a rapid increase of the hard X-ray intensity, the type III radio emission spreads to higher frequencies and a decimetric to metric continuum (type V) arises superposed to the type III emission (see figure 6). There is an excellent overall similarity between radio and hard X-ray time profiles. This similarity demonstrates that both emissions are produced by a common and continuous acceleration/injection of electrons. More precisely at the time where the radio emission shows this sudden evolution from type III's to type III/V emission, a new radio source appears and proceeds to fluctuate in phase with one of the preexisting radiosources (see figure 6). This behaviour has been interpreted as being caused by the rapid interaction of two systems of magnetic loops giving rise to the main energy release during the impulsive phase (Raoult et al, 1985) ; Hernandez et al, 1986). This interpretation has recently received direct confirmation from VLA observations at 21cm : for one event, a single unpolarized source is responsible for the radio emission during the preflare phase. This is followed by the sudden appearance of a new radio source with a bipolar structure (Willson et al, 1989). Finally, observations of the fine structure of radio emission leads to the following remarks :

- There exists a good temporal association between hard X-rays, decimetric spikes and type III emissions (Benz, 1985; see figure 7). Benz concluded that energy is released into thousand of microflares of typical dimension $3.20\ 10^9$ cm^{-3} (Benz and Kane, 1986).
- There exists microwave type III's with reverse drifts (Stahli and Benz 1987) which originate near the energy

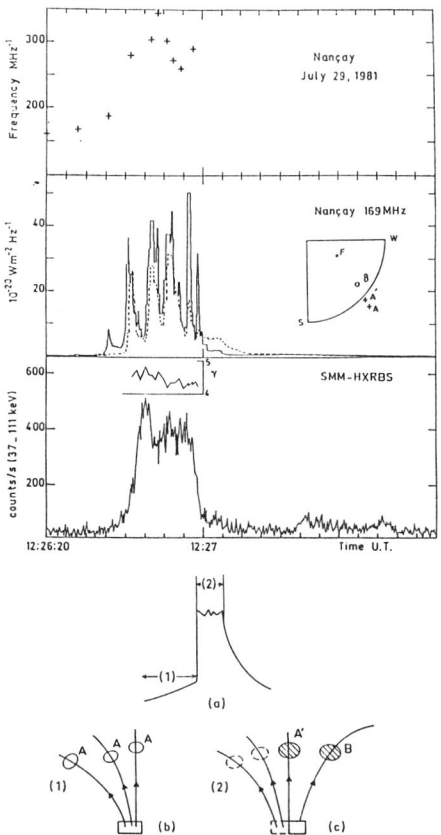

Figure 6. (adapted from Raoult et al. 1985)
Upper panel : *Top* : Starting frequency of the radio emission. *Middle* : Time profiles of the brightness from source A' (solid line) and source B (broken line) and location of sources A-A' and B. *Bottom* : Evolution of the hard X-ray power law spectral index and count rate.
Lower panel : *(a)* Schematic evolution of HXR burst showing pre-flash phase (1) and flash phase (2) ; *(b)* suggested magnetic field geometry during the pre-flash phase. The radio sources are labelled A ; *(c)* same as *(b)* during the flash phase. A' and B show the radio sources.

release region. The authors deduced that the typical electron density of this region ranges around 1.4 10^{10} cm^{-3}.

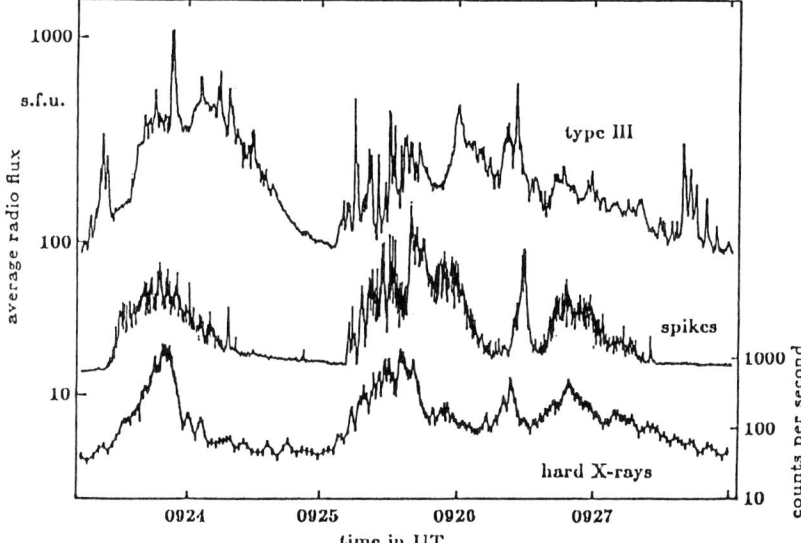

Figure 7. (from Benz 1985)
Time histories of the average spike flux in the frequency band 580 to 640 MHz (middle), type III emission in the 250 to 310 MHz band (top) and hard X-rays (bottom).

5. CONCLUSION

The quiescent as well as the active corona comprises a collection of large and small scale magnetic structures which evolve on different time scales. Radio observations provide continuous series of data which enable to study the evolution of the corona over time periods ranging from hours and less to the $\simeq$ 11 years solar cycle. Future multiple wavelength observations will bring new insight on different problems : corona topology, independant diagnostics of the temperature and density as a function of height, link with the interplanetary medium. This is also needed to study the coronal evolution through the solar cycle. Such radio observations represent a fundamental support to future space missions on the Sun or the interplanetary medium (ULYSSES, SOHO, CLUSTER).

REFERENCES

Achmedov, Sh. B., Borovik, V.N., Gelfreich, G.B., Bogod, G.B., Korzhavin, A.N., Petrov, Z.E., Diku, V.N., Lang, K.R., and Willson, R.F., (1986) Astrophys. J., 301, 460.

Alissandrakis, C., Kundu, M.R., and Lantos, P.,
  (1980), Astron. Astrophys. 82, 30.
Alissandrakis, C., and Kundu, M.R., (1984), Astron.
  Astrophys.179, 271.
Axisa, F., Avignon, Y., Martres, M.J., Pick, M., and
  Simon, P., (1971), Solar Phys., 19, 110.
Benz, A.O., Barrow, C.H., Dennis, B.R., Pick, M., Raoult,
  A., and Simnett, G., 1983, Solar Phys., 83, 267.
Brosius, J.W., and Holman, G.D. (1989), Astrophys.
  J.,342, 2, 1172
Brueckner, G.E., and Bartoe, J.D.P., (1983), Astrophys.
  J. 272, 329.
Fu, Q., Kundu, M.R., and Schmahl, (1987), Solar Phys.
  108, 99.
Gary, D.E., Hurford, G.J. (1987), Astrophys. J. 317, 522.
Habbal, S.R., Ronan, R.S., Withbroe, G.L., Shevgaonkar,
  R.K., Kundu, M.R., (1986), Astrophys. J. 306, 740.
Hernandez, A.M., Machado, M.E., Vilmer, N., and Trottet,
  G. (1986), Astrophys. J. 167, 77.
Kundu, M.R., (1985), Solar Phys. 100, 491.
Kundu, M.R. (1986), Adv. Space Res., Vol. 6, n°6, 93.
Kundu, M.R., Gergely, T.E., Schmahl, E.J., Szabo, A.,
  Loiacano, R., Wang Z., and Howard, R.A. (1987), Solar
  Phys.108.
Lang, K., Willson, R.F., and Trottet, G. (1988), Astron.
  Astrophys. 199, 325.
Lin, R.P., Schwartz, R.A., Kane, S.R., Pelling, T., and
  Hurley, K.C. (1984), Astrophys. J., 283, 421.
Mercier, C., Klein, K.L., and Trottet, G. (1988), Adv.
  Space Research. Vol. 8, n°11, 193 - 197.
Mercier, C. (1990), Solar Phys., CESRA Meeting, to be
  published.
Priest, E.R., Gaizauskas, V., Hagyard, M.J., Schmahl,
  E.J., and Webb, D.E. (1986), in Energetic Phenomena on
  the Sun, eds. M.R. Kundu and B. Woodgate, NASA-CP,2439.
Raoult, A., and Pick, M. (1979), Astron. Astrophys. 111,
  306.
Raoult, A., Pick, M., Dennis, B.R., and Kane S.R. (1985),
  Astroph. J. 299, 1027.
Roelof, E.C., and Pick, M. (1989), Astron. Astrophys.
  210, 717, 424.
Susuki, S., and Dulk, G.A. (1985), in Solar Radiophysics,
  McLean and Labrum eds. Cambridge University Press.
The Radioheliograph Group (1989), Solar Phys., 120, 193.
Trottet, G., Pick, M., House, L., Illing, R., Sawyer, C.,
  and Wagner, W. (1982), Astron. Astrophys. 111, 306.
Willson, R.F., Klein, K.L., Kerdraon, A., Lang, K.R.,
  Trottet, G. (1990), Astrophysical Journal (submitted).

The Solar Wind and the Winds from Cool Stars*

R. HAMMER

Kiepenheuer-Institut für Sonnenphysik, Schöneckstr. 6, D-7800 Freiburg

ABSTRACT. This review begins with a brief summary of observed characteristics of the winds from cool stars. The mechanisms that drive these winds are not yet known; but some of their basic properties can be inferred already from time-independent stellar wind theory. These properties include the size and spatial distribution of the energy input, and the question if the driving occurs via body forces or via heating the outer stellar atmosphere to high temperatures. It is shown that compressive waves are an unlikely mechanism to produce hot coronae, even for old, slowly rotating stars with inefficient dynamos. Recent time-dependent models of the winds from stars with and without coronae are reviewed.

1 INTRODUCTION

The mass loss associated with the wind from a star like the Sun is too faint to affect the evolution of the stellar interior. Nevertheless, even such a gentle wind is astrophysically very important – because it controls the circumstellar environment and, indirectly, the evolution of the stellar atmosphere. The latter results from the fact that the outflow implies a significant angular momentum loss, which is enhanced by magnetic fields that force the gas into corotation out to large radii. The angular momentum loss leads to the braking of the star's rotation, which in turn reduces the dynamo generation of magnetic fields and thus the activity level of the stellar atmosphere.

We are far from a detailed understanding of this central role that cool star winds play in stellar astrophysics. Indeed, the mechanisms that drive these winds have not yet been identified unambiguously – not even in the case of the Sun.

But the subject has received continued interest ever since the 1950s, when it was first proven that both cool supergiants and the Sun emit persistent winds (Biermann 1951; Deutsch 1956; Parker 1958; Neugebauer and Snyder 1962). This interest has even increased over the past few years. In 1987, for instance, at least three important international symposia were dedicated to the stellar wind phenomenon (Bianchi and Gilmozzi 1988; Pizzo et al. 1988; Stalio and Willson 1988), and several other meetings touched upon the subject (e.g., Linsky and Stencel 1987; Havnes et al. 1988;

---

* Mitteilungen aus dem Kiepenheuer-Institut Nr. 319

Schröter and Schüssler 1988). The proceedings of these conferences contain a number of comprehensive reviews on various aspects of the phenomenon. (Of particular relevance to some of the topics to be addressed in the present paper are the articles by Bowen 1988b; Drake 1988; Hammer 1988b; Holzer 1988; Leer 1988a,b; Reimers 1988; Willson 1988; and Withbroe 1988b.)

Against the background of this fairly broad coverage of the general subject, the present review emphasizes a selection of more recent developments, such as time-dependent calculations of the corona/wind problem (Sect. 5) and new insights into old heating theories (Sect. 4). This discussion is based upon an overview of the underlying physical principles (Sect. 3). The paper begins with a brief account of the empirical evidence (Sect. 2).

2 OBSERVATIONS

The main objective of the present paper is the comparison of the solar wind with the winds from other stars. We exclude, therefore, stars for which such a comparison does not make sense because the underlying physics is known to be too different. This applies, for example, to the radiation driven winds from early-type stars; and to pre-main sequence objects, which often show indications of a very complicated pattern of simultaneous wind outflow and accretion inflow (cf. the review by Bertout, these proceedings). Also, strongly (radially) pulsating stars like Cepheids and semiregular or long-period variables will be mentioned only briefly.

The stars in the remaining part of the HR diagram (cf. Fig. 1) can be divided into two groups that are separated by a dividing line. When we cross that line in the direction of later spectral type and increasing luminosity, the X-ray and UV emission from hot coronae and their transition regions disappears (Linsky and Haisch 1979; Ayres et al. 1981); chromospheres become geometrically more extended; and blue-shifted circumstellar absorption features indicate the onset of cool winds (Reimers 1977). This change of the structure of the outer stellar atmosphere is remarkable, although not necessarily abrupt. A number of hybrid stars (Hartmann et al. 1980) near the dividing line show both evidence of cool winds and emission from hot plasma that does not participate in the wind expansion.

Several possible explanations of this dichotomy have been suggested (for recent reviews see Antiochos 1987; Haisch 1987; and Hammer 1988a); but the final answer has not yet been found. This might be due to the fact that many fundamental stellar properties change rapidly near the dividing line: from left to right in Fig. 1, gravity and rotational velocity decrease; the photospheric density scale height and the size of convective elements increase faster than the stellar radius; and the stars become fully convective. It is well conceivable that *several* of these changes contribute to the observed phenomena.

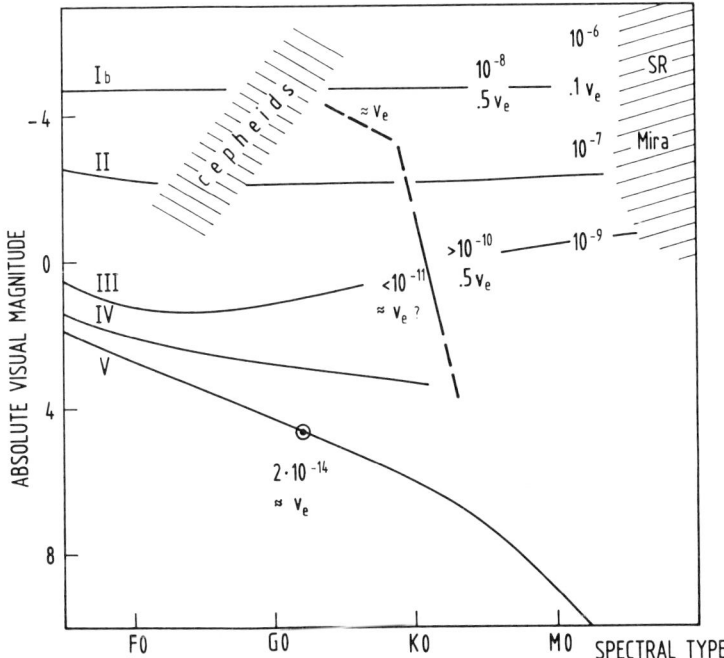

Figure 1 (*from Hammer 1988b*). A dividing line (drawn heavy) separates the stars in this schematic HR diagram into two groups. Coronal stars (to the left of the line and on the main sequence) are surrounded by hot coronae, which can produce hot, fast winds. Noncoronal stars (to the right) are characterized by extended chromospheres and cool, slow winds. Typical mass loss rates (in solar masses per year) and wind speeds (in units of the surface escape speed v_e) are indicated (after Drake 1986, 1988; and Reimers 1988).

2.1 Coronal Stars

Giants to the left of the dividing line and dwarfs of spectral type later than early F are often called "coronal" stars. Their outer atmosphere is commonly believed to be structured like that of the Sun: it consists of a warm chromosphere and a hot, multi-million K corona, separated by a thin transition region that is controlled by thermal conduction. The geometry and dynamics are complex, however. This is mainly due to the presence of magnetic fields, which completely fill the solar atmosphere everywhere above the mid chromosphere. Magnetic fields channel the flow of mass, momentum, and energy into the outer solar atmosphere. When the field lines form closed loops, they prevent the gas from escaping in the solar wind; hence, the energy that is dissipated must all be radiated away. In fact, most of the solar high-temperature emission originates from the hot, dense plasma in coronal loops.

The solar wind comes from magnetically open regions. Their footpoints probably cover only some 20% of the solar surface. Even in open regions the magnetic field is strong enough to dominate the gas: it guides the wind flow and forces it into corotation out to the so-called Alfvén radius, which lies typically at 10 — 20 solar radii, depending on the local wind properties. Magnetic fields are probably also responsible for the energy supply of the solar wind (cf. Sect. 4).

Most of our current knowledge of the solar wind comes from *in situ* measurements by various spacecraft, which explored the heliosphere near the ecliptic plane outside of 0.3 AU (see, e.g., Schwenn 1983). By contrast, our knowledge of the inner solar wind (in particular the wind acceleration region within the first few solar radii) and of the polar wind is rather rudimentary. This situation will improve considerably with the advent of the ESA/NASA mission SOHO, which will carry, among other instruments, potent UV- and X-ray telescopes and spectrographs specifically tuned to the investigation of the source region of the solar wind.

At 1 AU, the solar wind proton temperature is of the order of 10^5 K and highly variable. The proton flow speed varies between about 300 km s^{-1} and 800 km s^{-1}. It is comparable with the gravitational escape speed from the surface of the Sun, $v_e = 620$ km s^{-1}. High-speed wind streams can be traced back to long-lived coronal holes (e.g., Lindblad 1989), whereas the origin of the more variable low speed wind is not definitely known.

The proton flux in the ecliptic plane is known from spacecraft experiments. It corresponds to a mass loss rate $-\dot{M} \approx 2\,10^{-14}\,M_\odot$ yr^{-1}, less than a third of the mass loss associated with the photospheric radiation. In high speed streams the mass flux is somewhat lower than in the slow wind. The proton flux out of the ecliptic can be determined only indirectly, notably by studying the rate at which the protons react with interstellar neutral hydrogen atoms that penetrate into the solar system. Since these H atoms backscatter solar $L\alpha$ photons, it is possible to measure their spatial distribution and to infer the anisotropy of the solar wind proton flux. From such measurements, Berteaux et al. (1985) and Lallement et al. (1985) concluded that the average solar wind mass flux over the poles is smaller than at low latitudes. The reduction factor can be as large as 2 and appears to vary with the solar activity cycle (Lallement 1988).

The winds from coronal stars other than the Sun have not yet been detected spectroscopically. But it is reasonable to expect that these stars with hot coronae produce winds similar to the Sun.

Recently Drake (1986, 1988) derived upper limits to the associated mass flux. He assumed that within a group of stars of a given type, those with the minimum X-ray luminosity have coronae dominated by magnetically open regions. By making further assumptions on the average coronal temperature $\bar{T}$ and wind velocity profile $v(r)$, he could calculate the mass flux of a wind that produces the observed X-ray emission. Mass loss rates derived by this method represent *upper limits* to the true values, since even in the most quiet stars of a sample, the X-ray contribution from coronal loops is likely not completely negligible. A further source of uncertainty is that we do not know if the open regions of active and inactive stars are similar.

From his analysis, Drake concluded that for all coronal dwarfs and giants the total mass loss per unit area cannot be significantly greater than the solar mass flux; otherwise these stars would produce more X-rays than the observed minimum. The upper limit to the mass loss from coronal giants near the dividing line is of the order of $10^{-11} M_\odot \text{ yr}^{-1}$.

2.2 Noncoronal Stars

The mass loss involved in the cool winds from stars to the right of the dividing line can be estimated by a variety of methods, ranging from simple volume emission measure techniques to complex line profile modeling. (For extensive reviews on such methods see Drake 1986, 1988; Dupree 1986; and Reimers 1988.) All these methods need to make severe approximations. As a result, different authors often derive very different mass loss rates for the same star. Only for a few well-studied stars (listed in Table 1 of Drake 1986 and Table 2 of Reimers 1988) is $-\dot{M}$ currently known to within an uncertainty factor of less than about ± 5 (Drake 1986; Dupree 1986). Moreover, the mass loss rate of a given star may well be variable in time; and there might be a considerable scatter in the mass loss rate of stars at a given position in the HR diagram.

The most accurate mass loss data exist for binary systems, in which the companion star acts as a light source that probes the wind from a red giant. In order to provide useful diagnostics, the spectra from both stars must be separable. Therefore, the two stars must either have a sufficient angular distance (as in the visual binary α Her, Deutsch 1956); or they must have very different temperatures and thus spectral energy distributions (as in the chromospheric eclipsing binaries of the ζ Aur type, where the secondary is a B dwarf). Only for binary systems is it possible to derive some information on the wind velocity profile $v(r)$ (Schröder 1985) and fairly reliable values of the mass loss rate (cf. the review by Reimers 1988). On the other hand, the secondary may affect the wind from the red giant primary by its gravitation and by its ionizing radiation. The latter is particularly important for the ζ Aur systems, where the secondary is hot. Therefore, it is not clear if the winds from binary systems are typical of the winds from single stars at the same position of the HR diagram.

Despite the large uncertainty in the determination of $-\dot{M}$ for individual stars, a gross trend for its variation in the cool part of the HR diagram is evident (cf. Fig. 1): The mass loss rate increases more or less systematically towards later spectral type and higher luminosity, from somewhat more than $10^{-10} M_\odot \text{ yr}^{-1}$ for K giants (like α Boo, K1 IIIp) and a few times $10^{-9} M_\odot \text{ yr}^{-1}$ for early M giants, up to 10^{-7}—$10^{-6} M_\odot \text{ yr}^{-1}$ for early M bright giants (like α Her, M5 II) and supergiants (like α Ori, M2 Iab; and α Sco, M1.5 Iab). The mid K supergiant components of ζ Aur-systems have $-\dot{M} \approx 10^{-8} M_\odot \text{ yr}^{-1}$. (The sources and accuracy of these data are discussed in compilations by Drake 1986, 1988; and Reimers 1988.)

The mass loss rate of an early M supergiant is by 8 orders of magnitude larger than that of the Sun. Most of this increase, however, is due to a likewise drastic increase of the stellar surface area. When normalized to the surface, many characteristic properties of a stellar wind turn out to vary only little within the cool part of the HR diagram. This applies in particular to the energy flux density needed to drive the wind, $\Delta\Phi_w = \left(-\dot{M}/4\pi R^2\right)(v_e^2 + v_\infty^2)/2$, where $v_e = \sqrt{2GM/R}$ is the surface escape speed and v_∞ the terminal flow speed (cf. Eq. (11)). This total wind energy requirement is about 10^5 erg cm^{-2} s^{-1} for all stars including the spatially averaged Sun (Holzer and MacGregor 1985; Reimers 1988 – Table 2).

Flow speeds of the cool winds from noncoronal stars can be inferred from Doppler shifts of circumstellar absorption features. The measured speed is usually assumed to represent the final speed v_∞, which the wind attains beyond its initial acceleration zone. This assumption is debatable – as mentioned above, the only stars for which we have some limited information on the radial variation of the flow speed are the Sun and a few ζ Aur-type binaries. The observed speed is found to decrease towards later spectral type, down to 10—20 km s^{-1} for M supergiants. This value corresponds to only a small fraction (10—20%) of the gravitational escape speed v_e at the surface of these stars. By contrast, the observed wind speed from noncoronal K giants and supergiants (like ζ Aur systems) reaches some 50% of the escape speed, while hybrid stars – just as the Sun and perhaps other coronal stars – have fast winds of speeds comparable with the escape speed.

3 THEORY: BASIC CONCEPTS

The variation of wind speed and mass flux among cool stars, as discussed in the preceding section and summarized in Fig. 1, places important constraints on possible wind driving mechanisms. In order to illustrate these constraints and to discuss some basic ideas of stellar wind theory, it is sufficient to consider a simplified wind model governed by the equations,

$$\frac{p}{\rho} = c^2 = \frac{kT}{\bar{m}} \tag{1}$$

$$r^2 \rho v = const = \frac{-\dot{M}}{4\pi} \tag{2}$$

$$\rho v \frac{dv}{dr} + \frac{dp}{dr} + \rho \frac{GM}{r^2} = \rho D \tag{3}$$

$$\frac{1}{r^2}\frac{d}{dr}r^2\rho v\left(\frac{v^2}{2} + \frac{\gamma}{\gamma-1}c^2 - \frac{GM}{r}\right) + n_e n_H P_{Rad}(T) - \frac{1}{r^2}\frac{d}{dr}r^2\kappa(T)\frac{dT}{dr}$$

$$= \rho v D + \frac{1}{r^2}\frac{d}{dr}r^2 F_M, \tag{4}$$

where M is the stellar mass; $-\dot{M}$ the mass loss rate; v the wind flow speed; c the isothermal sound speed; p, ρ, T the gas pressure, density, and temperature; $\bar{m}$ the mean particle mass; n_e and n_H the electron and hydrogen densities; γ the ratio of specific heats; $P_{Rad}(T)$ the emissivity and $\kappa(T)$ the thermal conductivity. The source terms for radial momentum input (ρD) and mechanical heating ($\nabla \cdot \mathbf{F_M}$) depend on the assumed wind driving mechanism and will be discussed later.

In the above form, the equation of state (1) and the conservation equations for mass (2), momentum (3), and energy (4) describe a highly idealized model of a steady, spherically symmetric, and optically thin wind. Real stellar winds are more complicated. The flow geometry is not radial, neither in the Sun nor in red giants. In the latter stars, radiative transfer effects and chemical reactions (i. e., molecule and dust formation) under non-LTE conditions become important (cf. the review by Linsky 1987). In the solar corona, on the other hand, collisions are so rare that electrons and protons have different temperatures and deviate from a Maxwellian energy distribution, with important consequences for the efficiency of thermal conduction (cf. several review articles in Pizzo et al. 1988; in particular Shoub 1988).

Despite these limitations, such a simple model contains already the essential physical ingredients to discuss some fundamental theoretical problems (e. g., Parker 1958, 1965; Hundhausen 1972; Holzer 1988). The first, and most basic, question that needs to be answered is,

- Why do Stellar Winds Blow?

The momentum equation (3) describes the balance between the forces that act on a volume element: inertial, thermal pressure, gravitational, and other forces. The latter may be due to gradients of the radiation pressure, the turbulent pressure associated with disordered nonthermal motions, or the wave pressure associated with ordered wave motions. In Eq. (3), all these effects are symbolically represented by the source term ρD. If this term is much smaller than the thermal pressure gradient term, the stellar wind is called thermally driven.

In the inner part of such a thermally driven wind ($D = 0$), the flow speed is small, so the first term in Eq. (3) can be neglected. The force balance is then characterized by hydrostatic equilibrium,

$$\frac{dp}{dr} + \rho \frac{GM}{r^2} = 0. \tag{5}$$

The pressure gradient force is the only term left to balance gravity; consequently it decreases very rapidly from dwarfs to giants. If other forces are present ($D \neq 0$), and if they do not decrease so rapidly with g, their influence is much more important for giants than for dwarfs. Thus it is well possible that a given mechanism (e. g., Alfvén waves; Hartmann and MacGregor 1980) drives the winds from low gravity stars directly via body forces; whereas on high gravity stars these forces are small, and the mechanical energy must first be dissipated to heat up the atmosphere before a wind can be driven thermally.

In order to see why there needs to be a wind outflow, we use Eq. (1) and integrate the hydrostatic equilibrium equation (5) for an isothermal atmosphere. This gives the pressure as a function of radius. It is found to decrease outward, and to approach at large distances the constant value

$$p_\infty = p_0 \exp\left(-\frac{GM\bar{m}}{RkT}\right), \tag{6}$$

where p_0 is the pressure at the base. The value of p_∞ depends exponentially on the temperature T. If T is larger than a certain critical value T_*, the asymptotic pressure exceeds the interstellar pressure. Then the interstellar medium can no longer confine the atmosphere; and a wind begins to blow.

For a star like the Sun, the critical temperature T_* can be estimated to be about $5\,10^5$ K. Since the solar corona is hotter, it must produce a wind because the thermal pressure gradient alone would already suffice to drive it (Parker 1958). For low gravity stars, T_* is smaller – but in many cases not as small as the inferred wind temperature. In such stars, the thermal pressure gradient must be assisted by other forces in order to produce the observed wind (see also the discussion below).

- *Why are Stellar Winds Transonic?*

As soon as flows become important, the inertial term is no longer negligible in the momentum equation. Then we must consider the full Eq. (3), which can be combined with (1) and (2) to give the familiar stellar wind equation,

$$\frac{1}{v}\frac{dv}{dr} = \frac{\frac{GM}{r^2} - \frac{2c^2}{r} + \frac{dc^2}{dr} - D}{c^2 - v^2}. \tag{7}$$

When the flow speed approaches the sound speed, the denominator of the right-hand side vanishes. Then the velocity gradient remains finite only if the numerator also vanishes at this critical point $r = r_c$,

$$v^2 = c^2 = \frac{GM}{2r_c} + \frac{r_c}{2}\left(\frac{dc^2}{dr} - D\right). \tag{8}$$

In an isothermal corona with a thermally driven wind ($T = const;\ D = 0$), there exists a single critical point at a radius r_c that varies inversely with the coronal temperature. The wind can become supersonic only if it passes through this critical point. Parker (1958) proved that the solar wind indeed must become supersonic because only then has it a small enough pressure at large distances and can be matched to the tenuous interstellar medium. The so-called breeze solutions, which remain subsonic everywhere, could be realized only if the interstellar pressure were much larger. In more general cases ($T \neq const$; $D \neq 0$; nonspherical geometry) the solution topology is more complex, and several critical points may exist (Holzer 1977).

• *Comments on Wind Energetics*

The energy equation (4) describes the balance between various energy sources and sinks: wind, optically thin radiation, thermal conduction, energy sources associated with momentum sources, and heat sources with negligible contribution to the momentum balance. The latter term has been expressed in Eq. (4) as the divergence of a mechanical energy flux $\boldsymbol{F_M}$.

As long as the actual coronal heating mechanism is not fully understood, it is an acceptable and useful compromise to parametrize the heating law in the form,

$$F_M = \left(\frac{R}{r}\right)^2 F_{M_0} \exp\left(-\frac{r-R}{L}\right), \tag{9}$$

where a given energy flux F_{M_0} is absorbed in the corona over a characteristic damping length L. By varying the free parameters F_{M_0} and L, one can study how an open corona depends on the properties of the heating law (Hammer 1982) and try to determine those values of the heating law parameters that lead to an optimum fit of the available observational data (Withbroe 1988a,b). The results can then provide valuable constraints on possible coronal heating mechanisms.

It is very important that such corona/wind models are self-consistent in the sense that they take all essential energetic processes into account. In particular, they must include the whole outer stellar atmosphere from the base of the transition region out to the interstellar medium (Couturier et al. 1979, 1980; Hearn and Vardavas 1981; Hammer 1982; Souffrin 1982; Axford 1985; Hollweg 1986; Withbroe 1988a,b), because this part of the atmosphere is strongly coupled together by an extremely efficient thermal conduction. Many previous models, which approximated the coronal energy balance by a polytropic law and stipulated fixed boundary conditions at some point in the inner corona, are (at best) suitable for a discussion of the dynamics of stellar coronae, but cannot be used to describe the energy and mass balance.

The study of such complete energy balance models can help to resolve a number of puzzles that arose in the past. For example, Holzer (1988) noted that the relative constancy of the observed mass flux carried by high-speed solar wind streams is surprising, since according to Eq. (7) the mass flux should depend very sensitively on the coronal temperature. Self-consistent models, on the other hand, show that this sensitivity is largely offset by the *in*sensitivity of the coronal temperature with respect to the heat input: For situations typical of the open solar corona, the maximum coronal temperature varies roughly like $T_{max} \propto F_{M_0}^{0.2}$, and the mass flux depends only slightly stronger than linearly on F_{M_0} (Hammer 1982). Hence the relative constancy of the observed mass flux does not severely constrain the heat input.

- *How Much Energy is Needed to Drive a Stellar Wind?*

The wind energy flux density (i.e., the first divergence term in Eq. (4)) consists of the mass flux ρv times the sum of kinetic energy, enthalpy, and potential energy per unit mass. When normalized to the stellar surface, the wind energy flux reads,

$$\Phi_W = \left(\frac{r}{R}\right)^2 F_W = \left(\frac{-\dot{M}}{4\pi R^2}\right)\left(\frac{v^2}{2} + \frac{\gamma}{\gamma-1}c^2 - \frac{GM}{r}\right). \qquad (10)$$

The total energy requirements of a stellar wind per unit surface are thus

$$\Delta\Phi_W = \Phi_W(\infty) - \Phi_W(R) \approx \frac{1}{2}\left(\frac{-\dot{M}}{4\pi R^2}\right)\left(v_\infty^2 + v_e^2\right), \qquad (11)$$

where the initial flow speed and the enthalpy change have been neglected. Therefore, the total energy needed to drive a stellar wind consists primarily of the potential energy and the final kinetic energy (cf. Holzer and MacGregor 1985). The energy supply to the outer stellar atmosphere must balance these energy requirements of the stellar wind in addition to other energy losses (viz., radiation).

As mentioned in Sect. 2, the wind energy requirements appear to be roughly constant, $\Delta\Phi_W \approx 10^5 \, \mathrm{erg\,cm^{-2}\,s^{-1}}$, among all stars discussed here. However, different types of stars use this constant total wind energy budget in very different ways. On the Sun, the hybrid stars, and perhaps on all coronal dwarfs and giants (Fig. 1), the measured or inferred wind speed v_∞ is comparable to the surface escape speed v_e; hence comparable amounts of energy lift the gas out of the gravitational field and accelerate it to its final speed (cf. Eq. (11)). The slow winds ($v_\infty/v_e \approx 0.1$—0.2) from M supergiants, on the other hand, need essentially all of their energy (up to 99%) as potential energy.

- *Where is the Energy Needed?*

Clearly, a stellar wind picks up its potential energy over a characteristic length scale of the order of the stellar radius; while the kinetic energy must be supplied anywhere below the radius to which the observed ("final") wind speed refers. In a star with a hot corona, the driving energy must not necessarily be deposited where the wind needs it. Thermal conduction can efficiently redistribute the heat. Moreover, part of the enthalpy which the wind acquires below the coronal temperature maximum can be converted into potential or kinetic energy in the outer corona. In a star with a cool atmosphere, on the other hand, energy redistribution (in particular by thermal conduction) is much less efficient; hence the spatial distribution of the energy supply must match more closely the potential energy requirements (Wilson 1960).

Of particular interest is the location of the energy deposition with respect to the flow speed. Eq. (7) shows that energy addition to the lower atmosphere, either as momentum (increasing D) or as heat (increasing c) has two important effects: it reduces the average velocity gradient and the radius of the critical point (zero of the numerator in Eq. (7)). Hence, in order to reach sound speed at the critical point, the flow must start at the coronal base with a much larger speed, which normally translates into a larger mass flux (Hammer 1982). By contrast, energy that is deposited in the supersonic flow regime cannot significantly affect the mass flux, because the information on such energy deposition cannot easily be communicated upstream to the mass reservoir at the stellar surface. Consequently, when the flow energy (cf. Eq. (11)) changes due to energy addition to the supersonic regime, this tends to affect mainly the final kinetic energy rather than the mass flux (Leer and Holzer 1980). The mass flux is largely determined by the physics of the *inner* atmosphere.

Therefore, the small flow speeds observed in cool supergiants imply that most of the driving energy is deposited in the subsonic flow regime (Wilson 1960, Holzer 1988). On the other hand, solar high speed streams require some energy addition in the supersonic regime (typically 30% of the total energy; cf. Withbroe 1988).

- *What Kind of Energy – Heat or Momentum?*

As discussed above, the solar corona is hot enough to drive a wind by the thermal pressure gradient force (Parker 1958). However, possible coronal heating mechanisms do not only heat the plasma, but generate body forces as well (e. g., Jacques 1977), which assist in the acceleration of the outflow.

There has been substantial controversy about the importance of such nonthermal momentum sources (i. e., on the term ρD in Eq. (3)). Munro and Jackson (1977), for example, concluded on the basis of observations of a polar coronal hole that the solar wind cannot be thermally driven. Nearly a decade later, Lallement et al. (1986) reanalyzed these data in the light of their new results on the reduced proton flux over the poles of the Sun (see above). They found that an essentially thermally driven solar wind is well possible, if not likely. More recently, Withbroe (1988a,b) has calculated elaborate energy balance models of various types of open coronal regions. He found that the most reliable empirical data can be fitted surprisingly well by a thermally driven wind model in which the mechanical energy is dissipated over a characteristic damping length of a few tenths of the solar radius ($L \approx 0.25$–$0.8\,R_\odot$ in Eq. (9)). Such a heating law is consistent with the available empirical data on the coronal temperature, base density, and mass flux. In order to account also for the observed velocity in high speed wind streams, some energy input at large heliocentric distances is needed (e.g., Parker 1965). In Withbroe's models, this final acceleration of the plasma is achieved by a momentum source operating at $r > 15\,R_\odot$, which provides up to a third of the total energy requirements of the high speed solar wind.

From these results, it appears that the solar wind is basically thermally driven. In the inner corona, which determines the mass flux, the thermal pressure force dominates over nonthermal forces.

This crucial role of the thermal pressure gradient diminishes rapidly with decreasing gravity. Models of magnetically open coronal regions with a thermally driven wind can be scaled from the Sun to other stars. A remarkable property of this scaling is that for a given heating law the coronal temperature, base density, and mass flux decrease with increasing stellar radius. As a result, it turns out to be virtually impossible to produce a mass loss rate larger than the solar one on a coronal giant to the left of the dividing line by coronal heating alone, at least if the heating is confined to the subsonic part of the atmosphere (Hammer 1981, 1988b).

For the noncoronal giants on the other side of the dividing line, thermally driven winds can be excluded even more safely. The atmospheres of these stars are too cool to provide a sufficient thermal pressure gradient. The massive winds from late-type supergiants could be driven thermally only if the atmospheres were much warmer; but then they would produce enormous amounts of UV radiation, which are not observed and could not be balanced by any conceivable heating mechanism (e. g., Weyman 1977; Holzer and MacGregor 1985). Therefore, the cool winds from low gravity stars must be driven by nonthermal momentum sources such as wave or radiation pressure.

4 MECHANISMS

Over the past decades, a variety of mechanisms were proposed to heat the outer solar atmosphere. The oldest and best studied example is shock dissipation of acoustic waves that are generated by turbulent phenomena in the convection zone (Biermann 1948; Ulmschneider 1979, 1989; Jordan 1981; Bohn 1984). Until the mid 1970s, acoustic shock waves were considered a prime energy source for not only the chromosphere, but also the corona. This is no longer true for the Sun; the solar corona is now thought to be heated by other mechanisms, even in open coronal regions. This change of opinion was caused by stringent empirical limits on the energy flux in these waves, which could be derived from the nonthermal broadening and phase shifts of lines formed in the upper chromosphere (Athay and White 1978; Staiger 1987).

Unfortunately, the interpretation of such measurements is complicated by the nonlinearity of compressive waves at these heights. Moreover, observations alone do not explain why there is not enough energy left to heat the corona, even though such waves are certainly abundant in the photosphere. Therefore, a better theoretical understanding of this problem is desirable. This is particularly important since acoustic shock waves have recently been proposed to heat the chromospheres of old, slowly rotating stars ("basal flux stars"), in which the dynamo production of magnetic fields is inefficient (Schrijver 1983, 1987, and these proceedings; Oranje and Zwaan 1985).

Could such waves possibly produce coronae as well? This question can be addressed by combining (1) information on the *maximum* energy flux carried by compressive waves at a given point in the chromosphere, with (2), calculations of the *minimum* energy flux that is needed to sustain a transition region and corona beyond this point (Hammer and Ulmschneider 1989).

It is easy to see why periodic shock waves can transport only a certain amount of energy. The velocity amplitude as a function of height is determined by two competing effects: shock dissipation tends to reduce the amplitude, while the outward decline of the density tends to increase it. Ultimately, both effects balance each other, so that the waves reach a state of constant amplitude (cf. Ulmschneider, these proceedings). This general property of shock waves has been confirmed in full nonlinear hydrodynamic calculations (Schmitz et al. 1985; Bowen 1988; Cuntz and Ulmschneider 1988). The energy flux associated with such a constant amplitude wave can be estimated by various techniques (e. g., Ulmschneider 1970; Willson and Bowen 1985) and turns out to be proportional to the gas pressure p and to the square of the wave period P,

$$F_M \propto pP^2, \qquad (12)$$

but independent of the initial amplitude with which the wave was excited. Therefore, shock waves rapidly lose any knowledge of their generation mechanism and adjust their damping length to the pressure scale height. Self-regulation processes of this kind, in which the damping length is automatically adjusted to other scales, are probably operating in the (unknown) mechanisms that drive the winds from cool, low gravity stars (Wilson 1960; Holzer 1988).

Fig. 2 shows the energy flux carried by monochromatic shock waves that have reached the state of constant amplitude. The periods are specified in units of the photospheric acoustic cut-off period, $P_A = \frac{4\pi c_S}{\gamma g}$, where c_S is the adiabatic sound speed. Bohn (1984) has investigated the generation of acoustic waves in the convection zones of late-type stars. His spectra peak near $0.1P_A$ for solar-type stars, and at somewhat larger periods for dwarfs of later spectral type. Virtually no power is generated beyond about $0.5P_A$.

Also shown in Fig. 2 are the energy requirements of a corona. Hearn (1975) discovered that for a given base pressure the overall coronal energy loss must exceed a certain minimum value. This minimum exists because for large coronal temperatures the wind energy losses and the radiation from the transition region increase rapidly, while for small temperatures the radiative energy losses from the corona become large. Unfortunately, many authors speculated that all stellar coronae are of the "minimum flux" type; this assumption was later shown to be incorrect (Endler et al. 1979; Hearn 1982; Hammer et al. 1983). Nevertheless, Hearn's finding of a minimum possible energy loss is undisputed, and important for global energy balance considerations.

The curve "MFC" in Fig. 2 shows the results of an improved calculation of this minimum possible energy loss. By definition, the energy requirements of an open corona must always lie *above* this curve. On the other hand, the contribution of compressive waves of a given period to the coronal energy supply must lie *below* the corresponding dashed curve. This is because the shock waves may not yet have reached their asymptotic strength; and not all of the energy transported by the waves is available for coronal heating – part of it is reflected back from the transition region. Consequently, the intersection of the two types of curves determines *upper limits* for

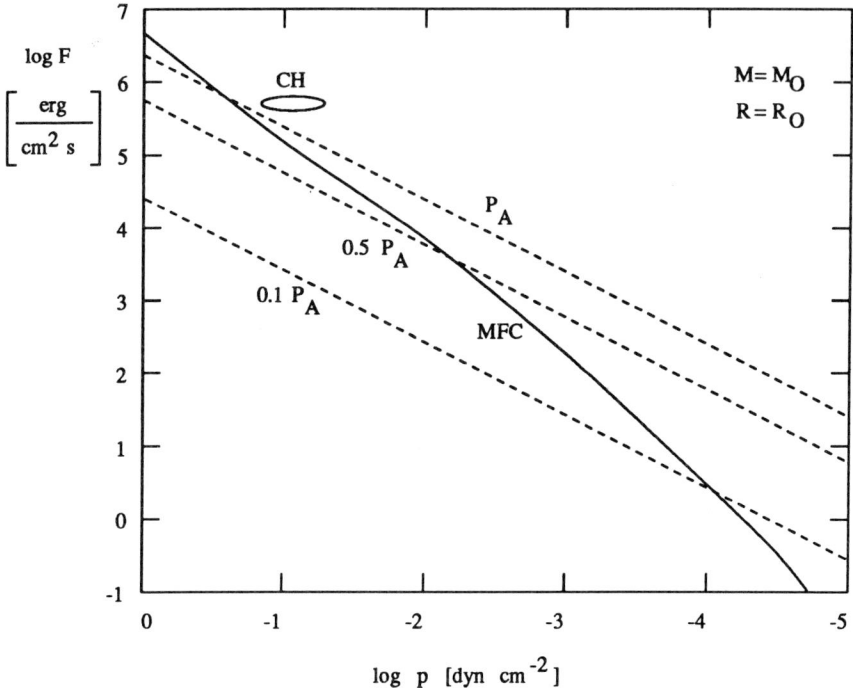

Figure 2 (*from Hammer and Ulmschneider 1989*). The dashed curves show the maximum energy flux carried by a compressive wave of a certain period (given as parameter in units of the photospheric acoustic cut-off period P_A). A wave can produce a corona only if its energy flux is sufficient to balance the coronal energy losses, which lie always above the (drawn) curve labeled MFC. Therefore, the intersection point between the two curves defines upper limits for the base pressure p_0 and total energy budget F_{M_0} of a corona that is heated exclusively by compressive waves of the given period. CH indicates empirical solar coronal hole parameters after Withbroe (1988).

the base pressure and for the total energy losses of a corona that is heated exclusively by such waves.

These theoretical limits are surprisingly severe. Waves of period $0.1 P_A$, which are most efficiently produced in the convection zone, cannot contribute more than 0.5% to the energy needed by actual solar coronal holes. If no other heating mechanisms were available, these short-period waves could produce a "corona" of their own only at pressures at least three orders of magnitude below the observed values; and the total energy budget of that corona would be smaller than that of real coronal holes by more than five orders of magnitude. It can be shown that the temperature of such a corona would be much lower than the critical temperature required for a Parker wind to resist the interstellar pressure. Therefore, short-period waves would lead to accretion rather than to a wind.

The situation improves with increasing wave period (cf. Eq. (12)). But even waves with the longest periods to be reasonably expected, $P \approx 0.5 P_A$, could produce only a very faint corona, with a base pressure and energy supply smaller than observed by one and two orders of magnitude, respectively. In order to balance the energy losses of realistic coronal holes we would need wave periods exceeding the acoustic cut-off period. Such waves cannot propagate through the photosphere as long as they are linear. There are in principle two ways out of this dilemma; however, both are not applicable to the Sun and solar-like stars. First, long-period waves might become nonlinear already *below* the photosphere, as in long-period variable stars. And second, long-period shock waves might be produced in the chromosphere out of *non*monochromatic short-period shock waves by the coalescence of individual shocks. Calculations illustrating this latter process will be discussed below. These calculations show, however, that shock mergence is too inefficient to allow for purely compressively heated coronae around solar-like stars.

If slowly rotating stars have coronae, it is extremely unlikely that these coronae are produced by compressive waves, unless such waves can be generated with periods $P > P_A$ by radial or nonradial pulsations. Coronae around basal flux stars must therefore be energized predominantly by other means, such as magnetic mechanisms. It is conceivable that even the most slowly rotating stars continue to produce magnetic fields at a slow rate, which can be energized by convective motions to provide coronal heating. The magnetic field need not be large. Even a 100 times smaller field than on the Sun dominates the atmosphere for $p \lesssim 10^{-3}$ dyn cm^{-2} – i. e., before compressive waves alone are able to produce a corona.

If it is not compressive waves – which other mechanisms heat the coronae of solar-like stars and drive the winds from noncoronal stars? Many suggestions were put forward in the past. The proposed mechanisms are so numerous that it is impossible to give a fair account of them in a single review – in fact, it is beyond the scope of the present paper to discuss even one of them in any detail. In the remainder of this section, I would rather like to refer the interested reader to some introductory and review papers on such mechanisms.

A recent and comprehensive compilation of heating mechanisms for the hot coronae of solar-like stars can be found in Narain and Ulmschneider (1989). Many articles (e.g., Castor 1981; Holzer and MacGregor 1985; Linsky 1987; Morris 1987; Hearn 1988; Holzer 1988; Willson 1988) review mechanisms that might drive the winds from noncoronal stars. These mechanisms include the wave pressure associated with Alfvén and large amplitude shock waves, and the radiation pressure on dust grains and molecules. Alfvén waves are also a popular candidate for heating the open solar corona and accelerating its wind to high speeds (e.g., Hollweg 1978, 1986; Leer et al. 1982; Leer 1988b). Finally, these quasi-steady wind driving mechanisms might well be assisted by discrete events like the ejection of diamagnetic plasma bubbles in an outward diverging magnetic field (Pneuman 1986) or reconnection processes (Parker

1988) that manifest themselves in the observed high velocity jets in the transition region (Brueckner and Bartoe 1983).

5 TIME-DEPENDENT EFFECTS

The previous two sections were mainly concerned with time-independent stellar wind theory, which is applicable as long as the temporal average over the conservation equations gives a reasonable description of the relevant physics. This approach is commonly chosen for models of the fairly steady high speed wind from coronal holes. The low speed wind is more variable. Moreover, wind variability is ubiquitous among cool giants (Reimers 1977, 1988) – from the vicinity of the dividing line (e.g., α Boo: Chiu et al. 1977; Judge 1987) up to the K and M supergiants (e.g., 32 Cyg: Schröder 1985; α Ori: cf. the review by Goldberg 1979). Finally, there has been some debate whether pulsating stars lose mass at higher rates than nonpulsating stars (cf. the reviews by Morris 1987; Drake 1988; and Willson 1988).

This increasing interest in temporal variability has stimulated the development of time-dependent theoretical stellar wind models. One group of these models addresses long-period variable stars. The winds from these stars appear to be driven by a combination of pulsation-induced shocks, which lead to an extended density distribution, and radiation pressure on dust grains, which ultimately drives most of the mass loss. The currently available models, reviewed by Bowen (1988a,b), are able to produce reasonable mass loss rates for these stars of extremely low gravity.

Cuntz (1987, 1990) has carried out acoustic wave calculations for noncoronal giants closer to the dividing line (in particular α Boo), for which dust driving is unimportant. The code (Cuntz and Ulmschneider 1988) is based on the method of characteristics, which makes it possible to treat shocks as discontinuities and to spatially resolve postshock radiation zones. Cuntz investigated three types of acoustic waves:

- Short-period ($P = 0.1 P_A$) monochromatic waves lead to an extended chromosphere, but no significant mass loss. This confirms earlier time-independent models of Hartmann and MacGregor (1980).
- Mass loss rates comparable to the empirical estimates ($-\dot{M} \approx 10^{-10} \, M_\odot \, \mathrm{yr}^{-1}$ for α Boo) can only be produced with *adiabatic* shock waves of *very* long periods ($P \approx 4 P_A$), corresponding to wavelengths of the order of the stellar radius. Only waves of such extremely long wavelengths carry their energy far enough to lift the wind plasma out of the gravitational field of the star (cf. Sect. 3). Atmospheric variations on large time scales have in fact been observed on Arcturus (e.g., Smith et al. 1987). However, the adiabaticity assumption overestimates the mass loss.
- In *non*monochromatic waves the shocks interact nonlinearly with each other (e.g., Gosling et al. 1976; Wood 1979; Bohn and Stein 1985). Occasionally individual shocks travel slightly faster than others, so they can merge with the preceding shocks, thus growing bigger and traveling even faster (cf. Fig. 8 of Ulmschneider, these proceedings). In this way, some of the shocks become very large and eject

matter from the star. However, the total mass loss associated with these sporadic events is still much smaller than the empirical value, although larger than for monochromatic short-period waves. Shock mergence leads to a reduction of the number of shocks that cross a given height level in a given time interval. Thus the average shock "period" increases with height and can eventually exceed the cut-off period. However, the calculations of Cuntz show that it takes a very thick slab to accomplish this increase. Therefore, this mergence process cannot remedy the fundamental difficulty of producing a corona with shock waves (cf. Sect. 4).

Models of this kind are extremely compute-intensive: in order to study the effects of waves in detail, it is necessary to resolve them in space and time; i.e., one needs certain minimum numbers of grid points per wavelength and time steps per period. This makes it virtually impossible to study the whole stellar wind zone out to its interface with the interstellar medium, and to follow its secular evolution in response to changes of the heat input or of the interstellar pressure.

The latter issues were addressed recently by Korevaar in a series of papers. He did not attempt to resolve individual waves, but rather described their combined effect by a time-independent heating law*. This allows him to extend the solution from the photosphere out to very large distances (up to 3000 stellar radii), so that in transonic solutions the outer boundary lies far in the supersonic flow regime, which cannot easily affect the inner regions of the wind. Numerical side effects of the outer boundary condition are thus minimized.

The numerical method (Korevaar and Van Leer 1988) is based on an implicit conservative upwind scheme and particularly suited to solve slowly varying problems in an efficient manner. Since the method is implicit, it is absolutely stable and permits large time steps when the solution becomes stationary. Moreover, the method is specifically designed to conserve the mass, momentum, and energy fluxes; thus it can even deal with shocks, for example with the shock at the transition from the stellar wind to the interstellar medium.

Korevaar applied this code to a number of problems. He investigated how the solution topology depends on the interstellar pressure and on the total coronal energy supply. By varying these parameters, he found not only transonic wind solutions, but (for large interstellar pressures) also stellar breezes (i.e., winds that remain subsonic everywhere), various types of accretion solutions, and even a static solution.

Korevaar and Hearn (1989a) studied the response of a corona to a change of the heat supply. They demonstrated that after the disappearance of transient wavefronts a hot corona normally finds a new stationary state. It has been known for some time, however, that this is possible only as long as the coronal heating flux does not exceed a certain limit, which depends on the damping length or wave period. When heated beyond this limit, a corona cannot be both extended and stationary, as was realized

---

* Wave momentum deposition was not taken into account.

independently by several authors (Hammer 1981, Hearn and Vardavas 1981, Souffrin 1982). In these "overheated" coronae, thermal conduction is no longer able to supply sufficient energy to the outer atmosphere. The only possible stationary solution consists of a rather peculiar sequence of hot coronal shells, separated by cool layers; but such a configuration can be shown to be convectively unstable (Hammer 1985). Hearn et al. (1983) suggested that an overheated corona might not be stationary at all, but rather undergo global relaxation oscillations, in which periodically the hot corona collapses and is then rebuilt. Korevaar and Hearn (1989b) were indeed able to reproduce such a behavior in their time-dependent calculations. Their oscillation period is very long, of the order of months to years, depending on the properties of the star. The period likely depends also on the assumed heating law. Korevaar and Martens (1989) speculate that the phenomenon of coronal relaxation oscillations might perhaps be relevant to F stars.

All of Korevaar's calculations were made for one star*; but the coronal portion of the models can be scaled to other stars (Korevaar and Martens 1989) with scaling laws found by Hammer (1984). When scaled to the Sun, the underlying heating law corresponds to acoustic shock waves of a period of 580 s, which is more than three times the cut-off period of the upper photosphere and therefore unrealistic. In this sense, the models are consistent with our previous conclusion that shock waves of *reasonable* periods cannot produce a hot corona around cool stars. Nevertheless, Korevaar's scaled models are valuable even for the Sun because they do not depend on the effects of waves in detail, so that this heating law is just a specific example for the as yet unknown heating mechanism of the open solar corona.

REFERENCES

Antiochos, S. K.: 1987, in *Cool Stars, Stellar Systems, and the Sun*, eds. J. L. Linsky and R. E. Stencel, *Lecture Notes in Physics* **291**, Springer, p. 283
Athay, R. G., and White, O. R.: 1978, *Astrophys. J.* **226**, 1135
Axford, W. I.: 1985, *Solar Phys.* **100**, 575
Ayres, T. R., et al.: 1981, *Astrophys. J.* **224**, 1064
Berteaux, J. L., Lallement, R., Kurt, V. G., and Miranova, E. N.: 1985, *Astron. Astrophys.* **155**, 1
Bianchi, L., and Gilmozzi, R. (eds.): 1988, *Mass Outflows from Stars and Galactic Nuclei*, Kluwer
Biermann, L: 1948, *Z. Astrophys.* **25**, 161
Biermann, L: 1951, *Z. Astrophys.* **29**, 274
Bohn, H. U.: 1984, *Astron. Astrophys.* **136**, 338
Bohn, H. U., and Stein, R. F.: 1985, in *Chromospheric Diagnostics and Modeling*, ed. B. W. Lites, p. 228
Bowen, G. H.: 1988a, *Astrophys. J.* **329**, 299

---

* in fact an O-type star, for historical reasons

Bowen, G. H.: 1988b, in *Pulsation and Mass Loss in Stars*, eds. R. Stalio and L. A. Willson, Kluwer, p. 3
Brueckner, G. E., and Bartoe, J. D. F.: 1983, *Astrophys. J.* **272**, 329
Castor, J.: 1981, in *Physical Processes in Red Giants*, eds. I. Iben, Jr., and A. Renzini, Reidel, p. 285
Chiu, H.-Y., Adams, P. J., Linsky, J. L., Basri, G. S., Maran, S. P., and Hobbs, R. W.: 1977, *Astrophys. J.* **211**, 453
Couturier, P., Mangeney, A., and Souffrin, P.: 1979, *Astron. Astrophys.* **74**, 9
Couturier, P., Mangeney, A., and Souffrin, P.: 1980, in *Solar and Interplanetary Dynamics* (IAU Symp. **91**), eds. M. Dryer and E. Tandberg-Hanssen, Reidel, p. 127
Cuntz, M.: 1987, *Astron. Astrophys.* **188**, L5
Cuntz, M.: 1990, *Astrophys. J.*, in press
Cuntz, M., and Ulmschneider, P.: 1988, *Astron. Astrophys.* **193**, 119
Deutsch, A. J.: 1956, *Astrophys. J.* **123**, 210
Drake, S. A.: 1986, in *Cool Stars, Stellar Systems, and the Sun*, eds. M. Zeilik and D. M. Gibson, Springer, p. 369
Drake, S. A.: 1988, in *Proceedings of the Sixth International Solar Wind Conference*, eds. V. J. Pizzo, T. E. Holzer, and D. G. Sime, NCAR/TN-306+Proc, p. 129
Dupree, A. K.: 1986, *Ann. Rev. Astron. Astrophys.* **24**, 377
Endler, F., Hammer, R., and Ulmschneider, P.: 1979, *Astron. Astrophys.* **73**, 190
Goldberg, L.: 1979, *Quart. J. R. Astron. Soc.* **20**, 361
Gosling, J. T., Hundhausen, A. J., and Bame, S. J.: 1976, *J. Geophys. Res.* **81**, 2111
Haisch, B. M.: 1987, in *Cool Stars, Stellar Systems, and the Sun*, eds. J. L. Linsky and R. E. Stencel, *Lecture Notes in Physics* **291**, Springer, p. 269
Hammer, R.: 1981, Ph. D. Thesis, Univ. Würzburg
Hammer, R.: 1982, *Astrophys. J.* **259**, 767
Hammer, R.: 1984, *Astrophys. J.* **280**, 780
Hammer, R.: 1985, in *The Origin of Nonradiative Heating/Momentum in Hot Stars*, eds. A. B. Underhill and A. G. Michalitsianos, NASA-CP 2358, p. 125
Hammer, R.: 1988a, in *Solar and Stellar Physics*, eds. E.-H. Schröter and M. Schüssler, *Lecture Notes in Physics* **292**, Springer, p. 77
Hammer, R.: 1988b, in *Pulsation and Mass Loss in Stars*, eds. R. Stalio and L. A. Willson, Kluwer, p. 51
Hammer, R., Endler, F., and Ulmschneider, P.: 1983, *Astron. Astrophys.* **120**, 141
Hammer, R., and Ulmschneider, P.: 1989, in preparation; and in *Cool Stars, Stellar Systems, and the Sun*, ed. G. Wallerstein, in press
Hartmann, L., Dupree, A. K., and Raymond, J. C.: 1980, *Astrophys. J.* **242**, 260
Hartmann, L., and MacGregor, K. B.: 1980, *Astrophys. J.* **242**, 260
Havnes, O., Pettersen, B. R., Schmitt, J. H. M. M, and Solheim, J. E. (eds.): 1988, *Activity in Cool Star Envelopes*, Kluwer
Hearn, A. G.: 1975, *Astron. Astrophys.* **40**, 355

Hearn, A. G.: 1982, *Astron. Astrophys.* **116**, 296
Hearn, A. G.: 1988, in *Mass Outflows from Stars and Galactic Nuclei*, eds. L. Bianchi and R. Gilmozzi, Kluwer
Hearn, A. G., and Vardavas, I. M.: 1981, *Astron. Astrophys.* **98**, 230
Hollweg, J. V.: 1978, *Rev. Geophys. Space Phys.* **16**, 689
Hollweg, J. V.: 1986, *J. Geophys. Res.* **91**, 4111
Holzer, T. E.: 1977, *J. Geophys. Res.* **85**, 4665
Holzer, T. E.: 1988, in *Proceedings of the Sixth International Solar Wind Conference*, eds. V. J. Pizzo, T. E. Holzer, and D. G. Sime, NCAR/TN-306+Proc, p. 3
Holzer, T. E., and MacGregor, K. B.: 1985, in *Mass Loss from Red Giants*, eds. M. Morris and B. Zuckerman, Kluwer, p. 229
Hundhausen, A. J.: 1972, *Coronal Expansion and Solar Wind*, Springer
Jacques, S. A.: 1977, *Astrophys. J.* **215**, 942
Jordan, S.: 1981, in *The Sun as a Star*, ed. S. Jordan, NASA SP-450, p. 301
Judge, P. G.: 1987, in *Circumstellar Matter* (IAU Symp. **122**), eds. I. Appenzeller and C. Jordan, Kluwer, p. 321
Korevaar, P., and Hearn, A. G.: 1989a, *Astron. Astrophys.* **220**, 177
Korevaar, P., and Hearn, A. G.: 1989b, *Astron. Astrophys.* **224**, 141
Korevaar, P., and Martens, P. C. H.: 1989, *Astron. Astrophys.*, in press
Korevaar, P., and Van Leer, B.: 1988, *Astron. Astrophys.* **200**, 153
Lallement, R., Berteaux, J. L., and Kurt, V. G.: 1985, *J. Geophys. Res.* **90**, 1413
Lallement, R., Holzer, T. E., and Munro, R. H.: 1986, *J. Geophys. Res.* **91**, 6751
Lallement, R.: 1988, in *Proceedings of the Sixth International Solar Wind Conference*, eds. V. J. Pizzo, T. E. Holzer, and D. G. Sime, NCAR/TN-306+Proc, p. 651
Leer, E.: 1988a, in *Activity in Cool Star Envelopes*, eds. O. Havnes et al., Kluwer, p. 297
Leer, E.: 1988b, in *Proceedings of the Sixth International Solar Wind Conference*, eds. V. J. Pizzo, T. E. Holzer, and D. G. Sime, NCAR/TN-306+Proc, p. 89
Leer, E., and Holzer, T. E.: 1980, *J. Geophys. Res.* **85**, 4681
Leer, E., Holzer, T. E., and Flå, T.: 1982, *Space Sci. Rev.* **33**, 161
Lindblad, B. A.: 1989, paper presented at this meeting
Linsky, J. L.: 1987, in *Circumstellar Matter* (IAU Symp. **122**), eds. I. Appenzeller and C. Jordan, Kluwer, p. 271
Linsky, J. L., and Haisch, B. M.: 1979, *Astrophys. J.* **229**, L27
Linsky, J. L., and Stencel, R. E. (eds.): 1987, *Cool Stars, Stellar Systems, and the Sun, Lecture Notes in Physics* **291**, Springer
Morris, M.: 1987, *Publ. Astron. Soc. Pacific* **99**, 1115
Munro, R. H., and Jackson, B. V.: 1977, *Astrophys. J.* **213**, 874
Narain, U., and Ulmschneider, P.: 1989, *Space Sci. Rev.*, in press
Neugebauer, M., and Snyder, C. V. W.: 1962, *Science* **138**, 1095
Oranje, B. J., and Zwaan, C.: 1985, *Astron. Astrophys.* **147**, 265
Parker, E. N.: 1958, *Astrophys. J.* **128**, 664

ERUPTIVE PHENOMENA ON THE SUN

Z. SVESTKA

SRON Laboratory for Space Research, Utrecht, The Netherlands

Abstract. This paper reviews pnenomena in which plasma is injected into preexisting coronal structures (surges, X-ray bright surges, flaring arches) and phenomena in which the magnetic field is disrupted (sprays, erupting filaments, eruptive (dynamic) flares). At some places the reader is referred to another recent review (Svestka, 1989) where dynamic flares were discussed in more detail.

1. INTRODUCTION

Active phenomena on the Sun represent gradual or sudden rearrangements of the local magnetic field in the solar atmosphere. A gradual rearrangement, for example, is responsible for the growth and decay of active regions and for X-ray brightness variations in coronal loops that interconnect them (Howard and Svestka, 1977). Examples of sudden rearrangements without disruption of the magnetic field can be seen also at some brightenings of interconnecting loops (Svestka and Howard, 1979) and this kind of rearrangement is typical for confined flares (Priest, 1981; Svestka, 1986). In all these cases the general magnetic configuration remains preserved: new loops emerge and old loops dissolve or submerge in an active region, thus making different active region loops bright; the magnetic field strength varies which intermittently makes different flux tubes visible in systems of interconnecting loops; an instability (or a set of instabilities) heats a preexisting loop configuration in a confined flare, but the configuration does not basically change.

Quite often flares also excite secondary loops, which sometimes may become more impressive than the flares that have triggered them, because the secondary loop may be bigger and cool slower. For an example the reader is referred to the "Queens' flare" analyzed by De Jager et al. (1983) where energetic electrons, accelerated in a small limb flare, diffused into a preexisting structure extending to 35000 km above the limb. While the flare lasted only 15 minutes in 10 keV X-rays, the large structure could be seen for 90 minutes.

There are other cases, however, when a sudden change of magnetic field leads to injections of plasma (with smaller or larger admixture of neutral gas) into preexisting magnetic structures. These events include *surges* and *flaring arches*. They are still injections into preexisting field configurations which stay preserved. In contrast to that, *sprays* and *erupting filaments* disrupt magnetic field and the most energetic process of this disruption is a *dynamic (eruptive) flare*. We will discuss all these various kinds of eruptive events in this review.

2. SURGES

When observed in the Hα line, surges, or surge prominences (cf. Svestka's (1976) book and references therein) are magnetically confined ejections of dense material from the chromosphere which ascend with velocities typically between 50 and 200 km s^{-1}. After reaching the maximum height of 20 up to 200 Mm, the material is usually seen to descend along the same, or very nearby, magnetic field lines. Very often there is a chromospheric brightening at the footpoint of an Hα surge, but in the majority of cases this brightening is too weak to be classified as a flare or subflare. But at least 20% of flares, and probably more, are associated with surges and then the flare apparently triggers the surge. Surges are usually repetitive phenomena and series of surges are often seen originating from the same place.

The sites of surges prefer some specific locations, as the outer edge of sunspot penumbrae (Giovanelli and McCabe, 1958), isolated areas of one polarity completely surrounded by the opposite polarity (Rust, 1968), or outer boundaries of growing Emerging Flux Regions (Kurokawa, 1988). Thus, according to Heyvaerts, Priest, and Rust (1977), surges may be caused by the reconnection between new evolving flux and the pre-existing ambient magnetic field. The estimated pressure gradient along a surge is sufficient to drive surge material outward, after an impulsive heating low in the chromosphere (Rust et al., 1980).

On the disk, surges are usually seen in absorption in the Hα line. In some cases the initial (lowest and densest) part of a surge may be in emission, but thereafter absorption prevails. Surges can be also seen in other low-temperature lines like C III (Kirschner and Noes, 1971) and C IV (Schmieder et al., 1983, 1984); Schmahl (1981) found as the upper limit of their visibility an Ne VIII line corresponding to $T \simeq 7 \times 10^5$ K. No surge emission could be detected in lines corresponding to higher temperatures (Schmahl) nor in soft X-rays (Rust, Webb, and MacCombie, 1977). However, the situation is different in the rare events when disk surges occur entirely in emission, and we will discuss these bright surges in Section 4.

3. FLARING ARCHES

Recently Martin and Svestka (1988) presented examples of magnetically confined injections of both cool gas and hot plasma into closed loop structures in the

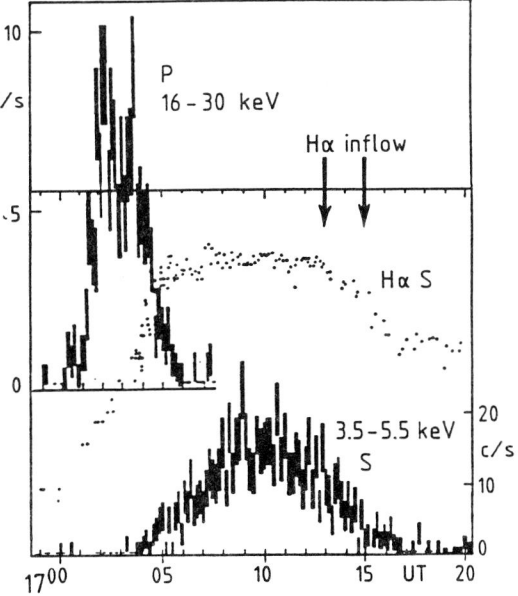

Fig. 1. Flaring arch of 12 November 1980 (length 260 Mm): the hard X-ray burst in the primary footpoint (P) is compared with the soft X-ray burst (S) and Hα brightening (dots, relative photometry) at the secondary footpoint; arrows show the time interval of the inflow of cool (Hα) material into the secondary footpoint. (After Martin and Svestka, 1988, by courtesy of Solar Physics.)

corona. These injections were observed as flows of bright emission in the Hα line and accompanying emission in >3.5 keV X-rays, propagating from a primary to a secondary footpoint of a coronal arch. The primary footpoint is the seat of a flare-like brightening, often a flare, and of a very steep and relatively long-lived hard X-ray burst (cf. example in Figure 1).

The secondary footpoint brightens immediately in Hα, and with some delay in X-rays (Figure 1). The immediate Hα brightening implies that the agent exciting the secondary footpoint, at least in the initial phase, must be energetic electrons. The propagation of the X-ray enhancement through the arches can be best fitted by conduction fronts moving with speeds between 1 and 2 Mm s^{-1} (Rust, Simnett, and Smith, 1985; Svestka et al., 1989b). The X-ray brightening at the secondary footpoint may be due to the conduction front as well, but the similarity of its time profile with that in Hα (Figure 1), as well as a hardening of the X-ray spectrum in flaring arches of smaller dimensions (Svestka et al., 1989b) rather indicate energetic particles as the excitation source: while electrons can explain all the observed characteristics in small arches, large structures, like that of Figure 1, would require protons replacing electrons in the main phase of the brightening.

The density in the arches must be very inhomogeneous: from $> 5 \times 10^{11}$ cm^{-3} in loops that emit in Hα (Svestka et al., 1987; Heinzel and Karlicky, 1987) to $< 4 \times 10^9$ cm^{-3} in loops through which enough energetic electrons can propagate (Martin and Svestka, 1988). The electron stream appears first, and a thermal front exciting the X-rays follows, while the bulk of the ejected plasma, at lower temperature, moves further behind. The densiest threads then emit in Hα. Similar phenomena were observed earlier, e.g. by Bruzek (1967), and Rust, Simnett, and Smith (1985), though they did not call them "flaring arches", and by Mouradian, Martres, and Soru-Escaut (1983) who already used this term, but in a somewhat different way.

Like surges, flaring arches are repetitive phenomena: the repeating events are of various intensity, but with the same basic characteristics and involving the same coronal structure (though different parts of the structure may be affected at different times). Svestka et al. (1989b) observed a series of 17 flaring arches on 6 November 1980 which showed a quasiperiodicity of about 18 minutes. These long periods might be due to oscillations of giant loops in the corona (Roberts, Edwin, and Benz, 1984; Harrison, 1987), though it is not clear how such oscillations can produce instabilities, near the loop footpoint, that repeatedly inject plasma into adjacent arches. Nevertheless, the observed quasiperiodical variations occurred below a giant X-ray-emitting coronal arch which stayed above the active region for several days. According to Roberts, Edwin, and Benz, the period of slow-mode oscillations of a coronal loop of length L and temperature T anchored in the photosphere is

$$P = 1.2 \times 10^{-4} \, LT^{-1/2} \, [1 + (s^2/v_A^2)]^{1/2} \text{ sec},$$

where s is the sound speed and v_A the Alfven speed. For a low-density loop with strong magnetic field $v_A \gg s$. Then, for $T = 10^7$ K, P = 18 min corresponds to L = 285 Mm. X-ray observations of the giant arch on 6 November and its magnetic modelling by Kopp and Poletto (1989) show that its loops extended to altitudes between 100 and 150 Mm, with lengths between 230 and 380 Mm.

4. X-RAY BRIGHT SURGES

It has been believed that, contrary to flaring arches, surges do not emit X-rays (cf. Section 2). However, very recently, several authors have reported soft X-ray emission associated with surges which are seen bright in projection on the solar disk in the Hα line (Harrison, Rompolt, and Garczynska, 1988; Schmieder et al., 1988; Simnett, Sotirovski, and Simon, 1989; Martin, 1989; Svestka, Farnik, and Tang, 1989). Figure 2 shows a series of X-ray bursts at footpoints of repetitive surges. All these surges were seen in absorption in the Hα line except the surge No. 2 which was completely, and for several minutes, in emission. This surge was also the only one that, in addition to the X-ray enhancement at its footpoint, clearly showed a temporary X-ray emission along its path (Figure 3).

As Figure 2 shows, the X-ray burst associated with this bright surge had the typical steep rise which characterizes flaring arches. Thus the flaring arches and X-ray bright surges are similar phenomena: in both of them cool gas and hot plasma are injected from a chromospheric source which brightens in Hα and

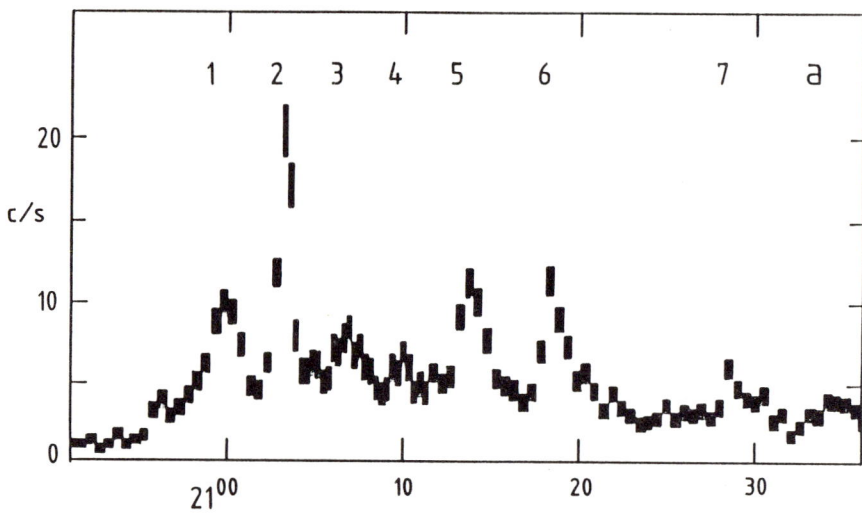

Fig. 2. Time development of the > 3.5 keV X-ray flux from AR 2550 during one orbit of the SMM on 8 July 1980 (HXIS data). All peaks 1 to 7 were associated with Hα brightenings accompanied by dark surges except No. 2 when the surge was in emission. (Svestka, Farnik, and Tang, 1989, courtesy of Solar Physics.)

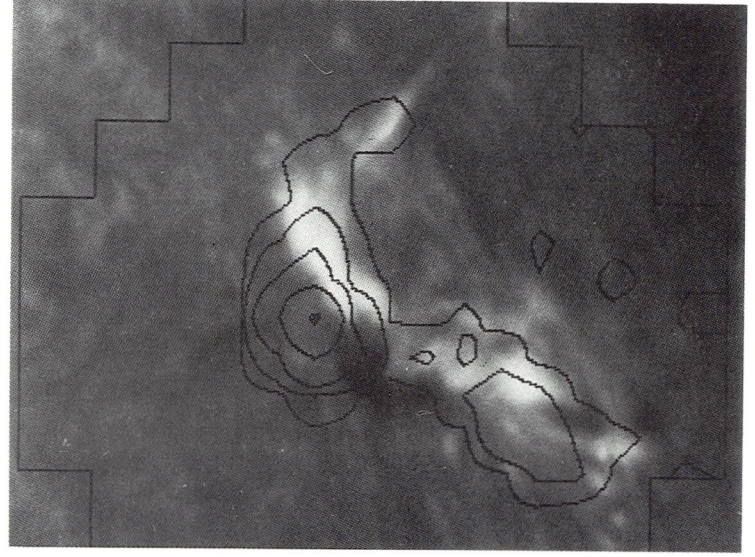

Fig. 3. HXIS fine-field-of-view image of AR 2550 and the surge No. 2 in Figure 1, superposed on the Hα photograph taken at Big Bear. The solar limb is in the upper right corner. (Svestka, Farnik, and Tang, 1989, courtesy of Solar Physics.)

X-rays and which produces a typical fast X-ray burst. While, however, in flaring arches the injection clearly follows closed magnetic structures, in X-ray bright surges the material seems to be injected along open field lines. It is likely, and often evident, that these field lines also form closed coronal loops, but of far larger dimensions: that one on 8 July 1980 (Figures 2 and 3) was at least 600 Mm long (Svestka, Farnik, and Tang, 1989). The much higher gradients of density and pressure in open or extended fields cause very rapid mass loss of the ejections, which explains the short duration of the detectable X-ray emission from bright surges when compared with flaring arches. The surge emission shown in Figure 3, for example, lasted less than 2 minutes, and its brightest phase only 20 - 30 seconds. This is apparently the reason why the X-ray emission from surges has been detected only so recently.

Another question is why only surges (and flaring arches) seen in emission in Hα yield detectable emission in X-rays. The Hα emission implies high density, close to 10^{12} cm^{-3} or higher. The density could be lower for large Doppler shifts, but as both dark and bright surges move with speeds of the order of 100 km s^{-1}, it is clear that the bright surges are denser than the dark ones. (Schmieder et al. (1984, 1988) estimate that dark surges have density of the order of 10^{11} cm^{-3}.) It is thus possible that a hot (i.e. about 10^7 K) plasma component is present in all ejections, but that it becomes detectable only in the densest events. A more likely interpretation, however, seems to be that only high-density ejections can give rise to thermal fronts which heat the plasma and thus produce the X-rays.

5. SPRAYS

All the phenomena that we discussed in Sections 2, 3, and 4, were injections into preexisting field configurations which did not disrupt. In contrast to that, *sprays* and *eruptive filaments* disrupt the magnetic field. A very appropriate definition was given by Sturrock (1980): the basic difference is that surges follow *unmoving*, preexisting magnetic field lines, while sprays and erupting filaments appear entrained in *moving magnetic fields*.

Both these erupting phenomena originate in preexisting filaments which lay above a zero line of the longitudinal magnetic field ($H_\parallel = 0$). Differences between them (on which opinions may differ), can be summarized as follows:

Sprays:	Erupting filaments:
accompany flares or can be considered to be flares themselves;	even quiescent filaments can erupt, without any chromospheric flaring;
powerful ejections right from their onset; speed frequently exceeds the escape velocity;	first slow rise, with speed growing while the eruption develops;
only a part of the filament erupts, in a random direction (sometimes along the surface);	the whole filament rises and erupts, in more or less radial direction;
in Hα very fragmented, though apparently material still is entrained on expanding loops;	in Hα more compact and clearly loop-shaped;
in X-rays emission rather chaotic (but essentially no X-ray images with good spatial resolution are available);	in X-rays emission is at tops of "post"-flare loops growing below the erupting filament.

Naturally, as in all classifications, there are eruptions which possess mixed properties of both these classes.

6. ERUPTING FILAMENTS

As mentioned above, filament eruptions can occur anywhere on the solar disk if a preexisting dark filament (i.e. a prominence projected on the disk) becomes subject of an instability. The typical helmet configuration in which filaments are embedded (Figure 4), is relatively stable and permits condensation and cool-

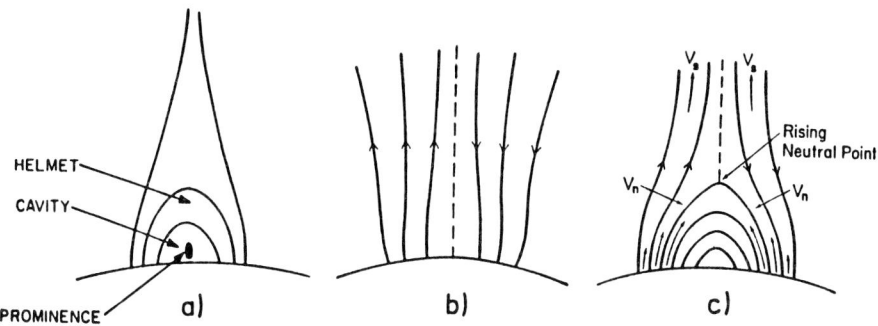

Fig. 4. (a) Carmichael's (1964) and Sturrock's (1968) helmet configuration above the $H_\parallel = 0$ line in which a filament (prominence) is embedded. (b) Opening of this configuration accompanying a filament eruption. (c) Subsequent reconnection and the creation of "post"-flare loops. (Kopp and Pneuman, 1976, and Pneuman, 1980; by courtesy of Solar Physics.)

ing of originally hot material along the $H_\parallel = 0$ line. However, either a loss of equilibrium or an MHD instability can cause the helmet to open and the filament to erupt. In the first case, for example, the filament can be considered as the coronal part of an electric current (Kuperus and Raadu, 1974; Van Tend and Kuperus, 1978; Kaastra, 1985; Martens and Kuin, 1989). If the current strength surpasses a certain threshold value, the force equilibrium ceases to exist and the filament starts erupting outwards. In the second case one can consider, for example, a horizontal flux tube rooted in the photosphere and held down by an overlying magnetic arcade (Sturrock, 1989). If the flux tube is sufficiently long and sufficiently twisted, an eruption of the flux tube will be energetically favourable: it moves to an untwisted state so reducing the magnetic energy along its entire length.

The equilibrium can also be destroyed, or the instability triggered, by external effects: a newly emerging flux (Rust, Nakagawa, and Neupert, 1975) or flux cancellation (Van Ballegooijen and Martens, 1989) (not necessarilly just below the filament; any magnetic field change, wherever the field lines supporting the filament are rooted, can unbalance the configuration), or slow-mode waves (Rust and Svestka, 1979). Figure 5 demonstrates that slow-mode waves, mostly originating in eruptive flares, with speeds decreasing from ~ 300 km s^{-1} at the onset

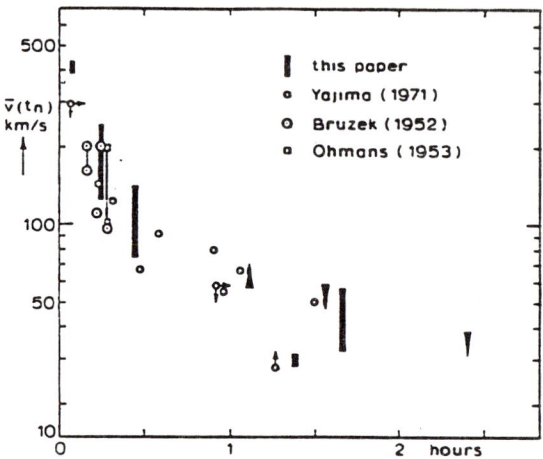

Fig. 5. Mean velocities of slow-mode waves deduced from the time difference between the onset of the wave in an eruptive flare and the onset of a distant filament eruption tentatively caused by the wave. (Rust and Svestka, 1979, courtesy of Solar Physics.)

to less than 50 km s^{-1} after two hours of propagation, fit very well the onset time of many filament eruptions observed in the Hα line on the Sun.

The consequences of a filament eruption in the solar atmosphere depend very much on the strength of the magnetic field in the filament's environment. In very old remnants of active regions, eruptions of quiescent filaments produce *disparitions brusques*, without any chromospheric brightenings along the $H_\parallel = 0$ line; in a somewhat stronger field a few Hα bright patches appear, which eventually merge into two bright Hα ribbons when a filament erupts in an (even spotless) active region. These ribbons, parallel to the $H_\parallel = 0$ line, become very bright in *two-ribbon flares* that appear in fully developed regions; the most energetic of them are the *proton* and *gamma-ray flares*. Because the basic process (field opening, filament ejection, and subsequent field line reconnection) is the same in all these phenomena, we call them collectively *eruptive* or *dynamic flares.*[*] All eruptive flares, from the disparitions brusques to the proton flares,

---

[*] *The term "dynamic" was proposed by Svestka (1986). Recently Priest (unpublished) proposed the term "eruptive" which seems to be used now more often.*

are capable to produce mass ejections in the solar corona.

It should be emphasized that the filament itself is not essential for the eruptive flare phenomenon. It is only a visible manifestation of the $H_\parallel = 0$ line, because that is the site of balanced forces where material can condense and cool. If there was enough time available for the condensation and cooling below 10^4 K, a filament becomes visible. However, this is not always the case. The active region may be too young, or an earlier eruption has not yet been restored so that no dark filament is visible in the Hα line. Nevertheless, the configuration and the physical process of the field opening and reconnection is still the same: only the manifestation of the eruption through the filament is missing.

7. ERUPTIVE (DYNAMIC) FLARES

Quite recently, in the IAU Colloquium No. 104 Proceedings, the present author published a review about the gradual phase of solar flares (Svestka, 1989) in which the properties of eruptive (dynamic) flares were thoroughly discussed. Therefore, the reader is referred to that paper (and references therein) and we will present here only a summary of the main items related to this topic.

A typical eruptive flare in a developed active region (a *two-ribbon flare* in Hα, a *long-decay event* in X-rays) can be briefly described as follows:

A pre-flare filament activation is followed by the filament eruption during the disruption of the field lines (Figure 4b). A coronal mass ejection is seen travelling through the corona and a shock develops, often giving rise to a type II burst on decametric and metric radio waves. According to the model first presented by Kopp and Pneuman (1976), plasma begins to stream upwards along the open field lines (towards the region of lower density) so that the magnetic pressure, originally in balance with the gas pressure, begins to dominate in the low atmosphere. This leads to field-line reconnection, starting first very fast low in the corona and proceeding with diminishing speed upwards (Figure 4c). Each reconnection heats the corona and accelerates particles which stream along the field lines to dense layers of the atmosphere, producing hard X-rays through thick-target bremsstrahlung and microwave bursts through interactions

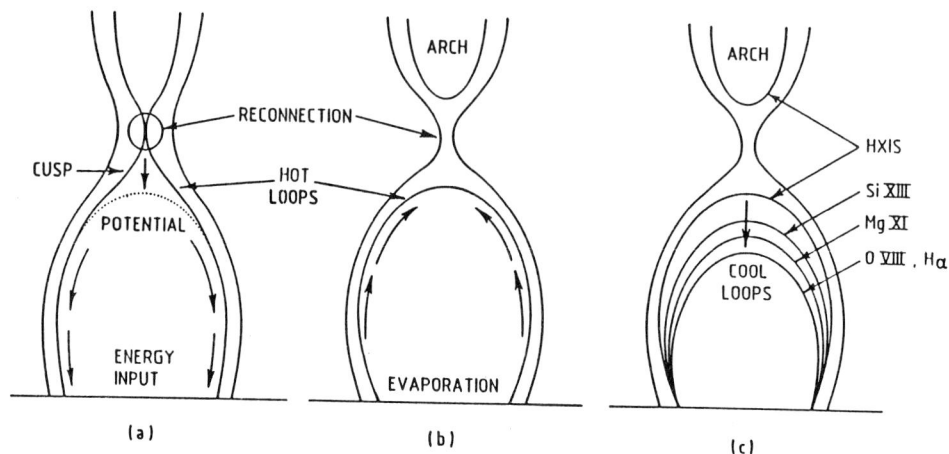

Fig. 6. *A scenario of the loop formation and heating (a), plasma evaporation (b), and the loop cooling and shrinking (c). (Svestka et. al., 1987, courtesy of Solar Physics.)*

with the strong magnetic field. Other accelerated particles diffuse upwards, are trapped in the corona, and produce decametric and metric type IV radio bursts.

Immediately after the reconnection the temperature at the top of each newly formed flare loop is very high (far in excess of 20×10^6 K) and makes the loop top visible in X-rays. This is the coronal X-ray emission observed after filament eruptions and typical for all kinds of eruptive flares. Particle streams (which dominate during the impulsive phase of the flare) and heat conduction (which prevails later) heat the chromosphere and evaporate chromospheric material into the loop. Thus the density of the loop grows, radiative cooling becomes faster, and the loop temperature sinks: the loop becomes visible in progressively cooler lines and eventually in Hα. During this process the loop shrinks: Figure 6 shows schematically the process of loop formation, heating, evaporation, and cooling.

The lower layers of the heated chromosphere which did not evaporate, become visible as two bright ribbons in the Hα line, at the footpoints of the hot loops emitting in X-rays (Figure 7). Thus the two Hα ribbons, extending parallel to the $H_\parallel = 0$ line, are connected by X-ray loops; below them are the cool Hα loops, visible in absorption when their density is below $\sim 10^{12}$ cm^{-3} and in emission when their density is higher. As the reconnection proceeds towards

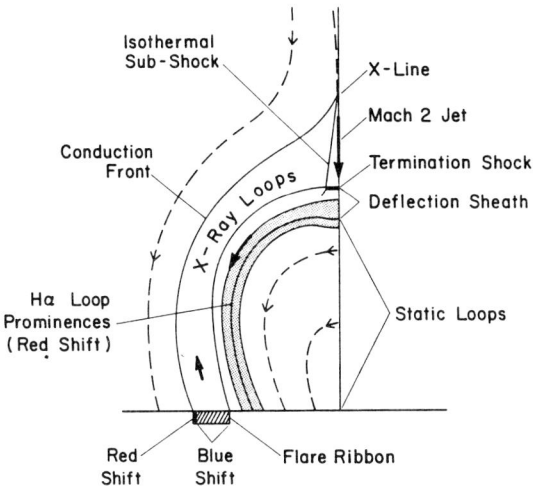

Fig. 7. Expected flow pattern in the reconnection model of an eruptive flare. (Schmieder et al., 1987, courtesy of Astrophysical Journal.)

higher altitudes and new progressively higher loops are formed, the loop system grows, both in X-rays (new hot loops) and in Hα (cooled, earlier loops) and the bright Hα ribbons gradually separate, their distance decreasing with a decreasing speed. The initial speed of the ribbon separation and the growth of the loops may be 50 km s^{-1} or more; it decreases rapidly, but even many hours after the flare onset the growth still continues, with a speed of a few hundred m s^{-1} (e.g., Moore et al., 1980).

Many aspects of this reconnection model of eruptive (dynamic) flares have been discussed in Svestka's (1989) earlier review (see additional references therein): the evidence of continuous energy release during the flare decay, indirect evidence for reconnection, heating of the loops (Cargill and Priest, 1983; Somov and Titov, 1985), chromospheric evaporation (Forbes and Malherbe, 1986), and shrinking of the loops. We will not repeat the discussion here, but one paper, by Schmieder et al. (1987), deserves particular attention, because it presented clear evidence for gentle chromospheric evaporation during the gradual phase of eruptive flares. Small blue shifts, lasting for several hours, were observed in the Hα flare ribbons of three flares and correspond to upwards chromospheric flows with speeds of 0.5 - 10 km s^{-1}. These upflows are sufficient to supply at least 10^{16} g needed to maintain a dense Hα loop system in the corona. In contrast to these upflows, the Hα loops exhibit large red shifts which are typical for material falling down in the cool loops. The expected flow pattern is shown in Figure 7. The narrow red-shifted region at the outer edge of the Hα ribbon in Figure 7 can be explained as due to downward compression of the lower chromospheric level, which occurs when it is suddenly heated at the footpoint of a newly reconnected loop.

Fisher (1986) distinguishes two types of evaporation: an explosive one, if the chromosphere is heated by particle flows, and a non-explosive evaporation produced by conduction. The maximum upflow velocities during an explosive evaporation can reach at most 2.3 sound speeds, whereas the non-explosive evaporation runs at speeds which are between 10 and 20 % of this upper limit. The explosive evaporation is to be expected during the impulsive phase. During the gradual phase we encounter the lower speeds which generally agree with the values obtained by Schmieder et al.

8. LARGE-SCALE STRUCTURES IN THE SOLAR CORONA

During recent years this picture of the eruptive flares has become somewhat complicated by discoveries of semi-permanent large-scale coronal structures above active regions in which eruptive flares occur (see again the earlier review by Svestka, 1989, for more details). These structures have been observed both in X-rays and on radio waves.

In X-rays, *giant post-flare arches* are seen for many hours after eruptive flares and their magnetic configuration apparently persists in the corona, because when another eruptive flare appears, essentially the same structure appears again in soft X-rays (Svestka, 1984). So far, using HXIS and FCS spectrometers aboard the SMM, altogether 11 such arches have been observed (Hick, 1988) and most of them were accompanied by a radio noise storm on metric waves. All of them, except one, followed eruptive (two-ribbon) flares. Similar structures were actually seen long ago on radio waves. Pick (1961) showed that metric type IV bursts, which as we know are typically associated with eruptive flares, consist of two parts. Nowadays we know that the first part is moving, and may be related to the mass ejection. The second part is stationary (see, as an example,

Wild, 1969, also reproduced in Svestka, 1989); it gradually becomes noisy and changes into a radio noise storm, and it clearly corresponds to the X-ray arch which forms the lowest component of the involved coronal structure.

Svestka et al. (1982) suggested that such an arch might be the upper product of the reconnection process during eruptive flares (as indicated in Figures 4b and c). A coronal mass ejection may form first, and the "arch" represents then a stationary remnant in the corona after the mass ejection had left. This would be in agreement with the radio data as well. But in many cases this does not seem to be true, because the new arch just "revives" the old one which apparently never disappeared. Thus, either no mass ejection was accomplished at all (which sometimes seems to be the case, cf. Svestka et al., 1989a), or the planes of the mass ejection and of the arch must be widely different, like in the event discussed by McCabe et al. (1986). Using radio data, also Gopalswamy and Kundu (1987) demonstrated that there can be a large difference between the direction of an outgoing shock (a type II burst in their case) and the position of stationary radio continua above an active region (Figure 8).

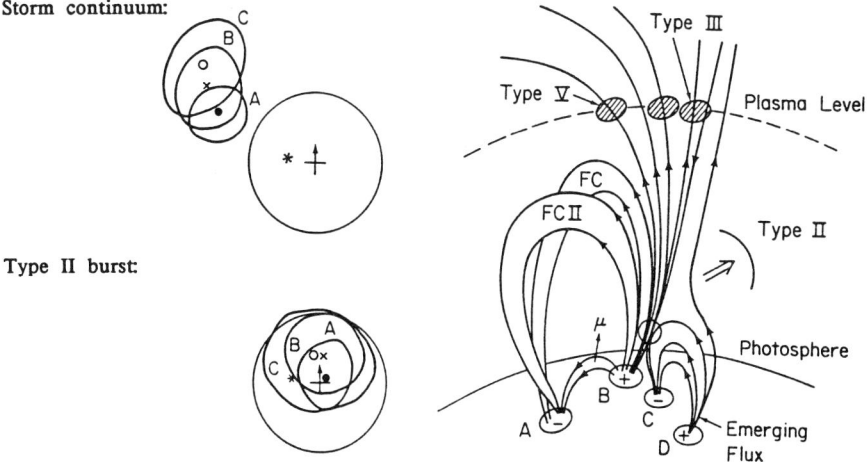

Fig. 8. Left: contours of peak brightness temperature at 73.8 (A), 50.0 (B), and 38.5 (C) MHz for the storm continuum (above) and type II bursts (below) originating in the flare marked by an asterisk. Right: model of the flare site based on radio images; FC loops emit the continuum and an arrow shows the direction of the type II burst (shock wave) propagation. (Gopalswamy and Kundu, 1987, courtesy of Solar Physics.)

Recently, Poletto and Kopp (1988; Kopp and Poletto, 1989), who could fit several giant post-flare X-ray arches with computed current-free magnetic fields, proposed a completely different model for this phenomenon. They agree that every arch brightening is the product of a magnetic reconnection process, like the "post"-flare loops seen at lower heights, but they suggest that this reconnection occurs in a separate, much more extensive magnetic structure, which is quite different from the configuration that creates the flare loops. That would explain why an arch behaves like a flare, with dimensions in time and space magnified by an order of magnitude (cf. Figure 5 in Svestka, 1989): at an altitude close to 100 Mm, the temperature peaks first, with ~ 1 hour delay, the X-ray brightness next, after more than 2 hours, and the emission measure last, with ~ 3.5 hours delay. However, the fact that the preceding arch does not seem to disappear during an arch revival, and the extremely close similarity of subsequent revived arches (cf., e.g., Figures 1 and 3 in Svestka, 1984) do not support this interpretation.

There are more observations in X-rays and on radio waves that indicate the existence of long-lived magnetic traps and large-scale coronal structures above the sites of eruptive flares. In a flare observed prior to the SMM imaging, Vilmer, Kane, and Trottet (1982) suspected that the source of the gradual X-rays was in a magnetic arch, in which the energetic electrons were trapped. Klein et al. (1983) observed a soft X-ray counterpart of the stationary Type IV burst for several eruptive flares. Kai et al. (1986) observed in other eruptive flares delayed bursts which appeared 0.5 to 1 hour after the strong impulsive phase had ended. These delayed bursts were observed from microwaves to meter waves. The wide difference in wavelengths at which this delayed radio burst appeared is again evidence that the delayed energy release occurs in a giant magnetic structure extending to at least 200 Mm altitude.

In many eruptive flares with long-lasting X-ray bursts the X-ray spectrum progressively hardens. Several such events were observed by Hinotori (e.g., Takakura et al., 1984) and by the SMM (e.g. Cliver et al., 1986). Takakura et al. observed three successive X-ray peaks in a flare event and each of them had harder spectrum than the preceding one. Cliver et al. observed flare events in which the power-law spectral index γ was steadily decreasing for tens of minutes (cf. Figure 7 in Svestka, 1989). This systematic hardening can be interpreted in terms of the energy-dependent collisional loss of electrons confined in a magnetic trap (Tsuneta, 1983; Bai and Dennis, 1985).

Observations of these large-scale and long-lived structures in the corona are still very incomplete: they have been rarely imaged and if so, then with bad spatial resolution, because good resolution is usually restricted to small fields of view. Hopefully more information about them, and about eruptive flares in general, will be obtained during the present solar maximum, in particular on board the SOLAR-A spacecraft. A special IAU Colloquium on eruptive flares is planned for the summer of 1991.

References

Bai, T. and Dennis, B.R.: 1985, Astrophys. J. **292**, 699.
Bruzek, A.: 1952, Z. Astrophys. **31**, 111.
Bruzek, A.: 1967, Solar Phys. **2**, 451.
Carmichael, H.: 1964, Proc. AAS-NASA Symp. on the Physics of Solar Flares, NASA SP-50, p. 451.
Cargill, P.J. and Priest, E.R.: 1983, Astrophys. J. **266**, 383.
Cliver, E.W., Dennis, B.R., Kiplinger, A.L., Kane, S.R., Neidig, D.F., Sheeley, N.R., and Koomen, M.J.: 1986, Astrophys. J. **303**, 920.
De Jager, C., Machado, M.E., Schadee, A., Strong, K.T., Svestka, Z., Woodgate, B.E., and van Tend, W.: 1983, Solar Phys. **84**, 205.
Fisher, G.H.: 1986, in D. Mihalas and K.H. Winkler (eds.), Radiation Hydrodynamics in Solar Flares, Lecture Notes in Physics **255**, 53.
Forbes, T.G. and Malherbe, J.M.: 1986, Astrophys. J. **302**, L67.
Giovanelli, R.G. and McCabe, M.: 1958, Australian J. Physics **11**, 130.
Gopalswamy, N. and Kundu, M.R.: 1987, Solar Phys. **111**, 347.
Harrison, R.A.: 1987, Astron. Astrophys. **182**, 337.
Harrison, R.A., Rompolt, B., and Garczynska, I.: 1988, Solar Phys. **116**, 61.
Heinzel, P. and Karlicky, M.: 1987, Solar Phys. **110**. 343.
Heyvaerts, J., Priest, E.R., and Rust, D.M.: 1977, Solar Phys. **53**, 255.
Hick, P.: 1988, Thesis, University of Utrecht.
Howard, R.F. and Svestka, Z.: 1977, Solar Phys. **54**, 65.

Kaastra, J.S.: 1985, Thesis, University of Utrecht.

Kai, K., Nakajima, H., Kosugi, T., Stewart, R.T., Nelson, G.J., and Kane, S.R.: 1986, Solar Phys. **105**, 383.

Kirshner, R.P. and Noyes, R.W.: 1971, Solar Phys. **20**, 428.

Klein, L., Anderson, K., Pick, M., Trottet, G., Vilmer, N., and Kane, S.: 1983, Solar Phys. **84**, 295.

Kopp, R.A. and Pneuman, G.W.: 1976, Solar Phys. **50**, 85.

Kopp, R.A. and Poletto, G.: 1989, Solar Phys., submitted.

Kuperus, M. and Raadu, M.A.: 1974, Astron. Astrophys. **31**, 189.

Kurokawa, H.: 1988, Vistas in Astronomy **31**, 67.

Martens, P.C.H. and Kuin, N.P.M.: 1989, Solar Phys. **122**, 263.

Martin, S.F.: 1989, Solar Phys. **121**, 215.

Martin, S.F. and Svestka, Z.: 1988, Solar Phys. **116**, 91.

McCabe, M.K., Svestka, Z.F., Howard, R.A., Jackson, B.V., and Sheeley, N.R.: 1986, Solar Phys. **103**, 399.

Moore, R., McKenzie, D.L., Svestka, Z., Widing, K.G., and 12 co-authors: 1980, in P.A. Sturrock (ed.), Solar Flares, Skylab Solar Workshop II, p. 341.

Mouradian, Z. Martres, M.J., and Soru-Escaut, I.: 1983, Solar Phys. **87**, 309.

Oehman, Y. and Oehman, N.: 1953, Observatory **73**, 203.

Pick, M.: 1961, Ann. Astrophys. **24**, 183.

Pneuman, G.W.: 1980, Solar Phys. **65**, 369.

Poletto, G. and Kopp, R.A.: 1988, Solar Phys. **116**, 163.

Priest, E.R.: 1981, in E.R. Priest (ed.), Solar Flare Magnetohydrodynamics, Gordon and Breach, London, p. 1.

Roberts, B., Edwin, P.M., and Benz, A.O.: 1984, Astrophys. J. **279**, 857.

Rust, D.M.: 1968, IAU Symposium **35**, 77.

Rust, D.M. and Svestka, Z.: 1979, Solar Phys. **63**, 279.

Rust, D.M., Nakagawa, Y., and Neupert, W.M.: 1975, Solar Phys. **41**, 397.

Rust, D.M., Webb, D.F., and MacCombie, W.: 1977, Solar Phys. **54**, 53.

Rust, D.M., Simnett, G.M., and Smith, D.F.: 1985, Astrophys. J. **288**, 401.

Rust, D.M., Hildner, E., and 11 co-authors: 1980, in P.A. Sturrock (ed.), Solar Flares, Skylab Solar Workshop II, p. 273.

Schmahl, E.J.: 1981, Solar Phys. **69**, 135.

Schmieder, B., Vial, J.C., Mein, P., and Tandberg-Hanssen, E.: 1983, Astron. Astrophys. **127**, 337.

Schmieder, B., Mein, P., Martres, M.J., and Tandberg-Hanssen, E.: 1984, Solar Phys. **94**, 133.

Schmieder, B., Forbes, T.G., Malherbe, J.M., and Machado, M.E.: 1987, Astrophys. J. **317**, 956.

Schmieder, B., Simnett, G.M., Tandberg-Hanssen, E., and Mein, P.: 1988, Astron. Astrophys. **201**, 327.

Simnett, G.M., Sotirovski, P., and Simon, G.: 1989, Astron. Astrophys., in press.

Somov, B.V. and Titov, V.S.: 1985, Solar Phys. **102**, 79.

Sturrock, P.A.: 1968, IAU Symp. **35**, 471.

Sturrock, P.A.: 1980, in P.A. Sturrock (ed.), Solar Flares, Skylab Solar Workshop II, Colorado Assoc. Univ. Press, p. 1.

Sturrock, P.A.: 1989, Solar Phys. **121**, 387.

Svestka, Z.: 1976, Solar Flares, D. Reidel Publ. Co., Dordrecht.

Svestka, Z.: 1984, Solar Phys. **94**, 171.

Svestka, Z.: 1986, in D.F. Neidig (ed.), The Lower Atmosphere of Solar Flares, NSO/Sacramento Peak, p. 332.

Svestka, Z.: 1989, Solar Phys. **121**, 399.

Svestka, Z. and Howard, R.F.: 1979, Solar Phys. **63**, 297.

Svestka, Z., Farnik, F., and Tang, F.: 1989, Solar Phys., submitted.

Svestka, Z., Stewart, R., Hoyng, P., Van Tend, W., Acton, L.W., Gabriel, A.H., Rapley, C.G., and 8 co-authors: 1982, Solar Phys. **75**, 305.

Svestka, Z., Fontenla, J.M., Machado, M.E., Martin, S.F., Neidig, S.F., and Poletto, G.: 1987, Solar Phys. **108**, 237.

Svestka, Z., Jackson, B.V., Howard, R.A., and Sheeley, N.R.: 1989a, Solar Phys. **122**, 131.

Svestka, Z., Farnik, F., Fontenla, J.M., and Martin, S.F.: 1989b, Solar Phys. **123**, 317.

Takakura, T., Ohki, K., Sakurai, T., Wang, J.L., Xuan, J.Y., Li, S.C., and Zhao, R.Y.: 1984, Solar Phys. **94**, 359.

Tsuneta, S.: 1983, Thesis, University of Tokyo.

Van Ballegooijen, A.A. and Martens, P.C.H.: 1989, Astrophys. J. **343**, 971.

Van Tend, W. and Kuperus, M.: 1978, Solar Phys. **59**, 115.

Vilmer, N., Kane, S.R., and Trottet, G.: 1982, Astron. Astrophys. **108**, 306.

Wild, J.P.: 1969, Solar Phys. **9**, 260.

Yajima, S.: 1971, Tokyo Astron. Bull. No. 207.

Radiative Transfer Problems in the Solar and Sun-like Atmospheres

J. TRUJILLO BUENO

Instituto de Astrofísica de Canarias, 38200 La Laguna, Tenerife, Spain

ABSTRACT: In this review we first provide a survey of certain fundamental radiative transfer (RT) problems, *i.e.* of problems which arise because of the need to develop improved RT theories capable of accounting for the complexity of radiation fields originating in Sun-like atmospheres. We then concentrate on studying both the diagnostic and energy balance problems of Sun-like atmospheres by means of various multi-dimensional RT calculations.

1. INTRODUCTION

As is well known, the theory of Radiative Transfer (RT) has a two-fold importance for Astrophysics. Through a detailed knowledge of RT we can develop diagnostic techniques to deduce, from the observed radiation field, not only the chemical composition, but also the thermodynamic and dynamic state of astrophysical plasmas. On the other hand, processes of transfer of energy and momentum by radiation play an important role in the structure and dynamical behaviour of many astrophysical fluids. Consequently, when aiming at giving a physical explanation of the observed phenomena, we are often obliged to couple the radiative transfer and the hydrodynamic equations, including also the influence of magnetic fields.

This double importance of RT is clearly seen in investigations dealing with stellar atmospheres (see *e.g.* Mihalas, 1978; Cannon, 1985). The atmospheres of stars, besides their particular chemical abundance, are characterized by the gravitational stratification of their constituent gas, by the presence of an effective surface through which photons escape, and by a net outward flux of energy. They thus represent open thermodynamical systems, where the variety of stellar atmospheric phenomena (radial and non-radial oscillations, magnetic flux concentrations, chromospheres, coronae, flares, winds, etc.) is the result of complex processes of exchange of energy and matter under non-equilibrium conditions.

There are many RT problems. Some of these problems are of a fundamental nature; others arise because of the difficulty in solving the RT equation obliges us to introduce unrealistic assumptions. Before going into details, it is first worth mentioning that if the atmospheres of the stars were in radiative and hydrostatic equilibrium, and if the role of magnetic fields were not so important, we would then be reasonably satisfied with our present treatment of RT. The physical properties of the atmospheric plasma would thus only vary along the radial direction of the star, and gas motions would not exist. Nowadays, efficient numerical methods are available to solve one-dimensional RT problems; such methods allow us to calculate both the emergent radiation from multi-level atoms, and the influence of millions of spectral lines on the temperature structure of the atmosphere under a global non-LTE context (see the papers by Cannon, Scharmer, Anderson and Werner in the books edited by Kalkofen 1984; 1987). The radiative transfer theoretician would, however, still be motivated to work on the problem of partial frequency redistribution, and on further improving the theories of spectral line shapes. On the other hand, those mainly interested in the quantitative spectroscopy of stars would probably concentrate on performing the RT calculations with improved basic atomic parameters (see *e.g.* Butler, 1987).

However, stellar atmospheres are generally not in hydrostatic and radiative equilibrium. In fact, they are often highly dynamic and spatially inhomogeneous systems (see *e.g.* Gustafsson, 1989 for a quick review survey of recent observations). Consider, in particular, the case of the solar atmosphere, where we can observe directly at high resolution not only temporal, but also spatial fluctuations of the radiation field at many frequencies. Through a careful analysis of these observations a rich variety of structural atmospheric patterns (exhibiting an intricate time-dependent behaviour) has been discovered on the Sun. Typical examples are the solar granulation pattern (Janssen, 1884; Muller, 1985), the radial and non-radial oscillations (Deubner and Gough, 1984), the small-scale magnetic flux concentrations (Stenflo, 1989), or the fascinating wealth of dissipative structures exhibited by the Sun's outermost layers, *i.e.* the chromosphere and corona (Gaizauskas, 1985; Vaiana and Rosner, 1978). In these examples, the horizontal structural lengths can be as small as only a few hundred kilometers and as large as the solar radius itself; the characteristic time scales may range from only a few seconds to hours, and the gas motions, either oscillatory or random, may reach velocities close to the speed of sound or even larger.

The fact that the plasma of the solar atmosphere, and more generally of stellar atmospheres, is highly dynamic and spatially inhomogeneous, implies the existence of a number of interesting RT problems. In this contribution we first provide information on certain fundamental RT problems; in particular concerning the transfer of polarized radiation, partial redistribution, theories of line broadening, and RT in stochastic media. We then concentrate on the issue of transfer of radiation in a highly inhomogeneous medium. This RT problem has turned out to be of crucial importance for both the diagnostic and the theoretical investigation of Sun-like atmospheres. We will thus study some physical effects of multi-dimensional radiative transfer (MRT), showing examples where a thorough knowledge of these MRT effects is decisive for a correct interpretation of observations. We will finally consider the problem of energy transfer by radiation under inhomogeneous situations typical of Sun-like atmospheres, demonstrating the importance of using reliable descriptions of the interaction between the hydrodynamic of the motion and the radiative energy transport in theoretical investigations of small-scale atmospheric processes. As will become evident from the discussion below, both the physical interest and the numerical challenge posed by most of these RT problems would disappear if the assumption of local thermodynamic equilibrium (LTE) were generally valid. Our emphasis here will be on the physical aspect of the problems, although we shall provide references on numerical methods with which to tackle the investigations suggested in the various sections of this review.

2. FUNDAMENTAL RADIATIVE TRANSFER PROBLEMS

The question posed here is the following: Do we have a sufficiently general theory capable of giving account of the radiation field originated in stellar atmospheres? This question leads to various fundamental problems related with the derivation of equations which govern the behaviour of the radiation field.

2.1 Transfer of Polarized Radiation

The interest for the derivation of a RT equation from the principles of Quantum Electrodynamics (QED) is not purely theoretical. On the contrary, the increasing need to interpret correctly solar polarimetric observations, *i.e.* spatio-temporal measurements of the Stokes parameters (see Shurcliff, 1962 for a definition of these four types of quantities), has led several authors to consider in great depth the problem of the derivation of the RT equation from first principles. The references of some of the key papers dealing with the quantum theory of transfer of polarized radiation can be found in Rees's (1987) introductory paper to polarized RT. After having provided such references, Rees (1987) writes: "Recently Landi Degl'Innocenti (1983a,b;1984;1985) has embarked on a grand synthesis of quantum line formation theory with a view to encompassing and extending all previous work. Here the algebraic minefield reaches formidable proportions and only deeply committed transfer theorists dare enter!". Given the great importance of achieving a rigorous understanding of the generation and transfer of polarized radiation for a correct interpretation of spectropolarimetric observations, it may be worthwhile to stimulate the reader to enter into this complex theoretical problem. To this end we shall comment here on the assumptions which lie behind Landi Degl'Innocenti's (1983) derivation of the RT and statistical equilibrium equations and, in particular, on how the same theoretical approach may be followed to derive equations of still greater generality. However, firstly it is helpful to begin by remembering the physical ingredients of this QED approach. (For more detailed information on QED ingredients see Cannon, 1985; and also Lamb and Ter Haar, 1971)

The quantum system under consideration is a volume element of a stellar atmosphere, which may be regarded as being composed of two interacting gases, a material gas and a photon gas. This interaction is due to the presence of microscopic interactions between material particles and photons (*i.e.* emission and absorption processes corresponding to bound-bound, bound-free, and free-free atomic transitions). There also exist microscopic matter-matter interactions (*i.e.* elastic and inelastic collisions), although photon-photon interactions are negligible. For the purpose of deriving equations which govern the behaviour of the radiation field (*i.e.* the behaviour of the photon gas), it is convenient to imagine the stellar atmosphere as being composed of three sub-systems: the atoms (A) which emit, absorb, or scatter the radiation; the material perturbers (P) capable of influencing the state of the atoms through collisions (*e.g.* free electrons), and the radiation field (R) itself. The quantum-mechanical Hamiltonian operator $\mathbf{H}$ of the whole system can be written, in the Schrödinger picture, in the form $\mathbf{H} = \mathbf{H_0} + \mathbf{V}$. Here $\mathbf{H_0} = \mathbf{H}_A + \mathbf{H}_R + \mathbf{H}_P$ is the unperturbed Hamiltonian made up of the sum of the energies of the atoms, radiation, and perturbers. $\mathbf{V} = \mathbf{V}_{RA} + \mathbf{V}_{PA} + \mathbf{V}_{RP}$ is the Hamiltonian which gives account of the various possible interactions among atoms, perturbers, and radiation.

Two reasons lead to the introduction of probabilities in the description of the state of a system like that defined above. One is the quantum-mechanical uncertainty related to the measurement process. The other is due (as in classical statistical mechanics) to our lack of complete information on the initial state of the system. This means that its description cannot be given in terms of a pure state $|\Psi>$, but through a statistical mixture of states (with the probability p_k of finding the system in the state vector $|\Psi_k>$). To incorporate into the quantum mechanical formalism the incomplete information we posses about the state of the system, the mixed state density operator (*i.e.* $\hat{\rho} = \sum p_k |\Psi_k><\Psi_k|$) was introduced (von Neumann, 1927; but see Fano, 1957; and ter Haar, 1961).

To appreciate the usefulness of the density operator, consider a complete set $\{|n>\}$ of eigenvectors of the atomic Hamiltonian (*i.e.* $\mathbf{H}_A|n> = E_n|n>$, with E_n the energy eigenvalue corresponding to $|n>$). To be able to make physical predictions about measurements bearing only on the atomic sub-system (A), the density operator $\hat{\rho}^A$ of this sub-system is introduced (*i.e.* $\hat{\rho}^A = \text{Tr}_P(\text{Tr}_R \hat{\rho})$, where the symbols Tr_P and Tr_R mean the traces over the perturbers (P) and the radiation field (R) coordinates respectively; note that the trace of a matrix is the sum of its diagonal elements). Calculating the matrix elements $\hat{\rho}^A_{ij}$ of $\hat{\rho}^A$ in the set $\{|n>\}$ one finds:

(*a*) $\hat{\rho}^A_{ii}$ is the population of the state $|i>$, since it represents the average probability of finding the system in the state $|i>$.

(*b*) $\hat{\rho}^A_{ij}$ with $i \neq j$ gives account of the interference effects between the states $|i>$ and $|j>$, which can appear simply because each state $|\Psi_k>$ of the statistical mixture is in general given by a linear superposition of the states $\{|n>\}$. These non-diagonal elements of $\hat{\rho}$ are called coherences.

The time evolution of the density operator $\hat{\rho}$ can be easily deduced from the Schrödinger equation. Using the usual symbol to indicate the commutator of two operators, one has

$$i\hbar \frac{\partial \hat{\rho}}{\partial t} = [\mathbf{H}_0, \hat{\rho}] + [\mathbf{V}, \hat{\rho}]. \tag{1}$$

Assume, for instance, that $\mathbf{H}_0 = \mathbf{H}_A$ and $\mathbf{V} = 0$. As can be easily deduced, the matrix element $[\mathbf{H}_A, \hat{\rho}]_{mn} = <m|[\mathbf{H}_A, \hat{\rho}]|n> = (E_m - E_n)\hat{\rho}_{mn}$. Therefore, from Eq. (1) we have

$$i\hbar \frac{d}{dt}\hat{\rho}_{mn}(t) = (E_m - E_n)\hat{\rho}_{mn}. \tag{2}$$

The solution of Eq. (2) is $\hat{\rho}_{mn}(t) = \hat{\rho}_{mn}(0)\exp[i(E_n - E_m)t/\hbar]$. Consequently, if $\mathbf{V} = 0$, the populations $\hat{\rho}_{nn}$ are constant, and the coherences oscillate at the Bohr frequencies of the system. However, if $\mathbf{V} \neq 0$, the matrix element $[\mathbf{V}, \hat{\rho}]_{mn}$ which results from Eq. (1) generally introduces an additional time-dependence for both the populations and the coherences.

One of the reasons which explains why the density matrix formalism is so suitable for dealing with the problem of generation and transfer of polarized radiation is the following. Through an emission process, polarization in spectral lines can be locally originated either by the presence of a splitting of the atomic levels (a splitting which can in turn be due either to the Zeeman or to the Stark effect), or by the presence of population differences and/or coherences among the sublevels. To obtain equations capable of describing the modification of this polarization due to transfer effects in a gas, it need only be remembered that the expectation value of an observable $\mathbf{O}$ is given by (see *e.g.* Fano, 1957):

$$<\mathbf{O}>(t) = Tr\{\hat{\rho}(t)\mathbf{O}\} = Tr\{\hat{\rho}_I(t)\mathbf{O}_I(t)\}, \tag{3}$$

where $\mathbf{O}_I(t) = \exp(i\mathbf{H}_o t/\hbar)\mathbf{O}\exp(-i\mathbf{H}_o t/\hbar)$ is the operator associated to the observable $\mathbf{O}$, expressed in the interaction picture, and $\hat{\rho}_I$ is the interaction representation density operator, whose expression is similar to the previous one, but with $\hat{\rho}$ in place of $\mathbf{O}$. Note that the expectation value of an observable has the same structural form in the two quantum-mechanical approaches (*i.e.* in the Schrödinger and in the interaction pictures). However, the interaction picture has the advantage of forcing the time dependence of wave functions to arise solely from the effect of the perturbing Hamiltonian $\mathbf{V}$, thus allowing a perturbation about the stationary state established by $\mathbf{H}_o$. From Eq. (1), one finds that the equation which governs the time evolution of $\hat{\rho}_I$ is given by

$$i\hbar \frac{\partial \hat{\rho}_I}{\partial t} = [\mathbf{V}_I(t), \hat{\rho}_I(t)]. \tag{4}$$

Substituting the mathematical expression for the direct solution which Eq. (4) gives for $\hat{\rho}_I(t)$ into the *rhs* of the same equation, and the ensuing expression into the equation which results when the time derivative is taken on both sides of Eq. (3), one finally obtains

$$\frac{d}{dt}<\mathbf{O}>(t) = Tr\{\frac{d\mathbf{O}_I(t)}{dt}\hat{\rho}_I(t)\} - \frac{i}{\hbar}Tr\{[\mathbf{O}_I(t), \mathbf{V}_I(t)]\hat{\rho}_I(0)\}$$

$$-\frac{1}{\hbar^2}Tr\{\int_0^t [[\mathbf{O}_I(t), \mathbf{V}_I(t)], \mathbf{V}_I(t')]\hat{\rho}_I(t')dt'\}. \tag{5}$$

This exact formula can be used to deduce the time evolution of physical quantities. This formula constitutes the starting point in Landi Delg'Innocenti's (1983) QED approach. One could obtain a very general RT equation provided that Eq. (5) could be solved exactly to deduce the time evolution of the polarization tensor $I_{\gamma\delta}(\nu, \vec{\Omega})$, with $\gamma, \delta = 1, 2$ being the sub-indices of the polarization unit vectors, ν the frequency, and $\vec{\Omega}$ the unit vector indicating the direction of propagation of the radiation beam; (see Landau and Lifshitz, 1962). Similarly, if one could solve Eq. (5) to deduce the time evolution of the diagonal and nondiagonal matrix elements $\hat{\rho}_{ij}^A$ of the density operator of the atomic system, one would end up with very general statistical equilibrium equations. Presumably, with this general set of coupled equations, a variety of possible atomic and RT processes could be accounted for in a self-consistent manner. For example, the expressions for the shapes of emission and absorption profiles would arise naturally, and this in turn would lead to expressions for redistribution matrices. Such matrices would give a consistent account of the correlation in frequency and angle occuring when the scattering process in multi-level atoms takes place in the presence of collisions, atomic motions, and the splittings due to magnetic and/or electric fields.

In what follows, we outline briefly the approximations which Landi Degl'Innocenti (1983) made to solve Eq. (5). These approximations imply the neglect of various physical processes. We shall comment on which of these physical processes should actually be included for a more rigorous description of the radiation field of stellar atmospheres.

The approximations used by Landi Degl'Innocenti (1983) are the following:

(*a*) $\mathbf{H}_0 = \mathbf{H}_A + \mathbf{H}_R$, *i.e.* $\mathbf{H}_P = 0$. Also, $\mathbf{H}_A$ is assumed to give account of the motion of the atomic electrons, but not of the motion of the atom. This implies that the basis vectors involve the specification of only the atom and radiation field states (e.g. $|i, n_1, n_2, ..., n_{\nu,\vec{\Omega},\delta}, ..., t>$, where i indicates the state of the atom, and $n_{\nu,\vec{\Omega},\delta}$ the number of photons in the $\nu, \vec{\Omega}, \delta$ mode, with $\delta=1,2$ characterizing the polarization state of the photons).

(b) $\mathbf{V} = \mathbf{V}_{RA}$, i.e. $\mathbf{V}_{PA} = \mathbf{V}_{RP} = 0$. $\mathbf{V}_{PA} = 0$ means that the effect of elastic and inelastic collisions of the perturbers with the atoms is not considered in the derivation of the RT and statistical equilibrium equations. $\mathbf{V}_{RA} = 0$ implies the assumption of no interaction between the radiation field and the perturbers.

(c) $\mathbf{V}_{RA} = \mathbf{V}_{RA}^{(elec.dipole)} + \mathbf{V}_{RA}^{(spin)} + \mathbf{V}_{RA}^{(two-photon)} = \mathbf{V}_{RA}^{(elec.dipole)}$. This implies: (1) the neglect of the interaction between the spin magnetic moment of the atomic electrons and the magnetic field (i.e. $\mathbf{V}_{RA}^{(spin)} = 0$), and (2) the omission of two-photon processes (either of absorption, emission, or diffusion of a photon from one mode to a different mode), i.e. $\mathbf{V}_{RA}^{(two-photon)} = 0$. The expression for the term retained ($\mathbf{V}_{RA}^{(elec.dipole)}$) gives account of one-photon processes of emission and absorption by nonrelativistic atomic electrons. This operator is given by a summation, over each mode ($\nu, \vec{\Omega}, \delta$) of the radiation field, of a linear combination of creation ($\mathbf{a}[\nu, \vec{\Omega}, \delta]$) and annihilation ($\mathbf{a}^+[\nu, \vec{\Omega}, \delta]$) operators of photons. It is easy to show that, whenever the Bohr radius is much smaller than the wavelengths of interest, these approximations concerning $\mathbf{V}$ are justified (see Heitler, 1954).

(d) $\hat{\rho}_I(t') = \hat{\rho}_I^R(t') \otimes \hat{\rho}_I^A(t')$ for $0 \leq t' \leq t$. This equation for $\hat{\rho}_I$ implies that the radiation field and the atomic system are supposed to be uncorrelated from the time when the interaction begins up to the time t. In terms of the wave function describing the whole system of atoms and radiation field, the above simplified expression for $\hat{\rho}_I$ means $|\Psi_{AR}> = |\Psi_A> |\Psi_R>$. However, this is strictly true only if $\mathbf{V}_I = 0$. The important point to keep in mind is that the present approximation for $\hat{\rho}_I(t')$, when substituted into Eq. (5), is equivalent to considering a second-order perturbative expansion to the solution of Schrödinger's equation (see Landi Delg'Innocenti, 1983).

(e) Optical-coherence phenomena (see e.g. Loudon, 1973) are supposed to play a totally insignificant role.

(f) $Tr\{\mathbf{a}^+[\nu, \vec{\Omega}, \delta]\mathbf{a}[\nu', \vec{\Omega}', \delta']\hat{\rho}_I^R\} = 0$ unless $\nu = \nu'$, and $\vec{\Omega} = \vec{\Omega}'$, i.e. only correlations between different polarization modes of the radiation field are allowed to occur. This approximation is restrictive, since it impedes the possibility of explaining partial frequency and angle redistribution effects.

With these approximations the commutators appearing in Eq. (5) can be evaluated, and a coupled set of equations composed of the rate equations for $\hat{\rho}_{ij}^A$ and the RT equation can be obtained. The resulting RT equation has the form:

$$\frac{1}{c}\frac{\partial}{\partial t}\mathbf{I} + \frac{\partial}{\partial s}\mathbf{I} = \mathbf{K}(\mathbf{S} - \mathbf{I}), \tag{6}$$

where $\mathbf{I}$ is the Stokes vector, $\mathbf{K}$ is a 4×4 matrix describing the modification of the Stokes parameters due to transfer effects along the ray-path coordinate s, and $\mathbf{S}$ is the source function vector in the four Stokes parameters. The various coefficients defining the vector $\mathbf{S}$ and the matrix $\mathbf{K}$ are given in terms of the diagonal (populations) and nondiagonal (coherences) density-matrix elements of the atomic system, which include magnetic and/or electric fields as parameters. Therefore, to find the solution for $\mathbf{I}$, the RT equation has to be solved consistently with the rate equations. Note that, in Sun-like atmospheres, the term equal to $(1/c)\partial \mathbf{I}/\partial t$ in Eq. (6) is usually omitted, since the time scales of variations of the thermal sources are generally much larger than the average transit time of photons.

The rate equations for $\hat{\rho}_{ij}^A$ were obtained by ignoring the effect of collisions between atoms and perturbers. Therefore they only contain radiative rates. The method used to account for the effect of collisions is to add the collisional rates in a purely phenomenological way. When doing this, and then particularizing to the unpolarized case, the non-diagonal elements of the atomic density-matrix vanish, and one finds the well-known phenomenological non-LTE equations describing the behaviour of the specific intensity $I_\nu(\vec{r}, \vec{\Omega}, t)$ (see Mihalas, 1978). The only difference is that, instead of an absorption profile with line-shape Φ, one has a Dirac delta function $\delta(\nu_{ij} - \nu)$, with $h\nu_{ij}$ the energy difference between the atomic levels i and j. This result is a consequence of having followed the second-order perturbation expansion specified in assumption (d). One could in principle obtain a Lorentzian line-shape due to the natural broadening mechanism. To achieve this goal, a suitable alternative to assumption (d) must first be found. This is, however, a difficult problem which must be carefully considered. One possibility might be to follow a higher-order perturbative expansion obtained by consecutively substituting the exact solution which Eq. (4) gives for $\hat{\rho}_I(t')$ into Eq. (5) and, once the desired order of the expansion is reached, to use the factorization of $\hat{\rho}_I$ given in assumption (d) (Landi Degl'Innocenti; private communication). Clearly, the obtaining of, say, a Voigt line-shape would additionally require us to take into account the Doppler effect of the atomic thermal motions. (The unperturbed wave functions would then additionally depend on the coordinate of the center of mass of the atom).

In practical applications one simply replaces the above-mentioned Dirac delta function by the desired line shape Φ (*e.g.* the Voigt profile). When the particularization to the non-polarized case is not made, one has Eq. (6) and rate equations for both the diagonal and nondiagonal density-matrix elements of the atomic system. In this more general situation one does not find the well known Faraday-Voigt function, which gives account of the so-called magneto-optical effects, but instead the principal part (in the distribution theory sense) of $1/(\nu_{ij} - \nu)$. This is again a consequence of having used assumption (*d*). In order to account for the effects of anomalous dispersion one simply replaces this latter expression by the Faraday-Voigt function. With the equations in this form, a variety of physical phenomena can be decribed, like resonance scattering, the Hanle (1924) effect, the Zeeman effect, and all the possible intermediate situations. As is well known, these effects should be taken into account when aiming at a correct interpretation of solar polarimetric observations. For those interested in working in this line of research, it may be informative to study the "polarized papers" edited by Kalkofen (1987). This will be helpful to appreciate the difficulties related with an outstanding RT problem, namely the development of efficient numerical methods for the solution of the coupled set of equations described above.

Although we are still far from a clear understanding of all the possible observational signatures which the above set of equations can produce (see results for the Hanle effect in two-level atoms in Bommier *et al*, 1989), it is still necessary to derive more realistic equations capable of giving account of the effects of partial frequency and angle redistribution in the presence of collisions. A recent review by Linsky (1985) shows that these effects are important for the diagnostic problem of stellar atmospheres.

2.2 Partial Redistribution

One possible method of deriving equations which include the effects of partial frequency and angle redistribution is to follow the QED approach outlined in the preceeding section (Landi Degl'Innocenti, private communication). The first task would be to evaluate the commutator expressions appearing in Eq. (5) without making use of assumption (*f*), including thermal motions of the atoms, but still retaining assumption (*b*). Success with such an evaluation would imply the derivation of equations which include the effects of redistribution of polarized radiation in the absence of collisions. The next task would be much more difficult: to derive the equations removing also assumption (*b*), *i.e.* considering the effect of collisions. One difficulty associated with such tasks is the necessity of finding a suitable alternative to assumption (*d*). As indicated earlier, only in this way can we expect to obtain expressions for the shapes of the probabilities of spontaneous emission (Ψ), stimulated emission (Θ), and absorption (Φ) of a photon of frequency ν travelling in the direction $\vec{\Omega}$ and, consequently, to give account of departures from the assumption of complete frequency redistribution (*i.e.* of deviations from the assumption $\Psi = \Theta = \Phi$). It is interesting to note that Oxenius (1965) has derived that $\Psi = \Theta$ from QED arguments for unpolarised light.

In addition to these difficulties, the investigation of partial redistribution effects in the presence of collisions requires the specification of the Hamiltonian V_{PA}. One possibility is to assume the impact and the binary approximations. In this respect, a classical work for the non-polarized case is that of Omont, Smith, and Cooper (1972). These authors investigated the effect of collisions on the redistribution of resonance radiation in the rest frame of the radiator, by using a formula (see Fiutak and Van Kranendonk, 1962) which gives the probability for absorption of a photon of frequency ν_1, and emission of a photon of frequency ν_2, accompanied by the corresponding change in the state of the atom. A more exact formulation of collisional redistribution, based on the binary approximation, has been developed by Burnett et al (1980) and Burnett and Cooper (1980-a,b). With the aim at facilitating quantitative calculations for Ly-α, Ly-β, and H-α, Cooper et al (1988) have simplified the above formulation. Following the philosophy of the papers by Cooper and coworkers (see also Omont, Smith, and Cooper, 1973), Domke and Hubeny (1988) have recently derived a redistribution matrix for scattering of arbitrarily polarized light by an atom undergoing collisions. Finally, we should mention that Streater, Cooper and Rees (1988) have applied a QED formalism to derive equations which govern the transfer of polarized light, including partial redistribution effects, but neglecting the influence of magnetic fields.

Before deciding which strategy to follow in solving some of the problems outlined here, it is highly desirable to achieve a thorough understanding of the various papers quoted above. To facilitate this enterprise, it may be useful first to digest the last three chapters of Cannon's (1985) monograph. Interesting discussions on various problems related to RT with partial redistribution can be found in the papers by Hubeny (1985), Oxenius (1985;1986), Simonneau (1985), and Frisch (1988).

2.3 Theories of Line Broadening

The point we wish to consider here is that, even in situations where the complete frequency redistribution approximation turns out to be acceptable, the problem of the specification of the line broadening in the atomic absorption coefficient is crucial for diagnostic investigations in Astrophysics, and in other areas of Physics as well. The general line-shape formula is given by (see e.g. Mihalas, 1978):

$$\Phi(\omega) = \frac{1}{\pi} Re\{\int_0^\infty F(s) e^{-i(\omega - \omega_0)s} ds\}, \qquad (7)$$

where $Re\{...\}$ means the real part of $\{...\}$, and $F(s)$ is the so-called correlation function. This function contains all the information necessary for the calculation of the dependence of the absorption probability Φ with angular frequency ω (see Margenau and Lewis, 1959). In order to evaluate $F(s)$ one has to choose the various line broadening mechanisms which one wishes to include (e.g. the Doppler effect, the effect of collisions, and the natural broadening), and to specify the desired approximations (e.g. either the quasi-static or the impact approximation for the treatment of the effect of collisions, etc). To proceed further one has then to select between a quantum mechanical or a classical path for the evaluation of the correlation function $F(s)$. In both cases the ergodic hypothesis is invoked in order to express $F(s)$ as an average over the statistical ensemble of quantities defining a function $f(t)$. If the classical approach of the radiation-matter interaction is chosen, the atom is represented by an atomic oscillator, whose vibration is described by $f(t) = \exp[i\omega_0 t + i \int_{-\infty}^{t} K(t')dt']$. Here ω_0 is the frequency at which the atom would oscillate were it not affected by the various possible causes of broadening which lead to the term $\eta(t) = \exp[i \int_{-\infty}^{t} K(t')dt']$. The final expression for the classical evaluation of the correlation function is given by (see *e.g.* Mihalas, 1978)

$$F(s) = < e^{i \int_0^s K(t')dt'} >, \qquad (8)$$

where the symbol $< ... >$ indicates an average over the statistical ensemble.

A successful quantum theory of line broadening in the impact approximation was developed by Baranger (1958 a,b,c). In this quantum theory $F(s)$ is defined by means of a statistical average of the expectation value of the interaction representation evolution operator. On the other hand, the main contributions to the broadening of lines of the hydrogen and helium spectrum in a plasma (*e.g.* in solar flares) is due to the Stark effect in the fields of electrons and ions (see Griem, 1974). A major problem arises here because the field of the slowly-moving ions is quasi-static, while the fast-moving electrons cause impact broadening. This in turn implies the necessity of achieving a unified line broadening theory capable of describing properly the Stark broadening of ions and electrons together with the Doppler broadening of the -generally nonthermal- motions of the ions. Important contributions to the problem of the broadening of hydrogen lines in plasmas have been made by Vidal *et al* (1973). However, much research in this area is still called for in order to develop a consistent unified theory of line-broadening, which could lead to the well known impact theory when the collision time is made negligible against the intercollision time, and to the quasi-static (or statistical) theory where the intercollision time is made much smaller than the collision time. The books by Griem (1974), Sobelman, Vainshtein, and Yukov (1981), and Breene (1981) are useful in achieving the required preparation to tackle this oustanding problem.

An example of the importance of developing unified theories is considered in the remaining part of this subsection. Most spectroscopic analyses of Sun-like atmospheres (*e.g.* for the determination of chemical abundances) make use of a Voigt line-shape for the description of the line broadening in the atomic absorption coefficient. The Voigt profile (Voigt, 1912) is often used to describe the combined effect of the following broadening mechanisms: (*a*) the natural broadening, which leads to a Lorentzian line shape (see Heitler, 1954), (*b*) the Doppler frequency shifts of the atomic motions, which lead to a Gaussian line shape if these motions are assumed to be described by the local Maxwellian velocity distribution function (Rayleigh, 1889; see also Mihalas, 1978), and (*c*) the effects of the random interruption of the atomic absorption of radiation by binary collisions of negligible duration, which also leads to a Lorentzian line shape (see *e.g.* Anderson, 1949; and note that such interruption processes are classically accounted for by assuming random phase shifts in the vibration of the atomic oscillator). The combination of these three causes of broadening is performed by assuming that they are statistically independent. This latter assumption implies (see Eq. 8) that the correlation function of the problem is simply given by the product of the individual correlation functions corresponding to each of these three broadening mechanisms. Consequently, the Voigt profile is simply given by the convolution product of the three line shapes mentioned above, *i.e.* by the convolution of a Lorentzian and a Gaussian (see Eq. 7, and recall that the convolution of two Lorentzians is also a Lorentzian).

However, as pointed out by Dicke (1953), the Doppler and collisional broadening mechanisms cannot be considered as statistically independent because the collisions between the absorber atoms and the neighbouring particles not only interrupt the radiation processes but also change the velocity and direction of motion of the absorber atom. Consequently, a correct evaluation of the Doppler broadening requires us to consider that the velocity component of the atom in the direction of observation is changing in a random way. Since the Doppler shift of frequency corresponding to an atom with velocity component u_z in the direction of observation is $\omega - \omega_0 = \omega_0 u_z/c$, the classical expression for the correlation function which describes the Doppler broadening in the presence of velocity changing collisions is given by (*cf.* Eq. 8)

$$F_D(s) = < e^{i\frac{\omega_0}{c} \int_0^s u_z(t')dt'} > . \qquad (9)$$

In order to evaluate this expression one has to average over all possible displacements $z(s) = \int_0^s u_z(t')dt'$ which can occur in the time interval s. As shown by Wittke and Dicke (1956), this statistical average reduces to an average over the distribution function of the initial velocity components, and over the distribution function of the net displacements which can occur in a time interval s for a given initial velocity component along the line of sight. This latter distribution function, which is characteristic of the Brownian motion (see Chandrasekhar, 1943), has a parameter β which is inversely proportional to the mean time t_D between collisions. Following these ideas, Galatry (1961) calculated the spectral line shape that results from the combination of the Doppler effect and the two effects (*i.e.* interruption of the absorption processes and change of motion) produced by collisions.

Galatry's formula for the line-shape may be considered the result of a unified line-shape classical theory within the framework of the impact approximation. The physical conditions under which the Galatry profile leads to the Voigt profile can easily be found by means of the following argument. As can be deduced from Eqs. (9) and (7), the conventional Gaussian function associated with the Doppler shifts in frequency of thermal atomic motions is obtained when the velocity component u_z is assumed to be time independent. However, if u_z is time dependent, the wavelength of the wavetrain of radiation which the absorbing atom 'sees' is changing in a random way. If such velocity component is constant, say, only during a time interval t_D, the absorbed radiation during this interval has a coherence time equal to t_D, and it contributes to the line-shape in a spectral range of width $1/t_D$ around the frequency $\omega_0(1 + u_z/c)$. As a result, the conventional Gaussian line-shape function is valid if $\omega_0 u_z/c \gg 1/t_D = u_z/L$, i.e. if $2\pi L/\lambda \gg 1$, where L is the mean free path of the atomic oscillator. However, if $\lambda \gg L$, the spectral profile is given by a Lorentzian (L_{Dicke}) whose width decreases with increasing gas density (Dicke, 1953). In conclusion, the Voigt profile is valid if the wavelength of the spectral line under consideration is much smaller than the atomic mean free path L.

In Fig. 1, we show schematically the limiting expressions of the Galatry profile, together with the parameters on which they depend. The Galatry line-shape itself depends on the following three parameters: (a) $\gamma_c = N < |\vec{v} - \vec{u}|\sigma_a >$, where N is the number density of perturbers, $\vec{u}$ and $\vec{v}$ the velocities of the absorbing atom and perturbing particle, respectively, and σ_a is a cross-section which depends on the relative velocity $|\vec{v} - \vec{u}|$. This parameter γ_c is the half-width of the Lorentzian function (L_{phase}) which describes the collisional broadening effect of the sudden interruption of the absorption processes. (b) β, i.e. the above-mentioned coefficient which appears as a parameter in the probability distribution characteristic of the Brownian motion. Note that β is inversely proportional to the mean time t_D between collisions, and that t_D is the larger the smaller the number density N. (c) A= $\frac{1}{2}\gamma_D^2/\beta^2$ (where $\gamma_D = \frac{\omega_0}{c}\sqrt{2KT/M}$, with M the mass of the absorbing atom, is proportional to the width of the Gaussian profile which results from the conventional Doppler broadening theory). In Fig. 1, we have used the series expansion expression given by James (1969) for the Galatry profile. Note that the number of terms of this series expansion which must be retained for a given value of N, essentially depends on A. Thus, for example, when N is so high that $L \ll \lambda$, only the first Lorentzian function in the series expansion need be considered. As expected, this Lorentzian is simply the result of the convolution product of L_{phase} with L_{Dicke}. On the other hand, if the number density N is sufficiently small so that $\lambda \ll L$, the well known Voigt line-shape of the conventional theory results.

As seen in the corresponding expression given in Fig. 1, the Voigt profile is obtained when the Lorentzian L_{phase}, describing the interruption of the absorption processes of an individual atom with velocity component u_z along the direction of observation, is averaged over the Gaussian profile which describes the Maxwell distribution of the velocity component u_z. However, this statistical thermal average is mathematically inconsistent because the half-width γ_c of the Lorentzian L_{phase} is proportional to the thermal average $< \sigma_a|\vec{v} - \vec{u}| >$. Consequently, it is not correct to average the Lorentzian L_{phase} over the Maxwell distribution without considering the explicit dependence of γ_c on the velocity of the atom that absorbs the radiation (Mizushima, 1971; but see Fowler and Sung, 1975). In fact, $\gamma_c(u) = N \int \int \int \sigma_a |\vec{v} - \vec{u}| W_m(v) d\vec{v}$, where $W_m(v)d\vec{v}$ represents the Maxwellian distribution function for the modulus of the velocities of perturbing particles of mass m. When this factor is taken into account, the correct mathe-

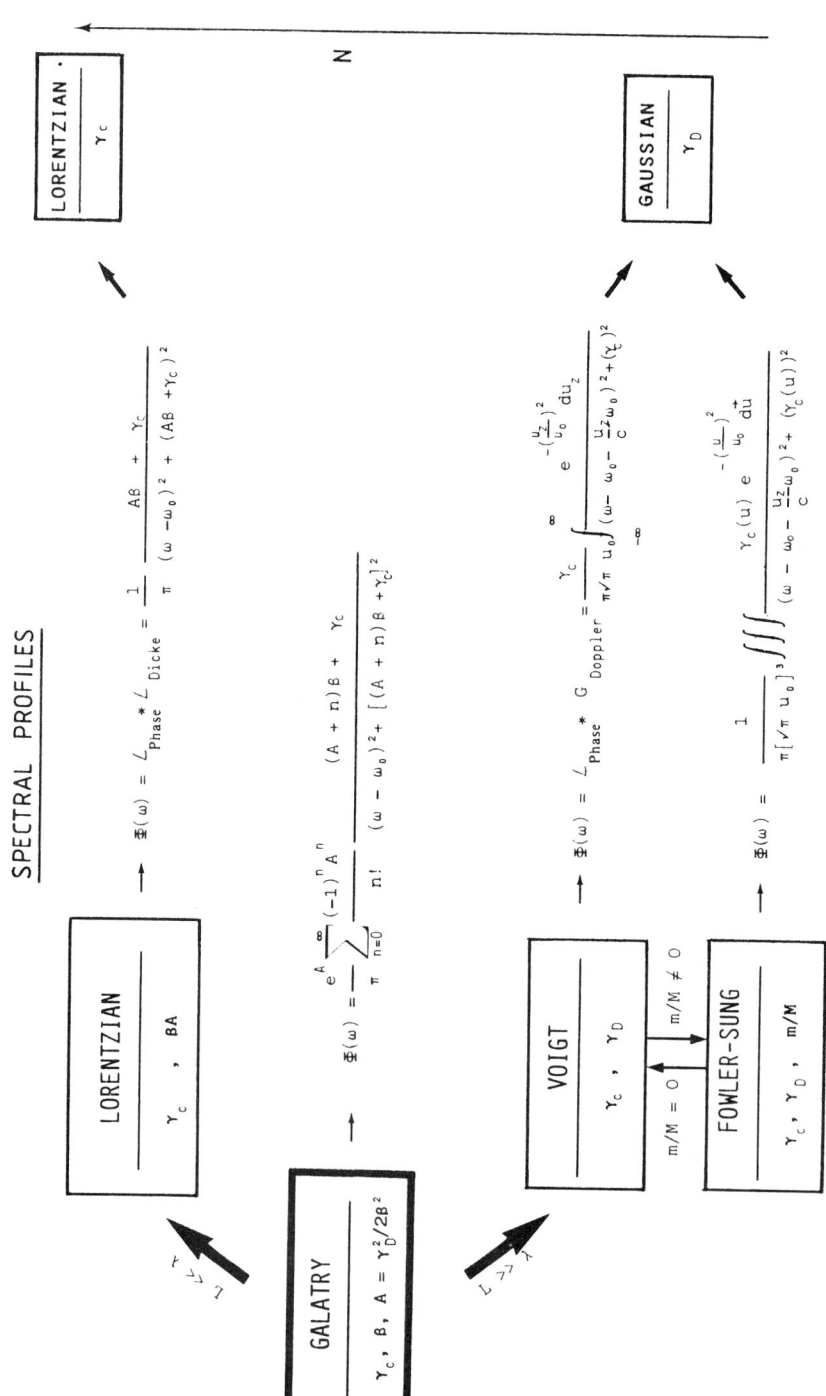

Fig. 1. The Galatry and the Fowler-Sung profiles, together with their limiting expressions. Note that $u_0 = \sqrt{2KT/M}$.

matical formula for the line-shape is the one shown in the lower part of Fig. 1. In general, this formula leads to an asymmetric profile. This asymmetry is due to the dependence of the cross section σ_a on the atomic velocity. Fowler and Sung (1975) solved the mathematical formula defining $\gamma_c(u)$ by assuming that the cross-section σ_a is a constant. By using this approximation one can show that, besides the Lorentzian parameter γ_c and the Gaussian parameter γ_D of the Voigt profile, the more refined profile includes a new parameter: the mass ratio (m/M) of the perturber particle and absorber atom. As indicated in Fig. 1, when the mass ratio is zero, the Voigt and Fowler-Sung profiles are identical. However, as the mass ratio rises, the width of the Fowler-Sung profile decreases and its height increases. This can be seen in Fig. 2, where we show a Voigt profile (*i.e.* a Fowler-Sung line-shape with $m/M = 0$; solid lines) and a Fowler-Sung profile with $m/M = 5$ (dotted lines), both corresponding to $\alpha = \gamma_0/\gamma_D = 5$, where γ_0 is the parameter γ_c for the case $m/M = 0$. More information is given in Fig. 3 (see also Trujillo Bueno, 1982), where we have plotted the half-width of the Fowler-Sung profile (in units of the Gaussian parameter γ_D) as a function of m/M for various values of α.

Both the Galatry and the Fowler-Sung profiles reflect the statistical dependence of the Doppler and impact broadenings. The Galatry profile reflects the interrelationship between the Doppler effect and the effect of collisions in changing the atomic velocity. The Fowler-Sung profile reflects the interrelationship between the Doppler effect and the effect of collisions in interrupting the atomic absorption processes. Clearly, a really unified theory requires a consistent evaluation of the global correlation function of the problem. This can be done either by following a classical approach (*i.e.* via Eq. 8), or by using a quantum-mechanical method (see Sobelman *et al*, 1981; Breene, 1981; and the original references therein).

As is well known, in stellar spectroscopic work usually $\lambda \ll L$, and the relevant perturbing particles are often lighter than the atoms which emit, absorb, or scatter the radiation. This suggests that a good approximation in dealing with the problem of partial redistribution in the presence of collisions is to ignore correlation effects between the Doppler and collisional broadening. This notwithstanding, since the averaging over velocities generally tends to reduce the interference effects which appear in the atomic rest frame, investigations on how the Doppler broadening influences the effects of level-crossing interferences should ultimately be carried out by considering the previously described correlation effects. On the other hand, as indicated in the introduction, Sun-like atmospheres are highly dynamic systems far from equilibrium. The ensuing complex gas motions lead to an additional broadening which is largely more important than the modification of the Voigt profile which results from the subtle correlation effects considered here. However, refined line-shapes theories, like those above, are important not only because they allow us to justify the use of approximate line-shape expressions like the Voigt profile, but also because they serve as a guide in choosing the relevant physical ingredients for the development of more realistic treatments of RT.

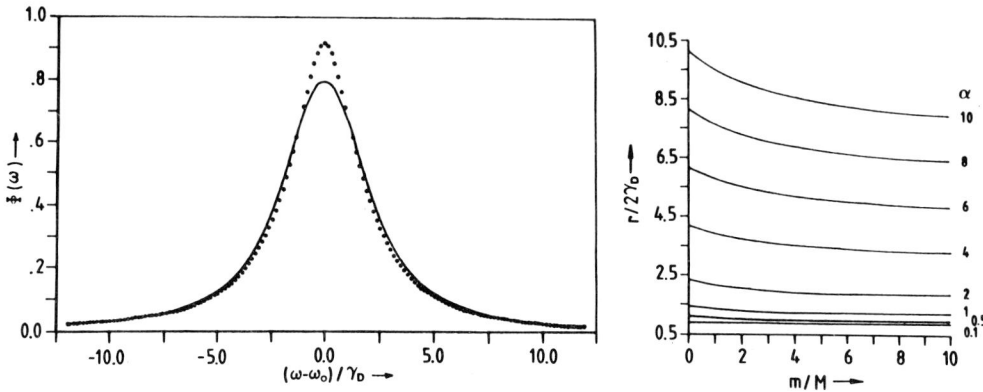

Fig. 2. Solid line: Fowler-Sung profile with $\alpha = 2$, and $m/M = 0$; dotted line: Fowler-Sung profile with $\alpha = 2$, and $m/M = 5$. The intensities have been calculated after multiplying the formula defining the Fowler-Sung profile by $\pi\sqrt{\pi}\gamma_D$. Frequencies distances from line centre are measured in units of γ_D.

Fig. 3. Halfwidths at half maximum (in units of γ_D) of Fowler-Sung profiles. The halfwidths are given against m/M for various values of $\alpha = \gamma_0/\gamma_c$.

2.4 Radiative Transfer in Stochastic Media

Consider the numerical simulations of stellar granulation performed by Nordlund and Dravins (1989). These hydrodynamic model atmospheres provide the temporal evolution of the three-dimensional structure of the temperature, density, and velocity of the stellar photospheric gas. With this information one can calculate the time variation of the values of the radiative properties (*i.e.* opacity χ and source function S) along any desired line of sight. Finally, the RT equation can be directly integrated to obtain spatially resolved line profiles across stellar surfaces as a function of time. Dravins and Nordlund (1989) carried out such calculations by assuming LTE, and concluded that the properties of averaged line profiles are not at all typical for individual points on the stellar surface, but *rather reflect the statistical distribution of photospheric inhomogeneities*. In fact, we would find a large number of possible variations of χ and S along a given line of sight if we fixed the time and moved around on the surface of the stellar model, or if we considered a given point on the surface, but let the time go by. We may thus say that Dravins and Nordlund (1989) calculated mean values of the emergent intensity (*i.e.* averaged line profiles) by first solving the RT equation for each particular realization of χ and S, and then averaging over the statistical ensemble of all such realizations.

The aim of stochastic RT theory is to derive an equation which allows the *direct* computation of the mean value $<I>$ of the emergent intensity, by assuming a given statistical description of the state of the medium under consideration, i.e. by assuming a given statistical specification of the physical variables (temperature, density, velocity, etc.) which lead to the determination of χ and S. A general statistical description of the state of a system would require us to give the hierarchy of n-point probability densities $P_n(\vec{r}_1, \vec{v}_1, T_1, \rho_1; ...; \vec{r}_n, \vec{v}_n, T_n, \rho_n)$ of finding at $\vec{r}_1$ the velocity $\vec{v}_1$, the temperature T_1, and the density ρ_1, and at $\vec{r}_2$ the velocity $\vec{v}_2$ ..., and at $\vec{r}_n$ the velocity $\vec{v}_n$, the temperature T_n, and the density ρ_n. Actually, a statistical description of the ensemble of the different flow situations, which one would encounter along different rays in a hydrodynamic model atmosphere would be very complicated. As pointed out by the above-mentioned authors the velocities in the inhomogeneous model photospheres do not have anything like a Gaussian amplitude distribution, and the correlated fluctuations in temperature, pressure, and velocity are beyond the classical turbulence model.

However, this expected result concerning the complexity of the flow situations in stellar atmospheres should not lead us to conclude that the only way of obtaining useful physical information from the interpretation of observed stellar profiles is via a physically consistent, hydrodynamic numerical simulation approach. What is clear is that we should abandon the use of the well known "micro-" and "macro-turbulence" classical fitting parameters. A possible alternative is the application of the stochastic RT theory developed independently by groups at Heidelberg and Nice (see Mihalas, 1978, and an extensive list of original references at the end of Sedlmayr's (1980) paper). Both approaches assume a statistical description based on a combination of both a Gaussian one-point and a two-point velocity distribution $P_2(v_1, z_1; v_2, z_2)$. The specification of this two-point probability density establishes the difference between the two proposed stationary Markovian stochastic processes: (a) a continuous process in space with a Gaussian P_2 (the Heildelberg approach), and (b) a discontinuous process in space with a non-Gaussian P_2 given by a linear combination of a completely uncorrelated part, and a part which is completely correlated (the Nice approach). An interesting comparison between the profiles which result from these two stochastic models has been carried out by Sedlmayr (1976) for various values of the mean turbulent velocity σ and spatial correlation length l, and by assuming LTE and plane-parallel geometry. Sedlmayr's calculations show how the two assumed stochastic models lead to increasing differences between the corresponding profiles as the value of the mean turbulent velocity σ increases. These calculations for selected oxigen and iron lines also show that the micro- and macroturbulent limits correspond to spatial correlation lengths $l < 1$ km and $l > 3000$ km, respectively. A similar conclusion for the behaviour of Ca II lines in a stochastic medium characterized by a two-point Gaussian probability velocity distribution was reached by Carlsson and Scharmer (1985), who firstly solved the non-LTE problem for each assumed realization of the velocity field, and then averaged over the assumed statistical ensemble. We note, however, that a non-LTE stochastic transfer equation has also been formulated, and that numerical solutions for two-level atoms in plane-parallel geometry have also been obtained (see Traving, 1976; Gail et al, 1975; and Sedlmayr, 1976).

A clarifying discussion concerning the question of how to relate the parameters l and σ to physically meaningful quantities can be found in Frisch and Frisch (1976). Clearly, in order to develop a stochastic RT theory suitable for the diagnostic problem, one should first try to isolate the basic parameters of the velocity field which are relevant to the line transfer problem, and then to derive a stochastic RT equation which depends on the relevant parameters only. In this respect Gail (1980) argues that weak lines, just like strong lines, do not contain any significant information on the structure of the velocity field extending beyond the probability density P_3. Therefore, the next theoretical step would be to derive a stochastic RT equation based on various particular forms of P_3. In any case it would be of great theoretical interest to derive a hierarchy of stochastic RT equations, which would allow us to investigate the influence, on the mean value of the emergent intensity, of increasing degrees of complexity in the structure of the velocity field. In this respect, it may be useful to mention that Magnan (1985) has also investigated the line formation problem in stochastic media by deriving stochastic RT equations based on the addition of layers formalism.

These stochastic RT theories assume that only the velocity is a random variable, that the velocity only depends on the height z in the atmosphere, and that the stochastic process is Markovian in z. Recently, Jefferies and Lindsey (1988) have proposed a numerical procedure for the calculation of the mean value of the emergent intensity from a LTE medium whose opacity and Planck function vary statistically along the line of sight. A three-dimensional non-LTE stochastic RT theory was developed by Rybicki (1976) by using a perturbation expansion approach, and then restricting the equations to the lowest significant order. Although this is a weak turbulence RT theory, it would be very interesting to obtain numerical solutions of the equations proposed by Rybicki (1976), in order to investigate the observational signatures of a three-dimensional stochastic medium under non-LTE conditions.

3. MULTI-DIMENSIONAL RADIATIVE TRANSFER

In Sun-like atmospheres the transfer of radiation is taking place in a highly inhomogeneous medium. The calculation of the radiation field which originates in such a medium therefore requires, in general, solving the multi-dimensional radiative transfer (MRT) equation under non-LTE conditions. Only in this way can one take into account not only that the transfer of radiation in small-scale atmospheric structures occurs in all directions, but also that fundamental transfer quantities like the opacity (χ) and source function (S) may also substantially depend, through their partial coupling to the radiation field, on the thermodynamic state of the surrounding medium. The extreme computational difficulties associated with the numerical solution of the MRT equation (see *e.g.* Cannon, 1976; Mihalas, Auer, and Mihalas, 1978; Kunasz and Mihalas, 1984) inevitably lead to the use of approximations which simplify the problem, thus making possible both theoretical and diagnostic investigations.

For instance, diagnostic investigations of Sun-like atmospheres are currently performed by assuming that the star's atmosphere is simply stratified in homogeneous plane-parallel layers. This is one of the approximations employed to derive semi-empirical models of solar atmospheric features (see *e.g.* Avrett 1989; Solanki, 1989). These models are then used to calculate, for example, the radiative energy losses per unit volume and time as a function of height in the atmosphere (see Vernazza *et al*, 1981; Anderson and Athay, 1989a). This semi-empirically determined quantity is then considered by theoreticians working in the field of chromospheric heating as the fundamental quantity which their theoretical models should ultimately reproduce (see Stein, 1985).

There are several reviews on MRT (see *e.g.* Jones and Skumanich, 1980; Jones, 1986; Kneer, 1986). In Cannon's (1985) monograph one can also find illuminating discussions of his many contributions to the subject; for example, Cannon (1985) shows why the failure to identify the effect of macroscopic *multi-dimensional velocity fields* on the transfer of spectral line radiation may lead to a complete misinterpretation of observations. Our purpose here is to examine the diagnostic and energy balance problems of a highly inhomogeneous medium. One of our aims is to show how a detailed knowledge of certain intricate effects of MRT, which have been only briefly considered in previous reviews, may lead to a better interpretation of various observations. On the other hand we aim at illustrating the *stabilizing* and *destabilizing* influence of the radiation-matter interaction, and at demonstrating the inadequacy of certain approximate descriptions of energy transfer by radiation.

3.1 Radiation Field Fluctuations

Some physical insight on the effects of MRT can be gained by considering the following question: How does the radiation field respond to fluctuations in the opacity and in the source function applied to an unperturbed medium with source function $\bar{S}_\nu$, opacity $\bar{\chi}_\nu$, and specific intensity $\bar{I}_\nu$? We may think of two extreme answers to this question: one useful and another realistic.

The Linear Response

The useful answer consists in assuming that the radiation field responds linearly to the perturbations. The time-independent RT equation for unpolarized radiation is

$$\frac{d}{ds} I_\nu = \chi_\nu (S_\nu - I_\nu). \tag{10}$$

Linearizing this equation by means of retaining only the first-order terms of the perturbations we have:

$$\frac{d}{ds}\delta I_\nu = \bar{\chi}_\nu(\delta S_\nu{}^{eff} - \delta I_\nu), \qquad (11)$$

where

$$\delta S_\nu{}^{eff} = \delta S_\nu + \frac{\delta\chi_\nu}{\bar{\chi}_\nu}(\bar{S}_\nu - \bar{I}_\nu). \qquad (12)$$

Eq. (11) governs the variation of the fluctuating specific intensity δI_ν along an arbitrary direction in a three-dimensional medium. δS_ν^{eff} (cf. Eq. 12) is the "effective source function" which determines the solution of Eq. (11). The notational convention used for an arbitrary function 'f' is that '$\bar{f}$' represents the unperturbed quantity, while 'δf' represents its first-order fluctuation. For harmonic perturbations, the symbol 'Δf' will be used below to indicate the *amplitude* of the fluctuating quantity 'δf'. Note that the unperturbed quantity $(\bar{S}_\nu - \bar{I}_\nu)$ has to be specified before being able to calculate the reponse of the radiation field to *opacity* perturbations. In this respect, the following example provides some interesting information.

Assume that the unperturbed medium is the diffusion region of grey stellar atmospheres in LTE. In such a medium the frequency-integrated specific intensity is $\bar{I}(\mathbf{r}, \overrightarrow{\Omega}) = \bar{B}(\mathbf{r}) + \frac{3}{4\pi}\overrightarrow{\Omega}.\bar{\mathbf{F}}$, where $\mathbf{r}$ is the position vector, $\overrightarrow{\Omega}$ the unit vector specifying the direction of propagation of the radiation beam, $\bar{B}$ the frequency-integrated Planck function, and $\bar{\mathbf{F}}$ the frequency-integrated radiation flux of the unperturbed medium (see *e.g.* Unno and Spiegel, 1966). Consequently, the frequency-integrated effective source function is given by (*cf.* Eq. 12):

$$\delta S^{eff} = \delta B - \frac{3}{4\pi}\frac{\delta\chi}{\bar{\chi}}\overrightarrow{\Omega}.\bar{\mathbf{F}} \qquad (13)$$

Equation (13) implies that a necessary condition for opacity fluctuations to have an effect is the existence of an average radiative flux. Thus, for example, if *small* B and χ perturbations are applied to systems in *thermodynamic equilibrium*, only B fluctuations can have an effect on the radiation field; for χ perturbations the change in extinction would be exactly cancelled by the altered re-emission, a result first found by Spiegel (1957) for an infinite homogeneous medium in LTE and with grey absorption.

Consider now the fluctuation of the mean intensity of the radiation field, *i.e.* $\delta J = \int d\Omega\delta I/4\pi$. In order to calculate δJ one has to introduce the formal solution of Eq. (11) for δI into the previous expression defining δJ, and to make use of (13) to obtain

$$\delta J(\mathbf{r}) = \int_V P(\mathbf{r},\mathbf{r}')\delta B(\mathbf{r}')d^3\mathbf{r}' + \int_V O(\mathbf{r},\mathbf{r}')\delta\chi(\mathbf{r}')d^3\mathbf{r}', \qquad (14)$$

where the kernels P for the Planck-function fluctuations and O for the opacity fluctuations depend only on the state of the unperturbed system, and are given by

$$P(\mathbf{r},\mathbf{r}') = \frac{\bar{\chi}(\mathbf{r}')e^{-\bar{\tau}(\mathbf{r},\mathbf{r}')}}{4\pi|\mathbf{r}-\mathbf{r}'|^2}, \qquad (15-a)$$

and

$$O(\mathbf{r},\mathbf{r}') = -\frac{3}{4\pi}\frac{(\mathbf{r}-\mathbf{r}').\bar{\mathbf{F}}}{4\pi|\mathbf{r}-\mathbf{r}'|^3}e^{-\bar{\tau}(\mathbf{r},\mathbf{r}')}. \qquad (15-b)$$

These kernels give the response of the mean intensity of the radiation field due to point-like Planck-function and opacity fluctuations, respectively. If we assume the z axis along the direction of the radiative flux, we would find that a point-like opacity disturbance at $\mathbf{r}' = 0$ would lead to negative values of δJ for $z > 0$, but to positive values of δJ for $z < 0$, with the magnitude of δJ depending on the angle between $\mathbf{r}$ and $\bar{\mathbf{F}}$. This is illustrated in Fig. 4a; in Fig. 4b we visualize how δJ responds to a point-like disturbance in the Planck-function.

We note that, while the validity of the Planck-function kernel P (cf. Eq. 15-a) is not restricted to the diffusion region of stellar atmospheres, the opacity kernel O (cf. Eq. 15-b) can only be used with Eq. (14) if the opacity disturbance $\delta\chi(\mathbf{r}')$ is confined to the diffusion region of a stellar atmosphere. A general expression for the opacity kernel O can be found in Rybicki's (1965) work on stochastic RT.

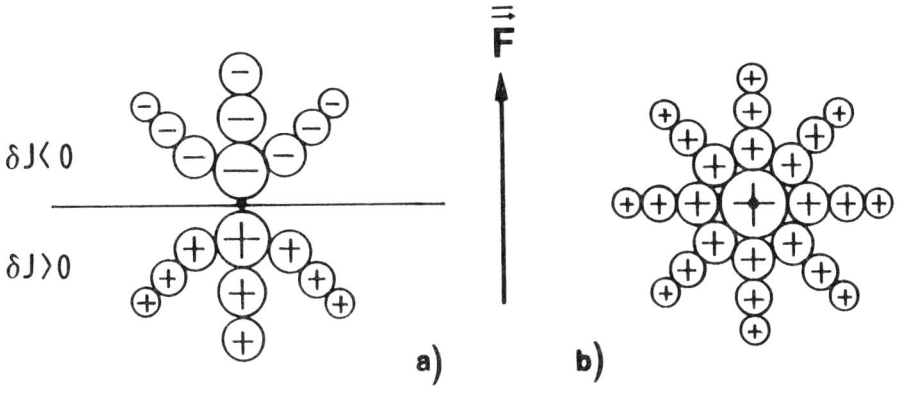

Fig. 4. The response of the mean intensity of the radiation field to a point-like disturbance in the opacity (a), and to a point-like perturbation in the Planck-function (b).

A linear analysis would be valid if the gas perturbations were *small* compared with the thermodynamic values characterizing the assumed unperturbed state. Since this is not generally the case, a *non-linear* analysis based on full solutions of the MRT equation becomes necessary.

The non-Linear Response

In a *non-linear* analysis δI has to be calculated by first fully solving the MRT equation for I, and then writing $\delta I = I - \bar{I}$. We thus have

$$\frac{d}{ds}(\bar{I}_\nu + \delta I_\nu) = \bar{\chi}_\nu(\bar{S}_\nu - \bar{I}_\nu) + \bar{\chi}_\nu\left[\delta S_\nu + \frac{\delta\chi_\nu}{\bar{\chi}_\nu}(\bar{S}_\nu - \bar{I}_\nu) + \frac{\delta\chi_\nu}{\bar{\chi}_\nu}(\delta S_\nu - \delta I_\nu) - \delta I_\nu\right]. \quad (16)$$

If LTE is assumed $\delta S = \delta B$. From Eq. (16) one then sees that if $\delta\chi = 0$ (i.e. if there are no opacity inhomogeneities) the radiation field responds linearly, while if $\delta\chi \neq 0$ the radiation field responds non-linearly. On the contrary, outside LTE the radiation field may respond *non-linearly* even if $\delta\chi = 0$. This can be easily seen by considering some expressions for the source function. In the coherent scattering problem the source function reads (see e.g. Mihalas, 1978)

$$S_\nu = (1 - \epsilon)J_\nu + \epsilon B_\nu, \quad (17)$$

while for the standard two-level atom line-transfer problem with complete frequency redistribution, the line source function is given by (see e.g. Mihalas, 1978)

$$S_l = (1 - \epsilon)\tilde{J} + \epsilon B_\nu, \quad (18)$$

where $\tilde{J} = \int \Phi_\nu J_\nu d\nu$ and ϵ is the well-known non-LTE parameter, which may be interpreted either as the probability that an absorbed photon is destroyed by collisional deexcitation, or as the probability that an excitation event is caused by collisions. Accordingly, even in the absence of opacity fluctuations we generally have non-linear terms in Eq. (16), which are due to products of the form $\delta\epsilon\delta B_\nu$ and $\delta\epsilon\delta J_\nu$.

3.2 The Diagnostic Problem

Due to the presence of inhomogeneities in the gas temperature, density, velocity, etc., the opacity χ and the source function S fluctuate horizontally and, consequently, the emergent specific intensity from a star fluctuates across its surface. If the star under consideration is other than the Sun we do not have spatial resolution, and the relevant question concerning whether the approximation of plane-parallel geometry leads to diagnostic errors is the following: *Are spatially averaged emergent intensities obtained from a MRT calculation different from the emergent intensity computed using the spatially averaged one-dimensional atmosphere?* If the radiation field were to respond *linearly*, the answer would be negative, and the plane-parallel approximation could be used safely. Since in general the radiation field responds *non-linearly*, the answer has to be affirmative; however, whether MRT calculations are actually necessary for a correct diagnostic of the observed spectrum depends on how important such a non-linear response is.

An example where the assumption of plane-parallel geometry leads to very important errors is due to Nordlund (1984), who investigated the problem of the solar iron abundance by using his three-dimensional numerical simulations of solar granulation. Nordlund first pointed out that the relative Fe I abundance is a very strong function of temperature (since $N_{FeII}/N_{FeI} \propto T^{10}$); he then showed that the iron abundance derived from his three-dimensional hydrodynamic model photosphere is about a factor two smaller than the one that would be derived from a one-dimensional model with temperature equal to the horizontally averaged temperature of the three-dimensional model.

If the star under consideration is the Sun, we also have to consider the question raised above, since our telescopes are still unable to fully spatially resolve the atmospheric features of interest. As mentioned earlier, semi-empirical models of solar atmospheric features are based, among other approximations, on the assumption of plane-parallel geometry. Therefore, the depth-dependent distribution of temperatures and densities which gives the best fit to the observed spectra may not represent at all the actual average state of the atmospheric feature of interest (see Cannon (1985) for further discussion on this point). In any case, we do observe horizontal intensity fluctuations on the Sun's surface, and with a spatial resolution which is being continuosly improved. Consequently, we also have to pose the following question: *Do we need MRT to interpret the observed radiation field fluctuations?* In order to answer this question it is first convenient to clarify some of the physical effects of MRT. To this end we begin by recalling an electrostatic textbook exercise, where the assumed geometry is identical to the next RT problem to be discussed.

The electrostatic Field of a Grid of Charged Wires

Consider the text-book problem of finding the equipotential surfaces above a grid of infinitely long charged wires parallel to the y axis, and arranged with a spatial periodicity L along the x axis (see *e.g.* Feynman et al, 1964). The solution to this problem can be found by substituting harmonic solutions of the form $\phi(x,z) = A_n(z)\cos(2\pi nx/L)$ in Laplace's equation (*i.e.* in $\nabla^2 \phi = 0$). The dependence with z of the amplitude A_n is found to be $A_n(z) = a_n \exp(-z/z_0)$, with $z_0 = L/2\pi n$. Note that, already for the Fourier component of harmonic $n = 1$, the amplitude A_1 falls by a factor $\sim 10^{-3}$ each time we increase z by one grid spacing L. The conclusion is clear: if we are only a few times the distance L away from the grid, the electric field is very nearly uniform, just as though the charge were uniformly spread over a plane.

Chandrasekhar's Searchlight Problem

Consider that the lower boundary of a plane-parallel grey atmosphere with exponentially stratified opacity is irradiated isotropically, as shown in the inner plot of Fig. 5. The problem of finding the ensuing distribution of the radiation field within the atmosphere under the constraint of radiative equilibrium may be considered

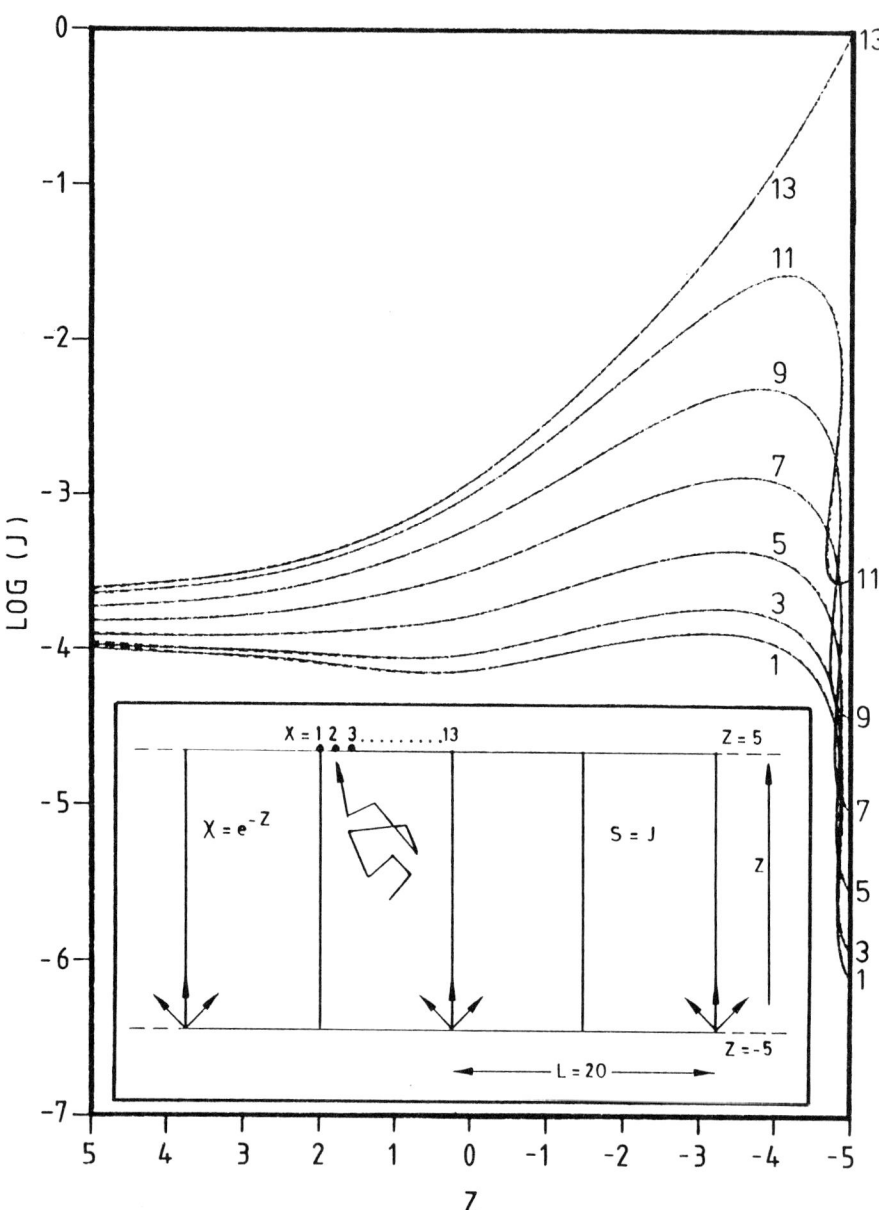

Fig. 5. The inner plot shows the geometrical configuration of the problem. The figure gives the height dependence of $J(x,z)$ at various horizontal positions indicated by a number on the curves. The opacity is $\chi(z) = \chi_0\exp(-z/H)$. Geometrical distances are measured in units of the opacity scale height H, and χ_0 is chosen such that the optical depth τ is unity at $z=0$, that is $\chi_0 H = 1$ and $\tau = \exp(-z)$. (Near z=-5, the curves are triple valued due to the smoothing procedure of the plotting routine.)

as a type of Chandrasekhar's (1958) searchlight problem. Chandrasekhar (1958) required to determine the intensity distribution in a plane-parallel scattering atmosphere whose boundary is illuminated at one point with an arbitrarily oriented radiation beam. Due to the property of completeness of the searchlight solutions, the solution to the particular inhomogeneous illumination problem we are considering here may be expressed by using the solution to the searchlight problem (see Rybicki, 1971). We have however calculated the spatial distribution of the mean intensity $J(x,z)$ given in Fig. 5 with an own-written code, which uses the numerical strategy of Mihalas, Auer, and Mihalas (1978). As can be seen in Fig. 5, the fluctuation of the mean intensity varies from 6 orders of magnitude at the bottom of the atmosphere ($z = -5$) to one order of magnitude at $\tau = 1$ ($z = 0$), and decreases to a factor 2.5 at the surface ($z = 5$). We have therefore to conclude that *it is indeed very difficult to distinguish the two head lights of a car, and thus to estimate the distance, on a foggy road.*

It is useful to note at this stage that if we had calculated the emergent intensity distribution for a *point source* illuminating the bottom of the atmosphere, we would have directly obtained the *point spread function* (PSF). Kneer (1981) used this PSF as an indicator for the lateral extent of RT. Note also that the Hankel transform of the PSF gives the *modulation transfer function* MTF (see *e.g.* Dainty and Shaw, 1974). With these clarifications we may now discuss Kneer's (1979;1981) work, where the optical transfer properties of the (solar) atmosphere are compared with those of a diffraction limited telescope.

Atmospheres with Inhomogeneous Distributions of Thermal Sources

The above example is a particular case of the general problem of finding diffuse radiation fields due to any inhomogeneous illumination of the atmosphere's lower boundary. We shall now consider an example in which inhomogeneous distributions of internal thermal sources are present. To this end we follow Kneer's (1981) paper, and investigate the coherent scattering problem (*cf.* Eq. 17) in an atmosphere with *constant* non-LTE parameter ϵ, and with opacity $\chi(z) = \chi_0 \exp(-z/H)$. We will always measure z and all other geometrical distances in units of the opacity scale height H; we will also choose χ_0 such that the optical depth at $z = 0$ is unity, that is $\chi_0 H = 1$. The response of the mean intensity of the radiation field to Planck-function fluctuations of the form $B = \bar{B}(z) + \Delta B \cos kx$ (with $k = 2\pi/\Lambda$, and Λ the horizontal wavelength) is $J = \bar{J}(z) + \Delta J(z,k) \cos kx$. Note that, since it is being assumed that both ϵ and χ do not fluctuate horizontally, the radiation field responds *linearly* (*cf.* Eqs. 16 and 17, with $\delta\chi = 0$ and $\delta S = (1-\epsilon)\delta J + \epsilon \delta B$). Our interest here lies in $\Delta J(z,k)$, for various values of the non-LTE parameter ϵ, and for the case of a height-independent ΔB.

Fig. 6 shows the results, which we obtained by using the above-mentioned MRT code. For plane-parallel perturbations (*i.e.* for perturbations with wavenumber $k = 0$ or $\Lambda = \infty$) we recover the correct surface value for the amplitude of the source

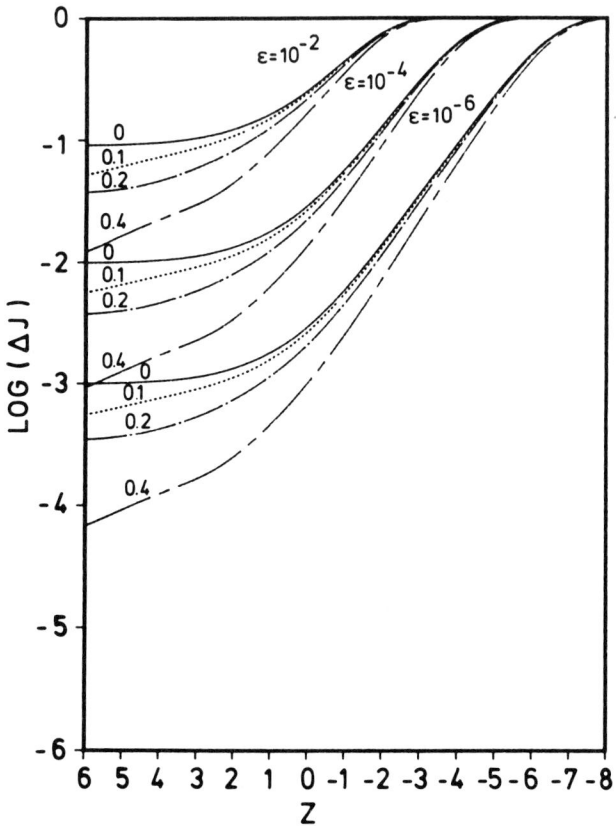

Fig. 6. The height dependence of ΔJ for the coherent scattering problem. Parameters are the wavenumber $k = 2\pi/\Lambda$ of the horizontal Planck-function fluctuations, and ϵ. Geometrical distances (*i.e.* z and Λ) are measured in units of the opacity scale height H. The optical depth $\tau = e^{-z}$.

function (note from Eq. 17 that $\Delta S = (1 - \epsilon)\Delta J + \epsilon \Delta B$; and recall from Mihalas (1978) that $\Delta S(\tau \approx 0) = \sqrt{\epsilon}$). We also have the correct value for the thermalization depth (*i.e.* $\Theta \approx 1/\sqrt{\epsilon}$; note that in Fig. 6 $\tau = e^{-z}$). However, in accordance with Kneer (1981), we note that with increasing values of the wavenumber k the *horizontal transfer of radiation effects* efficiently damp out the fluctuations of J. Note also that the larger the non-LTE effects (*i.e.* the smaller ϵ), the larger the damping.

In addition to the coherent scattering problem described above, Kneer (1981) also considered the standard line transfer problem, *i.e.* with line source function given by Eq. (18). He calculated the amplitude ΔI of the *normaly* emergent intensity, for various frequency distances $\Delta\nu$ from line centre, for several values of ϵ, and for various values of the horizontal wavenumber k of cosinusoidal Planck-function perturbations. As a result, he found that for a given value of k, ΔI decreases strongly when going from the line wing frequencies (at which we see layers where the source function has thermalized to the local Planck function) towards the line centre frequency (at which we observe layers where the source function is strongly decoupled from the thermal pool). From line transfer calculations in simple model structures with sizes well above the resolution limits of existing solar telescopes, he concluded that the effects of horizontal radiative transfer may diminish the contrast by an order of magnitude compared to that expected in a one-dimensional RT treatment. In order to demonstrate that the solar atmosphere itself has its own spatial resolution limits, Kneer (1979; 1981) calculated the *modulation transfer function* $M_I(k) = \Delta I(k)/\Delta I(k=0)$, pointing out that the MTF $M_I(k)$ shows how much of the signal (*i.e.* of the thermal fluctuation) is transmitted to an observer. Fig. 7 shows Kneer's (1981) results for M_I assuming observation at the line centre frequency. The dotted line in Fig. 7 is the MTF of a 1 m telescope at 5000Å. From this figure it becomes clear that *horizontal radiative transfer may compete in its effect with instrumental limitation.*

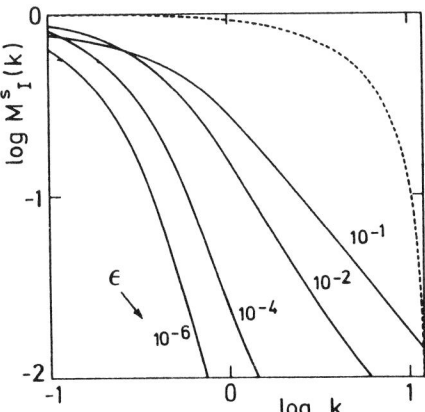

Fig. 7. The MTF $M_I(k)$ calculated by Kneer (1981) for the vertically emergent intensity at line centre, assuming complete redistribution and a Gaussian absorption profile.

Whenever solar physicists use earth-based telescopes to investigate the very small spatial features (*e.g.* granulation, chromospheric structures, high degree non-radial oscillations, etc), they have to take into account that the effect of seeing is to reduce the *amplitudes* of the high spatial frequency components seen in any solar image. As is well known, measurements of the atmospheric MTF are often used to correct the *amplitudes* of observed quantities from the effect of seeing (Deubner and Mattig, 1975; Collados and Vázquez, 1987; Kaufman, 1988).

Could we use a similar strategy to avoid the Sun's own spatial resolution limitation? In principle, yes. The problem is how to calculate, in a sufficiently realistic way, the function $M_I(k)$. Even assuming that we have at our disposal an efficient non-LTE MRT code, there remains an intrinsic difficulty, namely that we generally lack information on the height variation of the atmospheric inhomogeneities. Note that these inhomogeneities not only lead to Planck-function fluctuations, but also to *opacity* fluctuations, which were consciously ignored by Kneer (1981) for ease of analysis. As pointed out by Cannon (1970; 1985), *opacity* inhomogeneities tend to *channel* emerging radiation into lower opacity regions. This *radiation channelling* effect is particularly important for a correct diagnostic treatment of the thermodynamical properties of magnetic flux tubes on the Sun (Stenholm and Stenflo, 1977; 1978); the failure to identify this MRT effect may also lead to a complete misinterpretation of observations of atmospheres where multi-dimensional velocity fields are present (see Cannon, 1985). A useful investigation would therefore consist in solving the linearized equation (11), in order to examine the influence of horizontal *opacity* inhomogeneities on the function $M_I(k)$ (for numerical strategies see *e.g.* Kneer and Heasley, 1979; and Christensen-Dalsgaard and Frandsen, 1983).

There is a field of research in Solar Physics, where a more or less "realistic" calculation of $M_I(k)$ for given spectral lines should be possible. We are referring here to the field of Helioseismology, where numerical solutions of the linearized equations of non-radial oscillations in a semi-empirical solar model have succeded in providing the frequencies and eigenfunctions (see *e.g.* Christensen-Dalsgaard and Frandsen, 1983). A very interesting, but still unresolved problem in Helioseismology is the issue of the excitation mechanism that pumps energy into the solar p-modes of oscillation (see the review by Libbrecht, 1988). In order to compare the theoretical predictions of possible excitation mechanisms with observations, measurements of solar oscillation velocity amplitudes of high degree l are important (see Kaufman, 1988). Through calculations like the one we are suggesting here, we may expect to obtain information about the extent to which the Sun's own resolution limits (due to MRT effects in frequently-used spectral lines) may affect future seeing-free measurements (Scherrer *et al*, 1988) of the variation with l of the velocity power.

3.3 The Energy Balance Problem

In Sun-like atmospheres the radiation field influences the hydrodynamics of the motion through the appearence of the term $\nabla \cdot \mathbf{F}$ in the energy equation. This term gives the net energy which the atmospheric plasma gains (or looses) per unit volume and time through radiative processes. In order to gain physical insight on the stabilizing and destabilizing influence of the radiation-matter interaction, and also in order to demonstrate the inadequacy of some of the frequently-used approximations which render possible the calculation of $\nabla \cdot \mathbf{F}$ at each time step during the course of a numerical simulation, one may use the following strategy (Kneer, 1986; Trujillo Bueno, 1988a; see also Spiegel, 1957): to apply horizontal temperature perturbations to a grey RE solar model atmosphere in LTE, and to assume that the temporal evolution of the temperature is simply governed by the following radiative heat equation (*cf.* Unno and Spiegel, 1966):

$$\rho c_p \frac{\partial T}{\partial t} = -\nabla \cdot \mathbf{F} = 4\pi \int_0^\infty \chi_\nu (J_\nu - S_\nu) d\nu, \qquad (19)$$

where ρ is the density, and c_p the specific heat at constant pressure.

As a result of the imposed temperature perturbations, both χ and S deviate from their RE values, and therefore the radiative energy balance is altered. A crucial question is then: Does the system return to its initial RE configuration? If it does, one would then like to know how the *radiative relaxation times* depend on the geometry and structural lengths of the perturbations. If it does not, the conditions under which *radiative instabilities* may occur should be investigated. In order to gain knowledge on these issues both a *linear* analysis and a *non-linear* analysis are of interest (see Trujillo Bueno, 1988a).

Linear Analysis

The advantage of a *linear* analysis lies in that one can easily compare the relative importance of χ fluctuations with that of B fluctuations on the *radiative relaxation time* as a function of the wavenumber k of the perturbations, *i.e.* as a function of the characteristic length of the imposed atmospheric structures. These structures are assumed to be *small* horizontal temperature fluctuations of the form $T = \bar{T} + \delta T = \bar{T} + \Delta T \cos kx$.

Energy Transfer by Continuum Radiation

Consider, firstly, the problem of transfer of energy by continuum radiation, assuming grey absorption and LTE. Under these circumstances $\nabla \cdot \mathbf{F} = 4\pi\chi(B - J)$, where J is the frequency-integrated mean intensity and $B = (\sigma/\pi)T^4$, with σ the Stefan-Boltzmann constant. Linearizing Eq. (19) one has

$$\frac{\partial}{\partial t}\Delta T = -\big[n_{\Delta B}(z,k) + n_{\Delta\chi}(z,k)\big]\Delta T, \qquad (20)$$

where the growth rates $n_{\Delta B}$ and $n_{\Delta\chi}$ are due to the B and χ fluctuations, respectively. These growth rates are given by

$$n_{\Delta B} = \frac{16\sigma \bar{T}^3}{c_p} \frac{\bar{\chi}}{\bar{\rho}} (1 - \frac{\Delta J}{\Delta B}), \qquad (21)$$

$$n_{\Delta\chi} = \frac{4\pi}{c_p} \frac{\bar{\chi} \bar{F}}{\bar{\rho} \bar{T}} \frac{\partial \ln \bar{\chi}}{\partial \ln \bar{T}} (-\frac{\Delta J}{\alpha \bar{F}}), \qquad (22)$$

where $\alpha = \Delta\chi/\bar{\chi} = (\partial ln\bar{\chi}/\partial ln\bar{T})\Delta B/4\bar{B}$. Note that $\bar{B}$ decreases outwards (*e.g.* in the Eddington approximation $\bar{B} = \sqrt{3}(\sigma T_{eff}^4/4\pi)[1+\sqrt{3}\tau]$); note also that, in semi-empirical models of the solar photosphere, $\partial ln\bar{\chi}/\partial ln\bar{T}$ decreases both towards the interior and towards the surface from a value slightly larger than 10, which occurs approximately at continuum optical depth unity. Therefore, $|\alpha|$ should increase outwards, at least below $\bar{\tau} \approx 1$ in photospheric models of Sun-like stars.

In order to obtain these growth rates (*cf.* Eqs. 21 and 22), the *linear* response of the mean intensity to both B and χ perturbations has to be calculated separately by solving Eq. (11). On grounds of symmetry, J responds either in phase or π out of phase to *horizontal* fluctuations. Therefore, both growth rates are real, and the system can either be *locally* stable or unstable. Note that local radiative instabilities would occur if $n_{\Delta B} + n_{\Delta \chi} < 0$. As pointed out by Trujillo Bueno and Kneer (1989), the establishment of possible types of radiative instabilities follows from an evaluation of the possible ways of obtaining negative values of $n_{\Delta \chi}$. These authors showed that radiative instabilities may occur not only in atmospheric regions where $\partial ln\bar{\chi}/\partial ln\bar{T} < 0$, but even in those atmospheric regions where $\partial ln\bar{\chi}/\partial ln\bar{T} > 0$. For this latter type of instability to occur, the mean intensity J has to respond in phase to the opacity change (*cf.* Eq. 22). Trujillo Bueno and Kneer (1989;1990) demonstrated that, if $|\alpha|$ is an outwardly increasing function, J will always respond in phase to the opacity perturbation below a critical height which, though it actually *decreases* as the wavenumber k increases, lies close to $\bar{\tau} = 1$. They concluded that this particular *subphotospheric* radiative instability may explain the excitation of solar p-modes of oscillation which has been found by using the grey RE approximation for the assumed equilibrium model (Ando and Osaki, 1975). As mentioned earlier, the response of the radiation field to *opacity* fluctuations does depend on the state of the unperturbed medium (*cf.* Eq. 12). In this respect it is interesting to mention that Christensen-Dalsgaard and Frandsen (1983) have shown that, when the RE assumption is removed, the overstable behaviour found by Ando and Osaki (1975) for the solar p-modes disappears (see Trujillo Bueno and Kneer (1990) for further discussion on this point). We should also note that, in order to investigate whether stellar oscillations may extract energy from the radiation field, it is not only crucial which equilibrium model is chosen, but also the way by which the MRT equation is solved. Up to now, the possibility of a κ-like overstability mechanism for the excitation of *non-radial* stellar oscillations has been mainly investigated by means of diffusion-like descriptions of energy transfer by radiation.

In a diffusion-like description the radiative flux $\mathbf{F}(\mathbf{r}) = -C(\mathbf{r})\nabla G(\mathbf{r})$. There are two frequently-used diffusion-like descriptions of energy transfer by radiation: the MRT Eddington approximation ($C = 4\pi/3\chi$, $G = J$; see Giovanelli, 1959; Unno and Spiegel, 1966), and the optically thick or diffusion approximation ($C = 4\pi/3\chi$, $G = B$; see *e.g.* Rybicki, 1965). (Note that if the medium is in grey RE, and in LTE, both approximations become equivalent, since then $J = B$). A detailed comparison between the "exact" and approximate values of the growth rates $n_{\Delta B}$

and $n_{\Delta\chi}$ (*cf.* Eqs. 21 and 22) can be found in Trujillo Bueno (1988a). Here we simply summarize the results, providing also information on other investigations.

The MRT Eddington approximation is generally adequate for reproducing the *radiative damping* effects of B fluctuations in stellar atmospheres, *i.e.* in the presence of a surface and stratification. However Eddington's approximation fails to give the exact $n_{\Delta\chi}$ around and above $\bar{\tau}=1$; in particular, it does not reproduce the correct variation of $n_{\Delta\chi}$ with k (Trujillo Bueno, 1988a,b). Therefore, its use should be restricted to situations where the temperature sensitivity of the opacity is small enough to be sure that the radiative damping effects of B fluctuations play the dominant role (*cf.* Eq. 22). Since this is *not* the case near to optical depth unity in Sun-like atmospheres (note that at $\bar{\tau}=1$ in semi-empirical models of the solar atmosphere $\partial ln\bar{\chi}/\partial ln\bar{T} \approx 10$), we have to warn against the use of Eddington's approximation for the investigation of plasma atmospheric processes such as granular convection (*e.g.* Legait, 1986), magnetic flux concentrations (Hassan, 1988), or *nonradial* stellar oscillations (Ando and Osaki, 1975; note however that Christensen-Dalsgaard and Frandsen demonstrated that, for *radial* oscillations, the *plane-parallel* version of Eddington's approximation is generally suitable). Accordingly, Eddington's approximation should not be used to investigate whether a κ-like overstability mechanism may lead to the excitation of *non-radial* oscillations.

The optically thick or *diffusion* approximation can only be valid at depths where the photon mean free path is small compared to the scales over which B and χ vary (see *e.g.* Rybicki, 1965). Since now $\mathbf{F} = -(4\pi/3\chi)\nabla B$, the optically thick approximation gives a totally *local* expression for the radiative flux. A comparison of the "exact" results with those which the optically thick approximation gives for the height variation of $n_{\Delta B}$ and $n_{\Delta\chi}$ leads to the following conclusion (see Trujillo Bueno, 1988a): the smaller the characteristic length of the atmospheric structures, the greater the depth at which the asymptotic optically thick limit is reached. This limit occurs well below $\bar{\tau}=1$ for all spatial scales.

The optically thick approximation has been applied to obtain stationary models of magnetic flux concentrations through the use of a numerical relaxation approach. (Deinzer *et al*, 1984; Knölker and Schüssler, 1988). Recently such models have been re-calculated by solving the grey MRT equation in LTE (see the review by Schüssler, 1989; and more references therein). The energy budget in the models of magnetic flux concentrations is dominated by the balance of vertical radiative loss and lateral inflow of radiation into the *partially evacuated* magnetic region. However, while the optically thick approximation leads to models where the temperature of the gas situated inside the magnetic structures is lower than that of the surrounding medium at equal geometrical depth, in the upper layers of the refined models one finds gas temperatures which are a few hundred degrees larger than those of the surrounding plasma. This effect has been named the "illumination effect" (Schüssler, 1989). In our opinion, such a temperature enhancement may be considered to be the result of a well-known MRT effect, namely the *radiation channelling* effect (Cannon, 1970; 1985). As stated earlier, horizontal radiative transfer tends to *channel* emerging radiation into lower opacity regions. In order to illustrate that this *radiation channelling* effect may in fact lead to a temperature enhancement of the upper layers of a magnetic flux concentration, one only needs to carry out a MRT calculation with $\epsilon = 0$ in Eq. (17), assuming a prescribed deficit for the *opacity* of an embbeded structure in a grey atmosphere. Once $S(z) = J(z)$ is obtained, a temperature enhancement in the upper layers follows directly from the assumption $B = \sigma T^4/\pi = S = J$, i.e. from the LTE constraint. By imposing from the outset the constraint of grey RE in LTE for a slab with prescribed opacity, Kalkofen *et al* (1989) illustrated that the upper layers of a magnetic flux concentration may in fact be heated as a result of purely RT effects.

Energy Transfer by Spectral Line Radiation

Turning now to the problem of energy transfer by spectral line radiation, we ask: *How do radiative energy losses in spectral lines deviate from the plane-parallel case as the atmospheric horizontal structural lengths diminish?* To answer this question one may apply *small* temperature perturbations of the form $T = \bar{T} + \Delta T \cos kx$ to a plane-parallel model atmosphere, and calculate the *amplitude* of the radiative flux's divergence, *i.e.* the *amplitude* of the energy loss function L given by

$$\Delta L(z,k) = 4\pi \Big[\int_0^\infty \bar{\chi}_\nu (\Delta S_\nu - \Delta J_\nu) d\nu + \int_0^\infty \Delta \chi_\nu (\bar{S}_\nu - \bar{J}_\nu) d\nu \Big]. \quad (23)$$

Kneer and Trujillo Bueno (1987; see also Trujillo Bueno and Kneer, 1990) used simplified atomic and atmospheric models, and calculated the variation with wavenumber k of the *amplitude* ΔL due to a *single* spectral line of constant strength $\bar{\sigma}_l$ and non-LTE parameter ϵ. These authors investigated the influence of the MRT effects of the ensuing B and χ fluctuations, *i.e.* they calculated separately $\Delta L_{\Delta B}$ and $\Delta L_{\Delta \chi}$. Figure 8 shows an example where $\bar{T}(z)$ has been assumed to be identical to that of a grey RE solar model in LTE, and $\Delta T(z)$ such that $\Delta B_\nu = 1$. Note that $\Delta L_{\Delta B}$ (Fig. 8a) generally increases with increasing horizontal wavenumber k; note also that the critical height where $\Delta L_{\Delta \chi}$ (Fig. 8b) changes its sign largely depends on the atmospheric structural length. It is also worth noting that the dependence of ΔL with k is not significant within a *thermalization depth* from the stellar surface. However, at and below the height where thermalization in the plane-parallel case occurs (*i.e.* the case with $k = 0$), the dependence of ΔL with the horizontal structural length is important. In these atmospheric regions the radiative energy losses in LTE and non-LTE spectral lines are identical (see the physical explanation of this result in Trujillo Bueno and Kneer, 1987). From the observed dependence of ΔL on k it may be concluded that the *plane-parallel* approximation does not give proper account of the energetic role of spectral lines, if the atmospheric inhomogeneities have horizontal structural lengths *smaller* than 500...1000 km.

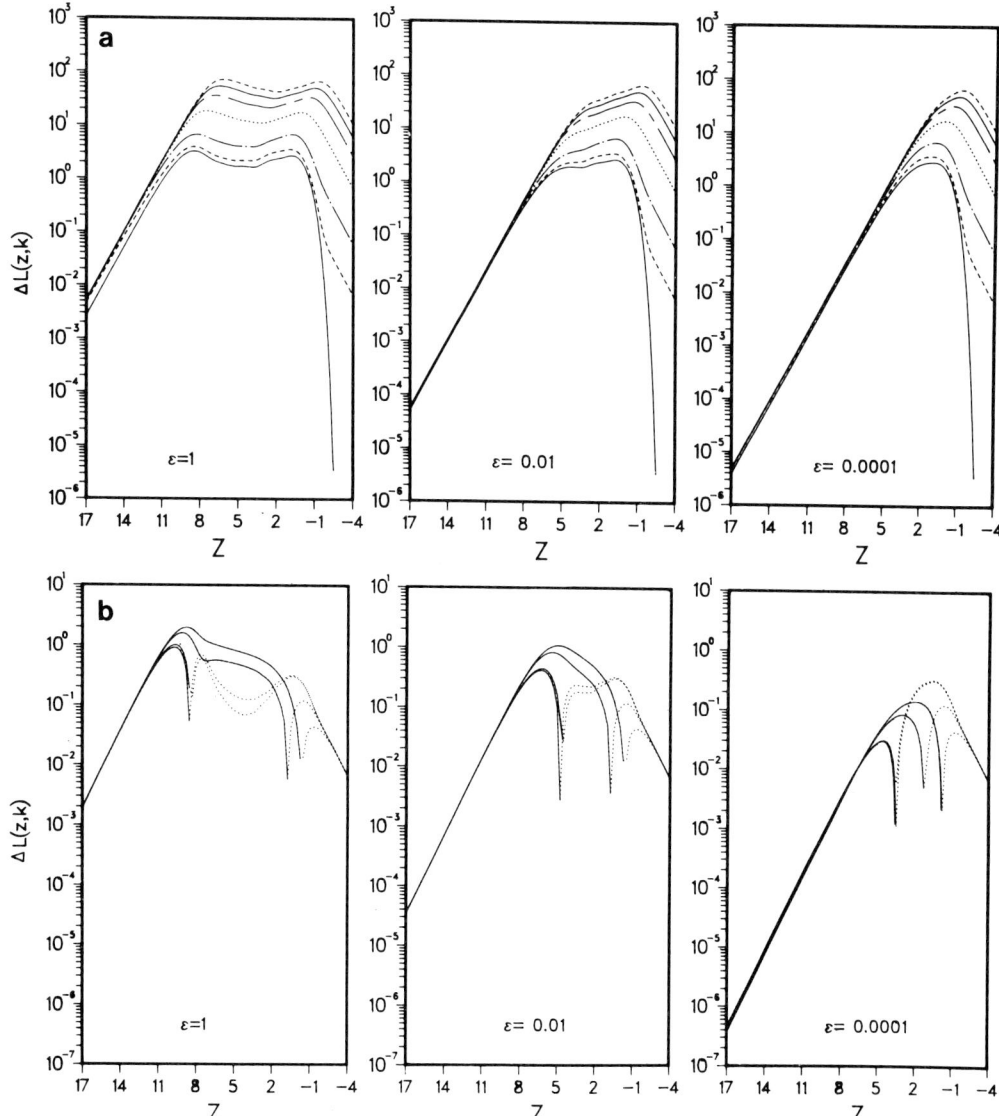

Fig. 8. $\Delta L_{\Delta B}$ (top of Fig. 8; *a*) and $\Delta L_{\Delta \chi}$ (bottom of Fig. 8; *b*) in units of $\chi_{c,o}\Delta\nu_D\Delta B_\nu$ due to a *single* spectral line with $\bar{\sigma}_l = 10^4$ and three constant values of ϵ. The absorption profile $\bar{\Phi}_\nu$ has been assumed to be Gaussian, with a constant Doppler width $\Delta\nu_D$. The line opacity $\bar{\chi}_l(\nu) = \bar{\sigma}_l\bar{\Phi}_\nu\bar{\chi}_c$, and the continuum opacity $\bar{\chi}_c = \bar{\chi}_{c,o}\exp(-z/H)$. Geometrical distances are measured in units of H. In Fig. 8*a* k=0, 0.2, 0.6, 2, 4, 6, 8 from bottom to top, whilst in Fig. 8*b* k=0, 0.2, 2, 6. In Fig. 8*b* solid lines indicate *positive* values, whilst dotted lines *negative* values. (These signs would have to be reversed if $\partial ln\bar{\chi}_l/\partial ln\bar{T}$ were -1 instead of 1). The results of Fig. 8*b* have been obtained by assuming that only the line-integrated opacity χ_l fluctuates.

To provide answers to questions like the above is important because, as is well known, in the outermost layers of stellar atmospheres, in chromospheres and coronae, where deviations from the assumption of LTE are most severe, the contribution of the radiation field to the energy budget stems mainly from spectral lines (see e.g. Avrett, 1985). From the linearization of Eq. (19) we may define not only *local* growth rates for the case of grey continuum radiation (*cf.* Eqs. 21 and 22), but also a *local* growth rate for *single* spectral lines through $n_l = 1/t_r = \Delta L/(\bar{\rho} c_p \Delta T)$. Here t_r is a *local radiative relaxation time* due to the effect of a *single* spectral line. Note that

$$t_{r\Delta\chi} = t_{r\Delta B} \frac{\Delta L_{\Delta B}}{\Delta L_{\Delta\chi}}, \qquad (24)$$

whith the sub-indices indicating whether reference to the influence of χ or B fluctuations is being made. By calculating $t_{r\Delta B}$ due to *single* spectral lines from two-level atoms, it is possible to illustrate that, at large atmospheric heights, transfer of energy in few spectral lines with $\sigma_l \epsilon \gg 1$ can compete with continuum processes (Kneer and Trujillo Bueno, 1987). In order to ascertain whether the MRT effects of the fluctuations in the *opacity* of spectral lines with $\sigma_l \epsilon \gg 1$ can also be important for the energy balance, one may compare $|t_{r\Delta\chi}|$ with $t_{r\Delta B}$. By using Fig. 8 and Eq. (24) one finds that, without requiring particularly large values of $|\partial ln \bar{\chi}_l / \partial ln \bar{T}|$, $|t_{r\Delta\chi}|$ can be as small as $t_{r\Delta B}$ *only* in atmospheric regions where the dependence of ΔL on the horizontal wavenumber k is unimportant, *i.e.* at heights larger than the height of thermalization. However, in order to have $|t_{r\Delta\chi}| \approx t_{r\Delta B}$ in regions where the k-dependence on ΔL is important, *i.e.* at depths greater than the thermalization depth, values of $|\partial ln \bar{\chi}_l / \partial ln \bar{T}| \approx 10$ or even much larger would be required. We should also point out that, for a given atmospheric height, $t_{r\Delta\chi}$ can be positive or negative depending on the sign of $\partial ln \bar{\chi}_l / \partial ln \bar{T}$ (*cf.* Fig. 8*b*); note that $t_{r\Delta\chi} < 0$ means that the MRT effects of line-opacity fluctuations play a *destabilizing* role.

The question thus arises as to whether the MRT effects of the fluctuations in the opacity of lines with large values of $|\partial ln \bar{\chi}_l / \partial ln \bar{T}|$ can give rise to radiative instabilities in the atmospheres of Sun-like stars. To answer this question is a much more complicated task than for the previously considered case of a grey continuum in LTE. Firstly, as with the grey case, one would have to choose an initially unperturbed configuration of the atmospheric plasma, since from Eq. (12) we know that the response of the radiation field to *opacity* perturbations does depend on the assumed equilibrium model. Secondly, one would have to calculate ΔL as given in Eq. (23), *i.e.* by taking into account not just the influence of a single spectral line, but the simultaneous influence of all the conceivable spectral lines under a global non-LTE context. One possibility is to select a number of absorbers and emitters, to construct a solar RE model and models with prescribed heating, and to investigate whether they are radiatively stable or unstable against perturbations. Recently, Anderson (1989) has shown that the *collective* effect of RT in multitudes of lines ensures the stability of his solar RE model against *plane-parallel* (*i.e.* $k = 0$) perturbations. Similarly stable behaviour has also been found (*cf.* Anderson and

Athay, 1989b) for Avrett's (1985) plane-parallel semi-empirical model (*cf.* Maltby et al, 1986). However, a CO-driven thermal instability as suggested by Kneer (1983;1985) has been indeed reported (*cf.* Anderson, 1989; Anderson and Athay, 1989b) for atmospheric models characterized by amounts of non-radiative heating substantially lower than the heating rate implied by Avrett's (1985) semi-empirical model. Through the application of efficient iterative algorithms (see Klein *et al*, 1989), one may expect in few years from now to be able to perform such global non-LTE calculations for increasing values of the horizontal wavenumber k. The results of the simplified MRT calculations we have discussed above do indicate that the investigation of such sophisticated models will in fact be worthwhile.

Non-Linear Analysis

On Fig. 9, we aim to illustrate the importance of the *non-linear* response of the radiation field. Here we show the time evolution of initially imposed *temperature* structures, by taking into account only the multi-dimensional transfer of energy by grey continuum radiation in LTE. At $t = 0$ seconds the atmospheric structures are assumed to be *slabs* of width W, and temperature enhancement $\Delta T(z)$ with respect to the surrounding unperturbed plasma. These slabs are embbeded with a horizontal periodicity P in a grey RE solar model atmosphere which is in LTE. The gravitational stratification of the unperturbed model is prescribed by assuming $\bar{\chi} = \bar{\chi}_0 \exp(-z/H)$ for the grey opacity, and $\bar{\rho} = \bar{\rho}_0 \exp(-z/2H)$ for the density. As in previous figures, we will measure all geometrical distances in units of the opacity scale height H (~ 60 km for H$^-$ absorption in solar-like photospheres). We note that the results of Fig. 9 have been obtained for $W = 8$, and $P = 16$ opacity scale heights (*i.e.* for $W \approx 480$ km, and $P \approx 960$ km); we also chose $\Delta T(z) = 1000$ K for $z > 0$ ($\bar{\tau} = \exp(-z) < 1$), and $\Delta T(z)$ slowly diminishing in value for $z < 0$ along an exponential decay until it virtually vanishes once the lower boundary is reached.

The time evolution given in Fig. 9a was computed (*cf.* Eq. 19) by solely taking into account the MRT effects of B perturbations, *i.e.* by maintaining χ and ρ fixed at their unperturbed values during the course of the numerical simulation. If only $B = \sigma T^4/\pi$ is allowed to change, Eq. (16) gives a *linear* response for the radiation field. (It is noteworthy that the times needed to reduce the initial amplitude ΔT to $\Delta T/e$ are similar to the *local* radiative relaxation times obtained by Kneer and Trujillo Bueno (1987) for cosinusoidal temperature perturbations with horizontal wavelength $\Lambda = 16$). Note that the time evolution is fastest near $\bar{\tau} = 1$. At this height in the atmosphere the temperature of the gas at $x = 8$ has decreased by about 250 K after 3 seconds, whereas the "walls" of the initially imposed structures (situated at $x = 4$ and $x = 12$) have been cooled down by almost 500 K. It is precisely around the horizontal boundaries of the atmospheric structures where horizontal energy exchange makes its most important contribution. For the case shown in Fig. 9 (*i.e.* for $W = 8$ and $P = 16$) horizontal radiative transfer does not essentially influence the time evolution of the temperature at $x = 0$ (or $x = P$) and at $x = 8$. From calculations with various values of W and P (see Trujillo

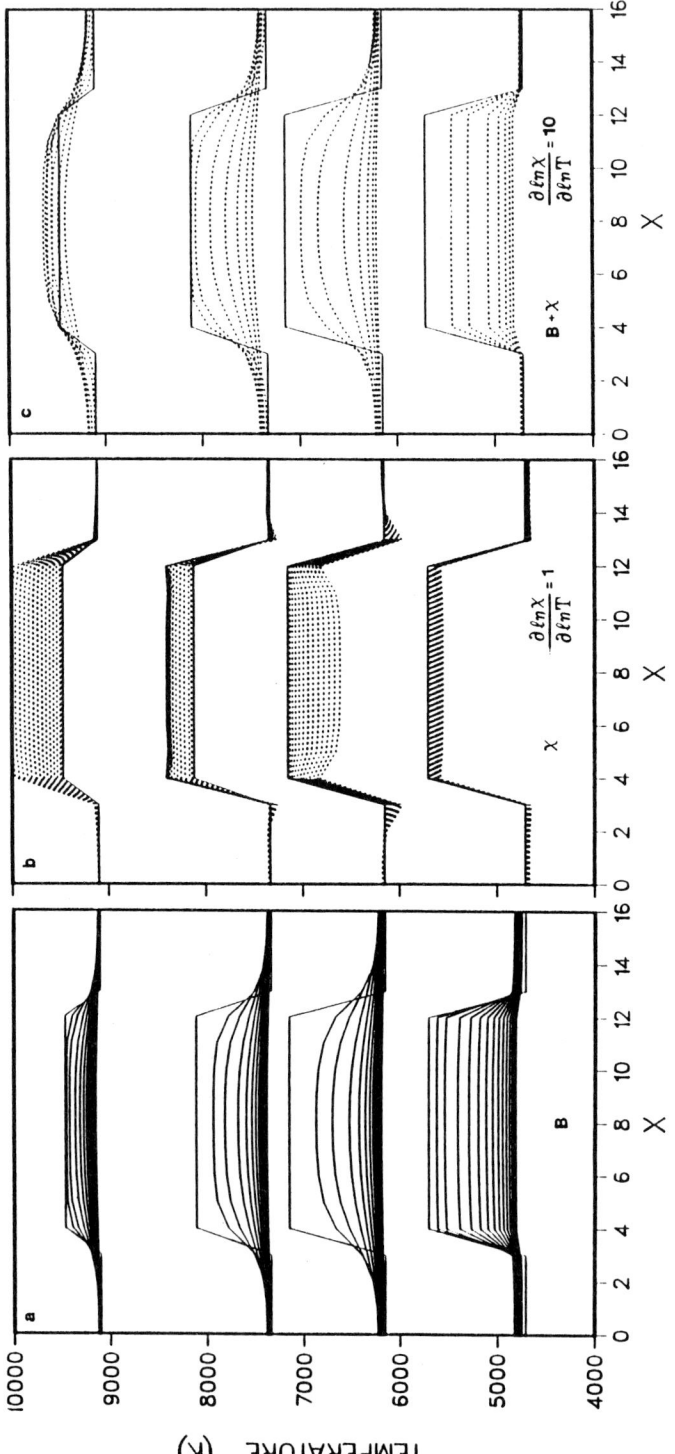

Fig. 9. Snapshots of the temperature field at various points in time, and atmospheric heights, for the case $P = 16$, $W = 8$. Times: 0 s, 3.5 s, 12.5 s, 18.5 s, etc. Heights: $z = 5$ ($\tau \approx 6 \times 10^{-3}$), $z = 0$ ($\tau = 1$), $z = -1$ ($\tau \approx 2.7$), $z = -2$ ($\tau \approx 7.4$); from bottom to top.

Bueno, 1988a), we may conclude that if $W \gtrsim 500$ km the temporal evolution of the temperature in the middle of the initially heated atmospheric regions is mainly dominated by *vertical radiative transfer*. However, for structures with $W \lesssim 250$ km *horizontal radiative transfer* is also important. These results also apply to the temporal behaviour of the temperature of the gas situated in the middle of the surrounding cooler structures (*i.e.* at $x = 0$ and $x = P$): as soon as $P - W \lesssim 250$ km horizontal radiative transfer is capable of efficiently influencing the thermal state of these atmospheric regions.

The results in Fig. 9b were obtained by taking into account *only* the MRT effects of χ fluctuations, *i.e.* by making $\delta S = \delta B = 0$ in Eq. (16). In order to introduce a temperature dependence for the opacity, we assumed that, after a temperature perturbation ΔT, the opacity responds as

$$\chi = \bar{\chi}(1 + \alpha) = \bar{\chi}(1 + \frac{\partial ln\bar{\chi}}{\partial ln\bar{T}} \frac{\Delta T}{\bar{T}}), \qquad (24)$$

where $\bar{T}$ is the temperature of the RE unperturbed model, and $\partial ln\bar{\chi}/\partial ln\bar{T}$ may be chosen at will. The results of Fig. 9b were obtained with $\partial ln\bar{\chi}/\partial ln\bar{T} = 1$. Note that, with $\partial ln\bar{\chi}/\partial ln\bar{T}$ constant, and with the already stated height dependence of ΔT, it follows that $|\alpha(z)|$ increases outwards. Although Eq. (16) now contains the non-linear term $\delta\chi\delta I$, we may still expect to find the MRT effects of *opacity* fluctuations producing *radiative amplification* of the gas temperature below a height close to $\bar{\tau} = 1$, but *radiative relaxation* in higher layers. The reason for this is that, as mentioned above, if $|\alpha|$ increases outwards in a grey RE atmosphere, the MRT effects of opacity fluctuations play a destabilizing role below $\bar{\tau} \approx 1$ (see Trujillo Bueno and Kneer, 1989). These expectations are confirmed with the results of Fig. 9b, where we have radiative amplification below $\bar{\tau} = 1$. Note however that in Fig. 9b the last dotted line plotted corresponds to $t = 72.5s$. After this time the MRT effects of B fluctuations have practically smoothed out the initially imposed temperature perturbation (see Fig. 9a). Clearly, $\partial ln\bar{\chi}/\partial ln\bar{T} = 1$ represents too small a value to expect a noticeable influence from the MRT effects of χ fluctuations.

The time evolution shown in Fig. 9c was obtained by assuming $\partial ln\bar{\chi}/\partial ln\bar{T} = 10$, and by taking into account the MRT effects of both B and χ fluctuations (see Eq. 16). The signatures of the effects of χ fluctuations may easily be noticed by comparing Figs. 9a and 9c. Note, for example, that after 3 seconds the RT effects of B fluctuations acting alone reduce the temperature of the gas situated at $z = -1$ and $x = P/2 = 8$ by 150 K, whilst in Fig. 9c the combined RT effects of B and χ fluctuations do not reveal any appreciable relaxation.

From these illustrative examples we see that for large values of $\partial ln\bar{\chi}/\partial ln\bar{T}$, the non-linear terms in Eq. (16) are important. Since in the photospheres of Sun-like atmospheres $\partial ln\bar{\chi}/\partial ln\bar{T} \approx 10$ near continuum optical depth unity, it follows that the non-linear response of the radiation field to typical photospheric inhomogeneities must be relevant there. Recently Kumar and Goldreich (1989; see also Libbrecht, 1988) concluded that the solar p-modes are probably not driven by a κ-like overstability mechanism, because there would be no way to stop the exponential growth of the p-mode energy. In Fig. 9c we have seen how the non-linear terms of Eq. (16) are capable of stoping the initial temperature amplification of the subphotospheric layers (see also Fig. 9b). We therefore propose the non-linear response of the radiation field itself as a possible non-linear mechanism which might arrest the growth of supposedly linearly overstable modes.

4. CONCLUDING REMARKS

There are certainly other RT problems which we have not discussed. For example, the influence of "history" on the state of excitation and ionization, and the problem of non-Maxwellian velocity distribution functions. The former problem is related to the commonly-made assumption of *statistical steady state*, which is not appropiate for application to the outer layers of stellar atmospheres, where ionization equilibrium is adjusted *slowly* compared with dynamical time scales (Kneer, 1980). The general expression for the line source function of a two-level atom obtained by Cannon and Cram (1974) shows clearly why the atmospheric gas may "remember" the physical conditions met at previous times. The physical reason for this effect resides in the fact that an atom may be excited under a certain set of physical conditions and thence de-excited under differing conditions. For this reason, to remove the assumption of statistical steady state is particularly important for shock wave analysis.

The book of Oxenius (1986) is particularly useful to appreciate the difficulty associated to the problem of non-Maxwellian velocity distribution functions. It is important to recall that, although the RT equation is the kinetic equation for the photon distribution function I_ν, the *energy* and *momentum* equations of the hydrodynamic description stipulate LTE for the *material particles* involved. In fact, if one wishes to investigate the reliability of the assumption of Maxwellian velocity distribution functions, one is obliged to formulate the problem in terms of the kinetic equations for each *particle type* present; for example, for a gas of two-level atoms one has to deal with the distribution functions f_u of excited two-level atoms, the distribution function f_l of non-excited atoms, the electron distribution function f_e, and the photon distribution function I_ν. Therefore all these functions are inter-related, and a self-consistent procedure must be applied to solve the problem. Oxenius (1986; see also Shoub, 1977a,b) illustrates how non-LTE populations of atomic levels may give rise to a non-Maxwellian f_e in a partially ionized hydrogen plasma.

As may be expected, both the *physical* interest and the *mathematical* challenge posed by most of the RT problems we have considered in this review would almost completely disappear if the assumption of LTE were to be generally valid. In fact, if the radiation field did not play a part in determining the state of the material gas, we would hardly find any physically interesting fundamental RT problem. This applies also to the previously considered problem as to whether we do need MRT to correctly interpret the spatial fluctuations of the solar radiation field. It is only outside LTE where the emergent intensity may not reflect the atmospheric parameters of the observed atmospheric structure itself. However, even under non-LTE conditions, many RT problems would vanish if the atmospheres of the stars were not highly dynamic and inhomogeneous systems. We may thus express our gratitude to the stars since, in attempting to understand their complexity, we indeed have the chance to learn Physics.

ACKNOWLEDGEMENTS: I would like to thank Franz Kneer and Egidio Landi Degl'Innocenti for many valuable and stimulating discussions. Finantial support provided by the spanish DGICYT, and also by the CICYT through project Nr PB87-0521, are gratefully acknowledged.

REFERENCES

Anderson, L. S.: 1989, *Astrophys. J.* **339**, 558
Anderson, L. S., Athay, R. G.: 1989a, *Astrophys. J.* **336**, 1089
Anderson, L. S., Athay, R. G.: 1989b, *Astrophys. J.* **346**, 1010
Anderson, P.: 1949, *Phys. Rev.* **16**, 647
Ando, H., Osaki, Y.: 1975, *Publ. Astron. Soc. Japan* **27**, 581
Avrett, E. H.: 1985, in *Chromospheric Diagnostics and Modelling*, B. W. Lites (ed.), National Solar Observatory, Sunspot, N.M. 88349, p. 67
Avrett, E. H.: 1989, in *Solar Photosphere: Structure, Convection, and Magnetic Fields*, J. O. Stenflo (ed.), Kluwer, Dordrecht, in press
Baranger, M.: 1958, *Phys. Rev.* **111**, 481
Baranger, M.: 1958, *Phys. Rev.* **111**, 494
Baranger, M.: 1958, *Phys. Rev.* **111**, 855
Bommier, V., Landi Degl'Innocenti, E., Sahal-Bréchot, S.: 1989, in Second Atelier Transfert du Rayonnement, H. Frisch and N. Mein (eds.), p. 44
Breene, R. G.: 1981, *Theories of Spectral Line Shapes*, John Wiley and Sons
Burnett, K., Cooper, J., Ballagh, R. J., Smith, E. W.: 1980, *Phys. Rev. A* **22**, 2005
Burnett, K., Cooper, J.: 1980a, *Phys. Rev. A* **22**, 2027
Burnett, K., Cooper, J.: 1980b, *Phys. Rev. A* **22**, 2044
Butler, K.: 1987, *Mitt. Astron. Ges.* **70**, 65
Cannon, C. J.: 1970, *Astrophys. J.* **161**, 255
Cannon, C. J.: 1976, *Astron. Astrophys.* **52**, 337
Cannon, C. J.: 1985, *The Transfer of Spectral Line Radiation*, Cambridge University Press
Cannon, C. J., Cram, L. E.: 1974, *J. Quant. Spectrosc. Radiat. Transfer* **14**, 93
Carlsson, M., Scharmer, G. B.: 1985 in *Chromospheric Diagnostics and Modelling*, B. W. Lites (ed.), National Solar Observatory, Sunspot, N.M. 88349, p. 137
Chandrasekhar, S.: 1943, *Rev. Mod. Phys.* **15**, 1
Chandrasekhar, S.: 1958, *Proc. Nat. Acad. Sci.* **44**, 933
Christensen-Dalsgaard, J., Frandsen, S.: 1983, *Solar Phys.* **82**, 165
Collados, M., Vázquez, M.: 1987, *Astron. Astrophys.* **180**, 223
Cooper, J., Ballagh, R. J., Hubeny, I.: 1988, in *Spectral Line Shapes*, J. Szudy (ed.), Elsevier Science Publishers, p. 275
Dainty, J. C., Shaw, R.: 1974, *Image Science*, Academic Press

Deinzer, W., Hensler, G., Schüssler, M., Weishaar, E.: 1984, *Astron. Astrophys.* **139**, 435
Deubner, F. L., Mattig, W.: 1975, *Astron. Astrophys.* **45**, 167
Deubner, F. L., Gough, D. O.: 1984, *Ann. Rev. Astron. Astrophys.* **22**, 593
Dicke, R. H.: 1953, *Phys. Rev.* **89**, 472
Domke, H., Hubeny, I.: 1988, *Astrophys. J.* **334**, 527
Dravins, D., Nordlund, Å.: 1989, *Astron. Astrophys.*, (paper IV), in press
Fano, U.: 1957, *Rev. Mod. Phys.* **29**, 74
Feynman, R. P., Leighton, R. B., Sands, M.: 1964, *The Feynman Lectures on Physics*, Vol. II, Addison-Wesley
Fiutak, J., Van Kranendonk, J.: 1962, *Canadian J. Phys.* **40**, 1085
Fowler, B. W., Sung, C. C.: 1975, *J. Opt. Soc. Am.* **65**, 949
Frisch, H.: 1988, in *Radiation in Moving Gaseous Media*, Eighteenth Advanced Course, Swiss Society for Astrophysics and Astronomy, p. 337
Frisch, H., Frisch, U.: 1976, in *Physique des Mouvements dans les Atmosphères Stellaires*, R. Cayrel and M. Steinberg (eds.), CNRS, p. 113
Gail, H. P.: 1980, in *Stellar Turbulence*, D. F. Gray and J. L. Linsky (eds.), Springer-Verlag, p. 183
Gail, H. P.: Sedlmayr, E., Traving, G.: 1975, *Astron. Astrophys.* **44**, 421
Gaizauskas, V.: 1985, in *Chromospheric Diagnostics and Modelling*, B. W. Lites (ed.), National Solar Observatory, Sunspot, N.M. 88349, p. 25
Galatry L.: 1961, *Phys. Rev.* **122**, 1218
Giovanelli, R. G.: 1959, *Australian J. Physics* **12**, 164
Gustafsson, B.: 1989, in *Modeling the Stellar Environment: how and why?*, P. Delache, S. Laloe, C. Magnan, and J. Tran Thanh Van (eds.), Editions Frontieres, p. 13
Griem, H. A.: 1974, *Spectral Line Broadening by Plasmas*, New York: Academic
Hanle, W.: 1924, *Zs. Phys.* **30**, 93
Hasan, S. S.: 1988, *Astrophys. J.* **332**, 499
Heitler, W.: 1954, *The Quantum Theory of Radiation*, Clarendon Press, Oxford
Hubeny, I.: 1985, in *Progress in Stellar Spectral Line Formation Theory*, J. E. Beckman and L. Crivellari (eds.), Reidel Publishing Company, p. 27
James, T. C.: 1969, *J. Opt. Soc. Am.* **59**, 1602
Janssen, P. J.: 1884, quoted by Bray, R. J. and Loughhead, R. E., in *The Solar Granulation*, Chapman and Hall, London, 1967, plate 1.4
Jefferies, T. J., Lindsey, C. A.: 1988, *Astrophys. J.* **335**, 372
Jones, H. P.: 1986, in *Small Scale Magnetic Flux Concentrations in the Solar Photosphere*, W. Deinzer, M. Knölker, and H. H. Voigt (eds.), Vandenhoeck and Ruprecht, Göttingen, p. 127
Jones, H. P., Skumanich, A.: 1980, *Astrophys. J. Suppl.* **42**, 221
Kalkofen, W.: 1984, *Methods in Radiative Transfer*, Cambridge University Press
Kalkofen, W.: 1987, *Numerical Radiative Transfer*, Cambridge University Press
Kalkofen, W., Bodo, G., Massaglia, S., Rossi, P.: 1989, in *Solar and Stellar Granulation*, R. J. Rutten and G. Severino (eds.), Kluwer Academic Publishers, p. 571
Kaufman, J. M.: 1988, in *Seismology of the Sun and Sun-like Stars*, E. J. Rolfe (ed.), ESA Publications Division, p. 31
Klein, R. I., Castor, J. I., Greenbaum, A., Taylor, D., Dykema, P.: 1989, *J. Quant. Spectrosc. Radiat. Transfer* **41**, 199
Kneer, F.: 1979, in *Future Solar Optical Observations, Needs and Constraints*, G. Godoli, G. Noci, and A. Righini (eds.), Tip. Baccini and Chiappi, Florence, p. 204
Kneer, F.: 1980, *Astron. Astrophys.* **87**, 229
Kneer, F.: 1981, *Astron. Astrophys.* **93**, 387
Kneer, F.: 1983, *Astron. Astrophys.* **128**, 311

Kneer, F.: 1985, in *Chromospheric Diagnostics and Modelling*, B. W. Lites (ed.), National Solar Observatory, Sunspot, N.M. 88349, p. 252
Kneer, F.: 1986, in *Small Scale Magnetic Flux Concentrations in the Solar Photosphere*, W. Deinzer, M. Knölker, and H. H. Voigt (eds.), Vandenhoeck and Ruprecht, Göttingen, p. 147
Kneer, F., Heasley, J. N.: 1979, *Astron. Astrophys.* **79**, 14
Kneer, F., Trujillo Bueno, J.: 1987, *Astron. Astrophys.* **183**, 91
Knölker, M., Schüssler, M.: 1988, *Astron. Astrophys.* **202**, 275
Kumar, P., Goldreich, P.: 1989, *Astrophy. J.* **342**, 558
Kunasz, P., Mihalas, D.: 1984, *Los Alamos Report*: LA-UR-84-3735
Lamb, F. K., Ter Haar, D.: 1971, *Physics Reports*, **2C**, 253
Landau, L. D., Lifshitz, E. M.: 1971, *The Classical Theory of Fields*, 3rd ed., Pergamon Press
Landi Degl'Innocenti, E.: 1983a, *Solar Phys.* **85**, 3
Landi Degl'Innocenti, E.: 1983b, *Solar Phys.* **85**, 33
Landi Degl'Innocenti, E.: 1984, *Solar Phys.* **91**, 1
Landi Degl'Innocenti, E.: 1985, *Solar Phys.* **102**, 1
Legait, A.: 1986, *Astron. Astrophys.* **168**, 173
Libbrecht, K.: 1988, in *Seismology of the Sun and Sun-like Stars*, E. J. Rolfe (ed.), ESA Publications Division, p. 3
Linsky, J. L.: 1985, in *Progress in Stellar Spectral Line Formation Theory*, J. E. Beckman and L. Crivellari (eds.), Reidel Publishing Company, p. 1
Loudon, R.: 1973, *The Quantum Theory of Light*, Clarendon Press
Magnan, C.: 1985, *Astron. Astrophys.* **144**, 186
Maltby, P., Avrett, E. H., Carlsson, M., Kjeldseth-Moe, O., Kurucz, R. L., Loeser, R.: 1986, *Astrophy. J.* **306**, 284
Margenau, H., Lewis, M.: 1959, *Rev. Mod. Phys.* **31**, 569
Mihalas, D.: 1978, *Stellar Atmospheres*, Freeman, San Francisco
Mihalas, D., Auer, L. H., Mihalas, B.: 1978, *Astrophys. J.* **220**, 1001
Mizushima, M.: 1971, *J. Quant. Spectrosc. Radiat. Transfer* **11**, 471
Muller, R.: 1985, *Solar Phys.* **100**, 237
Nordlund, Å: 1984, in *Small Scale Dynamical Processes in Quiet Stellar Atmospheres*, S. L. Keil (ed.), National Solar Observatory, Sunspot, N.M. 88349, p. 181
Nordlund, Å, Dravins, D.: 1989, *Astron. Astrophys.*, (paper III), in press
Omont, A., Smith, E. W., Cooper, J.: 1972, *Astrophy. J.* **175**, 185
Omont, A., Smith, E. W., Cooper, J.: 1973, *Astrophys. J.* **182**, 283
Oxenius, J.: 1965, *J. Quant. Spectrosc. Radiat. Transfer* **5**, 771
Oxenius, J.: 1985, in *Progress in Stellar Spectral Line Formation Theory*, J. E. Beckman and L. Crivellari (eds.) Reidel Publishing Company, p. 59
Oxenius, J.: 1986, *Kinetic Theory of Particles and Photons*, Springer-Verlag
Rayleigh (Strutt, J. W.): 1889, *Philos. Mag.* **27**, 298
Rees, D. E.: 1987, in *Numerical Radiative Transfer*, W. Kalkofen (ed.), Cambridge University Press, p. 213
Rybicki, G. B.: 1965, *Ph. D. Thesis*, Harvard University
Rybicki, G. B.: 1971, *J. Quant. Spectrosc. Radiat. Transfer* **11**, 827
Rybicki, G. B.: 1976, in *Physique des Mouvements dans les Atmosphères Stellaires*, R. Cayrel and M. Steinberg (eds.), CNRS, p. 189
Scherrer, P. H., Hoeksma, J. T., Bogart, R. S.: 1988, in *Seismology of the Sun and Sun-like Stars*, E. J. Rolfe (ed.), ESA Publishing Division, p. 375
Schüssler, M.: 1989, in *Solar Photosphere: Structure, Convection and Magnetic Fields*, J. O. Stenflo (ed.), Kluwer, Dordrecht, in press
Sedlmayr, E.: 1976, in *Physique des Mouvements dans les Atmosphères Stellaires*, R. Cayrel and M. Steinberg (eds.), CNRS, p. 157

Sedlmayr, E.: 1980, in *Stellar Turbulence*, D. F. Gray and J. L. Linsky (eds.), Springer-Verlag, p. 195
Shoub, E.: 1977a, *Astrophys. J. Suppl.* **34**, 259
Shoub, E.: 1977b, *Astrophys. J. Suppl.* **34**, 277
Shurcliff, W. A.: 1962, *Polarized Light*, Harvard University Press, Cambridge, Mass.
Simonneau, E.: 1985, in *Progress in Stellar Spectral Line Formation Theory*, J. E. Beckman and L. Crivellari (eds.), Reidel Publishingh Company, p. 73
Sobelman, I. I., Vainshtein, L. A., Yukov, E. A.: 1981, *Excitation of Atoms and Broadening of Spectral Lines*, Springer-Verlag
Solanki, S. K.: 1989, in *Solar Photosphere: Structure, Convection and Magnetic Fields*, J. O. Stenflo (ed.), Kluwer, Dordrecht, in press
Spiegel, E. A.: 1957, *Astrophys. J.* **126**, 202
Stein, R. F.: 1985, in *Chromospheric Diagnostics and Modelling*, B. W. Lites (ed.), National Solar Observatory, Sunspot, N.M. 88349, p. 213
Stenflo, J. O.: 1989, *Astron. Astrophys. Rev.* **1**, 3
Stenholm, L. G., Stenflo, J.: 1977, *Astron. Astrophys.* **58**, 273
Stenholm, L. G., Stenflo, J.: 1978, *Astron. Astrophys.* **67**, 33
Streater, A., Cooper, J., Rees, D. E.: 1988, *Astrophys. J.* **335**, 503
Ter Haar, D.: 1961, *Rept. Progr. Phys.* **24**, 304
Traving, G.: 1976, in *Physique des Mouvements dans les Atmosphères Stellaires*, R. Cayrel and M. Steinberg (eds.), CNRS, p. 145
Trujillo Bueno, J.: 1982, *Tesis de Licenciatura*, University of Zaragoza
Trujillo Bueno, J.: 1988a, *Ph. D. Thesis*, University of Göttingen
Trujillo Bueno, J.: 1988b, in *Seismology of the Sun and Sun-like Stars*, E. J. Rolfe (ed.), ESA Publications Division, p. 11
Trujillo Bueno, J., Kneer, F.: 1987, *Astron. Astrophys.* **174**, 183
Trujillo Bueno, J., Kneer, F.: 1989, in *Solar and Stellar Granulation*, R. J. Rutten and G. Severino (eds.), Kluwer Academic Publishers, p. 441
Trujillo Bueno, J., Kneer, F.: 1990, *Astron. Astrophys.* in press
Unno, W., Spiegel, E. A.: 1966, *Publ. Astron. Soc. Japan* **18**, 85
Vaiana, G. S., Rosner, R.: 1978, *Ann. Rev. Astron. Astrophys.* **16**, 393
Vernazza, J.E., Avrett, E. H., Loeser, R.: 1981, *Astrophys. J. Suppl.* **45**, 635
Vidal, C. R., Cooper, J., Smith, E. W.: 1973, *Astrophys. J. Suppl.* **25**, 37
Voigt, W. K.: 1912, *Bayer. Akad. München, Ber* **603**
von Neumann, J.: 1927, *Göttinger Nach.* **245** and **273**
Wittke, J. P., Dicke, R. H.: 1956, *Phys. Rev.* **103**, 620

Evolution of Solar and Stellar Rotation

SANTO CATALANO

Institute of Astronomy
University of Catania
Viale A. Doria, I-95125 Catania, Italy

ABSTRACT: The rotational velocity of stars is one of the key parameters that provide important constraints how stars form and evolve. The developements of new efficient instrumentation, new line profile analysis and observational thechnique, have allowed significant progress in the observation of rotation of low mass stars during the last decade. Observations on young clusters have shown that pre-main sequence stars have relatively high rotation velocity and for a given mass values are spred over a large velocity interval. Braking winds rapidly decrease the rotational velocity of F and G dwarfs to few tenth's of Km/sec in only few time 10^7 years, with a spin-down time scale smaller for higher stellar masses. This spindown is such that, as the stars seat on the main sequence the rotational velocity spread is strongly reduced.

On the main sequence the rotation is tightly defined by the stellar mass and age and, from about $1.2 M_\odot$ to $0.5 M_\odot$ it appears to obey a power-law relationship with the age: $\Omega \propto t^{-1/2}$. The Sun, in spite of the presence of a planetary system retaining the 98 % of the total angular momentum, has a rotation speed which is pefectly consistent with stars of its mass and age. This strongly suggest that only a defined amount of angular momentum can be retained by a star of a given mass and the excess of the initial angular momentum has to be left in a planetary system, like the solar one or in a protoplanetary disk as observed in β Pictoris. Recent models for rotational spindown of low mass stars seem to fit the solar interior fairly well but do not fit the surface rotational velocity of young stars. Implications for angular momentum transfer from the core to the envelope are also discussed.

1 Introduction

The measurement of the solar rotation has a rather old origin. The 25 day rotation period became obvious by the time Galileo discovered the solar spots and followed them on the solar disk. The stellar rotation has a more recent history which traces back to the end of last century when sir William Abney suggested that axial rotation might be responsible for the great widths of certain stellar absorption lines. However, it was only in the thirthies that measurements of the equatorial rotational velocity, $v \sin i$ of single stars have been reported (Struve 1931). These measurements, as well as those in subsequent works by Slettebak (1949, 1954, 1955) Abt (1961, 1965) and others relate only with stars considerable more massive than the Sun. A survey by Herbig and Spalding (1955) clearly showed that late type stars have small rotational velocity, hardly measurable or below their detection limit of about 15 Km/sec. The systematic work of Abt and Hunter (1962) showed a smooth decrease of rotational velocity going from early to late type stars. Late B and early A type stars have mean $v \sin i$ of 200 Km/sec which declines to about 100 Km/sec for late A type. However, they found also a sharp decrease of rotational velocity among F stars which drops steeply toward later subclass such that it becomes of order of 10 Km/sec for G0 main sequence field stars.

Observation down to the limit of the classical photographic technique allowed Kraft (1965, 1967) to measure $v \sin i$ values as low as 10 Km/sec in late type field star and to compare them with those of young cluster. He showed that while early type stars in the Hyades and Pleiades clusters have rotational velocities not very different than those of field stars, mid to late F in the Pleiades rotate systematically faster than the Hyades ones of the same spectral type and that old disk field F dwarfs rotate even slower. An evolutionary effect appeared the more obvious explanation rather than a local effect (i.e. star formed in different regions would have different intrinsic rotational velocity). The coincidence of stellar rotation drop with the appearance of chromospheric activity (Ca II H and K emission core) and the onset of predicted deep hydrogen convective zone around spectral type F4-F5 led O. C. Wilson (1966) to the suggestion of an intimate connection between convection, chromospheric emission and rotation braking. This connection was giving support to the idea that the rotational braking is due to angular momentum loss through magnetic stellar wind, as formerly suggested by Schatzman (1962). The bracking effect become more clear when the observed rotational velocity are converted to angular momentum (McNally 1965, Kraft 1967). Kraft showed that the angular momentum per unit mass for high mass stars $M > 1.5\ M_\odot$ follows quite well a power-law relation with the stellar mass of the form $J/M = K\ M^{0.57}$. This dependence has been approximated as $J/M \propto M^{2/3}$ by Dicke (1970) who shown that it could be the natural result of simple homology arguments. Low mass stars $M < 1.5 M_\odot$ have been found to have specific angular momentum smaller than the extrapolation of the high mass power law. By that time, two quite different behaviors of stellar rotation came out for high and low mass stars. A number of problems, many of which are still waiting for an answer were rised . Let me summarize here the main aspects

a) high mass stars, $M > 1.5\ M_\odot$:

i) they show a smooth decrease of $v \sin i$ with mass such that the specific angular momentum results to be $J/M \propto M^{2/3}$
ii) for a given mass the observed $v \sin i$ is Maxwellian or near Maxwellian
iii) no, or marginal time evolution of rotation is observed

b) low mass stars, $M < 1.5 M_\odot$:

i) have very small $v \sin i$, such that $J/M \ll M^{2/3}$
ii) the Sun specific angular momentum is near two order of magnitude smaller than $M^{2/3}$
iii) the solar system angular momentum agree fairly well with the extrapolation of $M^{2/3}$
iv) time evolution is apparent, younger stars rotate faster.

Because of the short contraction time and because the angular momentum loss is believed to be quite inefficient for high mass-stars, it was widely accepted that the power-law relation reflects the initial distribution of angular momentum. This view was enforced by the consideration that, if the angular momentum of our planetary system is included, the total angular momentum fits very well in the mean relation. This quite naturally led to speculations on the frequency of planetary systems around low mass stars (Huang 1965, 1969).

The time scale estimated from the solar wind torque (Dicke 1970) was found, with respect to the Sun life-time, to be too long to spin-down the whole Sun if it rotates as a rigid body and too short to spin-down only the convective zone. The suggestion that the Sun might have a rapidly rotating core (Dicke 1964, 1970) does not completely solve the problem of the initial angular momentum of the Sun.

Compilation of more complete catalogues and analyses on the rotational velocity distribution (Fucuda 1982, Wolff, Edwards and Preston 1982) have shown that the picture of rotation among high mass stars has not changed that much in recent years. However new progress and results have been achieved on the knowledge of rotation among low mass stars, during the last decade.

New instrumentation, new analysis techniques and metodology have allowed to measure rotational velocity as low as 2 Km/sec and to reach faint stars in young clusters. The spectroscopic method has highly improved both, by the use of new digital detectors, like Reticon and CCD which allow high signal to noise ratio and by the application of Fourier analysis technique to line profiles. This technique allows to extract rotational broadening from a line profile even when it is smaller than other broadening effects (Smith and Gray 1976). Important contributions to the determination of accurate spectroscopic rotational velocity, down to 2 Km/sec with accuracy for 0.6 Km/sec for solar type field stars have been given by Smith (1979), Soderblom (1983) and Gray (1984). But lower mass stars have been found to be below these limits. However, Wilson (1978) effort to study the activity cycle of solar type stars from the changes of the flux in the Ca II H and K emission reversals provided a new powerful tool to measure rotational velocity. Following Wilson suggestion that short-term fluctuations in HK measurement reflect modulation due to rotation of univen distribution of the solar analogue chromospheric Ca II active regions a systematic observational programme has been settled at Mt. Wilson. The HK project (Vaughan et al 1981, Baliunas et al 1983) has provided rotational period of near 100 star of spectral type F7 to M (Noyes et al 1984). Similarly rotational periods have been derived for a number of K-M stars using broad band photometry to measure light variation produced by large star spot groups (Krzeminski 1969, Bopp Evans 1973, Torres and Ferrar-Mello 1973, Vogt 1975, Bopp and Fekel 1977, Oskanyan et al. 1977,Blanco et al. 1979, Radick et al. 1987). These results are of fundamental importance in studying the rotational evolution of low-mass stars, because rotation periods can be determined with much higher precision, few per cent, than $v \sin i$ values, and are totally indipendent of the projection effect. Moreover the method has the further advantage that rotation period can be determined for stars that rotate at a rate that would be impossible to see any line broadening in the spectra (for istance, for 61 Cyg A a K 5 dwarf the observed rotation period of 38 days corresponds to an equatorial rotational velocity of 0.9 Km/sec). A further method of measuring the rotation period relay on the observed correlation between the HK emission and rotation at a give mass on the main sequence. This correlation is sufficiently good to estimate the rotational periods from the mean emission of Ca II HK, or other chromospheric lines, with an accuracy of 20% (Noyes et al. 1984, Marilli, Catalano, Trigilio 1986). A substantial data-base on rotational velocity of low mass stars which include data on T Tau stars, young clusters and field main sequence stars is now available. This allows us to attempt an evolutionary history of the Sun and Solar type star rotation and address a number of problems. Recent reviews (Gray 1982a, Stauffer and Hartman 1986, Hartman and Noyes 1987, Smith 1988, Stauffer and Soderblom 1989, Sofia et al 1988) have summarized some of the problems and models of stellar rotation. The present discussion will focus on the observational aspects which help to delineate the rotational history of the Sun and its behavior among stars of similar type.

2 The Angular momentum problem

The origin of stellar angular momentum is still an argument of debate. Because of galactic rotation, a star formed by condensation of interstellar medium of low density over a large volume would acquire a large amount of angular momentum, typically $J/M \simeq 10^{23} cm^2 sec^{-1}$. A number of observational results seem to contradict this seemingly convincing prediction. The observed specific angular momentum of single stars is only $J/M \simeq 10^{17} - 10^{18} cm^2 sec^{-1}$. In addition from simple consideration on stability it is possible to show that if the galactic angular momentum is conserved during contraction the surface rotational velocity greatly exceeds the break-up velocity for the star.

If, indeed stellar angular momentum is a localization of galactic angular momentum, the angular momentum vectors of stars should all be allined with the galactic vector. Analyses of rotation of single stars as well as of binary system indicate that rotation axes are randomly oriented (see, Huang 1969 for a summary). Recent observations on stellar association suggest that rotation axes of at least some young stars are aligned parallel to the local magnetic field direction. All these arguments rise the so called *angular momentum problem*.

If one assumes, as many theoretician do explicitly or implicitely that the stellar angular momentum comes from the galactic rotation of the pre-stellar medium, the questions arise: why stars have lost memory of their original angular momentum? When and how they loose the excess of angular momentum?

The most straight-forward way to determine when the angular momentum is lost is to measure the rotation of dense molecular clouds at different stages of contraction. Because the rotation rate for such clouds is small, measuring the rotation is difficult. In addition their age is very uncertain. Rotation rate measurements of dark clouds in Taurus with mass in the range $6 \; 10^3$ to $0.7 \; M_\odot$, and size from 17 to 0.1 pc lead to velocity gradients from 0.2 to 6 $Km \; s^{-1} pc^{-1}$ (Goldsmith and Arquilla 1985). Fragments with diameter of 0.01-0.02 pc and masses 1-2 $M_\odot$ in ρ Ophiuchi have rotation rate of 40 $Km \; sec^{-1} pc^{-1}$ (Wadiak et al 1985). NH3 observations in Orion shows cores with diameters 0.05 pc and masses 5 M having rotation gradients of 25-50 $Km \; sec^{-1} pc^{-1}$ (Harris et al 1983). These data show tendency for the smallest and densest regions to have the shortest rotation period. This apparent spin-up can be interpreted as due to contraction, however a considerable braking has already taken place because the specific angular momentum J/M of the various regions seems to decrease with decreasing size (Goldsmith and Arquila 1985). Anyway, such rotation rates lead to rotation velocity larger than the break-up velocity if the cloud collapse to stellar dimension without further angular momentum loss.

The ways how protostellar clouds can loose large amount of angular momentum include: hierarchical fragmentation (see Mestel 1965, Bodenheimer 1978, Zinnecker 1984) magnetic stress via ambipolar diffusion (see Mauschovias 1976, 1981), accretion disk and dipolar flow (see Shu et al. 1988). However, it is becoming increasingly evident that accretion disks play a major role in the star formation process (Strom, Strom and Edwards 1988, Hartman and Kenyon 1988, Adam, Lada and Shu 1987, Bertout 1988) and as an intermediate solution for the angular momentum problem (Larson 1983, Cassen, Shu and Tereby 1985). Typical size and masses estimated for the disks are 100 AU and of about 0.1 $M_\odot$, so the disks are clearly an important reservoir of angular momentum.

Theoretically, disks are likely to occur simply because of the angular momentum probably present in the pre-collapse interstellar material, and a substantial fraction of the mass of the star beeing accumulated through the disk. To allow the protostar to accumulate material of relatively high specific angular momentum, spin-up by the disk must be exactly balanced by spin-down by an extraordinarily powerful magnetized wind. Based on this condition, models of angular momentum evolution yeld an equatorial break-up velocity $V_o \simeq 140$ Km/sec, nearly independent of mass between 0.1 and 2.0 M at the *deuterium birth line* (Shu et al. 1988). However, it is not clear to what estend stars form such disks, the time scale and mechanism by which the disks dissipate, and how much mass and angular momentum is transferred between the disk and central star. This does not allow to confidently predict the initial angular momentum of low mass stars.

Based on the extrapolation of the high mass dependence and on simple consideration of homology Dicke (1970) suggests that low mass have specific initial angular momentum $J/M \propto M^{2/3}$. Recently Kawaler (1987) showed that for high mass the specific angular momentum is better represented by $J/M \propto M^{2.02}$ which correspond to an average velocity $V_e = V_{cr}/2.7$. A simple extrapolation of the high mass star relation does not seem to be able to predict the initial angular momentum of low mass stars, because it apply to stars already on the main sequence. Even if, the pre-main sequence evolution time of high mass stars is very short it is not possible to assume an evolution without angular momentum loss in that phase.

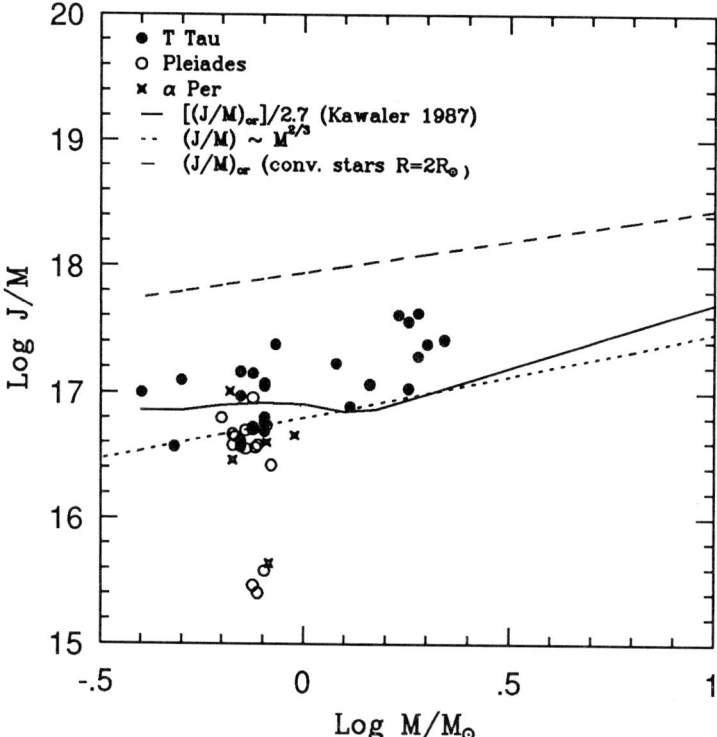

Figure 1. Specific angular momentum as function of the mass for T Tau stars and young K stars on the pleiades and α Persei clusters, based on photometric rotational periods. Different angular momentum dependences are shown for comparison.

Measurement of rotation at the very early phase of evolution, prior to the T Tau phase, i.e in FU Ori are very difficult because of the presence of disk. Rotational velocity at optical wavelength obtained for the three best-studied objects FU Ori, V1057 and V1515 Cyg are 65, 45 and 20 Km/sec respectively (Kenyon, Hartmann and Hewett 1988). This velocity is higher than the break-up velocity and it is interpreted as the Keplerian velocity of the inner part of the disk. Rotation velocity measurements recently made by several groups (Hartman et al. 1986, Bouvier et al. 1986, Hartmann, Soderblom and Stauffer 1987, Rydgren et al. 1987, Franchini, Magazzù, Stalio 1988) all agree with the previous results of Vogel and Kui (1981) that T Tau stars are not very rapid rotators. However, specific angular momentum values estimated from rotation periods and radii as collected by Covino et al (1989) do show that T Tau at an age between 10^6 and 10^7 years have J/M smaller than the critical J/M, but larger than both the Dicke (1970) $M^{2/3}$ and Kawaler (1988) $(J/M)_{cr}/2.7$ relation (Figure 1).

3 The Early Phases of Rotational Evolution: (Spin-up and Spin-down).

As discussed in the preceding section there is a large evidence that T Tau star rotate much slower than the break-up velocity. Typical values are 15-20 Km/sec. However an apparent differentiation exist between stars of different mass as demonstrated by Vogel and Kui (1981) and more recent surveys (Hartmann, Soderblom and Stauffer 1987, Hartmann and Stauffer 1989, Franchini, Magazzù, Stalio 1988). High-mass stars (M$>1M_\odot$) are generally found rotating considerably faster, v_e up to 50 Km/sec, than low-mass stars (M$<1M_\odot$) whose velocity rarely exceeds 30 Km/sec. This differentiation can be ascribed to the fact that high-mass stars are already on the radiative core phase of their pre-main sequence evolution, while most of the low-mass stars are still in the convective-track phase (Hartmann, Soderblom and Stauffer 1987).

The average slow rotation of T Tau stars is especially surprising in view of evidence for accretion of substantial amount of material, 5-10% of their mass, from circumstellar disks in that phase (Kenyon and Hartmann 1987, Bertout, Basri and Bouvier 1988). In addition, for stars on Hayashi track with mass less than $1M_\odot$, there are no obvious correlations between luminosity (and hence age) and rotational velocity. In particular the "Naked T Tauri" stars (stars without emission indicative of disk Walter (1986) do not show a rotational velocity distribution significantly different from the "classical" T Tauri stars or from the "continuum" T Tauri stars (stars with strong optical and IR continuum due to boundary-layer emission from rapid accretion disk) (Hartmann and Stauffer 1989). It is not clear whether Naked T Tauri stars do really form without disks, or they beeing on average older, have allowed their disks to have dissipated somewhat. Strom et al. (1989) from the IR properties of NTT suggest that there is simply a range of initial disk mass and that differences in the two populations indicate an age difference.

From the above consideration, it appears that a differentiated angular momentum loss is required to reconcile the accretion disk model of T Tauri stars with the low rotational-velocity observed. As low mass T Tauri stars do approach the main sequence are expected to increase significantly their rotational velocity due to the decrease in radius and increase in central concentration. The observed rotational velocity, if the evolution of pre-main sequence stars takes place without angular momentum loss lead to predicted maximum and mean rotational velocity of 150 and 75 Km/sec. The observed rotational velocity distribution for low mass stars in young clusters like, Alpha Persei (age 5 10^7 years) are remarkably close to this prediction (Stauffer and Hartman 1986) but there is also a large population of slow rotators, that require a similar population of slow rotator among T Tau stars that it is not yet identified (Hartmann et al. 1988, Bouvier et al. 1986). The large rotational velocities as well as the low ones are found for all the spectral type in Alpha Persei (Figure 2a), from F to M. Observation of the slightly older cluster, the Pleiades (age 7-10^7 years or 1.5 10^8 years according to recent evaluation (Mazzei and Pigatto 1989), show that one does find fast rotators,together with slow rotators only among the K and M dwarfs (Figure 2b)(Stauffer and Hartmann 1987). The Pleiades G dwarfs, however, are nearly all slow rotators. In the more older cluster, the Hyades (age 10^9 years) (Figure 2c) the G and K dwarfs are all slow rotators, while moderate rotational velocities can be found for many of the M dwarfs (Stauffer, Hartman and Latham 1987).

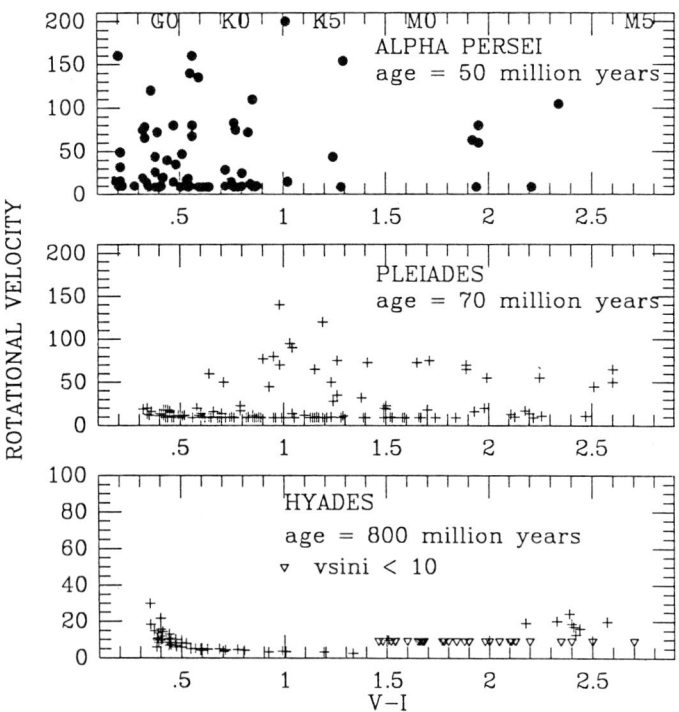

Figure 2. Rotational velocity distribution for low mass stars in the α Persei, Pleiades and Hyades. The disappearance of rapid rotators in early type stars with advancing age is apparent (from Stauffer 1988).

These results indicate that the time scale for the surface rotational velocity of late type stars to decrease from large to small values increases dramatically with decreasing stellar mass. For G stars, the last spin-down time is between 20 and 50 10^7 years, depending on the age assumed for the Pleiades cluster. For K dwarfs the time is, instead of the order of 10^8 years, and it is much longer ($\simeq 10^9$ years) for M dwarfs. The spread of rotational velocity of G stars in the Pleiades and, G and K stars in the Hyades is very small. This suggests that by the time a star sit on the main sequence should have a certain surface rotation determined by some self-regulating mechanism. A simple explanation for the different rotational velocities behaviour with cluster age is that low mass stars arrive on the main sequence as very rapid rotators, spin-down to relatively small rotational velocity in a period of time depending on the mass and therefore of the internal structure. Rotational evolution model by Endal and Sofia (1981) actually predict this brief period of rapid rotation upon arrival on the main sequence for solar mass stars. Their model assume essentially no coupling between the outer convective envelope and the core. The short spin-down time of G dwarfs derives by the fact that only the convective envelope which contains about the 10% of the total angular momentum of the star, is braked. The fraction of angular momentum contained in the outer convective envelope of main sequence stars increases with decreasing effective temperature, so that early K dwarfs have 30-40% of their angular momentum in the outer convective envelope. The spin-down time would be longer simply because a larger fraction of the star total angular momentum is involved. This simple conclusion is in contrast with the absence of a fast rotating core in the Sun inferred from recent helioseismology results (Harvey 1988 and references therein) which suggests some coupling between the core and envelope. Some of these problems are accounted for by new theoretical models for low mass stars which try to match the observed rotational velocity data including high angular momentum transfer (Kawaler 1988, Sofia et al. 1988).

Another putzling problem is posed by the numerous slow rotators v_e of about 10 Km/sec in young cluster (Stauffer et al 1985, Stauffer and Hartmann 1987). Their presence is not expected, because their precursor (i.e. the T Tauri and Naked T Tauri stars) do not show a large range in initial angular momenta or a peak at low rotation (Bouvier et al. 1986, Stauffer Hartmann 1989). If T Tauri stars evolved to the main sequence, without angular momentum loss, would have velocity, that nicely fit the rapid rotator population of young clusters, as we have seen before, but fail to explain the slowly rotating stars. This is clearly seen from Figure 1 were, the specific angular momentum of T Tauri Alpha Persei and Pleiades is plotted as a function of mass. The specific angular momentum values computed in the case of rigid rotation are based on photometric rotational periods and radii from Van Leeuwen and Alphenaar 1982, Covino et al. 1989, Stauffer et al. (1987). Fast rotator of Alpha Persei and Pleiades clearly mixed with T Tauri while slow rotators fall one order of magnitude below. Several explanations are possible:
a) there is a significant age spread among late type stars in these clusters; b) progenitors of the slowly rotating, open cluster stars are slowly rotating inactive PMS

stars not yet identified because they lack the intense chromospheric and coronal emission; c) different braking laws hold for high and low rotation rates (Stauffer and Hartmann 1987); d) fast rotating cluster stars are still acquiring angular momentum from accretion disk (Schatzman 1989), i.e. their progenitors are the classical T Tau or "continuum T Tauri while slow rotator progenitors could be the Naked T Tau stars.

4 The Steady Phase of Rotational Evolution . (The main sequence evolution)

At the end of the fast braking phase, a solar type star already on the main sequence rotates at about ten times the present solar velocity, i.e. $v \sin i$ about 20 Km/sec. The following evolution during the main sequence is characterized by a steady slowing-down, that Skumanich (1972) first posit as proportional to $t^{-1/2}$. More recent $v \sin i$ data on 1 $M_\odot$ by Smith (1978) Soderblom (1983), Baliunas et al. (1983), Benz Mayor and Mermilliod (1984) are consistent with Skumanich's power law relationship between rotation and age. The observed dependence of the Ca II chromospheric emission and transition region emission from the rotation period, the age and the mass, suggest that the rotation period of the main sequence stars later than F7 should be an unique function of the mass and age (Catalano and Marilli 1983, Simon et al. 1985). Moreover Duncan et al. (1983) shoved that the rotation periods of Hyades main sequence stars with B-V between 0.4 and 0.85 follow a nearly linear relationship as a function of the B-V (i.e., the mass). The relation appears to be well defined, with a scatter in P_{rot} of roughly 1-2 days in either directions. A more quantitative analysis of the rotation along the main sequence as function of the mass and age using observed rotation periods does confirm the unique dependence of rotation from mass and age (Trigilio, Catalano and Marilli 1986, Marilli, Catalano and Trigilio 1987, Catalano, Marilli and Trigilio 1988), as shown in Figure 3. Their results can be summarized as it follows:
- Stars of the same age lay along linear sequences (isochrones)
- the isochrones seem to converge towards a mass $M/M_\odot \sim 1.34$, corresponding to a spectral type F3 (the limit for thick convection zones?)
- the decay of rotation with the $t^{-1/2}$ holds for low-mass stars in the range of 1.1 to 0.5 solar masses (see also Rengarajan 1984);
- The rate of spin-down (i.e. the increase of rotation period) is larger for low mass stars;
- the rotation period for a given mass is tightly determined by its mass and age, expressed by the following relation:

$$P_{rot} = (1.32\ 10^{-3} - 0.92\ 10^{-3}\ M/M_\odot)\sqrt{t} + C(M) \qquad (1)$$

These results pose the important question: why lower mass stars spin-down faster?

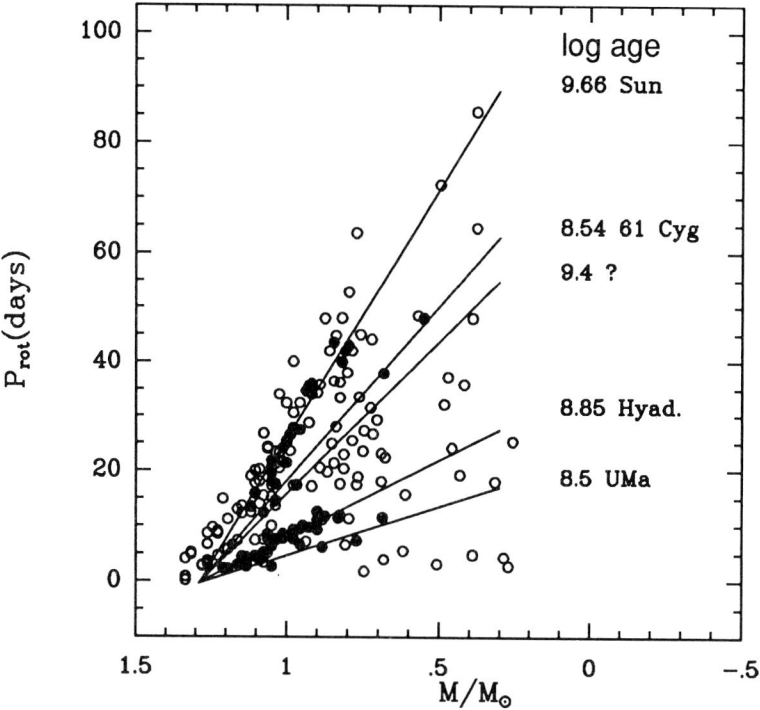

Figure 3. Rotation periods, observed and computed from Ca II emission, versus stellar mass. Filled symbols refer to stars of known age. Straight lines are fit to stars of equal age (adapted from Catalano et al. 1988).

This behavior is in the opposit sense to what appens during the approach to the main sequence. Since the total momentum of inertia along the main sequence decrease nearly as the square of the mass it seems unlikely that the whole star is spinning-down because this dependence is much faster than the dependence given by relation (1). Conversely the momentum of inertia of the convective envelope increases by about a factor of two as the mass decrease from 1 to 0.5 $M_\odot$. This trend is in the opposite direction with respect to the observed spin-down rate. Two possible explanation can be envisaged:

a) There is angular momentum transport from the radiative core to the convective envelope.

Higher mass stars have a larger radiative core and hence larger reservoir of the angular momentum left at the end of the fast spin-down, that can be tranferred to the convective evelope. The time scale for the angular momentum transport would be long enough to consider a partial coupling between the core and the envelope. An stimate for the time scale of the angular momentum transport of the persent Sun, comparing the the braking time scale due to the present solar wind and the time scale according to the $t^{-1/2}$ observed dependence, leads to $\tau \simeq 7 10^9$ years (Catalano, Marilli, Trigilio 1988). This time scale is not inconsistent with the internal rotation of the Sun deduced from helioseismology.

b). The magnetic wind braking is stronger for low mass stars.

Magnetic field mesurements of late type stars seem to support this possibility. Saar and Linsky (1986) found that the magnetic field intensity for main sequence stars is equal to the equipartition field (i. e. $B = B_{eq} = (8\pi P_{gas})^{0.5}$. This implies that the field strength increases as the temperature (and hence the mass) decrease along the main sequence. The total magnetic flux $B_{tot} = fB_{eq}$ (f the filling factor) has a much more complicated dependence, because the filling factor has been found to be proportional to the angular rotation Ω. A simple explanation of faster spin-down for lower mass stars due to the stronger magnetic field is not straightforward because the rate of angular momentum loss is highly dependent on the field geometry, which determine the lever-harm of the stellar wind. The fact that the $t^{-1/2}$ holds for lower as well as higher mass stars in the interval 0.5 to 1 $M_\odot$, does suggest a similar field geometry. The $t^{-1/2}$ relation follows from a radial field structure and B$\propto \Omega$ (Weber and Davis 1967, Mestel 1984).

The $t^{-1/2}$ relation does not seem to hold at the early phases of main sequence evolution. As shown by Benz, Mayor and Mermilliod (1984) between the Pleiades and the Hyades age the velocity of late F and early G dwarfs declines slower than predicted by the $t^{-1/2}$ relation (see also Trigilio Catalano and Marilli 1986). Since higher rotation rate would produce stronger magnetic field and therefore higher angular momentum loss Belcher and MacGregor (1976) argue that for rapid rotators the Alfven radius is smaller than for slow rotators, and this would reduce the efficeincy of the angular mometum loss. This could explain the observed slow braking of young F and G stars. However, the average rotation period of F-G Pleiades stars are longer (4-5 days) than the 2.5 days below which they found the angular momentum loss is less efficient. Gray (1982b) suggests that the character of the dynamo changes at a certain rotation velocity below which the magnetic field strength become very small and consequently the angular momentum loss rate is small. While this suggestion can explain the change from the fast spin-down between the Alpha Persei and the Pleiades age and the slower spin down after the Pleiades age, it is not consistent with the faster spin-down given by the $t^{-1/2}$ relation after the Hyades age when the stars rotate even slower. The more plausible explanation for the slow rotational decay of F-G dwarfs in young clusters is that there is a significant angular momentum transfer from the core to the envelope at this phase (Trigilio Catalano and Marilli 1986), that counterbalances a large portion of the wind braking. This implies that a large portion of the stellar interior is slowed-down during the early stage of main sequence evolution. Large angular momentum transport are possible due to hydrodinamical instabilities when the simultaneous spin-down of the convective envelope and spin-up of the radiative core produce large gradient in angular velocity at the base of convection zone (Sofia et al. 1988). However the problem of slower than $t^{-1/2}$ decay of rotation becomes less stringent if the new age of 1.5 10^8 years for the Pleiades (Mazzei and Pigatto 1989) is accepted.

5 Rotational future of the Sun.

There are no data to follow the rotational evolution of solar mass stars later than the solar age. Detecting rotation in oldest solar-type stars is very difficult; $v \sin i$ values are very small (< 1 Km/sec), chromospheric emission is generally too weak to produce measurable photometric variability and very uncertain to lead to confidently acceptable rotation period from emission rotation correlation (Marilli, Catalano Trigilio 1986). Even if, the angular momentum loss decreases or possibly stops because the dynamo process probably will switch off due to the slow rotation, the rotation period is expected to increase as the star leave the main sequence, due to the increase in radius. The apparent absence of any angular momentum reservoir in its core that can be drained when the convection zone deepen enforces the prediction for a very slow rotation *end* of solar and lower mass stars. This expectation appears to be fulfilled, even if some problems have been rised. Chromospheric emission of main sequence stars belonging to old moving groups (even older than the Sun) are consistent with yet slower than solar rotation, if the rotation activity relation is still valid (Catalano, Marilli, Trigilio 1989).

A possible contraddiction has been claimed to be provided by Arcturus (Stauffer 1989). The measured $v \sin i$ of about 2.5 km/sec lead to a rotation rate roughly of 50 Km/sec when it left the main sequence, if it is of one solar mass. However, recent analyses definitely settle the mass of Arcturus to be 2 $M_\odot$.

Peterson (1983) has shown that a large fraction of horizontal branch stars in some globular cluster have rotational velocities in the range 15-30 Km/sec. Since the progenitors of these one solar mass stars should have left the main sequence with surface rotational velocity of few Km/sec like the Sun, the high rotational velocity could be an indication of angular momentum drained from the core.

6 Conclusions

We can summarize the major phases of evolution of the angular momentum of a solar-type stars as a pre-main sequence high angular momentum loss, a phase of extraordinarily rapid rotation just as they arrive on the ZAMS, with a short-lived phase of angular momentum loss, and a long slow decline of rotation during and may be after the main sequence. Therefore, most of the life of a star like the Sun is spent rotating very slowly. This allows the star to adjust its internal equilibrium and redistribute the angular momentum. However, the fact that low mass stars spin-up due to contraction, in such rapid rates and then spin-down in an extremely short time (2-5 10^7 years) requires an high rate of angular momentum loss and presumibly considerable mass loss. Observable phenomena should be expected and studied to understand better the details of angular momentum loss in young stars. At present, only three young clusters have been observed well. This is not sufficient to sample the evolutionary time scale for different masses, and to probe the angular momentum loss process. The behavior of mass loss in the spectrum of HD36705, a very young field rapidly rotating star could be a good example of this process (Collier-Cameron 1989). On the theoretical side more sophisticated models of angular momentum loss are needed, especially models that incorporate treatment of coupling between the core and the convective envelope. Recent model of Yale group (Sofia et al. 1988) qualitatively agree with the average behavior of the observed rotation evolution but still need more refinement for a quantitative agreement.

Finally, in the study of rotational evolution of solar-type stars, the most basic question of all is: is the Sun typical for its mass and age? does the presence of a planetary make it different from other stars? can we look at these differences, if any, to dicovery other planetary systems? Unfortunately, the present answer is, no. There is no evidence that the Sun differs in any significant way in its rotation from old stars of similar mass and age (Soderblom 1983, Trigilio, Catalano,Marilli 1986, Catalano, Marilli, Trigilio 1987). This conclusion based mainly on rotation periods does confirm previous results on the H and K chromospheric emission: again the Sun looks like other old stars of similar age(Catalano and Marilli, 1983, Soderblom 1985). The thight dependence of the rotation period from the mass and age in Figure 3 does suggest that also stars of mass different from the Sun kip their own rotation wether or not they might have a planetary system. For the moment there is no way to discovery planetary systems from the rotational behavior of main sequence low mass stars. On the other hand these results show that rotation on the main sequence it is not a free parameter, and that stars adjust to an *equilibrium* rotation period probably throughout a process that Gray(1987) define a *rotostat* mechanism.

The pre-main sequece and main sequence values of rotation strongly suggest that only a defined amount of angular momentum can be retained by a star of a given mass and that the excess of initial angular momentum has to be left in a planetary system, like the solar one or in a proto-planetary disk as observed in β Pictoris. However, more detailed observations are needed on the rotation of very young stars (FU Ori, T Tau), and on the properties and evolution of disks to help identify the mechanism that can produce planetary systems.

Acknowledgements: this work has been supported by the Minister dell'Università della Ricerca Scientifica e Tecnologica through the University of Catania, The Catania Astrophysical Observatory and the CNR (Gruppo Nazionale di Astronomia) under contract No. 88.00349.02. The use of the Catania Astronet computing facilities it is also acknowledged. I would like to thank miss Cinzia Spampinato for kindly typing the manuscript, and the IAC for a partial support of the travel expenses.

7 References

- Abt, H.A.: 1961, *Astrophys. J. Suppl.*, **6**, 37.
- Abt, H.A.: 1965, *Astrophys. J. Suppl.*, **11**, 429.
- Abt, H.A., Hunter, J.H.,Jr.: 1962, *Astrophys. J.*, **136**, 381.
- Adams F., Lada C.J. and Shu F.: 1987, *Astrophys. J.*, **312**, 788.
- Baliunas, S.L., Vaughan, A.H., Hartmann, L., Middelkoop, F., Mihalas, D., Noyes,R.W., Preston,G.W., Frazier, J., Lanning, H.: 1983, *Astrophys. J.*, **275**, 752.
- Belcher, J.W., MacGregor, K.B.: 1976, *Astrophys.J.*, **210**, 498.
- Benz W., Mayor M. and Mermilliod J.: 1984, *Astron. Astrophys.*, **138**, 93.
- Bertout C.: 1988, in *NATO-ASI Formation and Evolution of Low Mass Stars*, A.K.Dupree and M.T.V. Lago eds., Kluwer Academic Publisher, p. **45**.
- Bertout C., Basri G., Bouvier, J.: 1988, *Astrophys. J.*, **330**, 350.
- Blanco, C.,Catalano, S., Marilli.: 1979, *Astron. Astrophys. Suppl. Series*, **36**' 297.
- Bodenheimer P.: 1978, *Astrophys.J.*, **224**, 488.
- Bopp, B.W., Evans,N.: 1973, *Mon. Not. R. Astr. Soc.*, **164**, 343.
- Bopp, B.W., Fekel, F.: 1977, *Astron. J.*, **82**, 490.
- Bouvier J., Bertout C., Benz W. and Major M.: 1986, *Astron. Astrophys.*, **165**, 110.
- Cassen P.M., Shu F.H., Tereby S.: 1985, *Protostar and Planets II* D.G. Black and M.S. Matthews eds., University of Arizona Press, p **448**.
- Catalano, S., Marilli, E.: 1983, *Astron. Astrophys.*, **121**, 190.
- Catalano, S., Marilli, E., Trigilio, C.: 1988, in *NATO-ASI Formation and Evolution of Low Mass Stars*, A.K.Dupree and M.T.V. Lago eds., Kluwer Academic Publisher, p. **377**.
- Catalano, S., Marilli, E., Trigilio, C.: 1989, (preprint).
- Collier-Cameron, A.: 1989, preprint.
- Covino E., Terranegra L., Vittone A.: 1989, *Mem. Soc. Astron. Ital.*, **60**, 111.
- Dicke, R.H.: 1964, *Nature*, **202**, 432.
- Dicke, R.H.: 1970, *Astrophys. J.*, **159**, 1.
- Duncan, D.K, Baliunas, S.L, Noyes, R.W., Vaughan, A.H.:1983, *Publ. Astron. Soc. Pacific*, **95**, 589.
- Endal A and Sofia S. 1981, *Astrophys. J.*, **210**, 184.
- Franchini, A. Magazzù, A., Stalio, R.:1988, *Astron. Astrophys.*, **189**, 132.
- Fucuda, I.: 1982, *Publ. Astron. Soc. Pacific*, **94**, 271.
- Goldsmith P.F. and Arquilla R.: 1985, in *Protostar and Planets II*, D.C. Black, M.S. Matthews eds., University Ariz. Press, Tucson p. **137**.
- Gray, D.F.: 1982a, *Mem. S.A.It.*, **53**, 931.
- Gray, D.M.: 1982b, *Astrophys. J.*, **261**, 259.

- Gray, D.M.: 1984, *Astrophys. J.*, **281**, 719.
- Gray, D.F.: 1987, *Lectures on Spectral-Line Analysis: F,G, and K Stars*, The Publisher, London Ontario.
- Harris A., Townes C.H., Matasakis D.N., Palmer P.: 1983, *Astrophys. J.*, **265**, L63.
- Hartmann L.W., Hewett R., Stahler S., and Mathieu R.D.: 1986, *Astrophys. J.*, **309**, 275.
- Hartman L., Kenyon S.: 1988 in *NATO-ASI Formation and Evolution of Low Mass Stars*, A.K.Dupree and M.T.V. Lago eds., Kluwer Academic Publisher, p. **163**.
- Hartmann L.W. and Noyes R.W.: 1987, *Ann. Rev. Astr. Astrophys.*, **25**, 271.
- Hartmann L., Soderblom D., Stauffer J.: 1987, *Astrophys. J.* **93**, 907.
- Hartmann L., Stauffer J.: 1989, *Astron. J.*, **97**, 873.
- Harvey, J.: 1988, in *Seimology of the Sun and Solar-Like Stars*, ESA SP-286, p. **55**.
- Herbig G.H. and Spalding J.: 1955, *Astrophys. J.*, **121**, 118.
- Huang S.S.: 1965, *Astrophys. J.*, **141**, 985.
- Huang S.S., 1969: *Vistas in Astronomy*, **11**, 217.
- Kawaler S.D.: 1987, *Publ. Astron. Soc. Pacific*, **99**, 1322.
- Kawaler S.D.: 1988, *Astrophys. J.*, **333**, 236.
- Kenyon S.J., Hartmann L., Hewett R.: 1988, *Astrophys. J.*, **325**, 231.
- Kraft R.P.: 1965, *Astrophys. J.*, **142**, 681.
- Kraft R.P.: 1967, *Astrophys. J.*, **150**, 551.
- Kraft R.P.: 1970, in *Spectroscopic Astrophysics*, ed G.H. Herbig,p. **385**.
- Krzeminski, W.:1969, in *Low Luminosity Stars*, S.S. Kumar ed., Gordon and Breach, London, P. **57**.
- Larson R.B.: 1983, *Rev. Mexicana Astron. and Astrophys.*,**7**, 219.
- McNally, D.: 1965, *The Observatory*, **85**, 166.
- Marilli, E., Catalano, S., Trigilio,C.: 1986, *Astron. Astrophys.*, **167**, 297.
- Marilli, E., Catalano, S., Trigilio,C.: 1987, in *Cool Stars, Stellar Systems, and the Sun*, M. Zeilik and D.M. Gibson eds, Springer-Verlag, p.**181**.
- Mestel L.: 1965, *J.R. Astr. Soc.*, **6**, 265.
- Mestel L.: 1984 in *Cool Stars, Stellar Systems, and the Sun*, S.L. Baliunas, L. Hartmann eds, Springer-Verlag, p.**49**.
- Mouschovias T. Ch.: 1976, *Astrophys. J.*, **207**, 141.
- Mouschovias T.Ch.: 1981, in *Fundamental Problems in the Theory of Stellar Evolution, IAU Symp. N. 93* D. Sugimoto, D.Q. Lamb, D.N. Schramm eds., Reidel, Dordrecht, Holland, p.**27**.
- Noyes R.W., Hartmann S.L., Duncan D.K. and Vaughan A.N.: 1984, *Astrophys. J.*, **279**, 763.
- Oskanyan, V.S.,Evans, D.S., Lacey, C., MacMillan,R.S.: 1977, *Astrophys. J.*, **321**, 459.
- Peterson
- Radick R., Thompson D., Lockwood G., Duncan D. and Baggett W.: 1987, *Astro-*

phys. J., **321**, 459.
- Rucinski S.M.: 1988, *Astrophys. J.*, **95**, 1895.
- Rydgren A., Vrba F., Chugainov P. and Shakhovskaya N. 1985, *B.A.A.S.*, **17**, 556
. - Saar, S.H., Linsky, J.L.: 1986, *Adv. Space Res.*, **6**, No. 8, 235.
- Schatzman E., 1962, *Ann. d'Astrophys.*, **25**, 18.
- Schatzman E.: 1989, XI ERAM in press.
- Shu F.H., Lizano S., Adams F.G., Ruden S.P.: 1988 in *NATO-ASI Formation and Evolution of Low Mass Stars*, A.K.Dupree and M.T.V. Lago eds., Kluwer Academic Publisher, p. **123**.
- Simon, T., Herbig, G.H., Boesgaard, A.M.: 1985, *Astrophys.J.*, **293**, 551.
- Skumanich A.: 1972, *Astrophys.J.*, **171**, 265.
- Slettebak,A.: 1949, *Astrophys. J.*, **110**, 498.
- Slettebak,A.: 1954, *Astrophys. J.*, **119**, 146.
- Slettebak,A.: 1955, *Astrophys. J.*, **121**, 653.
- Smith, M.A.:1978, *Astrophys. J.*, **224**, 584.
- Smith, M.A.: 1979, *Publ. Astron. Soc. Pacific*, **91**, 737.
- Smith, M.A.: 1988, in *Cool Stars, Stellar Systems, and the Sun*, J.L. Linsky, R.E. Stencel eds, Springer-Verlag, p.**192**.
- Smith, M.A. and Gray,D.F.: 1976, *Publ. Astron. Soc. Pacific*, **88**, 809.
- Soderblom, D.R.: 1983, *Astrophys.J.Suppl.*, **53**, 1.
- Soderblom, D.R.: 1985, *Astron.J.*, **90**, 2103.
- Sofia, S., Pinsonneault, M., Kawaler, S.D., Demarque, P.: 1988, in *Cool Stars, Stellar Systems, and the Sun*, J.L. Linsky, R.E. Stencel eds, Springer-Verlag, p.**192**.
- Stauffer, J.R.: 1988,in *Cool Stars, Stellar Systems, and the Sun*, J.L. Linsky, R.E. Stencel eds, Springer-Verlag, p.**182**.
- Stauffer, J.R., Hartmann, L., Burnham, N., Jones, B.: 1985, *Astrophys. J.*, **289**, 247.
- Stauffer, J. and Hartmann, L.W.: 1986, *Publ. Astron. Soc. Pacific*, **96**, 1233.
- Stauffer, J. and Hartmann, L.W.: 1987, *Astrophys. J.*, **318**, 337.
- Stauffer, J. and Hartmann, L.W.: 1989, *Astron. J.*, **97**, 873.
- Stauffer, J. and Hartmann, L.W., Latham, D.W.: 1987, *Astrophys. J. Lett.*, **320**, L51.
- Stauffer, J., Hartmann, L.W., Soderblom, D. and Burnham, J.N.: 1984, *Astrophys. J.*, **280**, 202.
- Stauffer, J.R., Schild, R.A., Baliunas, S.L., Africano, J.L.: 1987 *Publ. Astron. Soc. Pacific*, **99**, 471.
- Stauffer, J.R., Soderblom, D.R.: 1989, in, *The Sun in Time*, in press.
- Strom S.E., Strom K.M., Edwards S.: 1988, in *NATO ASI: Galactic and Extragalactic Star Formation*, R. Pudritz and M. Fich eds., Kluwer Academic Publisher p...
- Strom S.E., Strom K.M., Edwards S., Cabrit S., Skrutskie M.: 1989, preprint.
- Struve,O.: 1931, *Astrophys. J.*, **73**,94.
- Torres, C.A.O., Ferraz-Mello, S.:1973, *Astron. Astrophys.*, **27**, 231.

- Trigilio, C., Catalano, S., Marilli, E.: 1986, *Advances Space Res.*, **6**, 207.
- Van Leeuwen, F. and Alphenaar, P. 1982. *ESO Messenger*, No. **28**, p. 15.
- Vaughan A.H., Baliunas S.L., Middlekoop F., Hartmann L.W., Mihalas D., Noyes R.W. and Preston C.W: 1981, *Astrophys. J.*, **250**, 276.
- Vogel, S.N. and Kuhi, L.V.: 1981, *Astrophys. J.*, **245**, 960.
- Vogt, S.S.: 1975, *Astrophys. J.*, **199**, 418.
- Wadiak E.J., Wilson T.L., Rood R.T., Johnston K.J.: 1985, *Astrophys. J.* **295**, L43.
- Walter F.M. 1986, *Astrophys. J.*, **306**, 573.
- Weber, E.J., Davis, L.Jr. : 1967, *Astrophys.J.*, **148**, 217.
- Wilson O.C.: 1966, *Astrophys.J.*, **144**, 695.
- Willson, O.C.: 1978, *Astrophys. J.*, **226**, 379.
- Wolff, S.C., Edwards, S., Preston, G.W.: 1982 *Astrophys. J.*, **252**, 322.
- Zinnecker H.: 1984, *Mon. Not. R. Astr. Soc.*, **210**, 43.

Spectroscopy of Solar-Type Stars

P. E. NISSEN

Institute of Astronomy
University of Aarhus
DK-8000 Aarhus C., Denmark.

SUMMARY. Recent progress in spectroscopy of solar-type stars is reviewed with particular attention paid to the determination of stellar atmospheric parameters and the abundances of selected elements. Results obtained during the last few years are of great interest in connection with studies of nucleosynthesis of the elements and the evolution of the Galaxy. In some cases, however, errors in the analysis of the observed spectra make it difficult to draw any definitive conclusions.

1 INTRODUCTION

Solar-type stars, i.e. stars with about the same mass and luminosity as the Sun, play a very important role in astrophysics. They span a large range in age, some being as old as the Galaxy and some have newly been formed. Furthermore, due to the existence of an outer convection zone it can be assumed that the abundance ratios in the atmosphere of a solar-type star represent the original chemical composition of the interstellar gas out of which the star was formed. The close agreement between meteoritic and solar atmospheric abundances (Anders and Grevesse 1989) support this assumption. Therefore, by observing the chemical composition of solar-type stars, the nucleosynthesis of the elements and the chemical evolution of the Galaxy can be studied. If kinematical data such as radial velocity and proper motion are available, information about the dynamical evolution of the Galaxy can also be obtained. The ages and chemical compositions needed for such studies are based on spectroscopic determinations of stellar atmospheric parameters and abundance ratios. Thus, spectroscopy of solar-type stars is the key to a proper understanding of some of the most interesting problems in modern astrophysics.

During the last decade stellar spectroscopy has progressed significantly, the main reason being the construction of efficient high-resolution spectrographs and the application of highly efficient, linear RETICON and CCD detectors. Observations with S/N better than 200 and spectral resolutions in excess of 50.000 are now being carried out routinely for large samples of rather faint stars. This allows equivalent widths of absorption lines to be measured with errors less than a few mÅ. The data can be used to determine the basic atmospheric parameters, effective temperature T_{eff}, surface gravity g, and the abundances of nearly all elements. The major contribution to the uncertainties of these quantities are not any longer coming from errors in the observed spectra, as it was when photographic spectra were used, but from uncertainties in the model atmosphere analysis of the data caused by the lack of precise oscillator strengths, deviations from local thermodynamic equilibrium and inadequate models of stellar atmospheres.

In the case of solar-type stars uncertainties arising from the analysis of the observed spectra are generally smaller than for other groups of stars, due to the fact that it is possible to perform a *differential* analysis with respect to the comparatively well known solar atmosphere. It is a lucky circumstance that the Sun is of similar structure as the group of F and G main-sequence stars containing so much information about the evolution of the Galaxy.

In the following some highlights of recent years spectroscopy of solar-type stars are described and the reliability of the results obtained is discussed.

2 T_{eff} AND LOG G OF SOLAR-TYPE STARS

Accurate values of the fundamental atmospheric parameters, T_{eff} and surface gravity g, are needed to derive abundance ratios from stellar spectra and in order to determine ages and masses of stars. Thus the determination of these parameters is very essential for studies of galactic evolution.

In general T_{eff} is determined from a photometric colour index like $B-V$, $V-K$ or the Strömgren indices $b-y$ and β. Empirical calibrations of these indices have been carried out in recent years by the aid of the infrared flux method (Blackwell and Shallis 1977). Thus Saxner and Hammarbäck (1985) have calibrated $B-V$, $b-y$ and β for main-sequence F and early G stars ranging in metallicity from [Fe/H] = -0.8 to 0.2. Magain (1987b) has extended the calibrations of $B-V$ and $b-y$ to stars with lower metal abundances, and Arribas and Martinez Roger (1989) have calibrated $B-V$ and $V-K$ for somewhat cooler stars. The standard deviation in these calibrations is typically $\sigma(T_{eff})=50\,\mathrm{K}$ showing that differential values of T_{eff} can be determined for solar-type stars to that accuracy. However, the absolute values of T_{eff} may be subject to larger errors. A check of the zero-point in the calibrations would be possible if the colour indices of the Sun could be accurately measured but unfortunately this is very difficult. As discussed by Saxner and Hammarbäck (1985) various direct and indirect determinations of $(B-V)_\odot$ have resulted in values ranging from 0.62 to 0.68, whereas the calibration of Saxner and Hammarbäck corresponds to a value of $(B-V)_\odot=0.642$. If, par example, the true value is $(B-V)_\odot=0.67$ then the calibration of Saxner and Hammarbäck would be systematically wrong by 100 K. It cannot be excluded that the systematic error of the T_{eff} calibration for metal poor stars is even larger, say up to 200 K. Such an error would have quite a significant effect on some of the abundance ratios derived from stellar spectra (see discussion in Sect. 3) and on ages of stars determined by comparing their position in the $\log T_{\text{eff}}$-$\log g$ diagram with theoretical isochrones. A potential problem is that ages of metal-poor stars derived by this method become very high, 20 Gyr or even more (Hartmann and Gehren 1988, Schuster and Nissen 1989) in conflict with current values of the Hubble constant and also higher than ages of globular clusters, 14-18 Gyr, recently determined from the luminosity of the turn-off (see discussion by Schuster and Nissen 1989). The problem may be caused by a serious systematic error in the effective temperature calibrations for metal poor stars, but it could also be due to errors in the stellar models used to compute the isochrones, such as errors in the opacities or the mixing length parameter (see discussion by Hartmann and Gehren 1988).

Clearly it is very important to get some independent checks of the determination of T_{eff} for solar-type stars. Accurate methods do exist. Cayrel et al. (1985) and Perrin et al. (1988) have shown that high S/N observations of the profile of the H_α line when compared differentially to the solar H_α profile make it possible to determine T_{eff} with an accuracy of ± 35 K. As discussed by Cayrel de Strobel at this conference such spectroscopic determinations of T_{eff} for solar twins support the zero point of the calibration of Saxner and Hammarbäck (1985). Similar work for metal poor stars is much needed.

The surface gravity of solar-type stars may be determined from the Balmer discontinuity at 3650 Å as measured by e.g. the Strömgren c_1 index. According to the empirical calibration of c_1 by Nissen and Gustafsson (1978) differential values of $\log g$ can be determined to an accuracy better than ± 0.10 dex. However, the absolute values of g may be subject to larger errors, say ± 0.20 dex, especially for the metal deficient stars. Furthermore, it cannot be excluded that the c_1 index is affected by variations in other parameters than the three classical atmospheric parameters, T_{eff}, g and [Fe/H], like rotation velocity or strength of magnetic fields (Nissen 1988).

Often the surface gravities of solar-type stars is determined by an LTE model atmosphere analysis of the ratio between the equivalent widths of ionized and neutral lines of the same element. The correctness of the LTE-assumption has only recently been checked by non-LTE computations for the case of iron (see later) and it is clear that independent methods of determining the gravity parameter are very much needed.

An interesting spectroscopic method of determining the surface gravities of solar-type stars is based on the pressure broadening of metal lines. The profile of a strong metal line is measured together with the equivalent widths of a number of weak lines belonging to the same element as the strong line and having about the same lower excitation potential. The relative oscillator strengths of the lines involved should be known with high accuracy. From the weak lines the abundance of the element in question is determined both for the Sun and for the stars. From the wings of the strong line in the solar spectrum the van der Waals damping constant C_6 is determined, and from the wings observed in the stellar spectra the gravity can then be derived. The method was first applied by Blackwell and Willis (1977) for Arcturus. More recently it has been used for $\alpha\,Cen\,A$ and B by Smith et al. (1986), who have reached an accuracy as good as ± 0.05 dex in the $\log g$ determination. Smith and Drake (1987) have applied the method to two G dwarfs and Edvardsson (1988) to some G subgiants. In both cases accuracies of about ± 0.10 in $\log g$ have been obtained. Clearly the method should be extended to more solar-type stars in particular to the metal-deficient ones.

3 THE CHEMICAL COMPOSITION OF SOLAR-TYPE STARS

Many interesting results concerning abundance ratios of elements in solar-type stars have been obtained in recent years. Rather than giving a general review a few element ratios of particular interest in connection with galactic evolutionary models and Big Bang nucleosynthesis theories are discussed.

3.1 Oxygen

Oxygen is the most abundant element in the Universe after hydrogen and helium. It gives a major contribution to the opacity in stellar interiors and therefore stellar ages determined from comparing computed isochrones with observed parameters depend rather critical on the oxygen abundances assumed for the stars (Hesser et al. 1987). Furthermore, the behaviour of the O/Fe abundance ratio as a function of the Fe/H ratio gives information about such interesting problems as the relative production rate of oxygen and iron by supernovae of type I and II, the homogeneity of the proto-Galaxy and the time scale for the formation of the galactic halo (Wyse and Gilmore 1988).

Our knowledge about the O/Fe ratio in solar-type stars and K giants has recently been summarized by Wheeler, Sneden and Truran (1989). Although all spectroscopic investigations lead to greater O/Fe ratios in metal-poor stars than in the Sun the size of this oxygen overabundance is not well determined. Also the functional dependence of [O/Fe] on [Fe/H] is rather uncertain and it remains and open question if there is a significant scatter in [O/Fe] at a given value of [Fe/H].

An extensive study of the chemical composition of solar-type stars is presently being carried out on the basis of high resolution, high S/N spectra of about 200 disk stars observed at the ESO and McDonald observatories with their respective Coudé echelle spectrographs. Preliminary results from this survey have been discussed by the collaborators in this project: Nissen, Edvardsson and Gustafsson (1985), Andersen et al. (1988), Tomkin et al. (1989) and Lambert (1989). All spectra observed have been reduced and equivalent widths of up to 60 absorption lines, corresponding to O, Na, Mg, Al, Si, Ca, Ti, Cr, Fe, Ni, Y, Ba and Nd, have been measured in each star. Typical spectra are shown in Fig. 1 for spectral regions containing two faint oxygen lines at 6158 Å and the oxygen triplet at 7773 Å.

The 200 stars were selected by E. H. Olsen from his large catalogues of Strömgren $uvby$-β photometry for all F0-G2 stars brighter than $V = 8\overset{m}{.}3$. By means of the metallicity index m_1 these stars were divided into 9 metal abundance groups with [Fe/H] ranging from -1.0 to 0.3. Furthermore, the Balmer-discontinuity index c_1 was used to select stars that have evolved between $0\overset{m}{.}4$ and $2\overset{m}{.}0$ away from the zero age main sequence, so that their ages can be estimated from comparison with isochrones. In each metal abundance group the 25 brightest stars were observed. Thus the stars selected are about evenly distributed in [Fe/H]. Binaries, fast rotators ($V \sin i > 25$ km/s) and variable stars were excluded. The stars turn out to be confined to the ranges $5600 < T_{\text{eff}} < 6700$ K, $3.8 < \log g < 4.5$ and are thus of about the same type as the Sun.

From the equivalent widths measured abundances have been derived using standard LTE model atmosphere techniques. A plane-parallel, flux-constant model including convection in the mixing-length approximation is computed for each star using an updated version of the programme described by Gustafsson et al. (1975). The parameters T_{eff}, $\log g$ and [Fe/H] are determined from the Strömgren indices b-y, β, c_1 and m_1. The microturbulence is assumed to be independent of depth in the atmosphere and a function of T_{eff} and $\log g$, according to the formulae derived by Nissen (1981). The results are, however, not sensitive to possible errors in the microturbulence values, because most of the lines used are on the linear part of the curve of growth. The analysis is differential with respect to the Sun in the sense that gf-values of the lines are derived from solar equivalent widths as measured in the spectrum of the sunlight ESO 3.6m dome.

Preliminary results for the [O/Fe] ratio as a function of [Fe/H] are shown in Fig.2. Only stars observed at ESO are included because the McDonald observations did not include the spectral region around the oxygen triplet at 7773Å. As seen [O/Fe] decreases linearly from a value of about 0.6 dex at [Fe/H]=-1.0 to about -0.2 dex at [Fe/H]=0.3. This trend agrees well with the earlier study of Clegg et al. (1981) based on a smaller sample of dwarf stars. More recently Barbuy (1988) and Barbuy and Erdelyi-Mendez (1989) have derived oxygen abundances of K giants and G subgiants from the forbidden oxygen line at 6300Å. They find the same tendency as in Fig. 2, but [O/Fe] raises to about 0.35 dex only at [Fe/H]=-1.0. Furthermore, [O/Fe]

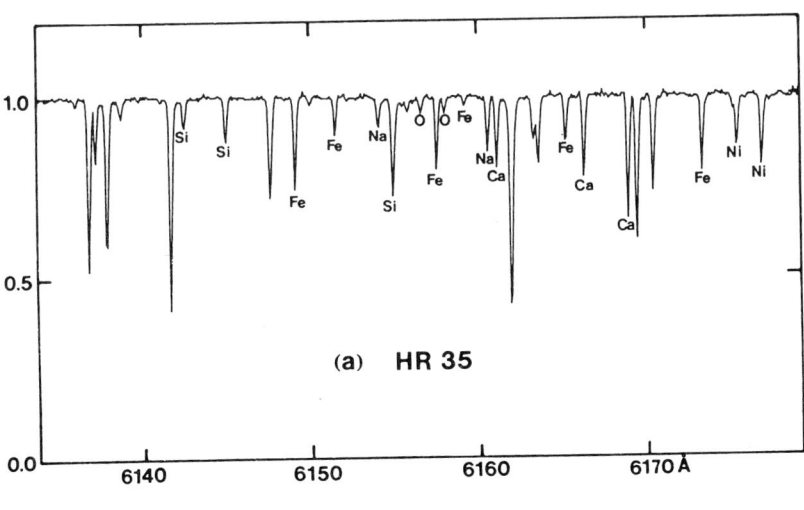

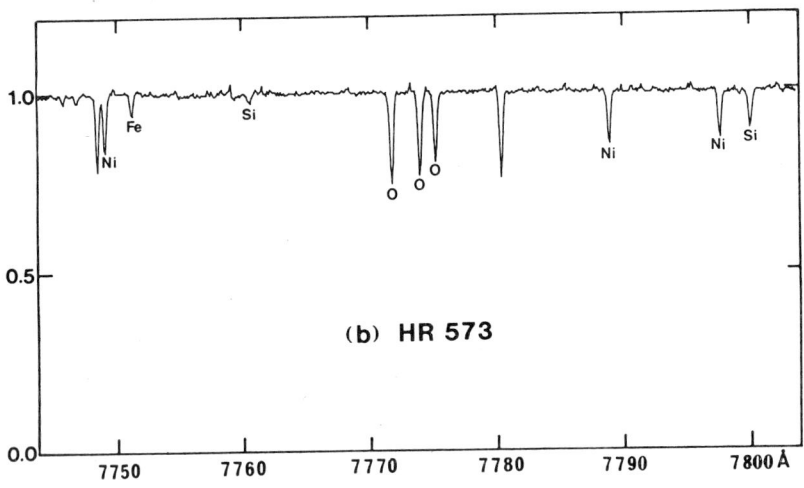

Fig.1. Typical spectra obtained with the ESO 1.4m CAT telescope and its Coudé Echelle Spectrograph with a Reticon as detector. Lines used in abundance determinations are marked with their corresponding elements. a) The 6155Å region for HR35 (V=5.25, 45m exp. time). Note the two weak oxygen lines at 6156.8 and 6158.2Å, having equivalent widths of 11 and 9mÅ, respectively. b) The region around the near infrared oxygen triplet at 7773Å for HR573 (V=6.10, 60m exp. time).

stays constant at this level when [Fe/H] decreases from -1.0 to -2.0. This is at variance with results from a study of 30 metal-poor solar-type stars by Abia and Rebolo (1989). They have derived oxygen abundances from spectra of the oxygen triplet at 7773Å observed with a resolution of 25.000 and a fairly high S/N ratio. The iron abundances were adopted from previous works. [O/Fe] is found to increase monotonically as [Fe/H] decreases below -1.0, reaching a value as high as 1.2 dex at [Fe/H]=-2.0.

In view of these rather discrepant results for the oxygen abundances in metal-poor stars it is of interest to have a closer look at the possible errors in the abundance determinations. Here we briefly discuss three major error sources: i) errors in the model parameters, ii) inadequate models, and iii) deviations from LTE.

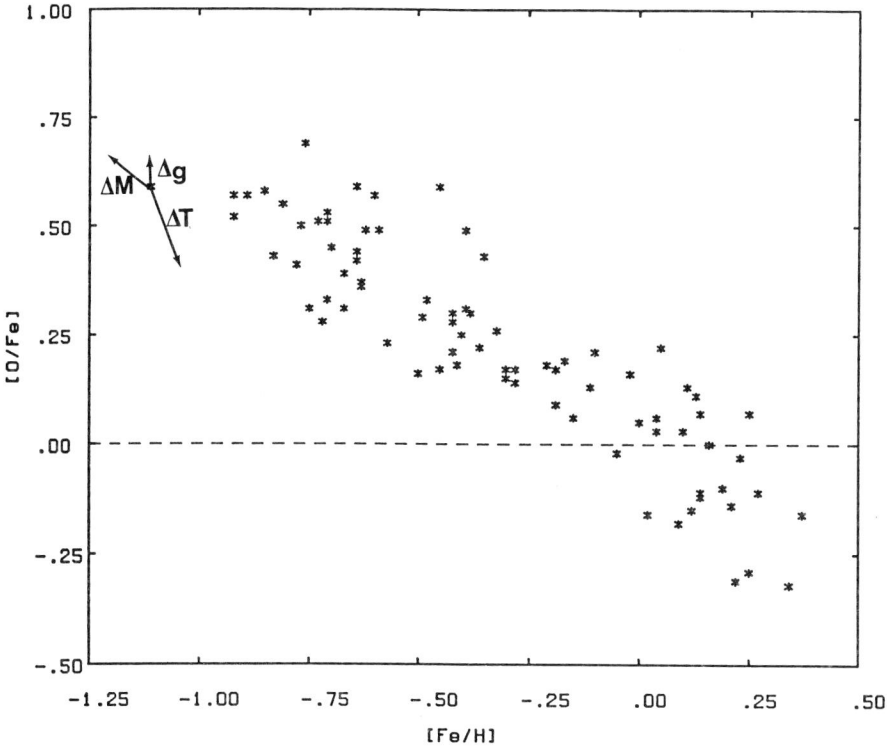

Fig.2. [O/Fe] versus [Fe/H] as derived for solar-type stars included in the large survey of chemical abundances in disk population stars described in the text. The sensitivity of the position of the most metal-poor star to errors in the analyses is indicated by arrows: ΔT, effect of increasing the effective temperature with 100K; Δg, effect of increasing $\log g$ with 0.2 dex; ΔM, effect of using the Holweger-Müller model for the solar atmosphere instead of using a theoretical model.

i) In Sect. 2 the maximum errors of T_{eff} and $\log g$ were estimated to be ±100K and ±0.2dex respectively. The corresponding shifts of a metal-poor star in the [O/Fe]-[Fe/H] diagram are shown in Fig. 2. It is seen that the [O/Fe] ratio derived is quite sensitive to errors in T_{eff}, which is due to the fact that oxygen is mainly in the neutral state and iron is in the ionized state in the atmospheres of solar-type stars. For K giants the uncertainty in the O/Fe ratios determined from the [OI] 6300Å line is even larger due to the possible errors in T_{eff} and $\log g$. For these cool stars one also has the problem of estimating the amount of oxygen bound in CO molecules.

ii) The use of theoretical, flux constant model atmospheres has often been criticized because they do not agree well with the observed flux distribution in the blue and ultraviolet part of the solar spectrum. The problem is probably due to the neglect of millions of faint absorption lines in computing the opacity. One may therefore expect that the use of flux-constant models introduces systematic errors in the derived abundance ratios as a function of [Fe/H]. In order to estimate the effect on abundances derived we have computed the gf-values of the lines using both a theoretical model and the Holweger-Müller model for the solar atmosphere (Holweger and Müller 1974). The corresponding shift of a metal-poor star in the [O/Fe] - [Fe/H] diagram is shown in Fig. 2. Although the shift is quite significant both in [Fe/H] and [O/Fe] it does not change the linear relation between the two quantities.

The existence of inhomogeneities, i. e. granulation structure created by convective motions, in the surface layers of solar-type stars constitutes a potential problem in connection with abundance determinations, which has been neglected in all abundance studies carried out so far. According to the recent hydrodynamical models by Nordlund and Dravins (1989) large temperature differences between hot, rising and cool, sinking elements do exist in the line forming regions. If these differences depend on the heavy element abundance in the atmospheres of solar-type stars then the use of homogeneous models may introduce systematic errors in [O/Fe] as a function of [Fe/H], because the O/Fe ratio is very sensitive to temperature changes.

iii) The determination of chemical abundances in solar-type stars is normally based on the assumption of local thermodynamical equilibrium, LTE. It has been suggested that the hot ultraviolet radiation from deep layers tends to over-ionize iron with respect to the Saha equilibrium and that the effect increases with decreasing [Fe/H], because of the decrease in line-blocking in the UV (Gustafsson 1987). The assumption of LTE then leads to an underestimate of the Fe abundance if neutral lines are used, especially in metal-poor stars, and spurious trends in O/Fe could be produced. According to detailed non-LTE computations by the Kieler group (Holweger 1988) the effect is very small for lines in the solar spectrum, typically 0.05 dex for low excitation neutral iron lines and 0.03 dex for high excitation lines, $\chi > 3.5$ eV. Recent

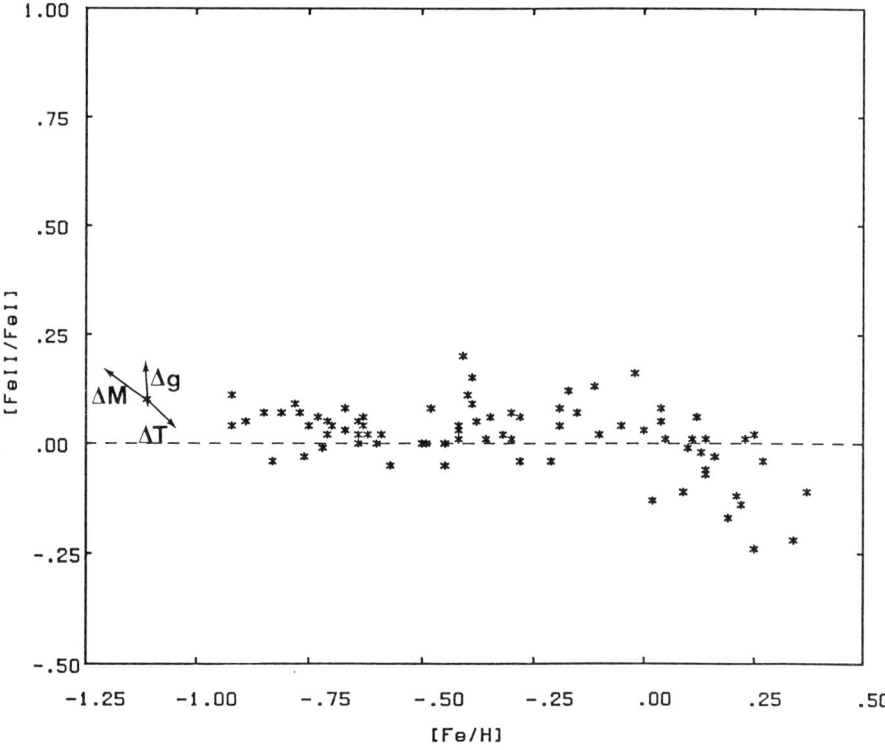

Fig. 3. [FeII/FeI] versus [Fe/H] for solar-type stars. The sensitivity of the position of the most metal-poor star to errors in the analysis is indicated by arrows in the same way as in Fig. 2.

computations by Lembke, Gustafsson and Holweger confirm that the degree of over-ionization of iron is larger in metal-poor, solar-type stars, but it is not more than 0.05 dex at [Fe/H]=-1.0 and 0.10 dex at [Fe/H]=-2.0 for the high excitation neutral lines used for abundance determinations. This agrees very well with Fig. 3, that shows the ratio of iron abundances derived from FeII and FeI lines as a function of [Fe/H] for the sample of stars plotted in Fig. 2. The scatter in [FeII/FeI] is very small and the average value is only slightly higher than 0.0 for [Fe/H]<0.0 in agreement with the non-LTE computations. As also seen this conclusion is not sensitive to the possible errors in T_{eff}, log g and the models. On the other hand there may be non-LTE problems for stars more metal rich than the Sun. [FeII/FeI] tends to decrease rather rapidly as a function of [Fe/H] for [Fe/H]>0.0, a curious effect that has not been explained yet.

On the basis of these results it seems safe to conclude that non-LTE effects in iron cannot be the cause of the apparent O/Fe overabundance in metal-poor stars. Non-LTE effects in oxygen itself is a greater problem. Thus Magain (1988) found the oxygen abundance of HD 76932 (T_{eff}=5815 K,

$\log g = 4.0$ and [Fe/H] = -1.0) derived from the triplet at 7773Å to be 0.6 dex higher than the oxygen abundance derived from the [OI] line at 6300Å. Barbuy and Erdelyi-Mendez (1989) found a corresponding difference of 0.15 dex for the somewhat cooler, metal-poor subgiant HD 10700. Although the forbidden line is very weak in these stars, W < 5 mÅ, these results seem quite significant and suggest that the overabundance of oxygen in metal-poor stars as derived from the near infrared triplet may be too large. The problem has been investigated in some detail by Abia and Rebolo (1989), who adopted a model of the oxygen atom consisting of the ground level, the continuum and seven excited states and solved the coupled statistical equilibrium and radiative transfer equations. The non-LTE effects on the abundances derived from the triplet at 7773Å were estimated to be less than 0.15 dex both in the Sun and in metal-poor, solar-type stars in disagreement with the large effect found by Magain (1988). Thus the situation remains unclear and new work, both observational and theoretical, needs to be done before the questions raised in the beginning of this section concerning trends and scatter in [O/Fe] can be answered safely.

3.2 The Light Metals.

The abundances of elements belonging to the group of light metals, Na, Mg, Al, Si, S, K, Ca, Sc and Ti, provide interesting information about nucleosynthesis in stars and the chemical evolution of the Galaxy. According to current theories the even-Z elements of this group are synthesized by α-capture processes and dispersed into the interstellar medium in connection with SNII explosions. The production of odd-Z elements, on the other hand, depends on the neutron excess, and thus on the initial heavy-element abundance at the site for the nucleosynthesis. The ratios of the abundances of these elements to that of iron as a function of [Fe/H] perform a critical test of these theories.

Our knowledge about the abundances of the light metals has recently been reviewed by Wheeler, Sneden, Truran (1989) and Lambert (1989). As an example of one of the interesting new results Fig. 4 shows the [Mg/Fe] data from the ESO survey described above. As seen the Mg/Fe ratio is solar for [Fe/H] > -0.5 with a remarkable small scatter in [Mg/Fe]. Below [Fe/H] = -0.5 magnesium becomes overabundant with respect to iron and [Mg/Fe] increases to a level of about 0.4 dex at [Fe/H] = -1.0. The work of Magain (1987a, 1989) shows that [Mg/Fe] is nearly constant at this level for stars with [Fe/H] in the range -1.0 to -2.5. The other α-elements Si, Ca and Ti show similar trends, whereas the odd-Z elements, Na and Al, have solar abundance ratios with respect to iron for all stars with [Fe/H] > -1.0.

As shown in Fig. 4 the Mg/Fe ratios derived are quite insensitive to possible errors in T_{eff}, g and the models. This is due to the fact that the two neutral

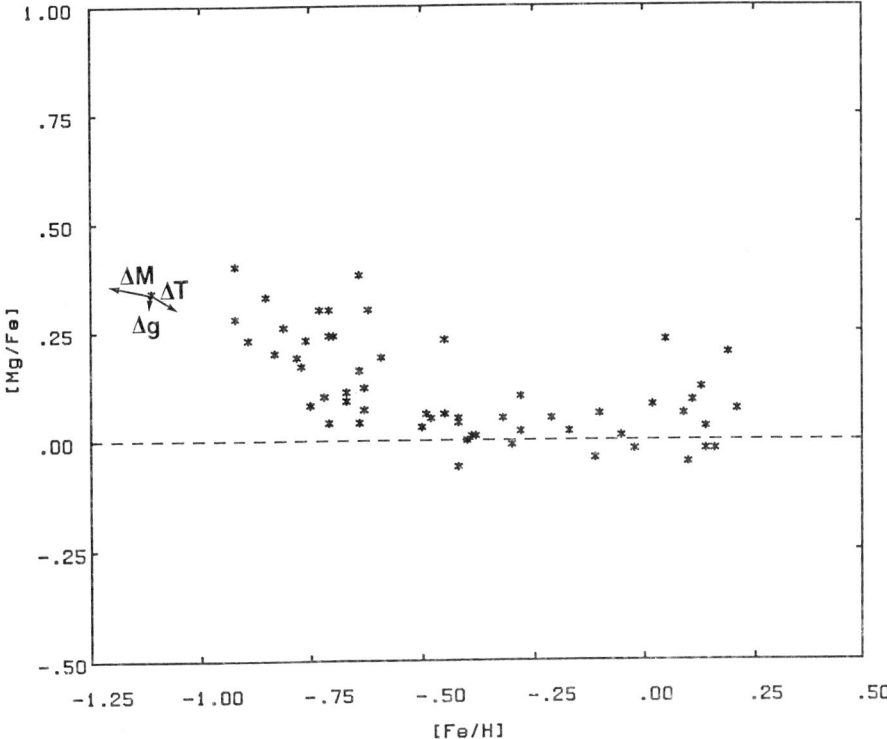

Fig.4. [Mg/Fe] versus [Fe/H] for solar-type stars. The sensitivity of the position of the most metal-poor star to errors in the analysis is indicated by arrows in the same way as in Fig. 2.

Mg lines at 8712.7 and 8717.8Å used for the abundance determinations have about the same strength and excitation potential as the iron lines used. Furthermore the ionization potentials of Mg and Fe are very similar. Errors in Mg/Fe are therefore much smaller than the corresponding errors in O/Fe, derived from the high excitation oxygen triplet at 7773Å. Possible non-LTE effects in Mg have not been studied in detail yet but are expected to be small like in the case of iron.

It is remarkable that the trend of [Mg/Fe] versus [Fe/H] is different from the trend of [O/Fe]. [Mg/Fe] reaches a plateau of 0.0 already at [Fe/H]=-0.5, whereas [O/Fe] continues to decrease even for [Fe/H]>0.0. According to current nucleosynthesis theories oxygen and magnesium are produced mainly by supernovae of type II, whereas iron is also produced by type I supernovae. Models of chemical evolution (Pagel 1989b, Matteucci and François 1989) predict the trends of [O/Fe] and [Mg/Fe] to be very similar and much like the trend for oxygen shown in Fig. 2. The result in Fig. 4 is clearly in disagreement with these predictions and shows that something is wrong with current nucleosynthesis theories and/or chemical evolution models.

3.3 Lithium and Beryllium

As described in detail by Boesgaard and Steigman (1985) the lithium abundances of solar-type stars provide important constraints on cosmological models and give interesting information on stellar structure and evolution.

The important work of Spite and Spite (1982) showed that the lithium abundance in metal-poor dwarfs with T_{eff} in the range 5500-6250K is remarkable constant at a value of $\log N_{Li} = 2.05$ on the usual scale where $\log N_H = 12.00$. New observations of the Li resonance line at 6707Å by Rebola, Molaro and Beckman (1988) have confirmed the near constancy of the Li abundance for a total of 30 stars with [Fe/H] < -1.5 and $T_{eff} > 5500$ K. The average abundance is $\log N_{Li} = 2.08$, but a slight trend towards 2.20 is found as T_{eff} approaches 6300 K. By comparing with the Li-T_{eff} relation for the Hyades Rebola et al. convincingly argue that $\log N_{Li} = 2.2$ is the primordial Li abundance. From this value narrow limits can be placed on the baryon density in a universe described by the standard (i.e. homogeneous) Big Bang model.

The Li abundance in solar-type stars derived from the 6707Å line is quite insensitive to errors in T_{eff}, g and the models. Altogether these error sources contribute with less than ± 0.10 dex to the error of $\log N_{Li}$. According to the work of Steenbock and Holweger (1984) the degree of non-LTE effects depends on how efficient the Li atoms are thermalized by collisions with neutral hydrogen atoms, but it is unlikely that the non-LTE correction of the Li abundance is larger than +0.10 dex. In view of the fundamental importance of knowing an exact value of the primordial Li abundance these errors should be further diminished. The sample of metal-poor stars with Li abundances measured should also be increased.

A better understanding of the mechanism of Li depletion in solar-type stars is needed before we can be sure that Li is not depleted in the metal-poor stars that provide the currently accepted primordial value. The surprising discovery by Boesgaard and Tripicco (1986) of the large Li depletion in Hyades stars with T_{eff} in the range 6500-6800 K shows that Li depletion is a complicated matter. Further observations of Li abundances in stellar clusters of different ages and metallicities, in addition to those reviewed by Boesgaard (1988), may reveal how the depletion depends on T_{eff}, g, [Fe/H] and stellar rotation.

Beryllium is as interesting an element as lithium. According to recent studies inhomogeneities in the early universe may arise in connection with the transition from the QCD phase to the hadron phase leading to neutron-rich and neutron-poor regions. In models describing this scenario a significant fraction of heavy elements ($A \gtrsim 12$) can be produced (Applegate, Hogan and Scherrer 1987) and the abundance of ^{9}Be may be enhanced by a factor of 10^4 relative to the abundance predicted by the standard Big Bang model. Thus Malaney and Fowler (1989) conclude that if a primordial ^{9}Be/^{1}H number ratio of 10^{-13} is found in very metal-poor stars then it would be a dramatic confirmation of inhomogeneities at the epoch of nucleosynthesis and would also leave open the possibility of an $\Omega_{Baryon} = 1$ universe.

A first attempt to measure the Be abundance in very metal-poor stars has been made by Rebolo et al. (1988). The BeII resonance doublet at 3130Å was observed with a resolution of 0.3Å and a S/N of about 30. Due to the severe blending by other lines only an upper limit of $2.5 \cdot 10^{-12}$ for the $^9Be/^1H$ ratio could be determined. Clearly one should try to push this limit by observing the BeII doublet with higher resolution and S/N.

3.4 Thorium

In a most interesting work Butcher (1987) has shown that high resolution observations of the ThII line at 4019.13Å and a nearby NdII line can be used to obtain information on the age of the Galaxy independent of stellar evolution theory, because thorium is unstable with a half-life of 14 Gyr. Butcher finds the Th/Nd ratio to be nearly constant in solar-type stars spanning a range of 20 Gyr in evolutionary ages. Assuming a constant production rate of Th and Nd this constancy sets an upper limit of about 10 Gyr for the ages of the stars in conflict with their evolutionary ages. As reviewed by Pagel (1989a) several attempts have been made to reconcile Butcher's data with the stellar evolutionary ages by assuming that the yields of the r- and s-process vary differently with age. However, as shown by Butcher (1988) such models do not agree with observations of the Eu/Ba ratio as a function of stellar age. The only model consistent with the observed constancy of Th/Nd and the large stellar evolutionary ages seems to be one in which all Th and Nd are synthesized in a single spike before the onset of star formation. However, such a model fails to explain that the abundances of r- and s-process elements have increased by an order of magnitude since the formation of the galactic disk (Lambert 1989).

As pointed out by Butcher (1987) the Th/Nd ratios derived are insensitive to errors in T_{eff}, g and the model atmospheres. Contamination of the Th line by other lines seems negligible because the solar thorium abundance derived from the 4019Å line is within 0.04 dex from the meteoritic value (Anders and Grevesse 1989). Probably the most critical part of Butcher's work is that only three very old stars are included in his determination of the Th/Nd ratio. The very high ages, 15-20 Gyr, determined for these disk stars from their T_{eff} and M_V values, may be rather uncertain as discussed in Sect. 2. Clearly, many more old, metal-deficient stars should be observed with very high spectral resolution in the region containing Th, Nd and Eu lines, and the parameters T_{eff}, $\log g$ and the abundances of the lighter elements should be carefully determined in order to improve the age determinations. Due to the relative faintness of metal-deficient stars a large statistical study probably has to await the large light collecting power of the next generation VLTs.

Nevertheless, Butcher's work is an excellent example of the incredible richness of information about galactic evolution and nucleosynthesis of the elements that may be extracted from the thousands of lines in the spectra of solar-type stars by the aid of accurate spectroscopic studies.

References

Abia, C., Rebolo, R.: 1989, *Astrophys. J.*, in press.

Anders, E., Grevesse, N.: 1989, *Geochim. Cosmochim. Acta* **53**, 197.

Andersen, J., Edvardsson, B., Gustafsson, B., Nissen, P.E.: 1988, Proc. IAU Symp. 132, *The Impact of Very High S/N Spectroscopy on Stellar Physics*, eds. G. Cayrel de Strobel and M. Spite, Kluwer, Dordrecht, p.441.

Applegate, J.H., Hogan, C.J., Scherrer, R.J.: 1987, *Phys. Rev. D* **35**, 1151.

Arribas, S., Martinez Roger, C.: 1989, *Astron. Astrophys.* **215**, 305.

Barbuy, B.: 1988, *Astron. Astrophys.* **191**, 121.

Barbuy, B., Erdelyi-Mendez, M.: 1989, *Astron. Astrophys.* **214**, 239.

Blackwell, D.E., Shallis, M.J.: 1977, *Mon. Not. R. astron. Soc.* **180**, 177.

Blackwell, D.E., Willis, R.B.: 1977, *Mon. Not. R. astron. Soc.* **180**, 169.

Boesgaard, A.M.: 1988, Proc. IAU Symp. 132, *The Impact of Very High S/N Spectroscopy on Stellar Physics*, eds. G. Cayrel de Strobel and M. Spite, Kluwer, Dordrecht, p. 273.

Boesgaard, A.M., Steigman, G.: 1985, *Ann. Rev. Astron. Astrophys.* **23**, 319.

Boesgaard, A.M., Tripicco, M.J.: 1986, *Astrophys. J.* **302**, L49.

Butcher, H.R.: 1987, *Nature* **328**, 127.

Butcher, H.R.: 1988, *ESO Messenger* **51**, 12.

Cayrel, R., Cayrel de Strobel, G., Campbell, B.: 1985, *Astron. Astrophys.* **146**, 249.

Clegg, R.E.S., Lambert, D.L., Tomkin, J.: 1981, *Astrophys. J.* **250**, 262.

Edvardsson, B.: 1988, *Astron. Astrophys.* **190**, 148.

Gustafsson, B.: 1987, Proc. ESO Workshop, *Stellar Evolution and Dynamics in the Outer Halo of the Galaxy*, eds. M. Azzopardi and F. Matteucci, ESO, Garching, p.33.

Gustafsson, B., Bell, R.A., Eriksson, K., Nordlund, Å.: 1975, *Astron. Astrophys.* **42**, 407.

Hartmann, K., Gehren, T.: 1988, *Astron. Astrophys.* **199**, 269.

Hesser, J.E., Harris, W.E., VandenBerg, D.A., Allwright, J.W.B., Shott, P., Stetson, P.B.: 1987, *Publ. Astron. Soc. Pac.* **99**, 739.

Holweger, H.: 1988, Proc. IAU Symp. 132, *The Impact of Very High S/N Spectroscopy on Stellar Physics*, eds. G. Cayrel de Strobel and M. Spite, Kluwer, Dordrecht, p. 411.

Holweger, H., Müller, E.A.: 1974, *Solar Phys.* **39**, 19

Lambert, D.L.: 1989, Proc. of Symp. *Cosmic Abundances of Matter*, ed. C.J. Waddington, Amer. Inst. Phys., New York, p. 168.

Magain, P.: 1987a, *Astron. Astrophys.* **179**, 176.

Magain, P.: 1987b, *Astron. Astrophys.* **181**, 323.

Magain, P.: 1988, Proc. IAU Symp. 132, *The Impact of Very High S/N Spectroscopy on Stellar Physics*, eds. G. Cayrel de Strobel and M. Spite, Kluwer, Dordrecht, p. 485.

Magain, P.: 1989, *Astron. Astrophys.* **209**, 211.

Malaney, R.A., Fowler, W.A.: 1989, *Astrophys. J.* **345**, L5.

Matteucci, F., François, P.: 1989, *Mon. Not. R. astr. Soc.* **239**, 885.

Nissen, P.E.: 1981, *Astron. Astrophys.* **97**, 145.

Nissen, P.E.: 1988, *Astron. Astrophys.* **199**, 146

Nissen, P.E., Edvardsson, B., Gustafsson, B.: 1985, Proc. ESO Workshop, *Production and Distribution of CNO Elements*, eds. I.J. Danziger, F. Matteucci and K. Kjär, ESO, Garching, p. 131.

Nissen, P.E., Gustafsson, B.: 1978, *Astronomical Papers dedicated to Bengt Strömgren*, eds. A. Reiz and T. Andersen, Copenhagen University Observatory, p.43.

Nordlund, Å., Dravins, D.: 1989, *Astron. Astrophys.*, in press.

Pagel, B.E.J.: 1989a, Proc. Adv. Study Institute, *Evolutionary Phenomena in Galaxies*, eds. J.E. Beckman and B.E.J. Pagel, Cambridge University Press, p.201.

Pagel, B.E.J.: 1989b, *Rev. Mex. Astr. Astrofis.*, in press.

Perrin, M. -N., Cayrel de Strobel, G., Dennefeld, M.: 1988, *Astron. Astrophys.* **191**, 237.

Rebolo, R., Molaro, P., Beckman, J.E.: 1988, *Astron. Astrophys.* **192**, 192.

Rebolo, R., Molaro, P., Abia, C., Beckman, J.E.: 1988, *Astron. Astrophys.* **193**, 193.

Saxner, M., Hammarbäck, G.: 1985, *Astron. Astrophys.* **151**, 372.

Schuster, W.J., Nissen, P.E.: 1989, *Astron. Astrophys.*, **222**, 69.

Smith, G., Drake, J.J.: 1987, *Astron. Astrophys.* **181**, 103.

Smith, G., Edvardsson, B., Frisk, U.: 1986, *Astron. Astrophys.* **165**, 126

Spite, F., Spite, M.: 1982, *Astron. Astrophys.* **115**, 357.

Steenbock, W., Holweger, H.: 1984, *Astron. Astrophys.* **130**, 319.

Tomkin, J., Lambert, D.L., Edvardsson, B., Gustafsson, B., Nissen, P.E.: 1989, *Astron. Astrophys.* **219**, L15.

Wheeler, J.C., Sneden, C., Truran, J.W.: 1989, *Ann. Rev. Astron. Astrophys.*, in press.

Wyse, R.F.G., Gilmore, G.: 1988, *Astron. J.* **95**, 1404.

SOLAR ANALOGS SEEN AT HIGH SPECTRAL RESOLUTION AND VERY HIGH S/N RATIOS

G. CAYREL DE STROBEL
Observatoire de Paris, Section de Meudon
F-92195 Meudon Cedex, France

SUMMARY Lists of photometric solar analogs based upon different authors are presented. A small number of candidates qualified to become real solar twins will be selected. With the help of the above sample of photometric solar twins candidates, the second part of this review will be devoted to the search of stars, whose fundamental parameters (mass, effective temperature, chemical composition, age and luminosity, equatorial rotation, velocity fields, chromospheric activity, etc...) are all very near to those of the Sun.

The tools employed for this research are :

i) High resolution, high S/N spectroscopy

ii) Appropriate models for stellar atmosphere analyses

iii) Evolutionary internal structure models computed with realistic physical input.

1 INTRODUCTION

"Is it possible to find, within a reasonable distance in our Galaxy one or several stars which are practically identical to our Sun ?" This was the question I have propound to my collaborators and to myself more than 10 years ago. And in asking this question I meant really, stars having all their physical parameters i.e. mass, chemical composition, age, luminosity, rotation, velocity fields, magnetic fields, chromospheric activity, the same Li-content etc., etc. equal, or nearly equal, to those of the Sun. Of course stars like the Sun are spread all over our Galaxy and in other galaxies. But this was not our concern : the question was whether stars previously classified as photometric solar analogs were remaining as such, when submitted to spectroscopic detailed analyses and discussed with the help of internal structure models.

The observations for the spectroscopic search of solar twins have begun in the late seventies at the coudé spectrograph of the 1.52 m telescope of the Haute-Provence Observatory. High dispersion photographic spectra have been obtained of four solar analogs. With the help of detailed analyses we have determined the atmospheric

Table 1

Photometric solar analogs

HARDORP			NECKEL			GRENON		
HD NAME	B-V	Sp	HD	B-V	Sp	HD	B-V	Sp
28099 VB64	.66	G2V	28099	.66	(G8V)	10307	.62	G1.5V
44594	.66	G3V	44594	.66	G3V	10800	.61	G2V
186427 16CygB	.66	G5V	186427	.66	G2.5V	30495	.64	G1V
191854	.66	G5V	191854	.66	(G5V)	52711	.60	G4V
1835 9Cet	.66	G2V	1835	.66	G2V	71148	.63	G5V
20630 KCet	.68	G5V	20630	.68	G5V	72905	.62	G1.5V
76151	.68	G3V	76151	.67	G3V	88742	.59	G1V
78418	.65	G5IV	78418	.66	G5IV-V	95128	.61	G0V
86728 20LMi	.65	G4V	--	--	--	115043	.60	G1V
89010 35Leo	.65	G1IV-V	89010	.67	G2IV	147513	.61	G1V
144585	.66	G4IV-V	144585	.66	G4IV-V	190406	.61	G1V
159222	.64	G5V	159222	.65	G5V	193664	.58	G3V
181655	.68	G8V	181655	.68	G8V	197076	.63	G5V
						207129	.60	G0V
B-V=0.661±.013			B-V=0.664±.009			B-V=0.611±.016		

parameters of the four stars: effective temperature, gravity, metallicity, microturbulent velocity, and discussed the same parameters taken from the literature for other four G-dwarfs. The comparison of these parameters to those of the Sun, together with the comparison of absolute bolometric magnitudes, evolutionary masses and ages of the eight stars form the subject of Paper I (Cayrel de Strobel et al 1981). The conclusion of Paper I was that we did not succeed, using our expression, to find a "real solar twin". Ten years later, we began to search again for real solar twins, stimulated by the fact that the observational material relying upon high S/N, high resolution solid state spectroscopy was by an order of magnitude better than the one we had used in 1981. This time, as it is seen in Paper II (Cayrel de Strobel and Bentolila 1989), we have been more successful : we have found that the photometric analog of Hardorp HD 44594, is a star for which most of the physical parameters are fairly close to those of the Sun.

Subsequently in our search for solar spectroscopic twins we shall follow the approaches described in Papers I and II, but at the same time lay stress on the interplay between photometry and spectroscopy in order to present at the end of this paper more than one good solar twin candidate.

2 WHAT KIND OF LISTS OF SUN - LIKE STARS HAVE WE CONSULTED IN OUR SEARCH FOR SOLAR TWINS ?

The problem of finding actual solar twins has been raised by Hardorp (1978). Using spectral scans at a resolution of 20 Å in the region $3650 < \lambda < 4060$ Å, he found that the solar energy distribution does not correspond to that of an average G2V star, but rather to later spectral types. Hardorp also found that the stars, which best match the solar spectrum at 20 Å resolution, all have (B-V) around 0.66. Both these statements: i) later spectral type and ii) (B-V) as high as 0.66 for stars having the same energy distribution as the Sun, cannot be taken lightly. If the continuous spectral distribution of the Sun matches those of G4 or G5 dwarfs, then stars classified as G2V are hotter than the Sun, and if we could finally be fixed on the exact (B-V) value of the Sun, this would be a great contribution to photometry, because the (B-V) index of the Sun plays an important role in the interpretation of the observational (color, luminosity) or (color, color) diagrams.

Recently Neckel (1986) has shown that the (B-V) index of the Sun must be very close to 0.65. Chmielewski (1981) using a new direct interpolation method has found 0.63 for $(B-V)_\odot$. In any case, since Gallouët (1964) who found 0.68, an impressive high value for $(B-V)_\odot$, this index has varied up and down from 0.62 to 0.68 between different authors in these last 25 years. The color indices in other photometric systems than UBV of the brightest object in the Sky, our Sun, are not well known, and in general experts in different photometries give informations about such indices very reticently.

Table 2

Standars of spectral type and colors around G2			Cousins of our Sun : The G stars		
Jaschek and Jaschek			**J.B. Kaler**		
HD NAME	B-V	Sp	HD NAME	B-V	Sp
115043	.60	G1V	212698 53AqrA	.61	G1V
10307	.62	G1.5V	128620 αCentA	.68	G2V
186408 16CygA	.64	G1.5V	1835 9Cet	.66	G2V
146233	.65	G2V	143761 ρCrB	.60	G2V
28099 VB64	.66	G2V	186408 16CygA	.64	G1.5V
186427 16CygB	.66	G2.5V			
140538 ψSer	.68	G2.5V	186427 16CygB	.66	G2.5V
B-V = 0.653 ± 0.019			B-V = 0.638 ± 0.033		

SUN   .625 ± 0.025   G2V

From now on we will frequently speak of : a) solar like stars , b) solar analogs , and c) real solar twins. Here below we give the definitions of these three classes of stars.

1. Solar like stars are stars of luminosity classes V or IV falling in a broad spectral interval centered at G2V and having U, B, V colors (Neckel 1986) :

$$0.48 \leq B-V \leq 0.80$$
$$-0.06 \leq U-B \leq 0.4202$$

2. Solar analogs are unevolved or slightly evolved Pop I disk stars having effective temperatures and photometric properties very similar to those of the Sun.

3. Real solar twins are not yet found stars having all their physical parameters identical to those of the Sun.

Following these definitions we see that compiling a list of solar analogs is harder than compiling a list of solar like stars. Indeed in this second case the photometric parameters of the solar analogs must all be very near to those of the Sun as well as their effective temperatures and we know that for only very few solar like stars their effective temperatures are known better than ± 50 K.

Lists of solar analogs are given in Tables 1 and 2. In the three columns of Table 1 we have included the best analogs of Hardorp (1978), of Neckel (1986_a, 1986_b, 1986_c) and of Grenon (1983) respectively. Table 2 contains two lists of solar type stars: they have been taken from Jaschek and Jaschek (1987) and Kaler (1986).

Table 1 of Hardorp (1978) contains 78 solar like stars and is divided in 8 sections. From this table we have chosen 13 stars, contained in column 1 of our table 1. The first 4 stars come from section 1 of Hardorp (1978) and their spectra are, according to Hardorp, "indistinguishable from solar", the other nine stars are contained in section 3 of the same table, headed "spectra very close to solar"; we omitted the second section containing spectra of stars "potentially equal to solar, but not observed enough". The remaining sections contain stars whose spectral absorption features were either to strong or to weak in respect to those of the Sun ; no solar analog candidates have been taken from these sections .

In his paper " The bright stars with UBV colors close to the Sun" Neckel (1986) provides a compilation of the main physical characteristics of all stars included in the "Bright Star Catalogue"(Hoffleit and Jaschek 1982), possessing B-V and U-B indices fulfilling the already mentioned conditions: $0.48 \leq B-V \leq 0.80$ and $-0.06 \leq U-B \leq 0.42$. The whole

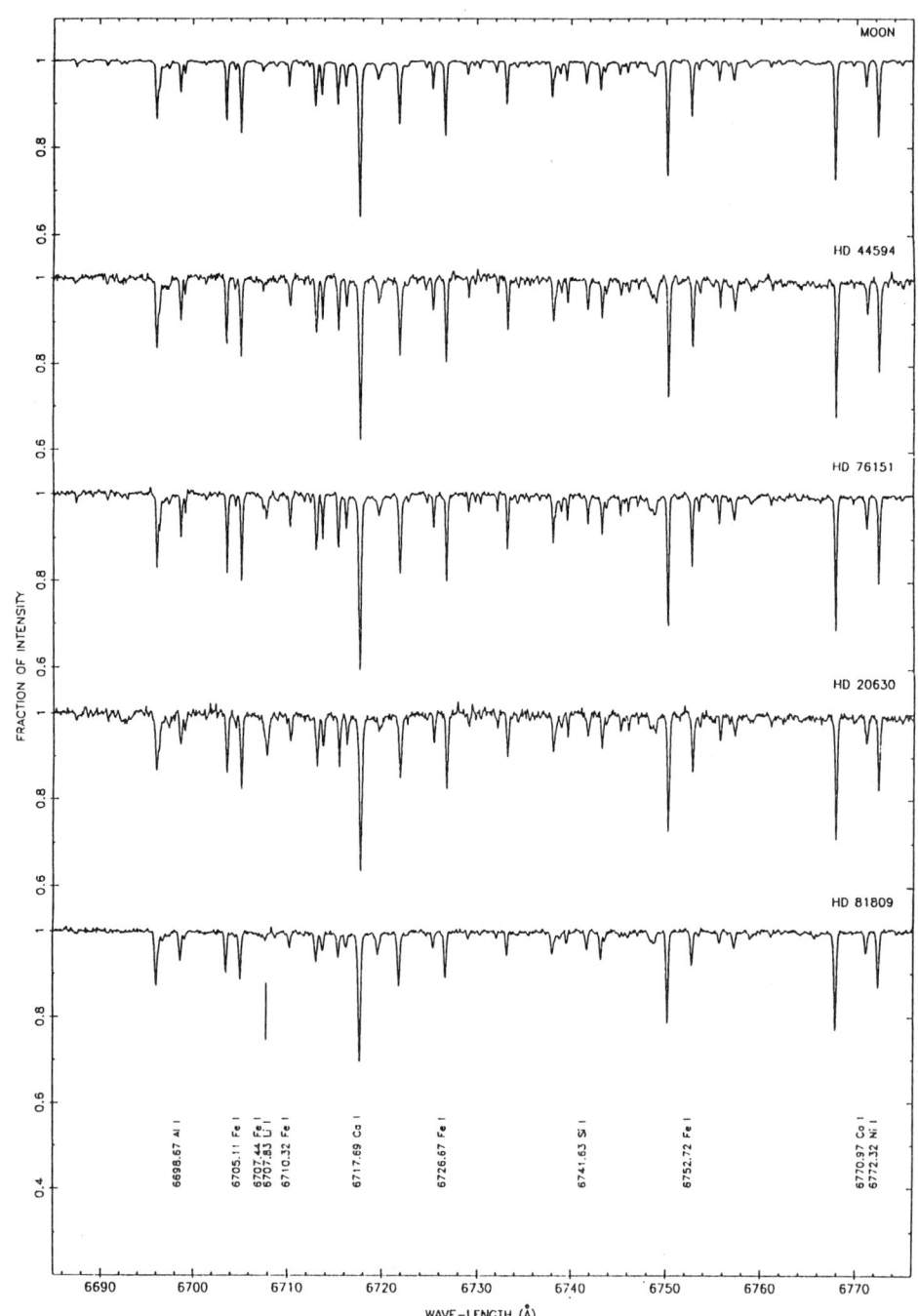

- Fig. 1 Examples of high S/N ratio, CFHT Reticon spectra of the 6730 Å region for sunlight (Moon) and for one solar analog (HD 44594) and three solar like stars (HD 76151, HD 20630, HD 81809).

sample of these stars has been plotted by Neckel on a (B-V, U-B) colour - diagram (see Fig.1 of Neckel 1986). In the same paper, Fig.2 represents the central part of Neckel's colour - diagram, the Sun occupying on it the position B - V = 0.650 and U - B = 0.195.. In this figure the stars neighbouring the Sun are mostly all the stars we have chosen in Hardorp's (1978) section 1 and 2 of his Table 1.Therefore we must not be surprised if the stars contained in columns 1 and 2 of Table 1 are the same.

The solar like stars contained in column 3 of Table 1 have been given to us by Grenon (1983). Grenon called this table "Some solar twins". A quick glimpse at Grenon list shows that there is no one star in common between the lists of Hardorp and Neckel on one side and that of Grenon on the other. Grenon's choice of candidates has been made with the help of the Geneva photometric boxes (Golay et al. 1977a, Golay et al. 1977b, Golay 1978). Indeed, Golay claims that stars contained in a same photometric box, i.e. stars having Geneva colors (UBVB$_1$ B$_2$ V$_1$ G) differing no more than 0.01 mag. from the colors of the central star of the box, in our case the Sun, have "more or less almost the same spectral type, the same absolute magnitude, the same chemical composition, the same v sin i, the same multiplicity, the same interstellar extinction etc..., as the central star of the box and , perhaps , almost the same mass, radius and age", (quoted from the above reference). Ultimately, the solar best photometric analogs have been selected by Grenon according to :

$(T_{eff})_{Geneva\ phot.}$ = 5780K ± 50K
$[Fe/H]_{Geneva\ phot.}$ = 0.00 ± 0.10
M_V Gliese1969 = 4.83 ± 0.15

To these three lists of solar "photometric analogs" we added, in our choice of reliable candidates, two lists of "spectral type solar analogs".These lists are reproduced in Table2, the first one is that of Jaschek and Jaschek (1987) the second one that of Kaler (1986). In considering the five lists all toghether,we see that, some stars are in common between three authors. Only one is in common between four authors : the star, 16 CygB. The B-V values written in bold letters at the bottom of each of the five columns of tables 1 and 2 are the mean of the (B-V)'s of all the stars contained in one of the columns.

Unfortunately not all the solar analog candidates, here presented, have been submitted to detailed analyses, and some of them have poorly determined atmospheric parameters still based upon photographic spectra and coarses analyses. In the next section we shall discuss which of the few remaining stars , having good and trustful determined physical parameters, present the best titles for becoming real solar twins.

3 Detailed spectral analyses of solar analogs

3 a Observations

Table 3 gives fundamental physical parameters of our Sun and of ten photometric solar analogs. They have been observed, with the exception of HD78418 and HD89010, by means of high resolution, high S/N solid state spectroscopy, mostly at CFHT and ESO. Eight of these stars are from Hardorp (1978, Table 1, section 1 and 3). Of the two remaining ones, αCenA is our nearest neighbour, and the second has been recommended to us by Mihalas (1981) as probably a good solar analog.Indeed HD81809 has the same activity cycle than the Sun: same mean Ca-index,same amplitude of Ca-index and same period (Baliunas and Vaughan 1985). This seemed to us a plus to consider HD81809 as an attractive solar twin candidate.Unfortunately, following the "Bright Star Catalog", this star is a single line spectroscopic binary.The difference of magnitude between the primary and the secondary is : $\Delta m=0.8$mag, not small enough to guarantee a negligeable contamination of the spectrum of the primary by the secondary.

Table 3
Fundamental Physical Parameters for the proposed Solar Analogs
by
Hardorp, Neckel, and Mihalas

HD Name	V	B-V	M_{bol}	T_{eff} K	log g	$[Fe/H]_\odot^*$	$\log N^7 Li$
Sun	-26.74	?	4.75	5770± 20	4.44	0.00±0.03	1.04
44594	6.60	0.66	4.60	5770± 40	4.50	+0.15± 0.06	<1.11
186427 16 Cyg B	6.20	0.66	4.52	5770± 50	4.40	+0.04±0.10	--
28099 VB64	8.12	0.66	4.80	5770± 50	4.50	+0.14±0.04	2.23
1835 9Cet	6.39	0.66	4.77	5770± 50	4.50	+0.17±0.04	2.48
128620 αCenA	F -0.01	0.69	4.31	5710± 25	4.00	+0.15±0.02	
	C -0.01	0.68	4.27	5800± 50	4.30	+0.20±0.04	1.34
78418	5.98	0.65	4.02	5730±100	4.20	-0.27±0.15	--
76151	6.00	0.68	4.89	5710± 40	4.50	+0.06±0.05	1.72
81809 (MIHALAS)	5.38	0.64	3.15	5630± 50	3.75	-0.31±0.07	<1.22
20630 κ^1Cet	4.85	0.68	4.94	5630± 40	4.50	0.00±0.06	1.95
89010 35Leo	5.97	0.67	3.56	5600±100	4.00	-0.03±0.15	--

Table 4 gives fundamental physical parameters of our Sun and of eight photometric solar analogs recommanded by Grenon (1983) The spectra of four of them (HD10307, HD71148, HD72905, HD115043) have been taken at CFHT with the Reticon detector. In order to show the quality of most of the spectra used in the search for solar analogs Fig.1 reproduces CFHT Reticon spectra of four solar analogs in the 6730 Å region.

3 b Data reductions

The equivalent widths have been determined by a detailed line profile fitting technique, with special attention to contamination by blends of weak lines. The method used in data reduction concerning 10 stars in Tables 3 and 4 (Cayrel et al. 1985, Cayrel de Strobel and Bentolila 1989) has been rigorously the same. In much the same way, have been reduced Reticon spectra of αCenA (Furenlid and Meylan 1989, Cayrel de Strobel unpublished) and 16 CygB (Branch et al.1980) . Hence for 6 stars the data are based upon photographic spectra alone.

3 c Results from Detailed Analyses

The spectra of all the stars contained in Table 3 and 4 have been interpreted with theoretical line computations using grids of model atmospheres of various effective temperatures, gravities and metallicities suitable for solar G dwarfs.

The most delicate operation has been the determination of the effective temperatures. They have been derived for most the program stars from the wings of $H\alpha$ or $H\beta$ lines (Fig.2).

Table 4
Fundamental Physical Parameters for the proposed Solar Analogs
by
Grenon

HD Name	V	B-V	M_{bol}	T_{eff} K	$\log g$	$[Fe/H]_\odot$	$\log N^7Li$
Sun	-26.74	?	4.75	5770± 20	4.44	0.00±0.03	1.04
10307	4.96	0.62	4.60	5860± 60	4.40	+0.14±0.25	<1.30
30495 58 Eri	5.49	0.64	4.86	6000±130	4.50	+0.10±0.25	--
52711	5.93	0.60	4.75	5860±130	4.50	-0.15±0.25	--
71148	6.36	0.63	4.65	5800± 60	3.75	+0.02±0.10	1.79
72905 π^1UMa	5.64	0.62	4.60	5850± 50	4.40	-0.09±0.08	2.70
95128 47 UMa	5.05	0.61	4.35	5860±100	4.31	-0.02±0.25	<1.60
115043	6.84	0.60	5.26	5830± 50	4.40	-0.03±0.08	--
193664	5.94	0.58	5.07	6000±100	4.64	+0.06±0.25	--

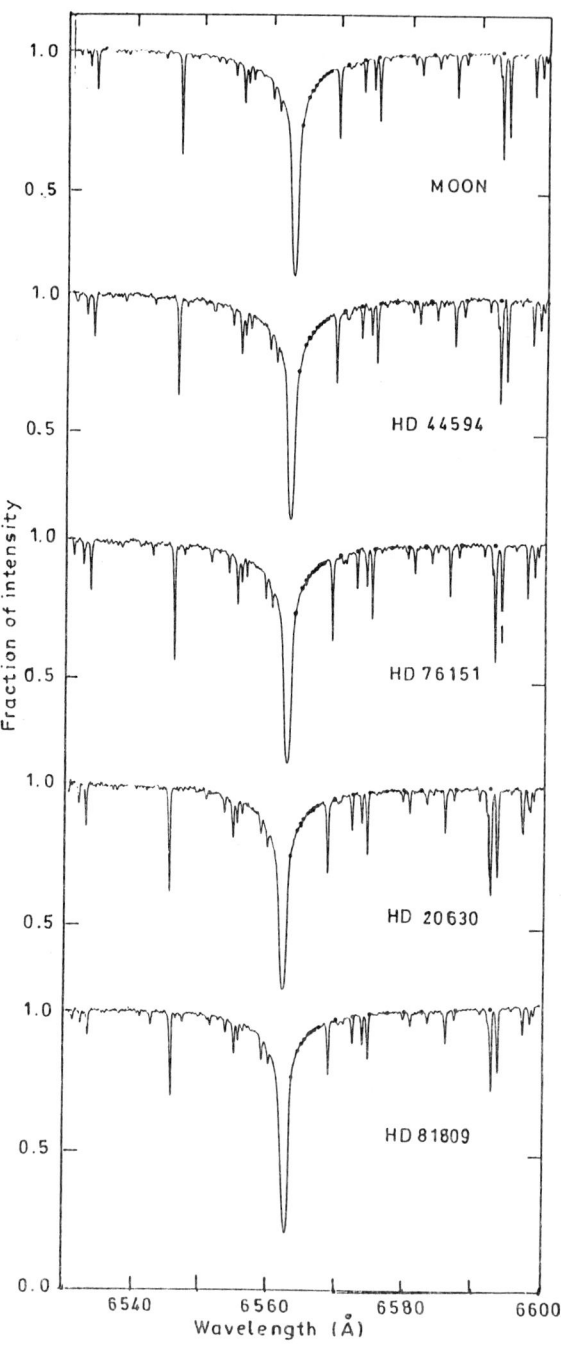

Fig. 2 The Hα region for the same stars than in fig. 1. Points represent computed Hα profiles.

The T_{eff} of some stars have been determined from the ratio of their Hα profiles to the Sunlight-profile (Cayrel de Strobel and Bentolila 1989). These observed ratios were then compared to the computed ratios (Fig.3). We note here that this method is the only one which can be applied to the ESO-CES Reticon spectra, because,unfortunately,their spectral intervals are to narrow to allow to set correctly the continuum in the Hα region and consequently to determine the effective temperature from the comparison between the observed and the theoretical Hα wing-profiles.

The metal abundances have been derived in matching equivalent widths in the observed spectra of solar analogs to those computed from a grid of appropriate model atmospheres.

The gravities have been determined from the ionisation equilibrium..

The "microturbulence" for the Sun must be determined from an absolute curve of growth with solar equivalent widths coming from the same observational material than that used for the program stars.There was no need changing the microturbulence from the solar value $\xi_t = 1$ Km s^{-1} for any star : the analysed lines were sufficiently

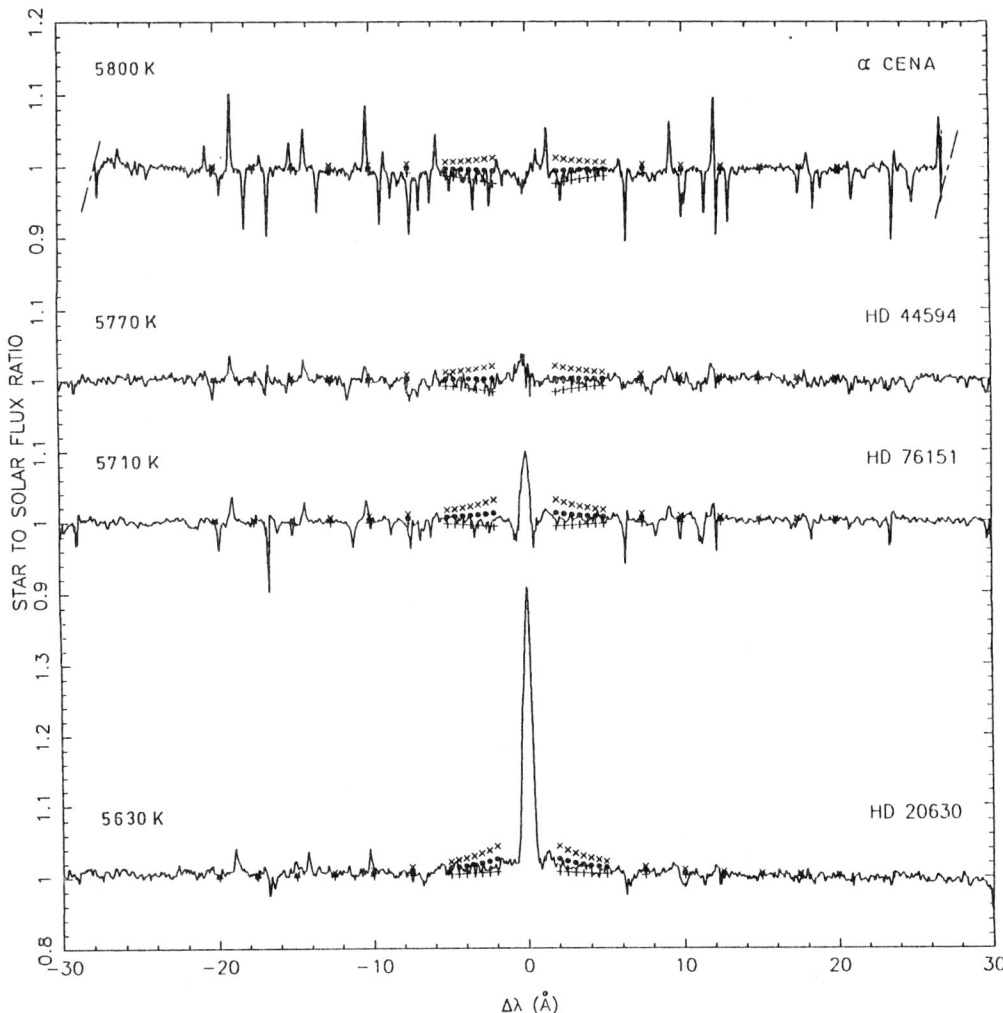

- Fig. 3 Observed ratio of ESO and CFHT spectra of two solar analogs, αCen A and HD 44594, and two solar like stars, HD 76151 and HD 20630, to Sunlight (Moon) spectrum near Hα. The black points represent the theoretical ratios for the best fit between observation and theory. The uncertainties of ± 100 K in T_{eff} is shown by crosses. The spikes are due to telluric velocity shifted lines.

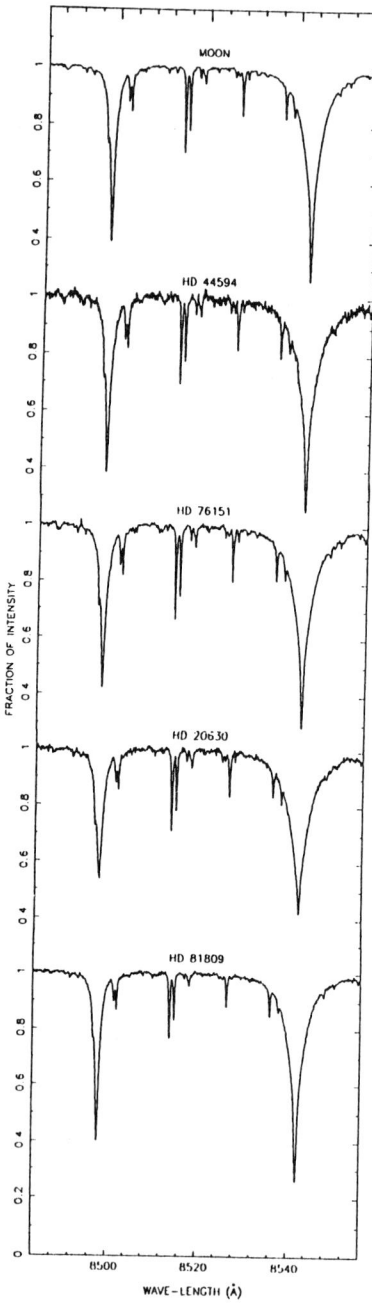

weak, so that a common curve of growth could be used.

The last column in Tables 3 and 4 gives the absolute abundance of Li with respect to H on the scale log N_H =12, in the program stars in which Li could be measured.

3 d Bolometric Magnitudes

The M_{bol}'s have been computed with the help of the trigonometric parallax Catalog of Gliese (1969), except for 16 Gyg B and HD76151 (1988, private communication). Equal bolometric corrections, BC = 0.07, have been applied to the absolute magnitudes of all the program stars.

3 e Chromospheric Activity

A crucial physical parameter for a star which may warrant it to be a solar twin, is its chromospheric activity. For this research, the chromospheric activity parameter has been estimated for a few stars from the central depths of the CaII infrared triplet lines. From Fig.4 we see, for instance, that the depth of the two visible lines : 8498.1Å and 8542.Å, are indistinguishable from that of the Sun in HD44594, HD76151, and HD81809.

This indicates, (Cayrel et al. 1983), that the chromospheric activity in these three stars must be very similar to that of the Sun. And, because, the chromospheric activity is tightly bound to the age of the star this indicates also that the three stars should have about the same age.

Fig. 4 CFHT Reticon spectra of the infrared CaII triplet region. Two triplet lines are visible: at 8498.06Å and 8542.14Å . except for HD20630 the depths of the lines of HD44594 , HD76151 and HD81809 indicate that their chromospheric activity is indistinguishable from that of the Sun.

4 How selecting the best spectroscopic solar analogs

For a discussion on the choice of the best spectroscopic solar analogs Tables 3 and 4, as well as the figures in the text, will be used as a working tool. With the exception of the V's and the B-V's the quantities contained in the other columns of the two tables are true physical parameters of the analysed stars. The bolometric magnitude gives informations on the degree of evolution and the internal structure of the star. With the help of the ($\log T_{eff}$, M_{Bol}) diagram (Fig.5) we see that, HD81809, HD89010 and HD78418 are already evolved and that the other stars are unevolved or, as the Sun, very mildly evolved.

Let us discuss, first, the stars contained in Table3. The effective temperature of each star deserves much attention. In Table 3, four stars have the same effective temperature as the Sun. The rest of the stars, with the exception of the T_{eff} found by Cayrel de Strobel for α Cen, are all by several tenths of degrees cooler than the Sun. The spectroscopic gravities of HD81809, HD89010 and HD78418 confirm what has been found from their bolometric magnitudes, i.e. that these stars are the most evolved of the sample. Only one star in table 3 is metal deficient by a factor of 2 : HD 81809. The other are metal normal or slightly metal rich. The Sun is the star having the smallest Li-abundance between the stars with measured Li-abundance.

The star α CenA has been analysed twice, once by Furenlid and Meylan (1989) and once by Cayrel de Strobel (1988 unpublished). There is a small, but significant difference in T_{eff}, in gravity and in $[Fe/H]_{\odot}^{*}$ between the two authors, although α Cen A has been observed by the two astronomers with the same equipement : ESO-Coudé-Auxiliary-Telescope(CAT), Coudé-Echelle-Spectrographe (CES), and Reticon detector. Both authors have used data of very high S/N ratio. The differences between the two authors are the stellar model atmospheres they have employed, and the method for the determination of the effective temperature of α CenA. The models used by Furenlid come from Kurucz (1979, 1984) and the ones used by Cayrel de Strobel come from Gustafsson (1981). Both Kurucz and Gustafsson's models are atomic line blanketed and assume LTE in the formation of lines and continuum. Therefore we do not think that the difference in the results of α CenA, in particular in T_{eff} and log g, come from the models, but rather from a different approach in the determination of T_{eff} and gravity. The excitation and the ionisation equilibria have been used by Furenlid for determining both, T_{eff} and log g. The wings of the Hα line have been used by Cayrel de Strobel for determining the T_{eff} and the ionisation equilibrium for determining log g. The first plot in fig.3 represents the observed ratio of the Hα wings of α CenA to the Hα wings of Sunlight as compared to the theoretical ratio. From this ratio we see that the temperature of α CenA is marginally higher than that of the Sun. and differs by 90K from that of Furenlid. With this temperature, (T_{eff}= 5800 K), the gravity of the star,

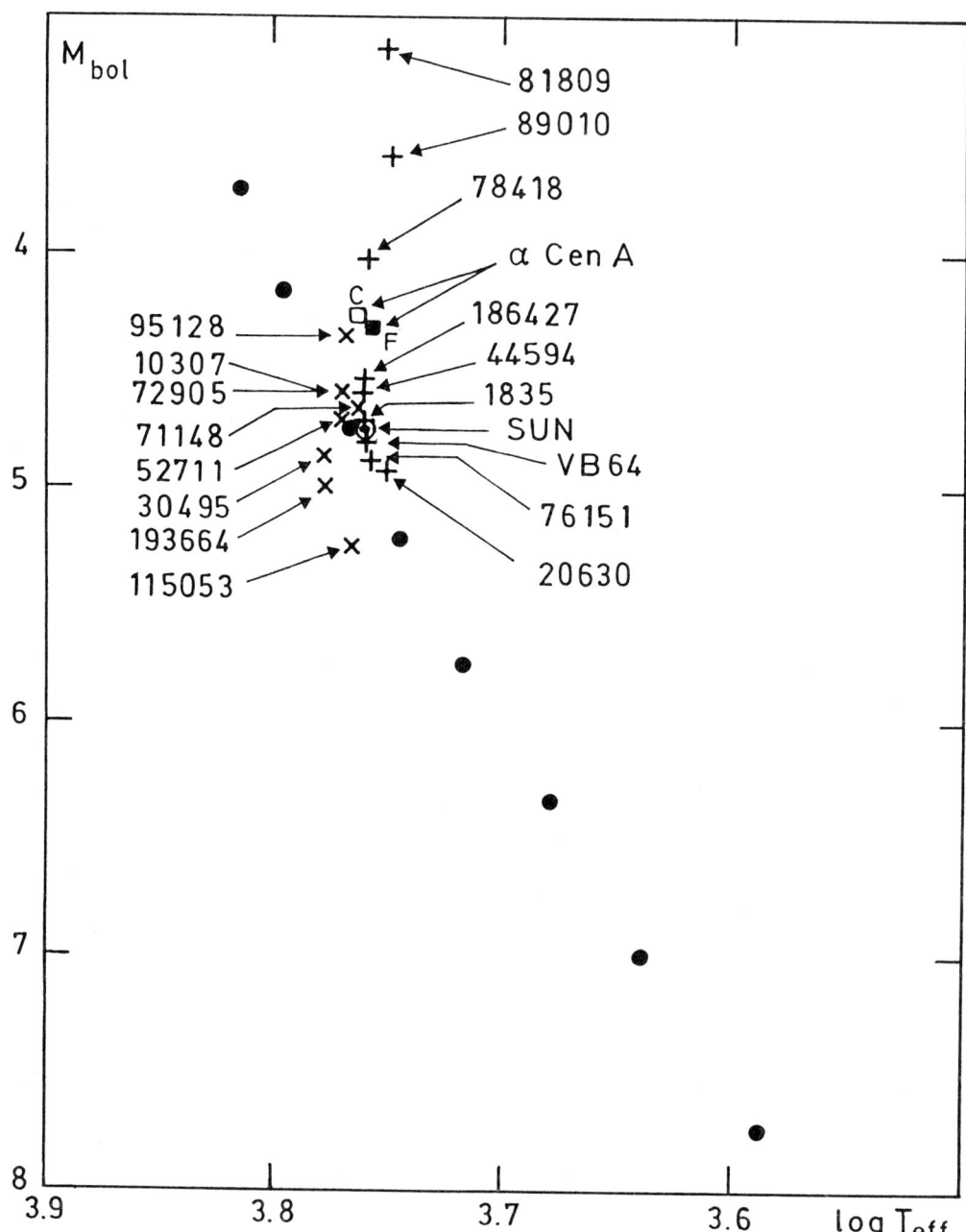

Fig. 5 Positions occupied by the stars, contained in Tables 3 and 4, in the theoretical (log T_{eff}, M_{bol}) diagram. Plusses are stars from Table 3 and crosses stars from Table 4. Black circles are evolutionary models computed by Lebreton (Cayrel de Strobel et al 1989) representing a theoretical ZAMS ($Z = 0.02$, $Y = 0.287$, $\alpha = 2.18$).

by Cayrel de Strobel is, log g = 4.30, which is higher by a factor of 2 than that of Furenlid. Indeed, the absolute magnitudes of the Sun and of α Cen A differ by about 0.45 mag or 0.18 in log. Neglecting the small difference in brightness due to a possible temperature difference, smaller than 90K ,and a difference in mass not exceeding 5%, one gets a log g-difference of 0.18, leading to a log g of 4.26 for α Cen A. The comparison between the absolute curve of growth of HD44594 and that of α Cen A, both computed with the atmospheric parameters found by G. Cayrel de Strobel, is given in Fig.6.

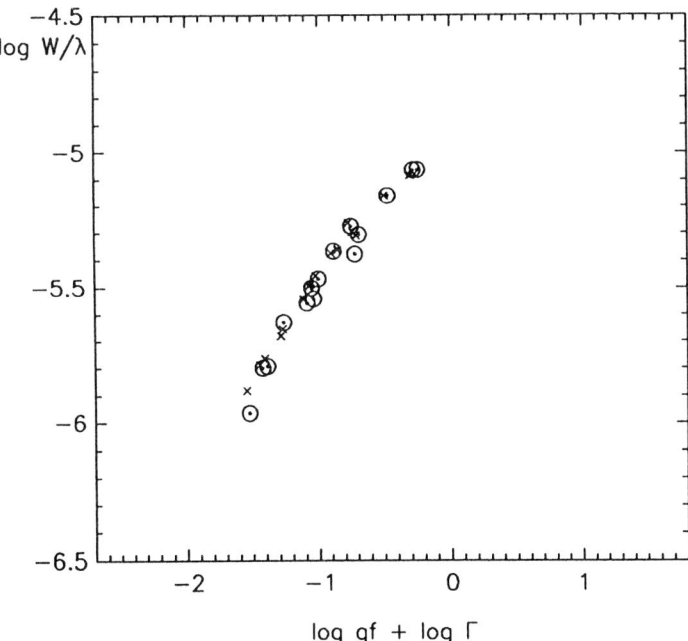

Fig. 6 Comparison between the absolute curve of growth of HD 44594 with that of αCen A. Here, the equivalent widths for both stars come from ESO observations, the absolute oscillator strengths gf from May et al (1974), and the thermodynamic term Γ has been computed assuming: T_{eff} = 5770, log g = 4.50, [M/H] = + 0.15 for HD 44594 (open dotted circles) and assuming T_{eff} = 5800, log g = 4.50, [M/H] = + 0.20 for αCen A (crosses)

The eight stars, found in Table 4, come from the list of solar analogs of Grenon (1983), contained in column 3 of Table1 of this paper. Four of these analogs, HD1030, HD72905, HD71148 and HD115043, have been observed with Reticon detectors and their analyses are based on high S/N, high resolution spectroscopic material. The atmospheric parameters of the other four stars have been taken out from the "Catalogue of [Fe/H] determinations" (Cayrel de Strobel et al 1985).The stars of Table 4 are all slightly hotter than the Sun, their B-V are smaller than those of the stars in Table3, their metallicities do not differ significantly from that of the Sun. The positions of three stars, HD30495, HD193664, and HD115053 do not follow the theoretical ZAMS (black circels)computed for normal solar helium and normal solar metal content .This can be attributed to: i) a bad parallax determination, ii) a bad effective temperature determination, or , iii) a very unlikely He - enrichment in the three stars.

5 The contribution of spectroscopic solar analogs to a better knowledge of the B-V of the Sun

Arribas and Martinez Roger (1988) have found that the dependence of the decrements in the (B-V) index due to metallicity is :

$$\Delta(B-V) = 0.227 \, ([Fe/H])^{2/3} \, (\Theta_{eff} - 0.48)$$

There are four stars in table 3 HD44594, 16 Cyg B, the Hyades solar analog, VB64, the Hyades moving group star, HD1835, which all have the same (B-V) equal to 0.66, and for all of which detailed analyses have found the same solar effective temperature, $T_{eff} = 5770K$. From Table 3 we can see that the mean of their [Fe/H] values is : [Fe/H] = +0.125, i.e. 33 % higher than the iron-content of the Sun. Introducing this value in the above relation, we obtain:

$$\Delta(B-V)_*^\odot = 0.022$$

From this its follows :

$$(B-V)_\odot = 0.64 \pm 0.015$$

Conclusion

This paper, is a short review of what has been done recently with the help of high resolution, high S/N spectroscopy techniques, on stars for which physical parameters are as close as possible to the Sun. None of the solar twins candidates, above presented, has all its physical parameters equal to those of the Sun. The spectral analyses have produced results more accurate by an order of magnitude than those obtained in the past on photographic material, therefore the non existence of a true solar twin is corroborated by such results. The two stars, in our list of solar analogs, which resemble the more to the Sun are 16Cyg B for the Northern Hemisphere and the photometric analog, found by Hardorp (1978), HD44594, for the Southern Hemisphere.

This work has been stimulated for many years by Johannes Hardorp, who died recently. To him this paper is dedicated.

REFERENCES

- Arribas, S., Martinez Roger, C. : 1988, Astron. Astrophys. **206**, 63
- Baliunas, S.L., Vaughan, H. : 1985, Ann. Rev. Astron. Astrophys. **23**, 379
- Branch, D., Lambert, D.L., Tombrin, J. : 1980 Astrophys. J. **241**, L83

- Cayrel de Strobel, G., Knowles, N., Hernandez, G., Bentolila, C. : 1981, Astron. Astrophys. **94**, 1
- Cayrel, R., Cayrel de Strobel, G.,Campbell,B.,Mein,N., Mein, P.,Dumont,S.: 1983, Astron. Astrophys.**123**,89
- Cayrel de Strobel, G., Bentolila, C., Hauck, B., Duquennoy, A. : 1985, Astron. Astrophys. Suppl. Ser. **59**, 145
- Cayrel de Strobel, G., Bentolila, C. : 1989, Astron. Astrophys. **211**, 324
- Cayrel de Strobel, G., Perrin, M.N., Cayrel, R., Lebreton, Y. : 1989, Astron. Astrophys. **225**, 369
- Chmielewsky, Y. : 1981, Astron. Astrophys. **93**, 334
- Dahn, C. : 1988, private communication
- Furenlid, I., Meylan, T. : 1989, Accepted Astrophys. J.
- Gliese, W. : 1969, Veröff. Astron. Rech. Inst. Heidelberg **22**
- Galouët, L. : 1964, Ann. Astrophys. **27**, 423
- Golay, M., Mandwewala, N. : 1977a, Geneva Obs. Publ. Ser.B, n° 3, Catalogue
- Golay, M., Mandwewala, N., Bortholdi, P. : 1977b, Astron. Astrophys. **60**, 181
- Golay, M. : 1978, Astron. Astrophys. **62**, 189
- Grenon, M. : 1983, private communication
- Gustafsson, B., Bell, R., Eriksson, K., Nordlund, A. : 1975, Astron. Astrophys. **42**, 407
- Gustafsson, B. : 1981, private communication
- Hardorp, J. : 1978, Astron. Astrophys. **63**, 383
- Hoffleit , D., Jascek,C: 1982,The Bright Star Catalogue, 4th ed., Yale Univ. Obs.
- Jaschek, C., Jaschek, M. : 1987, "The Classification of Stars" p. 280, Cambridge Univ. Press
- Kaler, J.B. : 1986, Sky and Telescope **72**, 450
- Kurucz, R.L. : 1979, Astrophys. J. Suppl. **40**, 1
- Kurucz, R.L. : 1984, Smithsonian Astrophys. Obs. , Special Report N° 362
- May, M., Richter, J., Wickelmann, J. : 1974, Astron. Astrophys. Suppl. **18**, 405
- Mihalas, D. : 1981, private communication
- Neckel, H. : 1986$_a$, Astron. Astrophys. **159**, 175
- Neckel, H. : 1986$_b$, Astron. Astrophys. **167**, 97
- Neckel, H. : 1986$_c$, Astron. Astrophys. **169**, 194
- Perrin, M.N., Spite, M. : 1981, Astron. Astrophys. **94**, 207

Surface Structures and Flares in Solar-Like Stars

B.H. Foing[1,2], S. Char[2], S. Jankov[2], E. Houdebine[2]

1 ESA/ESTEC Space Science Dept., P.O.Box 299, 2200 AG Noordwijk, NL
2 Institut d'Astrophysique Spatiale, IAS, BP10, 91371 Verrieres, F

SUMMARY: Diagnostics of active structures are possible from the comparison of active and quiescent stars, and from the spectroscopic variability in chromospheric or photospheric lines associated with the rotational modulation on late-type dwarfs and RSCVn binaries. Interpretative techniques have been developed based upon the rotational modulation, Doppler imaging or from chromospheric codes modelling, in order to diagnose the geometrical and physical properties of related magnetic structures. These results for different solar-like stars give constraints on chromospheric/coronal heating and on dynamo theories. Flares are the most violent phenomena occurring on these stars. Coordinated observations of dMe indicate continuous microflaring correlated in X-ray and in Balmer line spectroscopy, and complete multi-frequency coverage with photometry, optical spectroscopy radio, UV and X ray observations has proved to be crucial for studying the energy balance and dynamics of flares. Also coronal mass ejections can be detected during stellar flares. Coordinated multi-frequency multi-site observations from satellites or ground based telescopes with performant instrumentation will provide stimulating constraints on the physics of these activity phenomena.
Partially based on observations obtained at ESO, and with IUE and EXOSAT satellites
Keywords: solar/stellar connection, magnetic field, activity, chromospheres/coronae, flares

1 THE SOLAR-STELLAR CONNECTION

The solar-stellar connection allows now for the first time to intercompare directly activity phenomena on stars and on the sun, and to understand better solar processes by modelling them on stars with different parameters. On solar-like stars, spots, plages, intense chromospheric/coronal heating, winds, flares and other aspects of stellar activity are fundamentally magnetic in character. These phenomena manifest themselves as spectroscopic variations in lines and continua over a large wavelength domain and on timescales from seconds to years. The interest for studying chromospheres/coronae of late-type stars, apart from its own, arise also from the fact that they are excellent laboratories for using, testing and improving radiative transfer and modelling theories under conditions of non-LTE and non radiative equilibrium. It is also possible to diagnose from the chromospheric/coronal structure and dynamics the trace of subjacent phenomena associated with the existence of the magnetic fields in stellar surfaces, that are governed by the convective and internal properties of the stars. Some stellar situations allow to isolate better the physical processes at work in the various aspects of stellar activity. The activity relations with stellar parameters give also a tool for studying the global origin and consequences of magnetic field in solar-like stars.

1.1 Observations of small scale magnetic structures on the sun

High spatially resolved observations of the magnetic field in the photosphere have shown the existence of elementary field concentration in "flux tubes" of 1.5 kG in size less than the arcsec. These tubes, which originate from the coupling between convection and the magnetic field in the subphotosphere, are swept by the supergranular motions in the boundaries, and appear cospatial with the 0.2" filigree emission in white light (Dunn 1973) or the enhanced emission network observed in chromospheric spectroheliograms. The correlation between the chromospheric emission and the magnetic flux has been shown (Skumanich et al 1975). The vertical structure of these tubes can be diagnosed with high spatial resolution observations of emission excesses at different wavelengths corresponding to different temperatures. For instance 1" resolution pictures obtained with the Transition Region Camera (Bonnet et al 1980, Foing Bonnet 1984, Foing, Bonnet & Bruner 1986) in the 160 nm Ultraviolet continuum formed near the temperature minimum region, or in the Ly α line in the base of the transition region at 20000 K, or in the CIV transition region line at 10^5 K allow to span different altitudes for the diagnostic of these flux tubes and loops (cf Fig.1).

Figure 1, next page : Image at 1 arcsec resolution in the 160nm continuum obtained during the 2nd flight of the Transition Region Camera (TRC2). This filtergram shows at the same time the underlying photospheric dark spots (size 3-10Mm), surrounded by bright emission structures or chromospheric plages (100-200Mm). The chromospheric network structure (typical size 30Mm) corresponds to the concentration of magnetic flux tubes at the boundaries of supergranular convective cells (Foing, Bonnet 1984).

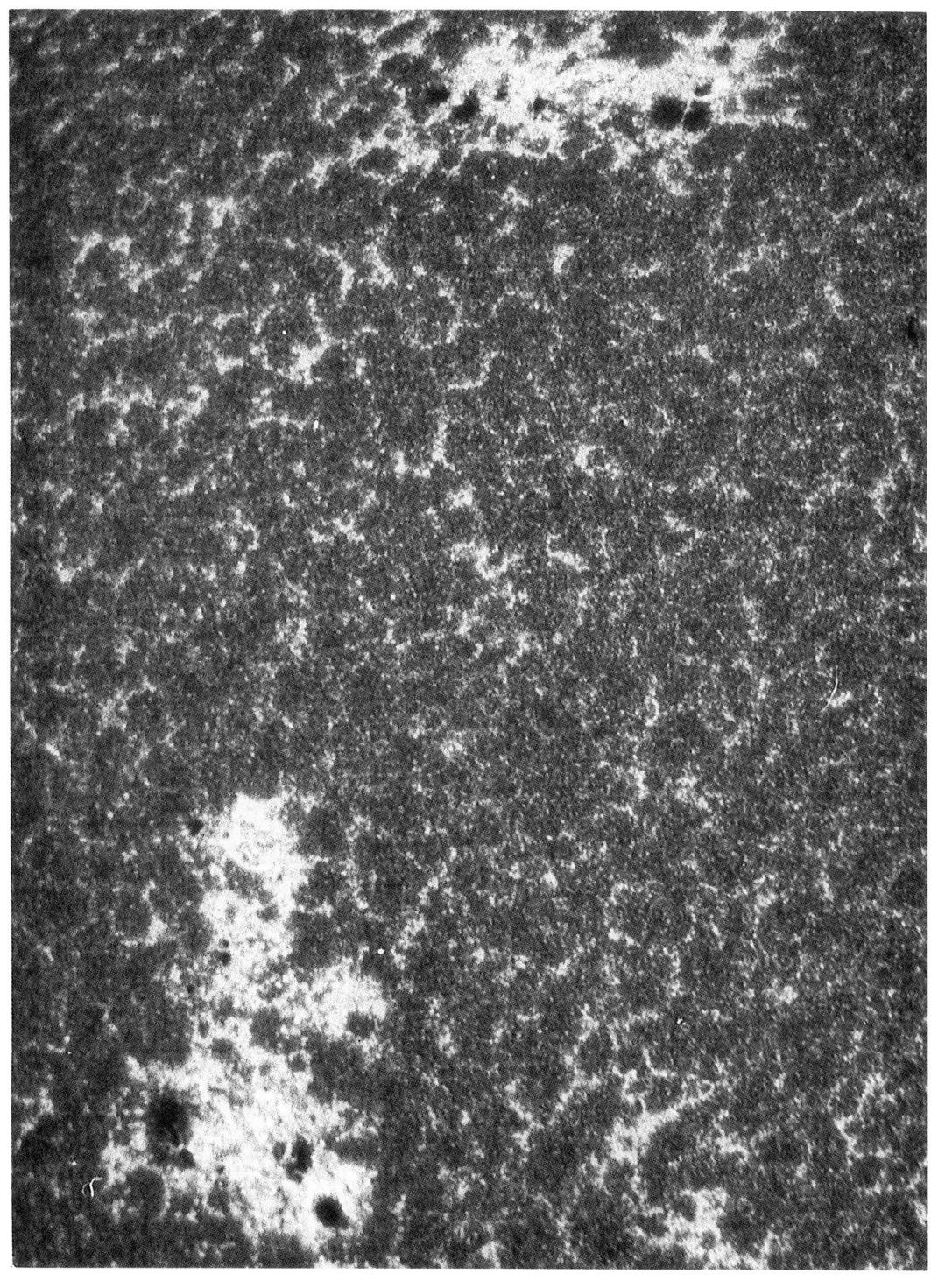

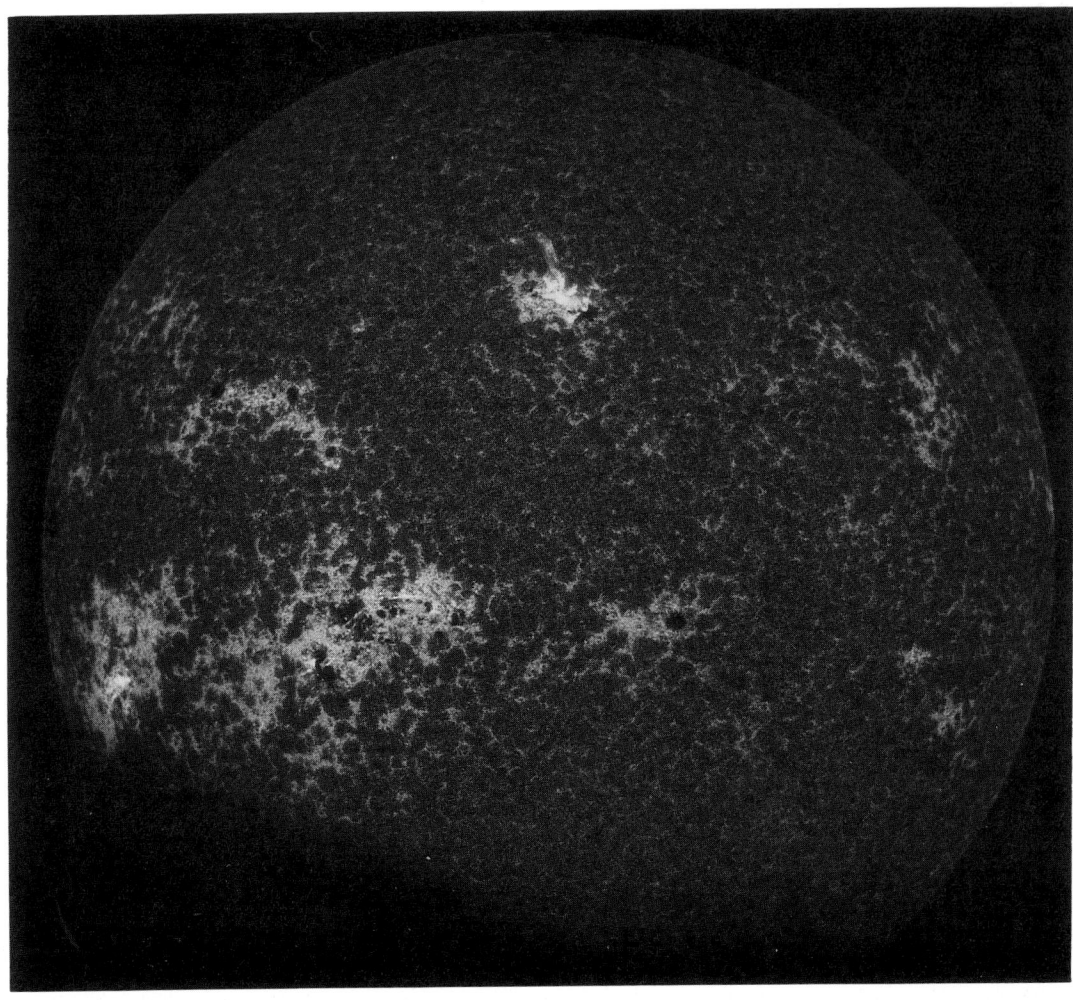

Figure 2 : Image at 1 arcsec resolution obtained by the TRC3 in 160nm over the solar disk (Foing, Bonnet, Bruner 1986b). The large scale emission structures due to extended complexes of magnetic regions concentrated along near equatorial bands can be distinguished. A flare started on the higher latitude active region and was also observed in real time during the rocket flight, leading to an ejected filament (Foing et al 1986a).

1.2 Semiempirical solar models

Basic and powerful tools for the study of the solar chromosphere are the semiempirical models, constructed by using observations either of EUV continua formed at different height in the chromosphere (as was derived by Vernazza, Avrett and Loeser, 1981 from Skylab observations, VAL81), or from the modelling of strong chromospheric lines as Ca II H and K observed from the ground, or Mg II h and k, Ly α from space. The current models rely on approximations of plane parallel geometry, homogeneity and hydrostatic equilibrium, with a very simple treatment of microturbulent velocities. Such models from the relation between temperature and height allow to calculate iteratively the pressure, ionization balance, population levels, mean intensity and the emergent spectra in different lines and continua. The predictions can then be compared to observational parameters or directly to the spectra, and the input model modified to allow a better adjustment to the data. Such a semi-empirical model (e.g. VAL81) indicates also the hierarchy of typical heights of formation of spectral features in the atmosphere.

However solar images of the sun obtained from Ca II spectroheliograms or EUV images (cf. Fig.2) show the inhomogeneity of the solar chromosphere (at large scale for the network and plages, and at small scale for the structure and properties of magnetic elementary flux tubes). With the advent of observations at high angular, spectral and temporal resolution on the solar chromosphere in different domains of the electromagnetic spectrum, it has been possible to analyse the effects due to the spatial heterogeneity, the magnetic field or the velocity fields. In the solar chromosphere, dynamical phenomena are observed, such as oscillations and waves. Models for the time resolved response of the chromosphere to waves and oscillations have been performed for calculating synthetic profile variations and compared consistently with the observations (Gouttebroze, Leibacher 1980). The shocks in the tenuous higher chromosphere involve nonlinear changes in the profile which keep an asymetry signature in a temporal average. Also systematic downflows are observed in network regions as an indication of a possible vertical circulation pattern of the material in the chromosphere. Multicomponent models from the very quiet sun to the enhanced network, (VAL 81) as well as models of active components (model of plage from OSO8 spectra by Lemaire et al 1981) or of flares (Machado et al, 1980) have been developed. From these models, the net radiative losses in the main line transitions and continua can be calculated as in Avrett (1985), showing the role of the Ca II, H alpha, H-, Mg II and Ly alpha losses in the chromospheric energy balance at different heights. These lines provide complementary and partially overlapping constraints on the temperature structure, and of the physical conditions (including the velocities) at the heights where they are formed.

2 SOLAR-LIKE CHROMOSPHERES

2.1 A general chromospheric programme of spectroscopic observations

On the base of this experience in the solar chromospheric research, and of the advent of instruments for stellar spectroscopy providing chromospheric profiles with a quality comparable to that of solar data, we have set up a network of collaboration between 3 institutes (IAS/LPSP Verrières, OAT Trieste and IAC Canarias), and defined a general program of observations and modelling of chromospheres in late-type dwarfs.

We have obtained for a sequence of active and quiescent dwarfs from spectral type F8 to K5, a series of spectra at high resolution and high signal to noise in the main chromospheric lines of Hα, Ca II infrared triplet and Ca II H and K from ESO with the CES, and h and k lines of MgII from the IUE satellite. The stars were selected to exhibit different degrees of chromospheric activity as measured as the enhanced chromospheric core emission in Ca II H and K. The signal to noise and the improved reduction techniques that allowed us reliable measurements and a quantitative comparison are discussed by Crivellari and Foing (1988). When comparing the Hα profiles of these pairs of active/quiescent stars, a difference was also found in the core due to the different chromospheric contribution over the background photospheric absorption profile due to the intrinsic S/N and to the precision of our registration procedures. A similar emission was also measured in the cores of the Ca II infrared triplet lines at 8498Å and 8542Å, and corresponding activity indices or fluxes were studied as a function of spectral type and compared with other chromospheric fluxes in Foing et al (1989).

2.2 Information from calibrated line profiles

The chromospheric spectra have been calibrated in order to estimate the absolute excess chromospheric fluxes either by using literature values for the normalisation of the continuum or pseudocontinuum. Another method was also used for Ca II H and K profiles (Castelli et al, 1988) by adjusting synthetic profiles computed under LTE and radiative equilibrium to the observed photospheric lines in the CaII H wings (cf Rebolo et al 1989).

Line core intensity indices or normalised chromospheric fluxes must be measured with a sufficient accuracy in order to derive a reliable relation with the stellar parameters (effective temperature, gravity, rotation) that can be compared to the predictions of theories of chromospheric heating and stellar dynamos.

A specific information about the radiative transfer opacity effects, about the dynamics at given heights in terms of microturbulent, macroturbulent, systematic flows and waves require a precise measurement of possible different spectral signatures of these processes in the line profiles. The measurements of line asymmetries, bisectors and multivelocity components fitting parameters are presented in Crivellari et al (1987), Vladilo et al (1988) in this framework.

The evidence for large scale structures (similar to solar plages) has been indicated by variations of activity indices along the rotational period of the star. Different techniques involved in the analysis of the spectroscopic variability associated with the chromospheric or photospheric activity can be used for different stellar cases including rotational modulation of fluxes, profiles, velocity shifts and Doppler imaging of the stellar surface inhomogeneities.

High signal to noise is thus required, either i) to define quantitative indices of activity in order to study their dependence with stellar parameters; ii) to obtain a differential spectrum between an active and quiescent star; iii) to compare spectra to synthetic profiles computed under different physical assumptions or iv) to measure the variability signature of activity phenomena.

3 SPECTROSCOPIC VARIABILITY ASSOCIATED WITH ACTIVITY

3.1 Magnetic activity of solar like stars

The magnetic activity of several classes of late-type stars has been particularly studied, both observationally and theoretically. In solar-like late-type dwarfs, the magnetic field seems to be amplified by the dynamo effect resulting from the coupling between convective motions and the differential rotation. Also, the modulation of the stellar flux and of some chromospheric indicators of activity along the stellar rotation indicates the presence of photospheric spots and chromospheric active regions. In analogy with the Sun, the appearance of spots could be attributed to intense sub-photospheric magnetic fields which inhibit the convective energy transport until the surface. Also long term periodical variations indicate solar-like activity cycles (Wilson,1978) on late-type stars. Some empirical relations were established (Mangeney, Praderie 1984, Noyes et al 1984) between the coronal or chromospheric activity and the rotational velocity through the Rossby number suggesting that the heating is related to the production of magnetic energy through a dynamo mechanism. However, little is known on the actual distribution of those magnetic fields and activity phenomena on these stellar surfaces. Some recent global magnetic field measurements (Saar et al 1987) have allowed to start deriving mean magnetic intensities and surface filling factor for active dwarfs.

3.2 Chromospheric flux rotational modulation

Flux variability of Ca II H and K in solar-like stars was studied by O.C. Wilson (1978) in his classic work on magnetic activity cycles. Short term H and K flux modulation was observed by Vaughan et al (1981) and used by them to compute rotational periods. The relation period activity-age using H and K was studied by Vaughan and Preston (1980). The possibility to use high S/N profiles instead of fluxes can give us three new types of information: 1) Proportion of active (i.e. plage + network) to quiet stellar cover, given a knowledge of the intrinsic profile emitted by each component; 2) The presence of velocity fields in three dimensions: the presence of an asymmetry or the position of a bump in the line profile, and the speeds with which these features change should enable us to distinguish vertical quasi-convective or wave motions from the projected rotational motions of plages; 3) Intensity and velocity can be combined via the technique of Doppler imaging to produce maps of stellar surfaces specifying active and quiet regions.

In Fig 3., we present the Ca II H rotational modulation for the K2 dwarf alpha Cen B.

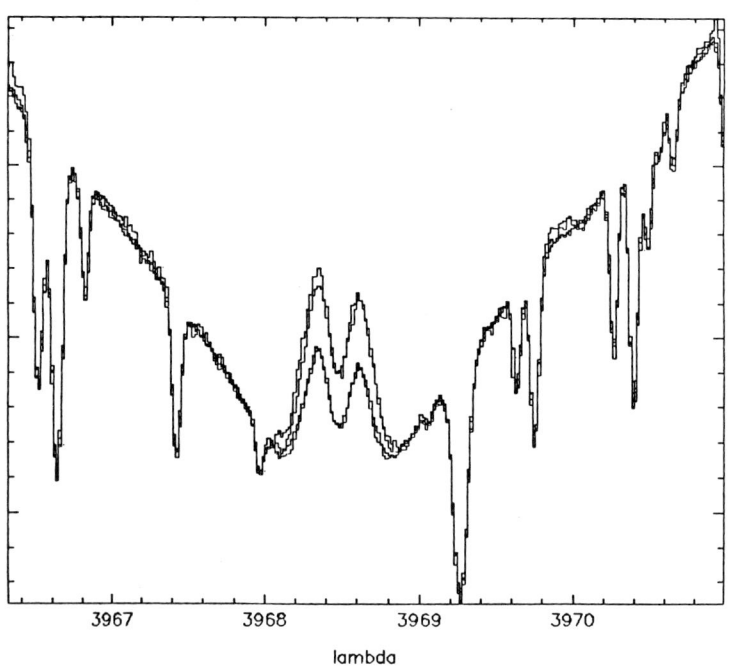

Fig.3 Observed variations in the Ca II H profile of alpha Cen B. The spectra taken respectively on 8, 9, 17 and 19 june 1985 show the rotational modulation of the profile by a long lived emitting structure (plage).

3.3 Solar-like components: quiescent, network, plage

The observed integrated flux is $F_l = \iint_\Omega I(M, \lambda - \lambda_0 - \lambda_{Rot}) \cdot (1-\varepsilon+\varepsilon\cos\theta) \cdot \cos\theta \, d\Omega$ where the center to limb factor ε may depend on the wavelength λ and on the component (quiet, network or plage) and $\lambda_{Rot} = y\lambda_0\Omega \sin i/c$ is the radial projection of the rotational velocity. The use of a strong rotational velocity to provide a Doppler image is described further in the paper, and the possibility for making velocity diagnostics on slower rotators in Crivellari et al (1987). At first, we then restrict ourselves to the variability of the profile due to rotational modulation by active structures. A three component description can be assumed, including a quiet stellar background virtually free of magnetic field, a network magnetic structure distributed homogeneously over the stellar surface and an enhanced emission from plages at given positions. If rotational velocity is neglected, the stellar flux can be expressed as $F = F_Q(1 - f_N - \Sigma \, f_{PN}{}^i I_Q/F_Q) + f_N F_N + \Sigma \, f_p{}^i I_p$ where $f_p{}^i = A_i \cos\theta_i \, (1-\varepsilon_i Q+\varepsilon_i Q\cos\theta_i)$ is an equivalent projected filling factor of the plage of area A_i specific intensity I_i, F_p being an average plage flux and $f_{PN}{}^i = A_i \cos\theta_i \, (1-\varepsilon_i Q+\varepsilon_i Q\cos\theta_i)$ the corresponding subtracted filling factor from quiet emission. If we compare two phases of observation, the only terms left in $F_2 - F_1$ are $\Sigma_2 - \Sigma_1 \, (f_p{}^i I_i - f_{PN}{}^i I_Q)$. If the center to limb effect difference is neglected $\varepsilon_i Q = \varepsilon_i P = \varepsilon_i$, and if we assume a fixed profile for the plage contribution $I_i - I_Q = a_i \, I_P(\lambda)$, then $F_2 - F_1 = I_P(\lambda) \cdot (\Sigma_2 - \Sigma_1 \, f_p{}^i a_i)$. Char and Foing (1989) simulated the rotational modulation of the Ca II H profile by plages to account for the observed variations. (cf Fig.4).

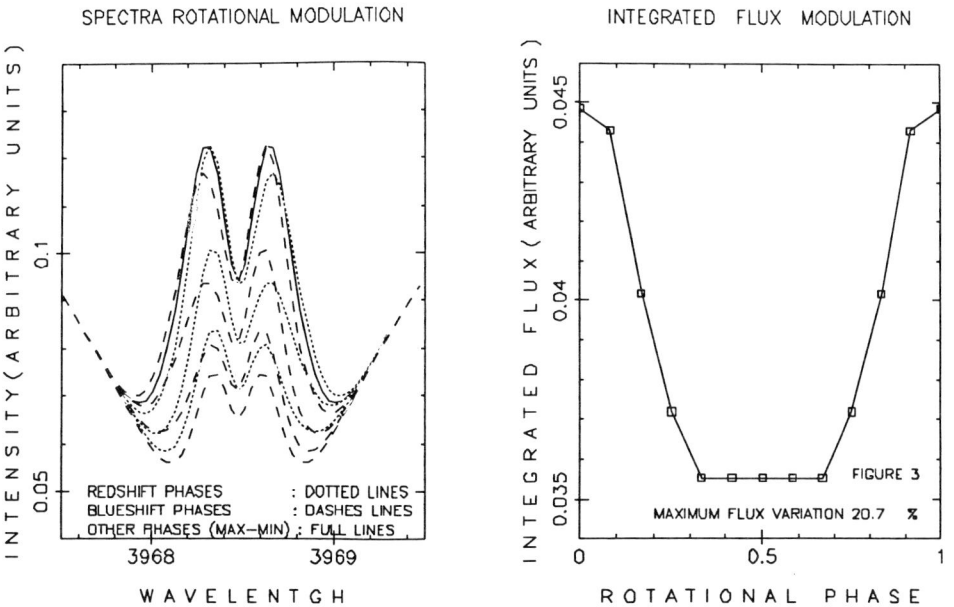

Fig. 4 Simulation of Ca II H profiles and 1A fluxes for different rotational phases of αCenB.

4 SOLAR-LIKE ACTIVE STRUCTURES ON STARS

The evidence collected from dedicated observations of active stars in the last ten years is strongly suggestive of a solar-type scenario with activity levels from the solar value to orders of magnitude higher. Activity phenomena are usually observed in red dwarfs, giants, T-Tauri and young stars, close late-type binaries because of i) contrast effects with the high temperature flare radiation and ii) deep convection zones and high rotation rates leading to differential rotation and efficient dynamos. They manifest themselves as flux variation in the continuum and emission lines over a wide range of wavelengths on timescales ranging from a few seconds, minutes (flares), hours and days (rotational modulation by active structures), to months and years (active region evolution, cyclic activity and differential motions).

4.1 Scientific interest for indirect imaging of RSCVn systems

RSCVn are binary detached systems with periods typically between 1 and 14 days, being synchronised by tidal effects, and generally composed with a subgiant primary (of type around K0 IV) and a dwarf secondary (around G5V). The review of properties for these systems can be found in Hall (1976). The most important photometric characteristic of these systems (Hall 1980) is the quasisinusoidal distortion of the lightcurve. Its slow migration towards decreasing orbital phases was discovered on RSCVn itself by Catalano and Rodono (1967). Among the proposed explanations, the spot model appear as the best established: the variations (except an eventual eclipse) are attributed to the rotational modulation of a nonuniform distribution of photospheric spots. The rotational modulation of chromospheric emission in the Ca II H and K lines was shown for several RSCVn systems in the Mt Wilson H and K variability survey (Vaughan et al 1981). Results from a joint IUE and ground based observations (Rodono et al 1987) have shown for the RSCVn system II Peg that the chromospheric and transition region emission reaches a maximum when the visible photometry is at minimum, thus indicating that chromospheric plages cover, in first approximation, an area correlated to the photospheric spots.

A further step in the study of the activity of these systems is to obtain the large scale distribution of the magnetic field on these active stars. Previous attempts to measure the magnetic field on these stars, using deconvolution of line profiles with different sensitivity to the Zeeman effect did not give any result until now. These measurements are made very difficult by the large rotational broadening (in general > 30 km/s) for these systems, and by the possibility that the magnetic field on the surface of the giant or subgiant may be lower than in the solar case. Thus, a crucial information relies on the knowledge of the spatial distribution of phenomena associated with these magnetic fields. The Doppler imaging method allows the localisation and reconstruction of the spots, active structures distribution, with constraints from the velocity dimension, in addition to the photometry. The observation of the rotational modulation of spectroscopic profiles and the application of reconstruction techniques must allow i) to obtain the configuration of the spots on these stars (polar, equatorial?), ii) to compare them with the signature of chromospheric, transition region or coronal emission, iii) to calculate extrapolated magnetic fields from these constraints and iv) to derive the vertical stratification and energy balance of magnetic structures. Also large scale magnetic structures can be followed over months and years to track differential motions, and constrain theories of internal rotation and dynamo.

4.2 T Tauri and young stars

Part of the surface activity of T-Tauri stars and young stars seems similar to the magnetic activity of other late type stars. For instance, rapid variations in X-rays, radio and in the optical are reminiscent of flares as observed on dMe or RSCVn stars. Also recently, repeated photometry (Bouvier et al., 1987) has shown a modulation of the light curve on time scale of a few days, that can be modelled with photospheric spots very similar to those observed on RSCVn stars. Also the measured rotational periods of T Tauri stars show a correlation with their coronal emission, that can be interpreted as the result of a coronal heating of dynamo origin. However those results suffer from the limits imposed by observing methods and interpretative models for adjusting photometric lightcurves. An additional information can be brought by Doppler imaging method in order to reconstruct an image of the temperature distribution, or of the chromospheric emission distribution on the stellar surface. Doppler imaging is well suited for stars with projected velocity between 30 and 70 km/s, which corresponds for T-Tauri stars to rotational periods from 2 to 6 days. Also spectral imaging methods in strong chromospheric lines such as CaII H & K, Mg II h & k, H α can provide the distribution of discrete chromospheric structures associated to the magnetic field. However T-Tauri stars being faint objects, such high resolution spectroscopy programmes require a major observing allocation on 2-4m class telescopes.

4.3 Circumstellar structures, coronal ejections and solar-like winds

Other young stars such as the Ae/Be Herbig stars present a rotational modulation in their wind, probably linked with the presence of a structured surface magnetic field. This poses the problem of the origin of such magnetic fields, and how this affects the stellar surface and circumstellar environment.

The interpretation of chromospheric rotational modulation in active stars can be complicated by the contribution of circumstellar regions to the line emission as observed in stellar winds (Basri, 1987). High resolution observations of T Tauri stars show a narrow component, of likely chromospheric origin and a wider component formed in a circumstellar region. Those contributions can be distinguished from their different temporal behaviour. The rotational modulation affects mainly the chromospheric component.

Also, circumstellar material can be observed in absorption due to ejected clouds of cool material transiting in front of a stellar disk, as described by Collier Cameron, Robinson (1988). The distribution of corotating clouds can also be derived from an inverse mapping technique of "skewing" technique. They deduced the structure of the clouds from continuous repeated observations at high resolution, in H alpha, and a more recent campaign showed also absorption transients in lines of Ca II, Mg II and Na I and even through H alpha photometry (Collier-Cameron et al 1990, Foing et al 1990).

5 INDIRECT IMAGING OF FAST-ROTATING STARS

5.1 Stellar active structures

The evidence for stellar photospheric spots can be obtained in the case of RSCVn or BY Dra systems from the photometric light curve periodic modulation. Migration of these photometric waves can also be observed as an indication of the change in the spot distribution over the surface. Modelling of the spotted image has been implemented for describing the photometric observations (e.g. Rodono et al 1987).

In particular the presence of plages and spots has been inferred from periodic low-amplitude photometric and spectral feature variations due to rotational modulation of spot-plage visibility . Systematic observations like those carried out at Catania Observatory for RSCVn and other active stars, have shown almost sinusoidal light curves to become multipeaked or even flat, suggesting variations in the spot number and distribution over the stellar surface. Spot and plage modeling (Byrne et al,1987) indicates that their physical characteristics are close to the solar ones, but they can cover up to 50% of the stellar surface. From simultaneous optical and IUE observations (Rodono et al 1986), a close spatial correlation between spots and chromospheric / transition region plages is apparent. High resolution spectroscopic observations of lines at different phases of the rotational period of active stars obtained with IUE or from the ground has allowed us to develop "spectral imaging" on IUE Mg II data (Walter et al, Neff Ph.D. 1987) , or "Doppler imaging" techniques for recovering the spatial distribution of surface activity over the star (Jankov, Foing 1987).

The objectives of such studies are to obtain the geometric distribution of activity phenomena, to understand the differences with their solar equivalent, to model the active and quiescent atmospheric regions, to study the correlation between the structures observed at different heights, and monitor the changes associated with active region behaviour, cyclic activity, dynamo phenomena and differential rotation. Basic considerations about the imaging of spotted stars from high resolution high signal to noise spectroscopy of photospheric lines have been given by Vogt and Penrod (1983).

5.2 Indirect imaging methods

The past decade has seen a very strong effort to understand and spatially resolve the surface structures and environment of late-type stars. As the techniques of interferometric imaging have still a limited angular resolution with the available baselines, indirect imaging techniques are necessary. An access to the information of quasistationary surface or extended structures is possible through the observation and interpretation of temporal photometric or spectroscopic variability along the rotational phase of the star, as well as during possible eclipses in binary systems. The intensity rotational modulation gives a one dimensional projection in longitude of the surface structures. In a method developed by Deutsch (1970) for chemically peculiar stars, the variation of line equivalent widths can be adjusted by parameters describing the development in spherical harmonics of the stellar surface inhomogeneities. This method did not make full use of the profile, mainly due to the low spectral resolution and low signal to noise ratio (S/N) available then on photographic plates spectra. The advent of high S/N observations with CCD and reticon detectors has stimulated the development of quantitative mathematical methods for studying the spectroscopic indirect imaging. With the Doppler imaging technique it is possible from the line profile disturbances observed at high spectral resolution and with adequate phase coverage, to obtain an information not only in longitude but also in latitude about surface stellar structures. Different formulation of the Doppler imaging method have been proposed or applied to various observations. Khokhlova (1985) calculates the line profile with the Doppler shifted contributions due to surface inhomogeneities; a Lagrange multiplier method is used to minimise an error functional between calculated and observed profiles, with a stabilisation Tikhonov functional. Jankov (1987), Jankov and Foing (1988) gave a mathematical formulation for the indirect imaging, and compared the reconstruction efficiency for different regularisation factors. Vogt et al (1987) express the relation between local surface intensities and the observed spectral profile using a matricial relation.

5.3 Comparison with observations

Several results have been obtained on Ap stars and RSCVn type stars by Vogt et al (1987, 1988), Khokhlova and collaborators (1984,1986). However, some aspects deserve further work, such as: i) the importance of the noise on the solution, ii) the possible biases in the reconstruction, iii) the position dependent resolution and accuracy for a characterised set of observations, and iv) the role of uncertainties in the stellar matrix of transformation parameters. A simple variant of the Doppler imaging method has been applied by Gondoin (1986), on observations of HR1099, by identifying bumps components in the profile and following their velocity changes with rotational phase. Recently, Neff (1987), Walter (1987) and collaborators have developed a spectral imaging method adjusting IUE MgII emission spectra of the system AR LAC with a minimal number of components.

5.4 Illustration of a method for Doppler imaging

Intensity and velocity can be combined via the technique of Doppler imaging to produce maps of stellar surfaces specifying active and quiet regions. Spectra of the RSCVn type binary HR1099 were obtained in december 1984 with the CAT + CES at ESO LA Silla in the 6430Å range at different phases, and around the H alpha line (cf Fig.5) showing the spectral changes due to the orbital motion of the system (Jankov, Foing 1987, Foing et al 1990). The rotational broadening of both components and the bump changes in the primary spectrum that are used for Doppler imaging can also be noticed.

The spotted image of the star is given in a sin B , L system coordinates with a number of pixels that is limited a priori by the resolution of the spectrograph versus the rotational broadening, and the number of observed phases (Jankov1987). Projection subpixels are also defined in order to keep the accuracy for the projection of equal radial velocity stripes. The apparent stellar surface can be divided in stripes of equal radial velocity (and then of equal Doppler position in the observed profile), which projections at a given phase are given on the stellar image defined previously. The position of a spot can be obtained from the intersection of these projections at different phases. Two methods can be distinguished, depending whether narrow band photometry of the continuum simultaneous to the spectroscopy , is available. The reconstruction of the image from such projections is an ill-conditioned problem, due to several sources: i) the deconvolution, ii) the matrix for the transformation contains systematic and random errors, iii) details observed close to the limb are a source of instability for the global reconstruction of the image. Then additional constraints, including a priori information, must be used to regularise the solutions.

5.5 Line profile synthesis and inverse Doppler imaging

We have produced first a test image of the primary star of the RSCVn system HR1099, with the spot positions given in Gondoin (1986), the elements of the system by Fekel(1983). Noise-free synthetic spectra were then calculated from the previous test image and system elements at different phases. We used the previous synthetic spectra in order to test our code for image reconstruction from spectra : different norms can be defined in the space of the solutions giving different methods and algorithms: the image reconstruction giving more ponderation to the CHI adjustment to the spectra is noisier and less representative of the input image, than the reconstruction giving more ponderation to the distance to an homogeneous distribution (Jankov, Foing 1987). The general method and algorithm for reconstruction of indirect images of rotating stars is presented in Jankov and Foing (1990).

The work is being followed : i) for finding an optimal mode of observation according to the instrument performances and stellar characteristics, ii) for using an optimal reconstruction method, iii) for getting information on the physical conditions used as constraints for the models, iv) for deriving the vertical structure of active regions localised by indirect imaging from multispectral observations.

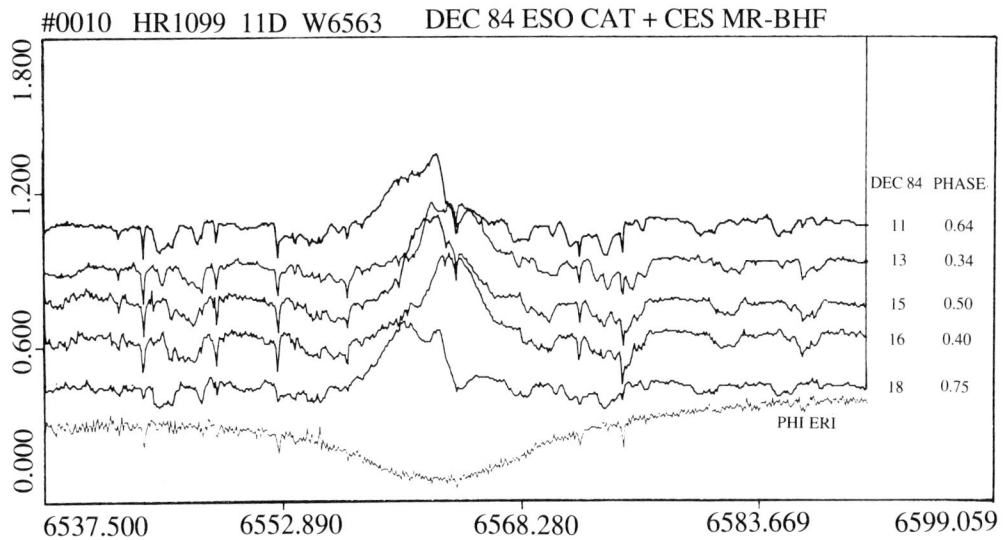

Fig.5: H alpha and photospheric line variation due to active regions of the RSCVn binary HR1099 vs the orbital phase . The comparison spectrum of the hot star phi Eri is shown for identification of H2O telluric lines.

6 FLARES ON SOLAR-LIKE STARS

6.1 Multiband observations of stellar flares

The flares are the most complex and violent phenomena occurring on these stars. Flare events have typical timescales of 10s for rise and 10^3s for decay. Simultaneous photometry and spectroscopy of flares have shown emission line enhancements of different species to take place as the response of the stellar atmosphere to the flare energy release. We refer here to the proceedings of the 104 th IAU Colloquium on
Solar and Stellar Flares (Haisch, Rodono eds, 1989) for an overview of the flare observations and corresponding theories. Fundamental issues for flare research concern the energy transport mechanism: particle beam vs heat conduction, the atmospheric response to flares; mass motions, ejected components and momentum balance; microflaring, flaring and the heating of coronae; statistics and recurrence of flares are adressed in Foing (1989), with relevant stellar flare spectroscopic diagnostics. The scientific objectives of this flare program are to determine: 1)what is the energy budget for a typical sample of flares, 2)what are the respective roles of radiation from the corona (as shown by the X rays), conductive losses through the transition region (EUV) and expansion (as indicated from velocity fields measurements), 3)what are the temperatures, densities and volumes of the hot flaring plasma. We shall here mention only some results obtained within multiwavelength observation campaigns. IUE observations of stellar flares were obtained , and soft X-ray flares were also detected by SAS-3, HEA01, Einstein satellites. We have undertaken several campaigns of coordinated X ray , UV, optical, infrared and radio observations of flare stars and RSCVn binaries, the more recent involving ESO and CFH in March 1984, in december 1984, in March 1985, in February and September 1987. An infrared decrease was for the first time detected during a flare, simultaneous to photometric, spectroscopic and radio (VLA) changes (Rodono et al 1985) . A correlation between ESO 3.6m spectroscopy of Balmer line H gamma and EXOSAT X-ray monitoring was found, suggesting the possible importance of microflares for the coronal heating of those stars (Butler et al 1986). The time behaviour of different lines and continua of several flare events has been analysed (cf Fig.6) (Foing 1989, Rodono et al 1990, Houdebine, 1990).

Our methods of investigation for these programs require, due to the transient behaviour of activity phenomena, especially for flare events that affect various atmospheric levels on a very short time scale, that we obtain simultaneous observations at all accessible wavelengths.

6.2 Need for Multiwavelength Multi Site Observations

The scientific need for multiwavelength multisite observations of these intrinsically variable active stars, has involved us in the organization of coordinated campaigns of observations, employing the various areas of expertise of our collaborating groups (Armagh and Catania observatories, IAS/LPSP, JILA and Lockheed), and their access to large telescopes and satellite observatories. Also, a project for a MUlti SIte COntinuous Spectroscopy network (MUSICOS) was set up to provide round-the-clock observations of such objects (Catala, Foing 1988). This work requires the analysis of different sets of data, the development and exchange of new methods and software for data handling, reduction, processing and archiving. Finally, we need to discuss the theoretical analysis of data corresponding to very different wavelengths and emission mechanisms, in relation with solar physics and astrophysics (MHD, stellar atmospheres, etc).

This collaborative research was set in the context of programmes accepted for ESA/NASA satellites (IUE, EXOSAT) and for large ground based telescopes at ESO, CFHT and US observatories. The longevity of the IUE satellite and the access to new ground-based telescopes, give a further perspective to the program. The future launch of the space telescope and of ROSAT (especially the HRS spectrograph) will give us instruments enabling to push research beyond present limits. In view of the competition for observing time on ST, it is necessary to prepare these scientific programmes well and to insure that the overall necessary expertise exists (coordination of space/ground based observations, data reduction, archival, analysis, scientific interpretation) from both solar/stellar aspects.

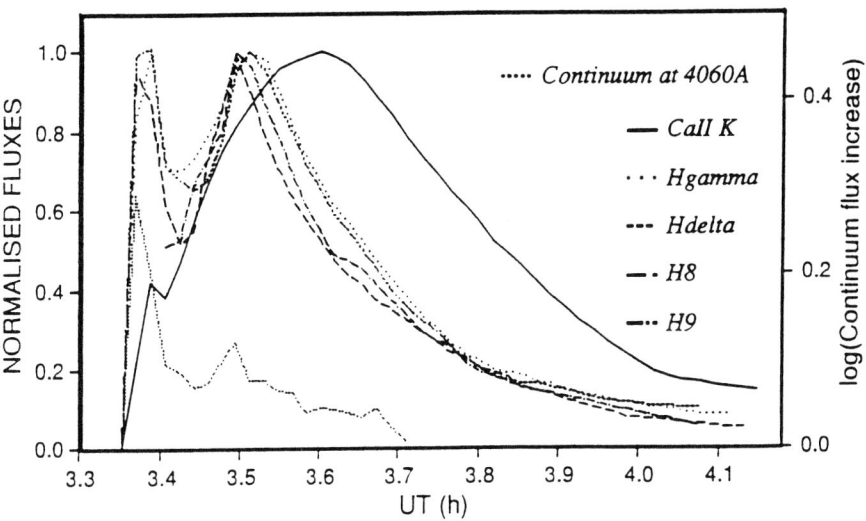

Fig.6 Time evolution of line fluxes and continuum of an AD Leo flare observed at ESO 3.6m.

7 CONCLUSION

Active regions and flares reflect the underlying stellar magnetic field and dynamo, and contribute a significant heating towards the upper stellar atmospheres. From high-quality spectra of the main chromospheric lines of Ca II H and K and their corresponding h and k in Mg II (observed by IUE), H alpha and the infrared triplet lines of Ca II, constraints on models incorporating velocity fields diagnostics and chromospheric inhomogeneities have been investigated. The rotational modulation of fluxes, asymmetries and spectral profiles allows to diagnose active regions on moderately rotating stars. For active fast rotating stars, the line profile changes can map the information on the geometrical surface structure distribution. For RS CVn type systems, profiles were calculated to account for the rotational broadening of immaculate and spotted stars. A line synthesis code including a model of the inhomogeneous temperature distribution of the primary star, and the binary system elements and parameters, allowed to calculate a series of spectra at different orbital phases. Different inversion algorithms were applied to test the image reconstruction.

From a series of flare campaigns from IUE, EXOSAT satellites, VLA, ESO, CFHT and other ground based observatories, we highlighted some results obtained (from the radio, infrared, visible, ultraviolet to the X-ray range) coordinated observations of flare events on dMe stars. We stress the need to coordinate future synoptic observations at all accessible wavelengths, and with multi-site networks of spectrophotometers, for these objects which are highly variable on all timescales from seconds to years.

Acknowledgements. We thank our colleagues R.M. Bonnet, L. Damé, M. Bruner, M. Martic (from the TRC collaboration), J. Beckman, L. Crivellari, R. Rebolo, G.Vladilo (from the IAS/IAC/Trieste collaboration), J. Butler, B. Byrne, J.G. Doyle, P. Panagi, M. Rodono, S. Catalano, G. Cutispoto, B. Haisch, J. Neff, J.L. Linsky (from the Cool Star Consortium) and C. Catala (and other collaborators from the MUSICOS project) for their participation or their kind advices and scientific discussions about some of the work reviewed here. We acknowledge also the support from different sources: CNES, ESA, NASA and SERC, ESO, CFH-INSU , NATO Grant for International collaboration in research; CNRS (IAS/LPSP and interlaboratory thematic GRECO funds) , CNR, Italian Ministry, Catania observatory, EEC (French contribution to EEC scientific programmes) for exchange of european scientists in the frame of a network of research laboratories on Stellar activity coordinated by B.H.F..

References

Avrett, E.H.: 1984. "*Small Scale Processes in Quiet Stellar Atmospheres*", Ed. S. Keil.
Beckman, J., Crivellari, L., Foing, B.H.: 1984. *ESO Messenger,* **38**, 24.
Bonnet , R.M. et al.: 1980. *Astrophys. J. (Letters)*, **237**, L47
Bouvier, J.: 1987, These de doctorat, Universite Paris VII
Butler, C.J., Rodono, M., Foing, B.H. et Haisch, B.: 1986. *Nature*, **321**, n° 6071, 679.
Butler, C.J., Doyle, J.G., Foing, B.H. et Rodono, M. : 1987 in *Activity in Cool Star Envelopes*, Tromso meeting
Byrne, P.B. et al.: 1987. *Astron. Astrophys.*, **180**, 172.
Castelli, F. et al.: 1988. "*The Impact of Very High S/N Spectroscopy on Stellar*

Physics", IAU Symp 102, Edts. G. Cayrel de Strobel and M. Spite, p. 153.
Catala, C., Foing, B.H. Eds: 1988, *1st Workshop on Multi-Site Continuous Spectroscopy*
Catalano, S.,Rodono, M.: 1967, *Mem. Soc.Astron.Ital.*, **38**, 395
Char, S., Foing, B.H.: 1989, in *"Modeling the Stellar Environment"*, Ed Frontieres, Gif
Collier-Cameron, A., Robinson, R.D.: 1989, *MNRAS* **236**, 57
Collier-Cameron, A. et al: 1990, submitted
Crivellari, L. et al. : 1987. *Astron. Astrophys.* **174**, 127.
Deutsch, A.J.: 1970, *Astrophys.J.* **159**, 985
Dunn R. : 1973. *Solar Physics* **33**, 281.
ESA SSD 1985 &1987, SOHO studies (Report on Phase A and A/O document).
Fekel, F. : 1983. *Astrophys. J.*, **268**, 274.
Foing, B.H. : 1983. Thesis Univ. Paris VII-Meudon-LPSP.
Foing, B.H. and Bonnet, R.M.: 1984a, *Astrophys. J.*, **279**, 848.
Foing, B.H. and Bonnet, R.M.: 1984b, *Astron. Astrophys.*, **136**, 133.
Foing, B.H. et al.: 1985. *ESO Messenger*, **41**, 18.
Foing, B.H.et al: 1986a, in *The Lower Atmosphere of Solar Flares*, Ed D.Neidig, p.319
Foing, B.H., Bonnet, R.M. and Bruner, M.E. : 1986b. *Astron. Astrophys.* **162**, 292.
Foing, B.H., Crivellari, L.: 1988, IAU SYMP 102, ibid
Foing, B.H.: 1989, *Solar Phys.* **121**, 117-133
Foing, B.H. et al: 1989, *Astron. Astrophys. Suppl.*, **80**, 189
Foing, B.H. et al: 1990 , in preparation
Gondoin P. : 1986. *Astron. Astrophys.* , **160**, 73
Gouttebroze, P. , Leibacher, J.: 1980. *Astrophys. J.* **238**, 1134.
Gouttebroze, P. et al. : 1986. *Astron. Astrophys.* **154**,154.
Haisch, B.M. and Linsky, J.L.: 1980. *Astrophys. J.(Letters)* **236**, L33
Haisch, B.M. and Rodono, M. Eds: 1989, *"Solar and Stellar Flares"* in *Solar Phys*.121
Houdebine, E.: 1990, PhD, in preparation
Jankov S. : 1987. DEA Univ Paris VII/LPSP.
Jankov S., Foing B.:1987 in *"Cool Stars, Stellar Systems and the Sun"*, Linsky, Stencel eds
Khokhlova, V.L., Rice, J.B., Wehlau, W.M.: 1986, *Astrophys. J.* **307**, 768
Khokhlova, V.L.:1985, Astrophys. Space Phys. Rev. **4**, 99
Khokhlova, V.L., Pavlova, V.M.: 1984 Sov. Astron. Lett. **10**, 158
Lemaire, P. et al.: 1981. *Astrophys. J. Suppl.* **45**, 350.
Machado,M.E., Avrett,E.H., Vernazza,J.E., Noyes,R.W.: 1980, *Astroph.J.* **242**, 336
Mangeney, A., Praderie, F.:1984, *Astron. Astrophys.* **130**, 143.
Marstadt, et al.: 1982, NASA Conf. Publ. 2238, 554
Neff, J. : 1987. PhD Thesis, JILA , Boulder.
Noyes, R.W. et al: 1984, *Astrophys. J.* **279**,763.
Rebolo, R. et al: 1989, *Astron. Astrophys. Suppl.*, **80**, 134
Rodono, M. et al.: 1984. *Publications ESA-SP* , **218**, p.247.
Rodono, M. et al.: 1985. *ESO Messenger*, **39**, 9.
Rodono, M. et al.: 1986. *Astron. Astrophys.* **165**, 135.
Rodono, M. et al.: 1987. *Astron. Astrophys.* **176**, 267.
Saar, S. , Linsky, J.L., Beckers, J.M.: *Astrophys. J.* **302**, 777.
Skumanich, A., Smythe C. and Frazier, E.: 1975. *Astrophys. J.* , **200**,747.
Vaughan , A.H. and Preston, G.W.: 1980. *P.A.S.P.* , **92**, 385.
Vernazza, J.E., Avrett, E.H., Loeser, R.: 1981. *Astrophys. J. Suppl.* **45**, 635.
Vladilo,G. et al: 1988, IAU SYMP 102, ibid
Vogt S. and Penrod H.: 1983. *P.A.S.P.* **95**, 565.
Vogt, S.S., Penrod, G.D., Hatzes, A.P.: 1987, *Astrophys. J.* **321**,496
Vogt, S.S.: 1988, IAU Symp. 132, eds. G. Cayrel & M. Spite, 253
Walter, F. et al.: 1987. *Astron. Astrophys.* **186**, 241.
Wilson, O.C.: 1978, *Astrophys.J.* **226**, 379

Quiescent X–ray Emission of Cool Stars

C.J. Schrijver

Space Science Department of ESA,
ESTEC, Mail Box 299, 2200 AG Noordwijk,
The Netherlands.

SUMMARY: X–ray observatories such as *EINSTEIN* and *EXOSAT* have shown that hot outer atmospheres are a common phenomenon of stars on the cool side of the HR–diagram. The coronal soft X–ray emission of these cool stars is often used as a diagnostic for stellar magnetic activity. Relations have been established between the X–ray emission and chromospheric and transition–region diagnostics, and stellar rotation rate. The detailed structure of stellar coronae cannot be studied because of the lack of spatial and spectral resolution. Nevertheless, the crude spatial structure of stellar coronae has been studied using rotational modulation of the X–ray emission of a few stars. A number of stars has been observed with limited spectral resolution, and correlations between coronal effective temperature and spectral type and rotation rate have been inferred. The higher spectral resolution of the *EXOSAT* transmission grating spectrometer allows tests of models of quasi–static coronal loops, and allows estimates of the loop geometry. Future missions, such as NASA's AXAF and ESA's XMM, promise the determination of coronal pressures, from which surface filling factors and loop lengths can be derived.

1. INTRODUCTION

The invention of optical narrow–band filters and spectrographs, which allowed high–resolution, wavelength–specific optical observations, has uncovered an unsuspected variety of effects related with atmospheric magnetic fields: active regions, cancelling magnetic features, emerging flux regions, faculae, filaments, moat cells, plages, spicules ... A long list of new terms was defined to describe the new discoveries. The X–ray telescopes of the late sixties and early seventies uncovered the effects of magnetic fields far above the photosphere: the corona, known to have temperatures of well in excess of 10^6K since the pioneering work of Grotrian and Edlén, appeared to have a highly inhomogeneous character. Magnetic loops were seen connecting opposite polarities observed in the photosphere. These fields, later demonstrated to confine and direct the flows of the tenuous coronal plasma, are often virtually potential field extensions of the photospheric fields (e.g. Poletto *et al.* 1975, Berton and Sakurai 1985).

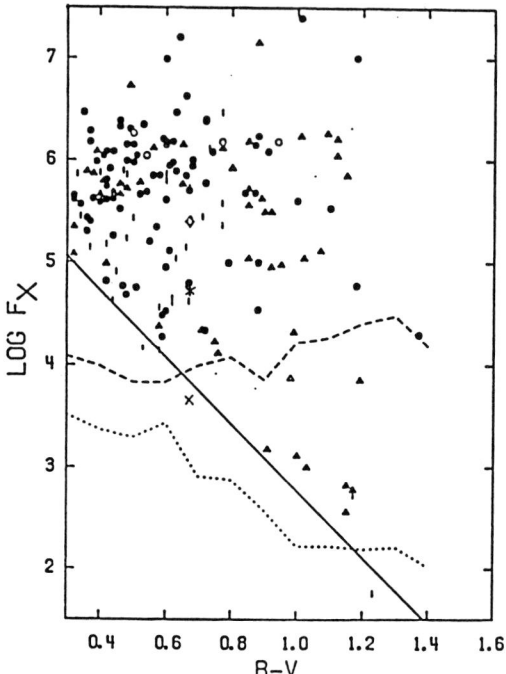

Figure 1. Soft X-ray flux densities F_X (as measured with the EINSTEIN IPC, 0.15–4. keV) vs. colour $(B - V)$ (figure adapted from Rutten et al. 1989b). Dwarfs are shown as circles and giants as triangles. Open symbols stand for less accurate flux detections, and the bars mark upper limits. The dashed and dotted lines show the intrumental detection limit for dwarfs and giants, respectively. Also shows are the mean flux density of a coronal hole $(\times)$, of a typical quiet region $(*)$, and the average value of an active region $(\diamond)$, cf. Section 4.

Stars which, like the Sun, have a convective zone immediately below the photosphere in general show signs of non–radiatively heated outer atmospheres. By analogy with the Sun, these outer stellar atmospheres are generally assumed to consist of magnetic loop structures. These loops are filled with a hot, X–ray emitting plasma. The soft X–ray emission of solar coronal loops and of their stellar counterparts are the topic of this review.

2. RADIATIVE DIAGNOSTICS OF STELLAR MAGNETIC ACTIVITY

One of the most direct measures of non–thermal atmospheric activity in cool stars is the stellar X–ray flux which originates entirely in the corona without being affected by a detectable photospheric contribution. Figure 1 plots the observed stellar X–ray fluxes against colour. At a given colour and luminosity class of a star, the observed stellar X–ray fluxes cover a large range, which implies that at least one additional parameter is required to describe stellar magnetic activity. Figure 1 shows that the detected X-ray fluxes go down to about the expected detection limits, except perhaps for the F-type stars. Flux-colour diagrams for radiative flux densities originating in cooler regions of the stellar outer atmospheres show a similar scatter in observed flux densities. The lowest detected values, however, increase relative to the detection limits and clearly exceed the detection limits for chromospheric diagnostics (Rutten et al. 1989a).

Solar-like activity has been shown to occur throughout the region in the HR-diagram where convective envelopes directly underlie the photospheres. Current theories of magnetic activity in cool stars therefore suggest that the origin of the magnetic fields is to be found in the interplay of convective motions and stellar rotation: magnetic fields are believed to be generated by a dynamo process of which at present incomplete knowledge exists. The dynamo process may feed on motions throughout the convective zone, or only on differential rotation in a thin region of rotational shear at the bottom of the convective zone. The dependence of stellar magnetic activity on rotation rate is widely accepted, but the precise functional relationship, or even the set of relevant parameters, has not yet been identified (Catalano, this volume). This relationship appears to depend on (at least) three parameters: effective temperature, surface gravity and rotation rate.

Several authors, starting with Ayres, Marstad, and Linsky (1981), have shown that the radiative losses from the outher atmospheres of the cool stars are highly correlated (e.g. Figure 2). The study of these diagrams suggests that two contributions to the observed stellar fluxes can be distinguished: a basal and an active component (Schrijver 1987b, Schrijver, Dobson, and Radick 1989, Rutten et al. 1989a). The active component is related with the stellar rotation rate and therefore, presumably, with a dynamo mechanism similar to that operating in the Sun. The basal component appears to be unrelated with rotation, depends strongly on colour, but may be a weak function of gravity.

If the observed fluxes are corrected for the basal components where necessary, the relationships between the radiative flux densities from stellar chromospheres, transition regions and coronae appear to be simple power laws which are independent of the fundamental stellar parameters (except for TTauri stars, e.g. Lemmens, Rutten and Zwaan 1989, and M-type dwarfs with their very dense chromospheres, Schrijver and Rutten 1987, Rutten et al. 1989b). Hence only a single flux needs to be observed from which all others can be computed, provided that the basal flux levels can be determined from the stellar colour.

3. MAGNETIC AND NON-MAGNETIC HEATING OF OUTER ATMOSPHERES

One of the earliest suggestions for a mechanism capable of heating the outer layers

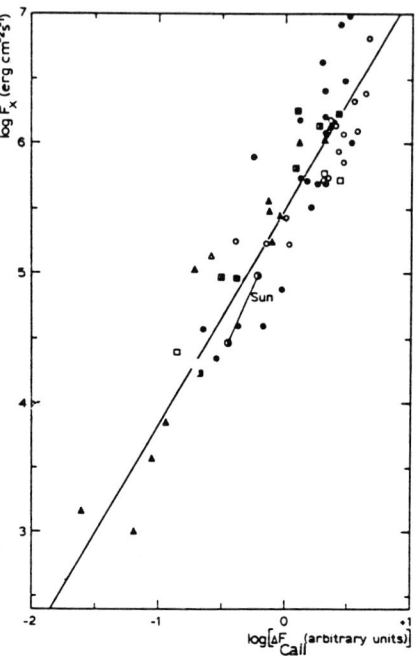

Figure 2. Soft X-ray flux F_X vs. the Ca II H+K flux in excess of the observational lower limit (from Schrijver 1983). Symbols: △ giants, □ subgiants, ○ dwarfs. Open symbols are used if $B - V < 0.6$, filled symbols if $B - V > 0.8$, and half-filled symbols for intermediate values.

of stellar atmospheres was the dissipation of acoustic waves. Motions in the convection zone generate noise that can travel upward into the atmosphere provided the wave periods lie below the acoustic cut–off period. Stepien and Ulmschneider (1989) and Hammer and Ulmschneider (1989) show that heating by pure acoustic waves can produce at best only a very weak corona (see also the papers by Hammer and by Ulmschneider in this volume). Hence, it appears that the soft X-ray emission of cool stars can only originate in plasma heated to coronal temperatures by processes associated with magnetic fields. Narain and Ulmschneider (1989) review the proposed coronal heating mechanisms, among which we find mode coupling, phase mixing, resonant absorption of Alfvén waves, and current heating by reconnection perhaps caused by microflares or the quasi–continuous removal of an azimuthal field component introduced by the random walk of the footpoints of the flux tubes. It should be noted that acoustic modes may be indirectly responsible for (part of) the radiative losses associated with magnetic activity because of various forms of mode coupling (see Ulmschneider, this volume).

Although pure acoustic waves are not responsible for the bright and highly structured

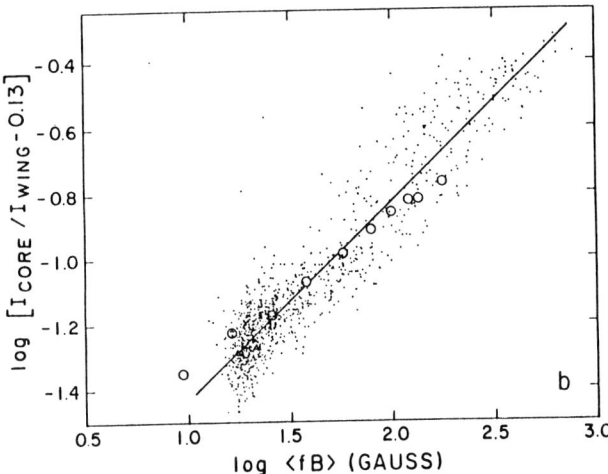

Figure 3. The Ca II K line-core excess intensity over a line-wing intensity (i.e. after subtraction of a minimal line-core intensity ratio) vs. absolute value of the magnetic flux density $\langle fB \rangle$ measured with 2."4 resolution (figure from Schrijver et al. 1989). The power-law fit has an exponent of 0.6. The open circles show results of Skumanich, Smythe and Frazier (1975).

corona of Sun and stars, they may contribute significantly to the chromospheric heating. It has been suggested that the basal components of the chromospheric radiative losses are directly caused by the dissipation of pure acoustic waves. Verification of this hypothesis is difficult: the amount of acoustic energy that is generated in the convection zone is certainly large enough to account for the chromospheric basal flux, but the large effects of radiation damping (see Ulmschneider, this volume) are not known with sufficient accuracy to allow a conclusive test.

The direct evidence for a link between the active component in radiative diagnostics of stellar magnetic activity and the magnetic field is growing both for solar and stellar data. Recently Schrijver et al. (1989) extended the earlier work of Skumanich, Smythe and Frazier (1975) and demonstrated a direct link between the Ca II K line-core intensity and the magnetic flux density in both quiet and active solar regions (Fig. 3). Saar and Schrijver (1987) presented preliminary results for the relationship between radiative losses and the magnetic flux for stars (Fig. 4).

4. SOLAR CORONAL EMISSION

Solar coronal structures can be divided into three global categories: open field structures, quasi-stationary closed coronal loops and flaring structures. Flares and coronal holes are discussed in the reviews by Svestka and Hammer in this volume. Quies-

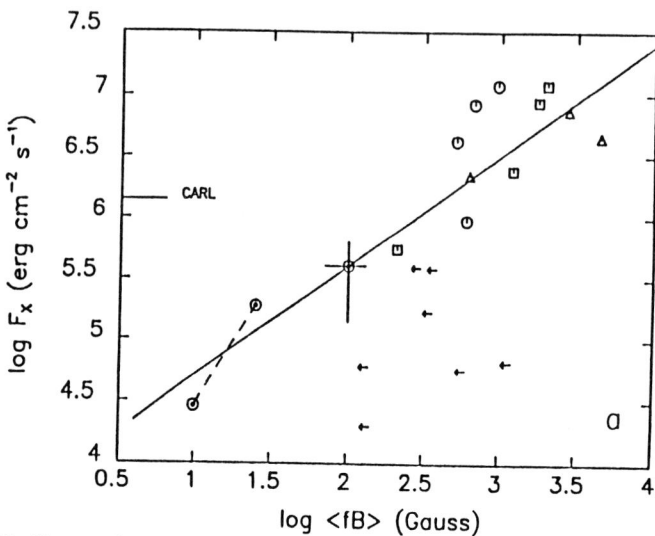

Figure 4. Stellar soft X-ray flux density F_X vs. the hemisphere-averaged absolute magnetic flux density $\langle fB \rangle$ (from Saar and Schrijver 1987). The estimates for the Sun at minimum and maximum activity are connected by a dashed line. The full-drawn line shows the best fit. The mean value and range for solar active regions are shown for comparison. The X-ray flux of one of the brightest non-flaring regions on the Sun, the Compact Active-Region Loop (CARL), is also shown. Symbols: ○ G-dwarfs, □ K-dwarfs, and △ M-dwarfs. Arrows mark upper limits.

cent, closed coronal features, the topic of this review, are much brighter than the coronal holes mainly because of their higher plasma density: the typical soft X-ray flux density of coronal holes is about $4\,10^3\,erg\,cm^{-2}\,s^{-1}$, over quiet regions about $5\,10^4\,erg\,cm^{-2}\,s^{-1}$, and over active regions about $2\,10^5\,erg\,cm^{-2}\,s^{-1}$ (from Schrijver et al. 1985). Loop lengths and pressures range from $L \approx 3\,10^9\,cm$ and $p \approx 5\,dyne\,cm^{-2}$ for compact active-region loops (CARL), through $L \approx 9\,10^9\,cm$ and $p \approx 1.5\,dyne\,cm^{-2}$ for extended active-region loops (EARL, matching the average of a solar active region), to $L \approx 2\,10^{10}\,cm$ and $p \approx 0.2\,dyne\,cm^{-2}$ for large-scale loops (LSL) which interconnect active regions (Pallavicini et al. 1981). The mean temperature of coronal condensations over active regions is about 3 MK, ranging from about 2 MK up to 6 MK, increasing with the Ca II K brightness of the plage (Schrijver et al. 1985).

The geometry of the plasma volume in the upper parts of the loops is prescribed by the magnetic field. The field is expected to diverge upward, away from the photospheric bipolar active region (see also Section 1). This is in fact indirectly observed on the Sun: the coronal condensation over solar bipolar regions has a projected area roughly an order of magnitude larger than the area of the underlying photospheric plage (e.g. Schrijver 1987a) suggesting that the ratio Γ of the loop cross section at the apex to the cross section at the footpoint (high in the chromosphere) is approximately $\Gamma \simeq 10$, or an increase in diameter by a factor of about 3 (*cf.* Section 1).

5. STELLAR CORONAL STRUCTURE

The above sections describe in some detail the arguments why we may assume that the X-ray emission from most of the cool stars during and after their main-sequence phase is primarily associated with stellar magnetic activity. A possible acoustic component exists in chromospheric diagnostics, but after correction for this component the relationships between radiative fluxes from different regimes of the atmosphere are simple power laws: there is little or no colour dependence, no clearly observed dependence on luminosity class, age, binarity, ... In fact, the data suggest that once energy is pumped into the magnetic field, it is distributed over the different temperature regimes in such a way that it is impossible for us to tell different types of stars apart. With this in mind, X-ray spectroscopy of quiescent stellar coronae might be expected to be a rather uninteresting field: should one not expect magnetic structures, plasma densities and electron temperatures similar to those observed for the Sun? Below I show that reality is in stark contrast to this naive expectation: stellar coronae do differ from star to star, depending on luminosity class and rotation rate.

5.1 Medium-resolution soft X-ray spectroscopy.

It may appear strange to start with the observations for which generalization to stars in general is very difficult: only three stars – Capella (G6III+F9III, the candidate X-ray emitter is underlined), σ^2 CrB (F8V+G1V), and Procyon (F5IV-V+DF) – have been observed with the highest spectral resolution achieved to date over a large wavelength range. These observations, however, set the stage for the interpretation of results obtained with low-resolution telescopes and broad-band photometry.

The spectra (Figure 5) of the three stars mentioned above have been observed with the *EXOSAT* Transmission Grating Spectrometer (De Korte et al. 1981, Brinkman et al. 1980, Mewe 1984). The *TGS* observes the wavelength range between 10 Å and 200 Å with approximately 3 Å resolution. Lemen et al. (1989) use the spectra to derive differential emission measure distributions, which are defined as follows. The line and continuum spectral intensity $f(\lambda_j)$ (counts s^{-1}) measured at Earth in a bin at wavelength λ_j for a multi-temperature plasma is given by:

$$f(\lambda_j) = \frac{1}{4\pi d^2} \int F(\lambda_j, T) n_e^2 \, dV, \tag{1}$$

where n_e is the electron density (cm^{-3}), V the plasma volume (cm^3), $F(\lambda_j, T)$ is the spectral emissivity (counts cm^5 s^{-1}) for the line plus continuum emission as a function of temperature T integrated over wavelength bin λ_j, convolved with the instrumental response function. The distance to the source is d (cm). With $\varphi(T) = n_e^2 \frac{dV}{dT}$ (cm^{-3} K^{-1}):

$$f(\lambda_j) = \frac{1}{4\pi d^2} \int F(\lambda_j, T) \varphi(T) \, dT = \frac{1}{4\pi d^2} \int F(\lambda_j, T) T \varphi(T) \, d(\ln T), \tag{2}$$

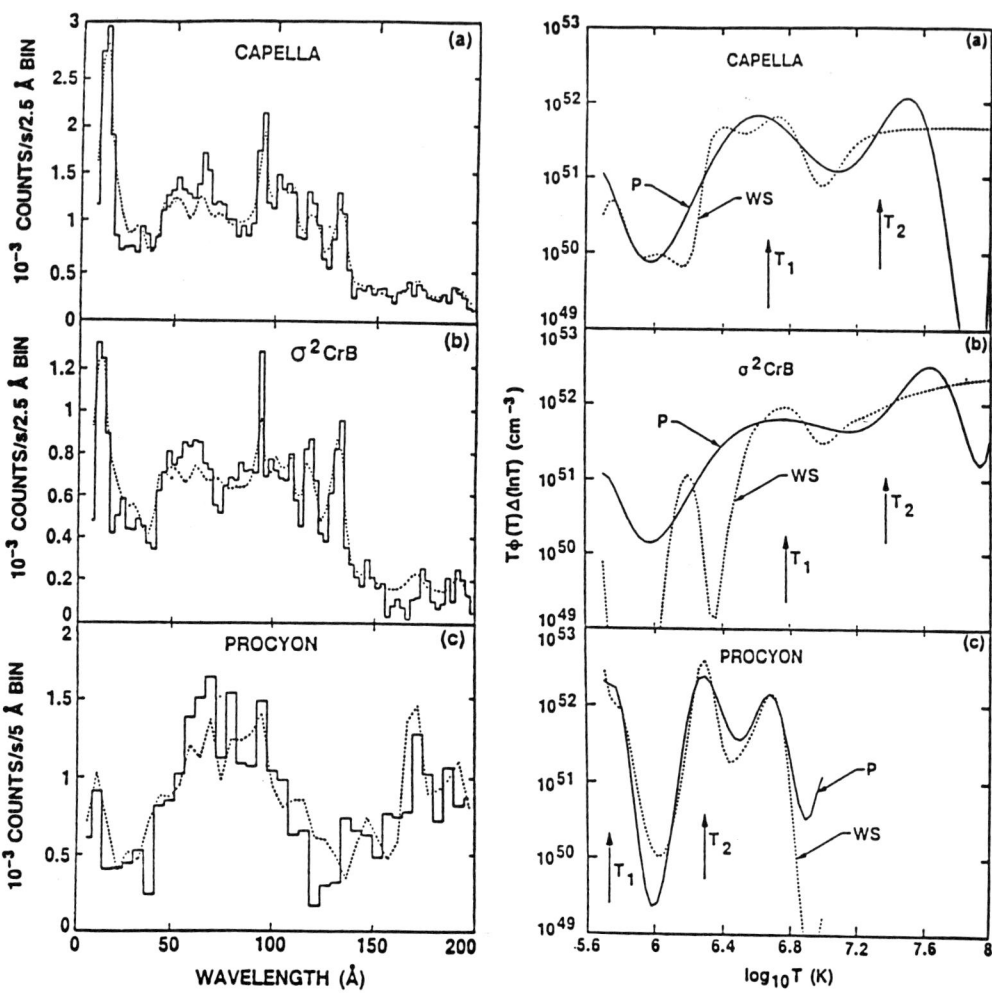

Figure 5. (left) EXOSAT 500 lines/mm TGS spectra of (a) Capella, (b) σ^2 CrB, and (c) Procyon compared with the "Polynomial" (P) best fit. (from Lemen et al. 1989). The spectra in panels a and b are given in 2.5Å bins, in panel c in 5Å bins.

Figure 6. (right) The differential emission measure curves $(T\varphi(T)\Delta \ln(T))$ (with $\Delta \ln(T) = 0.081$) for (a) Capella, (b) σ^2 CrB, and (c) Procyon (figure from Lemen et al. 1989). Each panel contains the result of the "Withbroe-Sylwester" (WS) method (dotted) and of the "Polynomial" (P) method (solid). Arrows in each panel mark the results of two-temperature fits.

The total emission measure is given by $\varepsilon = \int n_e^2 dV = \int T\varphi(T)d(\ln T)$.

Lemen et al. (1989) use two different algorithms to derive the differential emission measure distribution $T\varphi(T)$. The characteristic features of the DEM distributions are remarkably similar for the spectra of Capella and σ^2 CrB(Fig. 6): a very steep increase with temperature up to 3 MK, followed by a "depression" around 10 MK, and another increase towards temperatures about 20 MK. Lemen et al. (1989) demonstrate that the depression in the DEM around 10 MK is real and significant.

Large discrepancies between the $T\varphi(T)$ curves from the two different algorithms occur at temperatures below 3 MK (Figure 6). The spectral range between 160 Å and 200 Å contains line features which are sensitive to temperatures down to 0.5 MK. The observed intensities in that wavelength interval, however, are small and have a poor signal-to-noise ratio. Hence the differences at low temperatures between the two different DEM curves are not significant.

Strong disagreements between the two $T\varphi(T)$ models are also seen at temperatures above approximately 20 MK. This is caused by the absence of strong spectral lines characteristic of these high temperatures, and by the fact that the continuum changes only weakly as a function of temperature. Hence, the shape of the DEM curve contains little significant information for temperatures above 20 MK. In contrast to the *TGS*, the *EXOSAT* Medium Energy (ME, see Turner, Smith, and Zimmermann 1981) experiment is sensitive to emission from plasmas at temperatures above about 10 MK, although the spectral resolution is much lower. Lemen et al. show that if the *ME* and *TGS* spectra are analyzed jointly, the DEM at temperatures above ≈ 40 MK is reduced for both algorithms and the results of the two methods are in better agreement than in the analysis of the *TGS* spectra only.

The DEM derived for Procyon (Figure 6c) is uncertain because of the poor signal-to-noise ratio of the spectrum. A relatively strong contribution is suggested from plasma at temperatures around 0.6 MK and at temperatures in the range between 2 and 6 MK. The suggested separation of the peaks in the DEM between 2 and 6 MK is probably not statistically significant: a single, broad feature in the $T\varphi(T)$ distribution would yield an equally acceptable fit. This double peak may correspond to the 5 MK components seen in the spectra of Capella and σ^2 CrB.

The 5 MK component is seen on all three stars, although Procyon may have a somewhat lower value of 3 MK. We tentatively identify this component with common solar features. The fact that the quoted typical temperature of the coronal condensation of solar active regions (Section 4) is slightly lower than the 5 MK observed for stars could be caused by to the use of different spectral diagnostics and atomic data.

The stellar 25 MK component on stars may not have a quiescent solar counterpart, although Schadee (1983) reports that observations made with the HXIS instrument on *Solar Maximum Mission* show very weak, small, but long-lived features with temperatures in excess of 10 MK. The nature of these hot features is still unclear. With this one exception, temperatures as high as 25 MK are observed on the Sun only in flares. The stellar 25 MK component may be related to flares, in which case the generally featureless X-ray lightcurves of Capella and of σ^2 CrB (with the exception of a bright flare which was excluded from the analysis) would require a large number of small flares.

The 0.6 MK component of Procyon (F5IV-V), absent on Capella and σ^2 CrB, also lacks a clear solar counterpart. Such a cool coronal component may be a common phenomenon for early F-type stars (Walter, 1987). Perhaps the properties of the thin F-star convective zone result in a distinct, cool coronal component.

The existence of three dominant temperatures in the coronae of cool stars may reflect different heating mechanisms for the three components, or contain information on the stability of the corresponding loops, or both. Lemen *et al.* note that the dominant temperatures correspond with temperature intervals in which the coronal radiative loss $\Lambda(T)$ increases with increasing temperature. If radiation is the dominant cooling mechanism, the apex loop temperatures may tend towards values that make the loop relatively insensitive to heating fluctuations. The avoidance of decreasing sections of $\Lambda(T)$ may be more significant: the gradients in $\Lambda(T)$ in these intervals are stronger, so that the coronal plasma may be unstable to fluctuations.

5.2 Low-resolution X-ray spectroscopy and broad-band measurements.
Prior to *EXOSAT* the Imaging Proportional Counter (*IPC*, Gorenstein, Harnden, Fabricant 1981) of the HEAO-2 *EINSTEIN* observatory has observed the soft X-ray emission of a number of stellar coronae. The crude spectral resolution of the IPC instrument ($E/\Delta E \approx 1$, cf. Fig. 7) gives some indication of the temperatures of these coronae, but a detailed analysis of the differential emission measure structure is impossible. Given the limited capabilities of the IPC, single and two-temperature fits are usually sufficient to describe the observed spectra adequately. Figure 8 presents some of the results of single-temperature fits to IPC spectra of dwarfs and giants (Schrijver, Mewe and Walter 1984). The stars appear to define two bands in the diagram in which the emission measure per unit area $\zeta = \varepsilon/4\pi R_*^2$ is plotted against the "characteristic" coronal temperature T_X. At the high-activity end the two bands meet. The high-temperature band contains mainly giants, the low-temperature branch only dwarfs. Note that values of T_X of about 7 MK are found only for the earliest type stars in the sample.

The interpretation of such single-temperature fits is difficult without further information on the temperature structure of stellar coronae, leaving it unclear whether the resulting best-fit temperatures represent approximations of the actual coronal

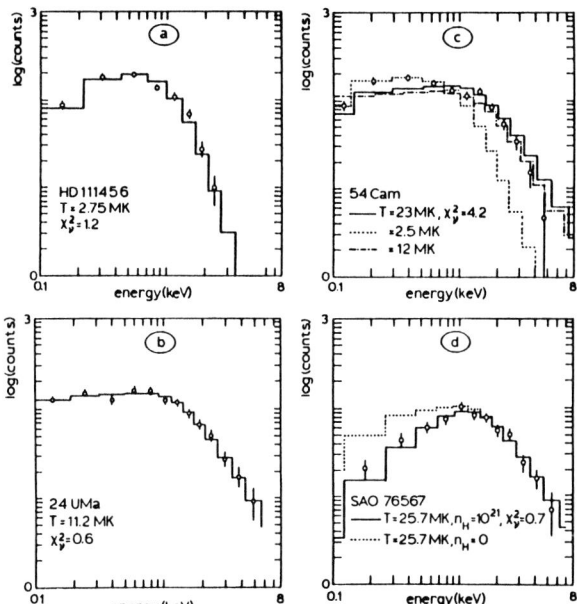

Figure 7. Examples of IPC spectra and best single-temperature fits (from Schrijver, Mewe, and Walter 1984). In panel c the single temperature fit is inadequate, and a pair of theoretical spectra is shown which reproduces the observed spectrum after scaling to the correct emission measures. Panel d shows the strong cutoff at low energies caused by a high hydrogen column density.

temperatures or whether they are largely artifacts of the instrumental response functions (e.g. Majer et al. 1986). If stellar coronae are indeed dominated by emission in two (or perhaps three) relatively narrow temperature intervals, as suggested by the analysis of medium-resolution spectra described in Section 5.1, a two-temperature description will be a useful approximation of reality. The spectral resolution of the IPC, however, is very low, and the use of *single*-temperature fits may be as useful if one has a good understanding of the limitations of the IPC. Schrijver and Bookbinder (1989) simulate stellar coronal sources with plasma at only two temperatures but with different mix ratios. They convolve these spectra with the *IPC* response function, and then use the standard IPC software to make 1-T and 2-T fits. They show that the single-T best-fit temperature is a weighted mean lying between the temperatures of the two components. This mean temperature can be interpreted properly only if the fractional emission measure in the high-temperature component exceeds approximately 0.6, because the single-temperature fit reponds only weakly to an increase in the fractional emission measure in the hot component for smaller relative contributions of the hot component. Conversely, if a single-temperature fit to an IPC spectrum yields an acceptable χ^2 value, a high-temperature component with as much as 60% of the total emission measure could be hiding in the spectrum

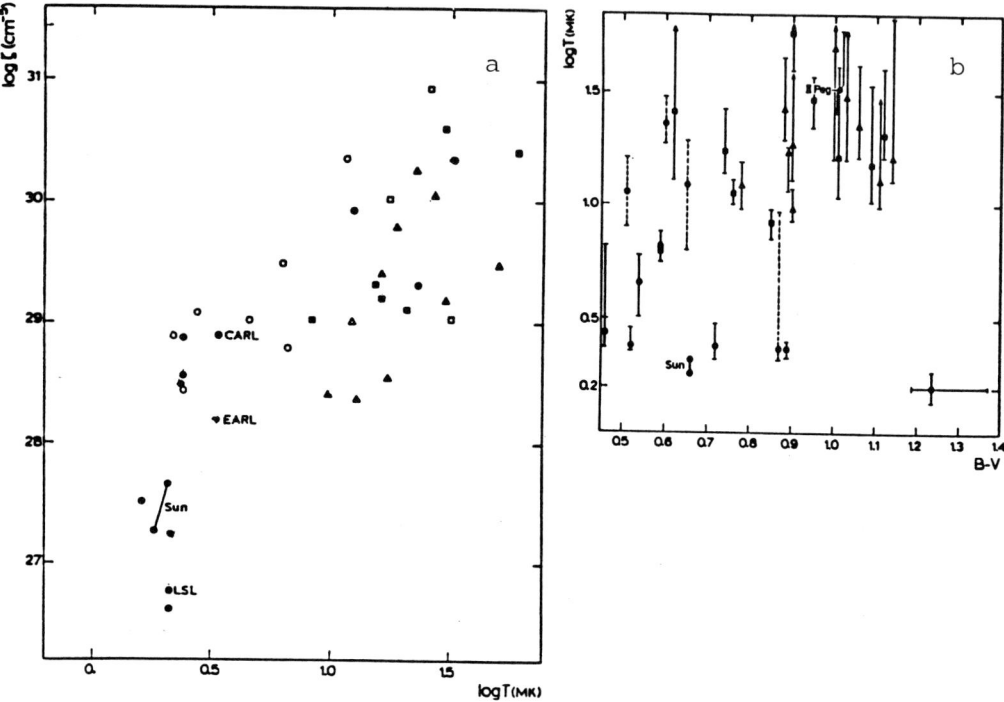

Figure 8a. Specific emission measure $\zeta = \varepsilon/4\pi R_*^2$ vs. coronal temperature T_X (Figure from Schrijver, Mewe, and Walter 1984). Typical values for solar compact (CARL) and extended (EARL) active-region loops and for large-scale loops (LSL) are shown (see Section 4). Symbols: △ giants, □ subgiants, ○ dwarfs.

Figure 8b. X-ray temperature T_X vs. $B - V$. Symbols as in panel a.

(compare, for instance, the 1-T IPC temperature of ≈ 12 MK for σ^2 CrB even though a significant fraction of the plasma has temperatures around 5 MK, cf. Fig. 6). They confirm that the IPC is virtually blind for plasmas with temperatures below ≈ 1 MK.

With these conclusions, we infer from Figure 8a that most of the coronal plasma in inactive to moderately active dwarf stars has a temperature of about 5 MK. In the most active dwarfs and in the giants most of the plasma is at about 20–50 MK, but we cannot exclude the possibility that a cool ≈ 5 MK component is present with an emission measure that can be comparable to that of the hot component.

The *EINSTEIN* Solid State Spectrometer (Swank and White 1980) had a higher resolution than the *IPC*, but only a relatively small wavelength window of about 3–30Å.

Several authors have shown that single–temperature spectra cannot to explain the
SSS spectra. Two–temperature models can reproduce most of the observed spectra,
but since temperatures derived from *IPC* and *SSS* spectra differed, the results were
interpreted as indicative of a continuous temperature distribution in stellar coronae
(e.g. Majer *et al.* 1986). Broad–band filter observations, with even less energy resolution, can only confirm that coronae are not isothermal (e.g. Pallavicini *et al.*
1988).

6. CONSTRAINTS ON LOOP GEOMETRY

The steep increase with temperature below 3 MK of the DEM $T\varphi(T)$ derived for
Capella and σ^2 CrB (Fig. 6) is incompatible with the model for static magnetic
loops with constant cross section as developed by Rosner, Tucker and Vaiana (1978).
Schrijver, Lemen and Mewe (1989) tested whether the steep increase of $T\varphi(T)$ could
be caused by an increase of the loop cross section with height: the relative contribution of plasma at temperatures below the maximum temperature decreases strongly
with increasing expansion, so that the loops will appear to be nearly isothermal if
the expansion is large. They performed a two–component analysis of the *EXOSAT*
spectra using computed loop spectra covering a range of apex temperatures T_m and
expansion factors Γ (defined as the ratio of the cross sectional area at the loop apex
and that at the footpoints in the upper chromosphere, i.e. well above the canopy
height). The best-fit DEM distributions are shown in Figure 9. Figure 10 shows χ^2
contours in the Γ_1–Γ_2 plane, where Γ_1 and Γ_2 are the expansion factors for the cool
and hot component.

The fits for Capella and σ^2 CrB (see Table 1) suggest that the low-temperature
component (T_m around 5 MK) originates in loops expanding significantly with height
($\Gamma > 7$ or 5, respectively, at 1σ, and $\Gamma > 5$ or 2, respectively, at 99% confidence). The
high-temperature ($T_m \approx 30$ MK) component of Capella requires loops with $2 < \Gamma < 5$
at 1σ, or $\Gamma < 10$ at 99% confidence). For the hot component on σ^2 CrB a value of
$\Gamma = 4$ is found, but this value cannot be constrained at the 99% confidence level.

The main constraint on the amount of plasma with temperatures below 3 MK is
formed by the weak signal between 160Å and 200Å. This limits the sum of the
emission measures of the plasma below 3 MK in the loop ensembles, but not of the
individual components. Hence, an increase in the DEM below 3 MK of the cool
loop component (by a smaller Γ) could be compensated for by a decrease in the
DEM of the hot loop ensemble (by a larger Γ). This possibility is reflected in the
χ^2 contour maps in Figure 10: if a small expansion factor is chosen for the cool
loop component, the best corresponding value for the hot loop is large, and vice
versa. The corresponding change in the amount of matter between 5 MK and 15 MK
mostly affects the continuum level in the spectra, without significantly changing the
predicted line ratios. Although the best-fit expansion factors differ quite strongly for

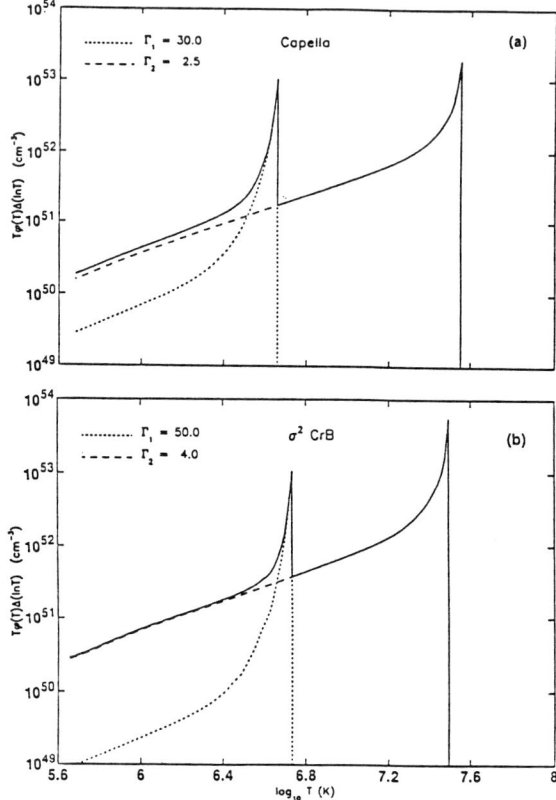

Figure 9. Panels a and b show the differential emission measure distributions $T\varphi(T)\Delta\ln(T)$ (with $\Delta\ln(T) = 0.081$) for Capella and σ^2 CrB, respectively, corresponding to the two-loop fits (from Schrijver, Lemen, and Mewe 1989). The dashed and dotted lines show the DEM for the two individual loop ensembles (see also Table 1).

the hot and cool loop ensembles, they could be the same within the 90% confidence contour of the best fit for Capella and the 67% confidence contour for σ^2 CrB. In that case, the expansion factors would be approximately 10.

Stern, Antiochos and Harnden (1986) analyzed *EINSTEIN* IPC spectra of the most active cool stars in the Hyades cluster. Their single-component fits suggest that the coronal loops of the Hyades stars have expansion factors smaller than approximately four, and apex temperatures between 10 MK and 15 MK, which is comparable with our results for the hot coronal components.

7. LOOP LENGTHS

Walter, Gibson and Basri (1983) used IPC eclipse spectra of AR Lac to determine typical loop lengths. They show that both stars in this close binary have a coronal component with a characteristic height of ≈ 0.02 stellar radii and an apex temperature

Table 1. Results of two-component loop fits. Shown are the apex temperature T_m for the low- (1) and the high-temperature (2) component, the expansion factor Γ, the ratio of the apex cross-sectional area A and loop half length L, the total emission measure EM, as derived from an integration along the half loop from T_m to $T_m/2$, and the ratio of the two components, the reduced χ_ν^2 value, the number of degrees of freedom ν, and the acceptance level α of the fit.

Source	T_m (MK)	Γ_i	A_i/L_i (10^{12})	EM_i (10^{51}cm^{-3})	EM_1/EM_2	χ_ν^2	ν	α (%)
Capella	4.6	30	44	120	0.35	1.13	74	20
	35.8	2.5	0.076	340				
σ^2 CrB	5.4	50	13	110	0.13	0.65	74	99
	31.2	4	0.30	850				

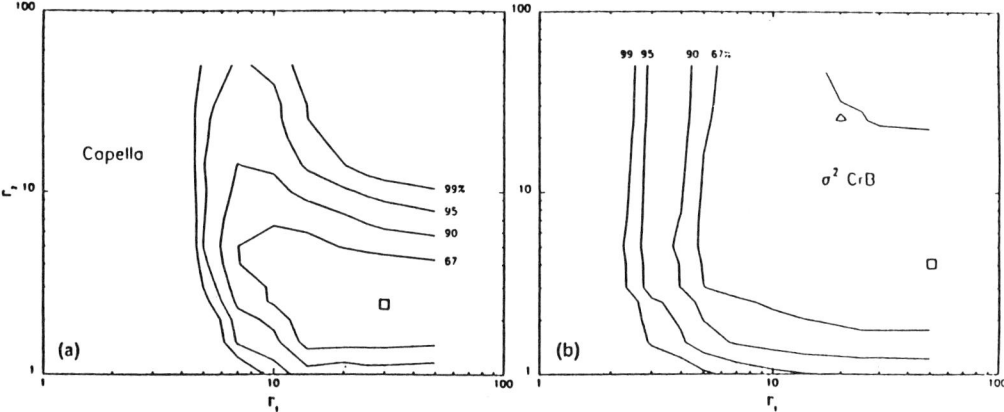

Figure 10. Chi-square contours in the Γ_1–Γ_2 plane for a two-component fit to the observed EXOSAT spectra of (a) Capella and (b) σ^2 CrB (from Schrijver, Lemen, and Mewe 1989). The model corona comprises two ensembles of loops with different maximum temperatures and different geometries. The factor Γ is the ratio of the cross section at the top of the coronal loop and at the footpoints of the loop: Γ_1 is the expansion factor for the low-temperature (~ 5 MK) loops, and Γ_2 is the expansion factor for the high-temperature (~ 30 MK) loops. Contours are drawn at different confidence level (expressed in percentages). The best fit is marked by a box.

of $T \approx 8$ MK. The K-type subgiant has an additional extended coronal component with a height of about one stellar radius and $T \approx 30$ MK.

White et al. (1986) observed the Algol system with the *EXOSAT* observatory for a period of 35 hours around the occultation of the K IV star by the B star primary. The quiescent spectrum in the 1–10 keV band can be attributed to a thermal plasma with a temperature of 25 MK. Since no obvious X-ray eclipse was seen, the scale height of this component is presumed to be comparable to or greater than the radius of the K IV star. Earlier observations with the *EINSTEIN* Solid State Spectrometer indicated an additional component with a temperature around 5 MK. Although the *EXOSAT* observations did show evidence for its existence, they could not constrain the dimensions of this cooler plasma component.

The two-loop fits discussed in the previous section provide an estimate of T_m, Γ, and ε for each component, but they do not constrain the loop length because no information is available on the plasma density. A crude estimate, however, can be made. Coronal loops can be divided into two categories: loops shorter than half the pressure scale height ($H_p/2$) which emit over their entire length, and longer loops which emit predominantly in their lower parts. The total brightness of compact loops with $L < H_p/2$ remains virtually unchanged if loop length and cross-sectional area are scaled proportionally (At a given apex temperature the product of electron density and loop length is constant – Rosner, Tucker, and Vaiana 1978 – so that longer loops are less dense. Since the brightness of the loop scales with the square of the electron density times volume, longer loops require a larger cross section to obtain the same total brightness). The total fractional area covered by the loops, however, cannot be larger than unity. This constrains the length of the cool (5 MK) loops to at most $0.3\,H_p/2$, i.e. much smaller than the stellar radius. The length of the hot (30 MK) loops cannot be similarly constrained: these loops may be longer than $H_p/2$ with a large filling factor, or shorter than $H_p/2$ with a small filling factor.

If AR Lac and Algol are used as typical examples, the 5 MK component on Capella and σ^2 CrB can be identified with compact, dense loops, much like solar loops. The hot 30 MK component can be identified with extended loop structures, with a length close to the pressure scale height and the stellar radius.

8. NEW WINDOWS TO THE UNIVERSE

As discussed above, the detailed temperature and density structure of stellar coronae, with few exceptions, has evaded study because the X-ray telescopes of the past had only a limited spectral resolution and sensitivity. NASA's Advanced X-ray Astrophysics Facility (*AXAF*) and ESA's X-ray Multi-Mirror Mission (*XMM*) will change that situation drastically. The high spectral resolution of these future space missions allows observations of resolved emission lines in soft X-rays with a high signal to noise ratio. Their large light-collecting areas allow exposure times for spectral observations as short as a few hours for many bright sources.

The wavelength region 4–140 Å as covered by the *AXAF* transmission grating spectrometer ($\Delta\lambda \simeq 0.05$Å) is rich in spectral lines from various ions in different ionization stages that can be used as diagnostics for temperature, density and velocity, allowing accurate studies of structure, energy balance and heating rates. Mewe, Lemen

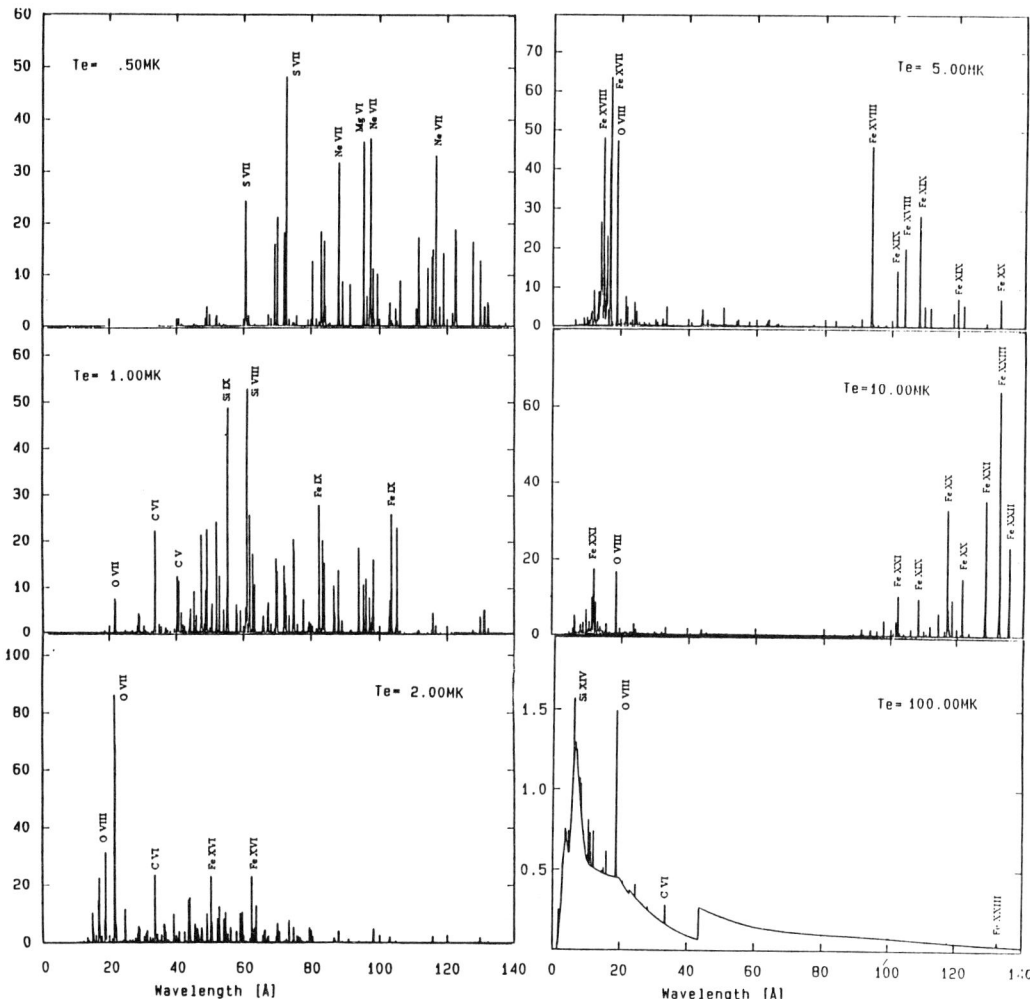

Figure 11. Spectra convolved with the AXAF–LETGS response (fluxes in phot cm^{-2} (0.05 Å)$^{-1}$, integrated over $t_{obs} = 1$ hr) for an optically thin plasma for electron temperatures $T = 0.5$–100 MK, reduced emission measure $\varepsilon/d^2 = 10^{50}$ cm^{-3} pc^{-2}, and hydrogen column density $N_H = 3 \; 10^{18}$ cm^{-2}. Several prominent lines from the H- and He-like ions and from a number of consecutive ionization stages of iron, are labelled. Figure taken from Lemen, Mewe and Schrijver (1989).

and Schrijver (1989) study the diagnostic capacities of the AXAF–LETGS. Figure 11 shows some calculated spectra convolved with the AXAF–LETGS response. Figure 12 illustrates the variations in some of the pressure sensitive lines that can be observed with the AXAF–LETGS.

Mewe, Lemen and Schrijver (1989) demonstrate that the AXAF–LETGS allows ac-

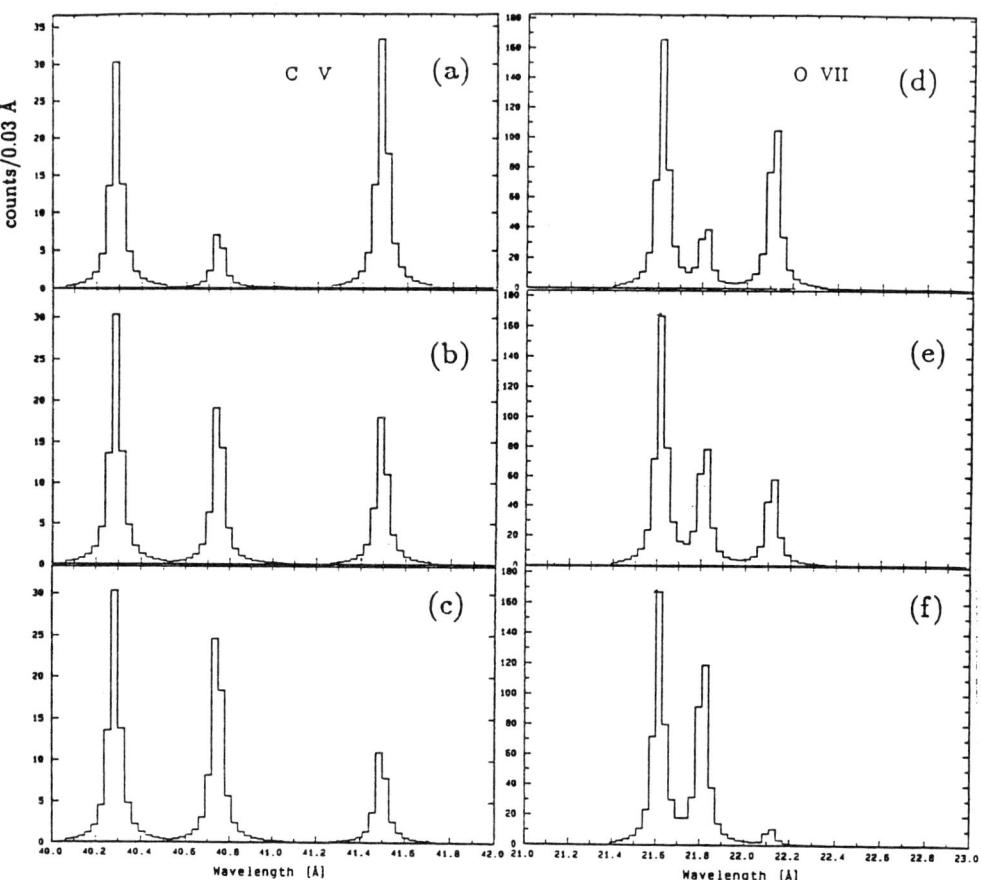

Figure 12. Intensities of the C V (41 Å) triplet at $T = 1$ MK and electron densities $n_e = 10^9$, 10^{10} and 10^{11} cm^{-3} (panels a,b,c, respectively) and of the O VII (22 Å) triplet for $T = 2$ MK and $n_e = 10^{10}$, 10^{11}, and 10^{12} cm^{-3} (panels d,e,f, respectively). We assume a reduced coronal emission measure of $\varepsilon' \equiv \varepsilon/d^2 = 10^{50}$ cm^{-3}, and $t_{obs} = 10^4$ s). The forbidden line of C V is blended with an Ar IX line which contributes about half of the total intensity at low electron density. The spectra are binned in 0.03 Å bins. Figure taken from Lemen, Mewe, and Schrijver (1989).

curate determinations of the DEM for temperatures below approximately 20 MK. The accuracy with which the input–DEM can be recovered from the simulated spectra leads to a high sensitivity to the geometry of the loops. Together with the study of density diagnostics, this will allow sensitive tests of proposed loop models, even though realistic coronae are likely to be more complicated than the currently simulated two–component models.

The large number of spectral lines that can be resolved allows the selection of sets of lines of a given element so that abundances can be determined. Abundance effects in the DEM analysis can thus be isolated and corrected for.

The high spectral resolution allows the study of velocity shifts of ≈ 100 km/s at long wavelengths. In short-period binaries the orbital motion "separates" the emission from the binary components. In large stellar flares plasma flows may be observable.

The large sensitivity and high resolution of *AXAF-LETGS* require relatively short exposure times, which allows studies of stellar flares and of stellar rotational modulation. Combination with optical and ultra-violet observations may allow an estimate of the surface distribution of active regions.

9. CONCLUSION

The information available to date on coronae of cool stars supports the following hypothesis. Stellar coronae are dominated by only a few temperature intervals: 5 MK and 25 MK and possibly 0.6 MK for F-type stars. The relative brightness of the components depends on the stellar rotation rate, gravity and possibly the stellar effective temperature, but the total X-ray flux from stellar coronae is related through a power law with the chromospheric or transition-region excess fluxes which is independent of effective temperature and gravity. The coronal loops expand with height, probably by a factor of 30–50 for the 5 MK component, and a factor of 2–5 for the 25 MK component, although they may be about 10 for both. The cool component must be more compact than half the pressure scale height, while the length of the hot loops may somewhat exceed the pressure scale height. The performance of *XMM* and *AXAF* will enable high-resolution spectroscopy in the fascinating realm of coronal soft X-rays, allowing not only a test of the above hypothesis, but also much more detailed studies of temperatures, densities, temporal variability, flows and abundances.

REFERENCES

Ayres, T.R., Marstad, N.C., Linsky, J.L.: 1981, *Astrophys. J.* **247**, 545
Berton, R., Sakurai, T.: 1985, *Solar Phys.* **96**, 93
Brinkman, A.C., Gronenschild, E.H.B.M., Mewe, R., McHardy, L., Pye, J.P., 1980, *Adv. in Space Research*, Nr. **3**, 65
Gorenstein, P., Harnden, F.R. Jr., Fabricant, D.G., 1981, IEEE Trans. on Nucl. Sc. NS **28**, 869
Hammer, R., Ulmschneider, P.: 1989, *Astron. Astrophys.* , in press
de Korte, P.A.J., Bleeker, J.A.M., den Boggende, A.J.F., Branduardi-Raymont, G., Brinkman, A.C., Culhane, J.L., Gronenschild, E.H.B.M., Mason, I., McKechnie, S.P., 1981, *Space Science Rev.* **30**, 495
Lemen, J.R., Mewe, R., Schrijver, C.J., Fludra, A.: 1989, *Astrophys. J.* **341**, 474
Lemmens,A.F.P., Rutten,R.G.M., Zwaan,C.: 1989, in prep. for *Astron. Astrophys.*
Majer, P., Schmitt, J.H.M.M., Golub, L., Harnden, F.R. Jr., Rosner, R.:

1986, *Astrophys. J.* **300**, 360
Mewe, R.:1984, in "Proc. 8th Int. Coll. on EUV and X-ray Spectr. of Astrophys. and Lab. Plasmas (IAU Coll. 86)," Naval Res. Lab., Washington D.C., p 59
Mewe, R., Lemen, J.R., Schrijver, C.J.: 1989, in preparation for *Astrophys. J.*
Narain, U., Ulmschneider, P.: 1989, *Space Science Rev.*, in press
Pallavicini, R., Peres, G., Serio, S., Vaiana, G.S., Golub, L., Rosner, R.: 1981, *Astrophys. J.* **247**, 692
Pallavicini, R., Monsignori-Fossi, B.C., Landini, M., Schmitt, J.H.M.M.: 1988, *Astron. Astrophys.* **191**, 109
Poletto, G., Vaiana, G.S., Zombeck, M.V., Krieger, A.S., Timothy, A.F.: 1975, *Solar Phys.* **44**, 83
Rosner, R., Tucker, W.M., Vaiana, G.S.: 1978, *Astrophys. J.* **220**, 643
Rutten, R.G.M., Schrijver, C.J., Lemmens, A.F.P., Zwaan, C.: 1989a, in preparation for *Astron. Astrophys.*
Rutten, R.G.M., Schrijver, C.J., Zwaan, C., Duncan, D.K., Mewe, R.: 1989b, *Astron. Astrophys.* , in press
Saar, S.H., Schrijver, C.J.: 1987, in "Cool Stars, Stellar Systems, and the Sun," J.L. Linsky, R.E. Stencel (eds.), Springer-Verlag, p 38
Schadee, A.: 1983, *Solar Phys.* **89**, 287
Schrijver, C.J.: 1983, *Astron. Astrophys.* **127**, 289
Schrijver, C.J.: 1987a, *Astron. Astrophys.* **180**, 241
Schrijver, C.J.: 1987b, *Astron. Astrophys.* **172**, 111
Schrijver, C.J., Bookbinder, J.A.: 1989, in preparation for *Astrophys. J.*
Schrijver, C.J., Coté, J., Zwaan, C., Saar, S.H.: 1989, *Astrophys. J.* **337**, 964
Schrijver, C.J., Dobson, A.K., Radick, R.R.: 1989, *Astrophys. J.* **341**, 1035
Schrijver, C.J., Lemen, J.R., Mewe, R.: 1989, *Astrophys. J.* **341**, 484
Schrijver, C.J., Mewe, R., Walter, F.M. 1984, *Astron. Astrophys.* **138**, 258
Schrijver, C.J., Rutten, R.G.M.: 1987, *Astron. Astrophys.* **177**, 143
Schrijver, C.J., Zwaan,C., Maxson,C.W., Noyes,R.W.: 1985, *Astron. Astrophys.* **149**, 123
Skumanich, A., Smythe, C., Frazier, E.N.: 1975, *Astrophys. J.* **200**, 747
Stepien, K., Ulmschneider: 1989, *Astron. Astrophys.* **216**, 139
Stern, R.A., Antiochos, S.K., Harnden, F.R. Jr.: 1986, *Astrophys. J.* **305**, 417
Swank, J.H., White, N.E.: 1980, in "Cool Stars, Stellar Systems, the Sun," A.K. Dupree (ed.) Cambridge, p 47
Turner, M.J.L., Smith, A., Zimmermann, H.V.: 1981, *Space Science Rev.* **30**, 513
Vesecky, J.F., Antiochos, S.K., Underwood J.H.: 1979, *Astrophys. J.* **233**, 987
Walter, F.M.: 1987, private communication
Walter, F.M., Gibson, D.M., Basri, G.: 1983, *Astrophys. J.* **233**, 987
White, N.E., *et al.*: 1986, *Astrophys. J.* **301**, 262

Duplicity of Solar-Like Stars in the Solar Neighbourhood

A. Duquennoy and M. Mayor

Geneva Observatory, CH-1290 Sauverny, Switzerland

ABSTRACT

The duplicity of solar-like stars in the solar neighbourhood is revisited, using a complete sample of 210 stars with primaries in the spectral range F7 to G9 IV-V, V, VI taken in Gliese's catalogue. With the help of about 4400 CORAVEL radial-velocity measurements obtained in 12 years with a precision better than 0.3 km/s, we derive significant new results on the orbital elements and mass-ratio distributions.

Such new distributions should be useful as constraints on stellar formation processes.

1. INTRODUCTION

Studies of stellar duplicity have a bearing on various matters such as:

i) The constraints on possible scenarios of stellar formation, that can be derived from:

– the distribution of the orbital elements such as eccentricity e, orbital period P, mass ratio $q = \mathcal{M}_2/\mathcal{M}_1$, and their possible dependence on the age of the systems.

– the correlation between orbital elements such as $(e, \log P)$, (q, P);

– the frequencies of singles: doubles: triples: quadruples systems, and the hierarchy observed in the plane ($\log P_{inner}/\log P_{outer}$) in multiple systems.

ii) The constraints on the evolution and interaction of the binary components evidenced by:

– tidal circularization of short period binaries;

– mass exchange between components in very close binaries filling their Roche lobe;

– chromospheric activity in short period binaries;

– peculiar surface element abundances (S, Ba, CH stars).

iii) The search for very low mass companions. A direct estimation of the masses can be derived from joint sets of visual and spectroscopic orbital elements, with the possibility to detect hypothetical brown dwarfs or giant planets.

iv) The improvement of the knowledge of the stellar mass-luminosity relation.

2. SOME RECENT ANALYSES ON STELLAR DUPLICITY

The distributions of orbital elements of solar type spectroscopic binaries have only been systematically investigated since slightly more than a decade. But major discrepancies still exist today among these different studies. The theory of stellar formation is in progress, and perhaps is about to be able to use the distribution functions of orbital elements as constraints (Boss 1988). So prior to expose the results of our survey, we wish to recall some of the most important recent steps.

Abt and Levy (AL, 1976) studied the stellar multiplicity in a sample of 135 F3-G2 IV or V bright field stars ($m_v \leq 5.5$), on the basis of 20 radial velocity measurements per star. The study of AL was certainly the most systematic effort to obtain a comprehensive view of the duplicity among solar-like stars. Their main results are:

i) The observed frequencies of singles: doubles: triples: quadruples are 42:46:9:2.

ii) The period distribution has a single maximum, with a median period of 14 days.

iii) The secondary masses distribution depends on the orbital period: for $P > 100$ yrs it fits the van Rhijn distribution, while for $P < 100$ yrs it varies as $\mathcal{M}_2^{-1/3}$. They conclude on the existence of two binary formation processes: fragmentation for the former, fission for the latter.

iv) there are really 1.4 companions for each primary star, and the total mass in the companions is just half of the total mass in the primaries.

However, Morbey and Griffin (1987) discussed the 25 new spectroscopic orbits derived by AL. They find that 24 are not supported by a statistical analysis, and that in at least 21 cases the evidence does not show that the stars are binaries at all. It is to be noted that these 24 (probably erroneous) SB represent about half of the totality of the SB in AL's sample, and this implies that the inferred multiplicity among G dwarf stars probably has been overestimated.

Besides, the distribution of the mass ratios $q = \mathcal{M}_2/\mathcal{M}_1$ in binaries has been studied by Halbwachs in two samples: one using the visual binaries (Halbwachs 1986), the other using the spectroscopic binaries (Halbwachs 1987) published in catalogues. In these studies the numerous selection and detection biases affecting these catalogues have been carefully taken into account. His main results are:

i) There is no maximum in the mass ratio distribution for $q = 1$, but a possible maximum around $q = 0.3$.

ii) There is no difference in this distribution between long and short period binaries, while AL found a dividing period of 100 yrs.

Trimble (1987) also studied the q- distribution in two samples: after taking into account the selection effects, she finds that only the cpm pairs could have come from a power-law distribution rising toward small q, while the distribution for the

VB is, at best, flat in q, not rising toward small values.

3 THE CORAVEL SURVEY OF SOLAR-LIKE STARS IN THE SOLAR NEIGHBOURHOOD

3.1 The CORAVEL spectrometer

The CORAVEL spectrometer is a high precision photoelectric radial velocity scanner adapted to late spectral-type ($\geq F5$) stars. It is described in detail by Baranne, Mayor and Poncet (1979) and we just recall here the main characteristics. The technique used is the optical cross-correlation of a mask with the stellar spectrum, following the pioneering instrumentation of Griffin (1967). The optical mask is computed from a high resolution spectrum of Arcturus and contains about 1500 metallic absorption lines in the wavelength range 3600-5200 Å. The spectrometer has a dispersion of 2Åmm^{-1}.

Two versions of the instrument have been built, one for each hemisphere. The first CORAVEL is mounted since 1977 on the 1-m Swiss telescope at the Haute-Provence Observatory (France) and records about 7000 stellar radial velocities per year. The second CORAVEL is mounted about 25% of the available time since 1981 on the 1.54-m Danish telescope at the ESO-La Silla Observatory (Chile).

The achieved radial velocity precision is about 0.3 and 0.2 km/s for the northern and southern CORAVEL respectively, for a typical G dwarf star with $m_v \simeq 7$ and a 3 min exposure. In fact, we can show that the precision can reach 0.1 km/s for bright stars by increasing the exposure time up to 10 min for a 7-mag star. However for higher magnitudes, efforts to increase the precision (mainly by decreasing the photon shot noise) would rapidly lead to prohibitive exposure times. The limiting magnitudes are $V \simeq 12.5$ and $V \simeq 14.0$ for the northern and southern CORAVEL respectively.

3.2 The radial velocity survey

The programme concerns all stars contained in the Gliese (1969) catalogue with spectral types later than F5 and declination above $-15°$. From this programme, the so-called 'nearby G-dwarf star sample' has been extracted in order to study the duplicity of nearby solar-type stars. The sample is defined by all primary stars with spectral types F7 to G9, luminosity classes IV-V, V, VI, declination above $-15°$, and trigonometric parallax greater than 0.045 arcsec (or distance $r \leq 22$ pc). 166 primaries fit the definition, plus 44 independently measurable secondaries (cpm or VB components). A total of about 4400 CORAVEL radial velocities have been obtained for the whole 210 stars, among which 90% with the northern CORAVEL. In fact, an important fraction of these measures (29%) concerns a small number of stars (10) which are IAU velocity standards.

The advantages of our G-dwarf star sample are the following:
  i) It is parallax-limited, instead of magnitude-limited in the case of AL's study of bright field stars, the latter introducing a bias favouring the SB2s.

ii) The spectral type of the blue limit (F7V) avoids large rotators for which the CORAVEL radial velocity precision decreases.

iii) The high precision of the derived velocities ($\epsilon \leq 0.3$ km/s).

iv) The large time coverage of the observations ($\Delta T \simeq 11$ yrs). This is a necessary step to overlap the orbital periods of binaries detected by various techniques (spectroscopy, astrometry, interferometry, speckle and visual).

v) The sample belongs to one of the most studied catalogues of stars (Gliese 1969), including knowledge of spectral types, photometry, kinematics, activity, age, and an almost complete detection of visual and common proper motion pairs.

vi) The extremely well defined determination of the error on each individual measurement allows a reliable statistical detection of radial velocity variables. With that aim, we used the distribution of the probabilities $P(\chi^2)$ to observe a given χ^2 for each star, where χ^2 depends on the number of measures and on the instrumental precision (I) (see Duquennoy and Mayor 1990a). For constant velocity stars, the shape of this distribution is expected to be flat. As an example, Fig.1 shows the $P(\chi^2)$ distribution for the G-dwarf sample for various values of (I) of the northern CORAVEL. In that way, we are able to infer the two values of 0.18 and 0.26 km/s

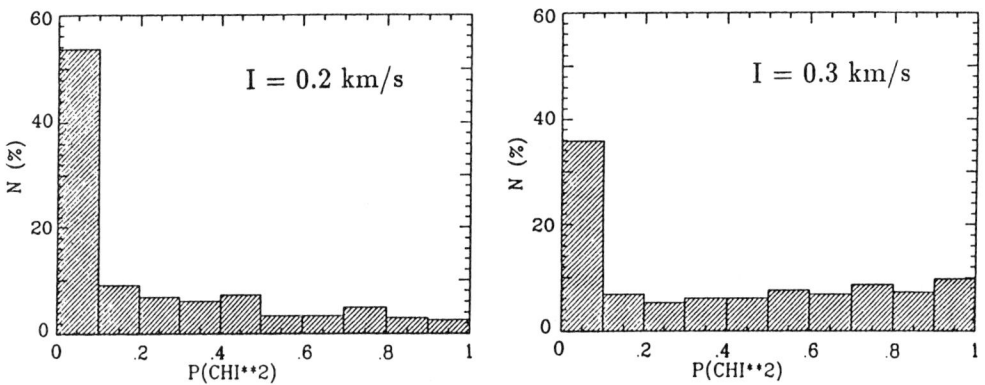

Figure 1: $P(\chi^2)$ *distribution for two values of the instrumental precision (I) of the northern CORAVEL and for the sample of nearby G-dwarf stars. The correct value to be adopted to obtain a flat distribution is between 0.2 and 0.3 km/s.*

for the standard error on one measurement respectively obtained by the southern and northern CORAVEL for our programme stars.

vii) The ability of CORAVEL to detect small amplitude SB2s through the variation of the shape of the cross-correlation function. As an example of such a case, Fig.2a shows the cc-dip of HD 137107 at three different observing dates, while Fig.2b shows the modelisation of these variations in excellent agreement with the ephemeris predicted by Dommanget (1982). Note the contrast in velocity dispersion around the computed curve between CORAVEL and earlier measures from Yerkes

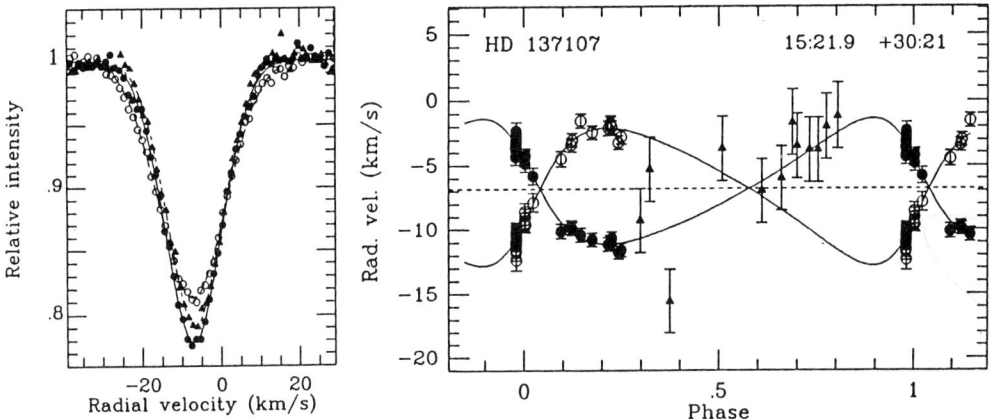

Figure 2: *2a) Shape of the CORAVEL cross-correlation dip for the visual binary HD 137107 ($P_{orb} \simeq 42$ yrs, $a \simeq 0.9$"), at three different dates:* ▲ = *1978 Apr. 16,* ● = *1979 Feb. 12,* ○ = *1985 May 29. 2b) Radial velocity curve for the same star: dots are CORAVEL observations, triangles are early Yerkes observations.*

made in 1929, which were subject to line blending.

3.3 The orbital period distribution

We collected all known spectroscopic, visual and cpm pairs in our G-dwarf sample of 166 primaries, as well as the spectroscopic pairs newly discovered with CORAVEL during the present survey. The orbital periods, when unkown, were approximated under appropriate assumptions to the nearest decade of days. The periods of the cpm pairs were determined statistically from their observed separation, and we adopted a cut-off for the gravitational bounding of the pairs similar to that used in AL's study. The resulting period distribution for 114 systems is shown in Fig.3. It presents a single maximum for $P_{max} \simeq P_{median} \simeq \overline{P_{orb}} \simeq 260$ yrs. This value is to be compared with the much smaller value of 14 yrs found by AL. The difference between the two results may have two sources:

i) The magnitude-limited sample of AL tends to favour the inclusion of SB2s, brighter than single stars of same type, and these SB2s have apparently often short periods because of the difficulty to resolve the components of low amplitude (i.e. usually long period) binary.

ii) The overinterpretation of the velocity data by AL, which led them to include a number of spurious short period binaries (Morbey and Griffin 1987, Duquennoy and Mayor 1988).

The distribution of the orbital periods in our sample may be approximated by the Gaussian-type relation (see representation on Fig.3):

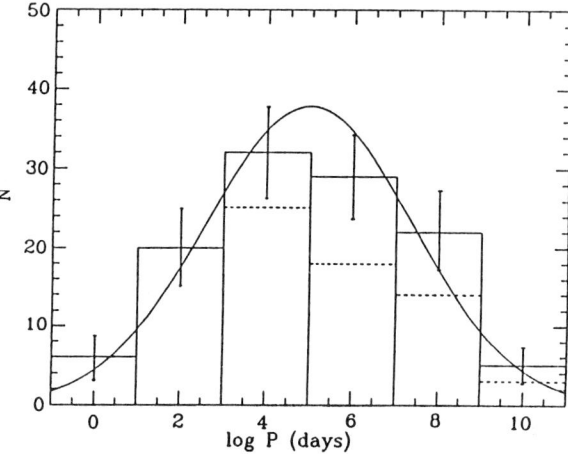

Figure 3: *The distributions of the orbital periods (in days), for the nearby G-dwarf binaries: observed (dashed line) and corrected from detection biases (continuous line), with a Gaussian curve fit.*

$$f(\log P) = Cst \times \exp\left\{ \frac{-(\log P - \overline{\log P})^2}{2\sigma_{\log P}^2} \right\}$$

where $\overline{\log P} = 5.0$, $\sigma_{\log P} = 2.4$, and P is in days.

It is interesting to compare this ditribution with that obtained by Griffin (1985) for two different samples. The first one, the sample of binaries among giant stars, agrees well with the above distribution. The second one, the sample of red dwarfs in the Hyades field, seems to show an excess of short period binaries compared to our G-dwarf sample. This comparison suggests a possible evolutionary effect in the distribution of the periods.

3.4 The eccentricity distribution

The eccentricity distribution is found to be strongly dependent on the orbital period. Following the idea of dichotomy of the orbital periods in three classes, as suggested by Mayor and Mermilliod (MM, 1983, 1984) and Burki and Mayor (BM, 1985) for the distribution of the eccentricities in open clusters, we find for the sample of nearby G-dwarfs:

   i) For P less than a certain value identified as the circularization period P_{circ}, all the orbits are circularized due to tidal interactions that occurred either in the pre-main sequence stage, or on the main sequence where P_{circ} would then depend on the age (see e.g. Zahn 1977, 1989, Zahn and Bouchet 1989, and with a different approach Tassoul 1987, 1988). The distribution of the nearby G-dwarf orbits in the plane $(e, \log P)$ given in Fig.4 shows clearly that $P_{circ} \simeq 11d$, which is to be compared with the values of 5.7 days for (young) open clusters (MM, BM), of 10-11 days for M67 (Mathieu and Mazeh 1988) and of 12-18 days for the halo stars (Jasniewicz and Mayor 1988, Latham et al. 1988). The $P_{circ} \simeq 11d$ found here is statistically in agreement with the mean age of the galactic disk. Indeed two kinds of stars could have polluted this result due to the age mixing in our sample:

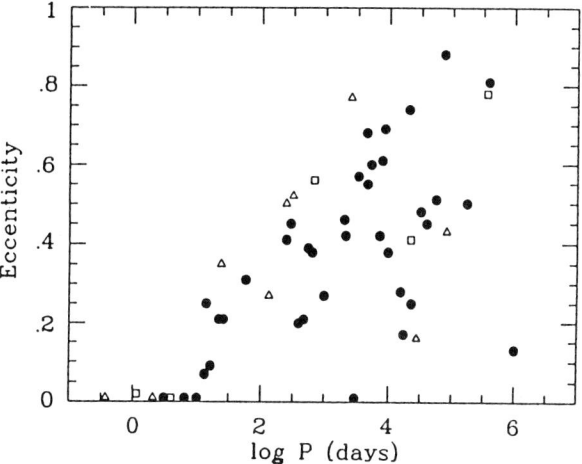

Figure 4: *Distribution of the orbital elements of the nearby G-dwarf binaries in the plane $(e, \log P_{orb})$, illustrating the circularization processes. Symbols follow the multiplicity of the system to which they belong:* • *double,* △ *triple,* □ *quadruple.*

– young stars (e.g. with age $\simeq 5 \times 10^8$ yrs, statistically about 20% of our sample) could have exhibited a binary with $P < 11d$ not circularized yet;

– old stars could have produced a SB1 with $P \simeq 1000d$ and circular orbit, provided the present secondary is a degenerate star (White Dwarf). Such a secondary was initially a primary B or A star, which during the giant phase of its evolution has circularized the orbit of its solar-type companion, now observed as SB1 primary.

ii) For $P_{circ} \leq P_{orb} \leq 1000d$ (see Fig.5a), called tight binaries, the eccentricity distribution is 'bell shaped' with a maximum for $e \simeq 0.3$, like in the case of open clusters (MM, BM). The mean eccentricity is $\bar{e} = 0.35 \pm 0.04$ for the nearby G-dwarfs, to be compared with $\bar{e} = 0.33 \pm 0.03$ for young open clusters and also $\bar{e} = 0.33 \pm 0.03$ for halo SB stars in the same period range. Moreover, in advance of the discussion presented in Sec. 3.6 we find (this work) $\bar{e} = 0.35 \pm 0.06$ for tight binaries with very low mass secondaries ($M_2 \sin i \leq 0.1 M_\odot$). These striking results show that the mean eccentricity of non-evolved tight binaries ($P_{circ} \leq P_{orb} \leq 1000$ days):

– is independent of the population (halo or disk);
– is independent of the mass of the companion.

Thus, keeping in mind the possible influence of age mixing evoked in i), this class of binaries may reflect the initial binary formation process. In this hypothesis, binary stars formed with a near-zero eccentricity would be rare.

iii) For $P_{orb} \geq 1000d$ (see Fig.5b), and provided detection biases for high eccentricities are taken into account, the eccentricity distribution tends toward $f(e) = 2e$ which is expected if that distribution is a function of energy only (Ambartsumian 1937). We note in particular the scarcity of low-eccentricity binaries in that period range. The same result is obtained by MM for young open clusters. Such a distribution can result from the dynamical disruption of small stellar systems (Van Albada 1968, Harrington 1975).

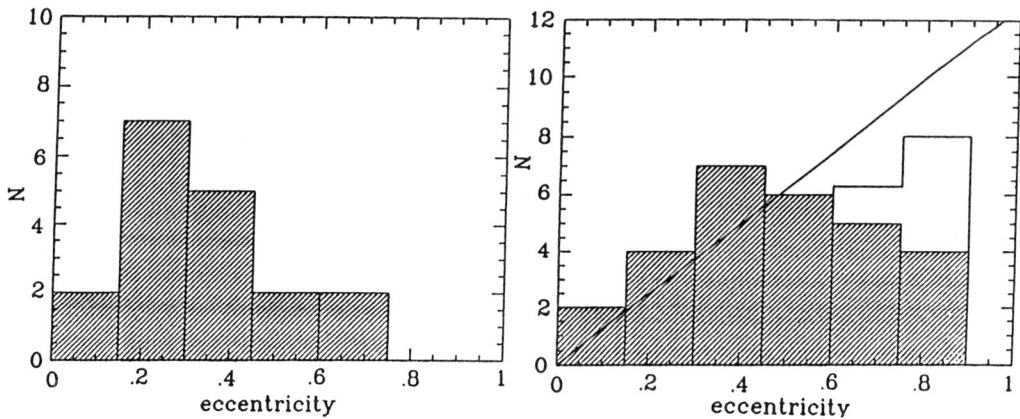

Figure 5: *Distribution of the eccentricities: 5a) for $P_{circ} < P < 1000$ days, 5b) for $P > 1000$ days. The empty region corresponds to the correction from observational biases.*

3.5 The mass ratio distribution

Several previous results on this distribution have been recalled in Sec. 2. Considering our G-dwarf sample, the mass ratio distribution corrected from observational biases (see Duquennoy and Mayor 1990b) is shown in Fig.6. The relevant results are:

    i) There is no (significant) maximum for $q = M_2/M_1 = 1$.

    ii) The shape of the distribution is remarkably fitted by one of the models derived by Kroupa and Tout (1989) for the field mass function of low mass stars. They used the result of a recent study on the stellar luminosity function and a mass/luminosity relation derived from both theoretical models and observational points obtained from binary stars. The mass function called 'GS' in their paper, with the same value of the adjusted parameters (μ, σ,) is represented by a dashed line in Fig.6:

$$\xi(q) = k \times \exp\left\{ \frac{-(q-\mu)^2}{2\sigma_q^2} \right\}$$

where $q = M_2/M_1$, $\mu = 0.23$, $\sigma_q = 0.42$, and $k = 18$. We note that the distribution below $q = 0.23$ could also be flat or even increasing after a plateau, the observed distribution in that range being still very uncertain. For comparison, the dashed curves correspond to the models called 'MS' (Miller and Scalo law) and 'SL' (Salpeter power law).

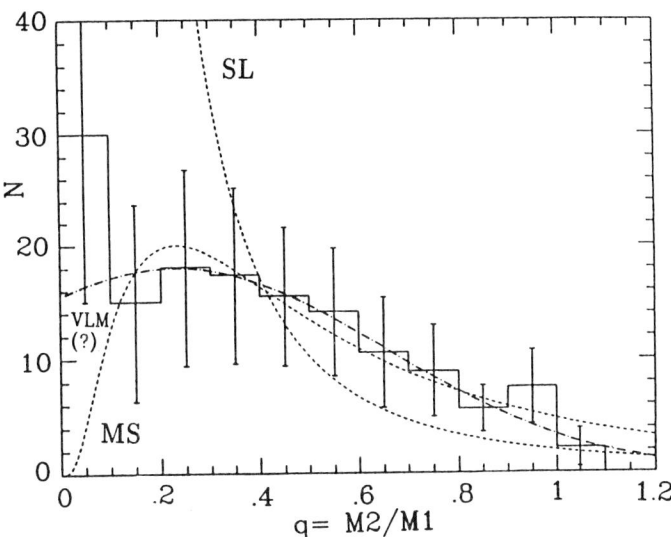

Figure 6: Distribution of the mass-ratios for all kinds of nearby G-dwarf binaries and corrected from observational biases. The fitting functions are those derived by Kroupa and Tout (1989) for low-mass field stars (see text).

The similarity of the mass functions for low-mass field stars and for secondary masses of nearby G-dwarf stars agrees with the idea of a binary star formation by random association of two stars formed with the same IMF. If such a scenario is acceptable for very long period binaries, probably for tight binaries the agreement betwen the observed $\xi(q)$ and the hypothesis of random association is fortuitous.

We have examined the possibility of a difference in $\xi(q)$ as a function of the orbital period. With the dividing period of 100 yrs suggested by AL applied to our G-dwarf sample, we found only a marginal 25% for the probability that the difference between the two populations of binaries is due to chance. Besides, the same probability is found to be 1% for the bright field stars, according to the data given by AL, and also 25% for K dwarf stars studied by Halbwachs (1987). This difference with AL's result may come from the selection biases already discussed. However, we find a statistically much more significant change (to a 95% confidence level) in the distribution $\xi(q)$ for $P < 1000d$: if for $P > 1000d$ the distribution remains close to the mean observed $\xi(q)$, on the contrary for $P < 1000d$ the distribution seems to be quite constant for all values of q.

3.6 The very low mass companions

By very low mass (VLM) companions we mean here $M_2 \leq 0.1 M_\odot$ or $M_2 \sin i \leq 0.1 M_\odot$. Some fundamental astrophysical questions in this domain are:

i) the behaviour of $\xi(q)$ for $q \leq 0.1$;
ii) what fraction of solar type stars are really single ?
iii) is there a lower limit to the mass M_2 of the companion ?
iv) the study of the transition from stars to brown dwarfs and to giant planets.

We first have to see if CORAVEL can contribute to these topics with a standard radial velocity error of 0.2-0.3 km/s. We made numerical simulations of a sample of binaries with various orbital elements (including sine-distribution of the inclinations, and distributions of eccentricities such as those found in Sec. 3.4), and with the same observing dates and velocity errors as in the real sample of G-dwarf stars. In fact we restricted the sample to a subset of well measured stars ($N \simeq 50$ measures, such as IAU standard stars). A binary is declared detected if $P(\chi^2)$ (see definition in Sec. 3.2) is less than or equal to 0.05.

The result of these simulations (see Fig.7) is that we have more than a 50% chance to detect secondary masses of 0.01 to 0.08 $M_\odot$ with orbital period ranging from 3 months to 16 years around a solar type primary star, provided the latter is observed with CORAVEL about 50 times within 12 years.

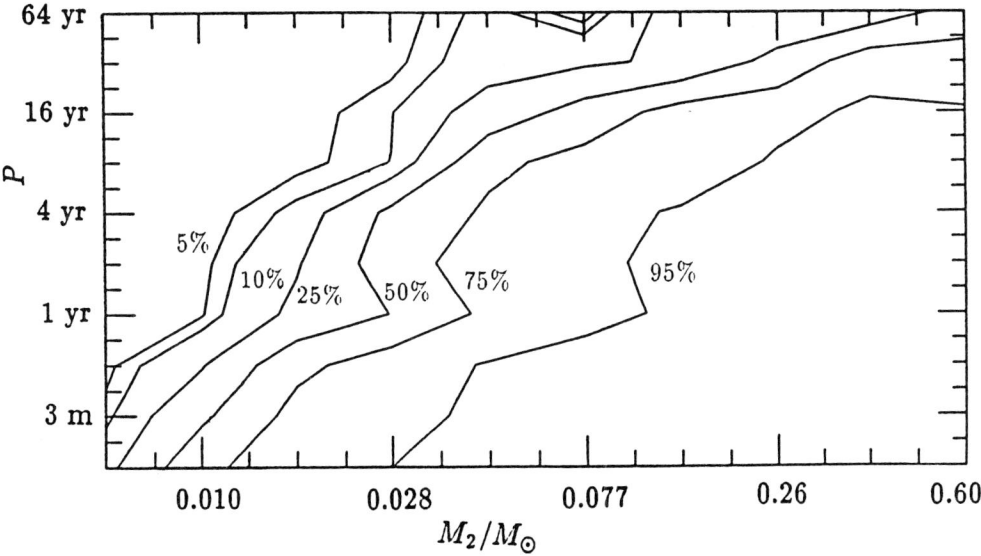

Figure 7: *Probabilities of detection with CORAVEL of a secondary component of mass M_2 with an orbital period P, assuming a $1M_\odot$-primary, measured about 50 times in 4000 days with a precision of 0.3 km/s.*

Now we consider the following sets of stars:
i) The IAU (and other) velocity standard stars observed with CORAVEL provide more than 100 stars with 7000 velocity measurements, a $\Delta T \simeq 4000$ days and a precision of 0.2 to 0.3 km/s. These stars have typical masses close to one solar mass. From this extremely well measured sample, we extracted 6 SB with derivable orbits (to be published) and $M_2 \sin i \leq 0.10 M_\odot$.

ii) Among published astrometric studies, we found seven astrometric binaries with orbits and $M_2 \leq 0.10 M_\odot$.

iii) Inspecting a sample of dM stars observed with CORAVEL since 12 years, we found another 6 SB with derivable orbits and $M_2 \sin i \leq 0.10 M_\odot$.

These three sets give a VLM sample of 19 binaries with $\overline{M_2 \sin i} = 0.057 M_\odot$. Their companions may thus be considered as probable 'soft brown dwarfs', in the sense that they statistically appear below the usually admitted hydrogen burning mass limit, but not too far from it. Some of the corresponding CORAVEL radial velocity curves, in phase with their orbital period, are shown in Fig.8. HD 3346 has already been published by Mc Clure et al. (1985), the illustrated orbit is only based on CORAVEL measurements. In a plane eccentricity versus $\log(M_2 \sin i)$ (see Fig.9) we plotted the nearby G-dwarf sample (SBs with $P \geq P_{circ}$) together with the VLM sample, and the 4 giant planets of the solar system. The mean eccentricities for the three sets are 0.33 ± 0.04, 0.35 ± 0.06 and 0.04 ± 0.01 respectively.

This result adresses the problem of the transition from stars and brown dwarfs to planets. The striking result on the eccentricities strongly suggests similar formation processes for stars and 'soft' brown dwarfs. Keeping in mind that the present lower limit to our secondary mass detection ($M_2 \sin i \simeq 0.01 M_\odot$) is similar to Boss' (1987) lower mass limit for fragmentation, the following questions arise:
 i) Can we find any companion in the mass range 0.001-0.010 $M_\odot$?
 ii) If yes, what would be the eccentricity of its orbit ?

Now returning to the G-dwarf sample, what is the fraction of low mass ($M_2 = 0.01$ to $0.10\ M_\odot$) companions to G-dwarf primaries ? We have no certain detection yet (i.e. with orbit), but only candidates. For a reliable variability criterion of $P(\chi^2) < 0.01$ and taking into account a statistical number of 1 false alarm detection, we still observe 5 candidate detections, which must be displayed in a range of three units of $\log P$. Assuming a mean detection rate of 50% in the considered mass and period ranges (from simulations similar to that in Fig.7 but adapted to the whole G-dwarf sample), that VLM secondaries follow the same period distribution as the others, we expect about 30 VLM companions or $(18 \pm 9)\%$ of the primaries.

This result is not necessary in contradiction with the study of Campbell (1988) who find no brown dwarf candidates in a similar type of stars, regarding the small number of stars (16) observed in his sample. Moreover we can compare with the proportion of VLM that can be expected in the sample of IAU standard stars: this sample can be considered as being extracted from a larger sample (say $\sim$ 130 stars) of any luminosity class objects but having roughly one solar-mass, whose large amplitude binaries (in same proportion $\sim$ 30% as that observed for the nearby G-dwarf stars) have been removed. With 6 certain detections (i.e. with orbits) among 100 stars, and rejecting 1 probable false VLM due to low $\sin i$, we have again 5 candidates in three units of $\log P$. Assuming the distribution found in Sec. 3.3 is valid and using the detection probabilities in Fig.7, we expect 20 VLM or 15% of the initial sample, which is comparable to the ratio estimated for the nearby

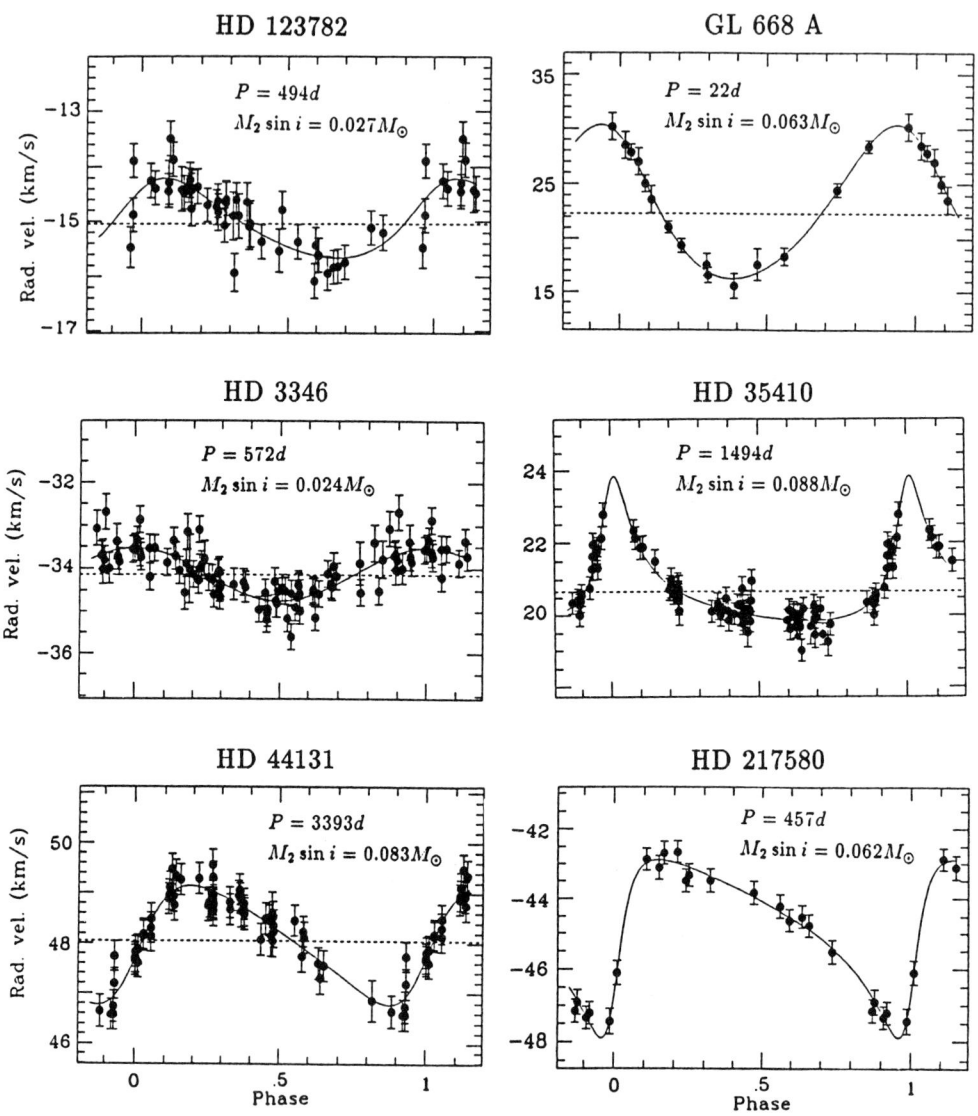

Figure 8: *Radial velocity curves obtained with CORAVEL for several binaries with very low mass ($\mathcal{M}_2 \sin i \leq 0.1 \mathcal{M}_\odot$) secondaries.*

Figure 9: The distribution of the eccentricities versus secondary masses, from stars to planets through brown dwarfs, or the missing link in the mass range 0.001 - 0.010 $M_\odot$.

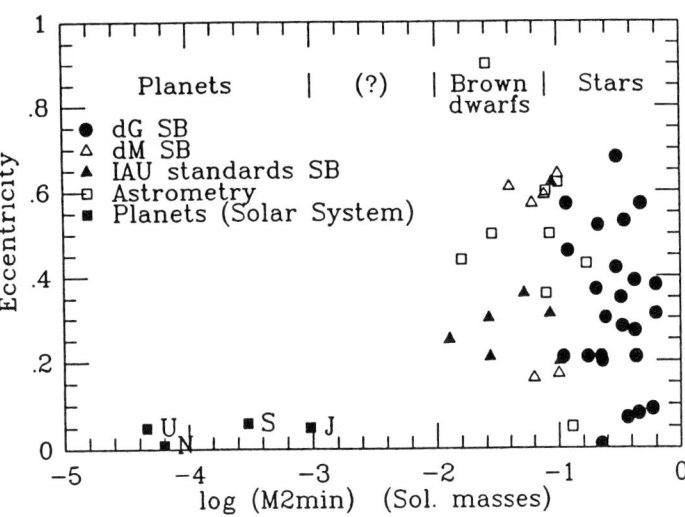

G-dwarfs.

4. CONCLUSIONS

We have studied an unbiased sample of 166 primary G-dwarf stars in the solar neighbourhood. We found 65% of them having a companion with $M_2/M_1 \geq 0.1$. These companions represent 32% of the total mass of the primaries. Among the 35% apparently single remaining stars, a large proportion, maybe about 50%, or $(18 \pm 9)\%$ of the total sample, could have a VLM companion in the mass range 0.01-0.10 $M_\odot$. Consequently, less than one quarter of the G-dwarfs may appear as real single stars.

Concerning the systems with $M_2/M_1 > 0.1$ in the nearby G-dwarf sample, we find the following results:

i) The orbital period distribution is unimodal and can be approximated by a Gaussian-type relation with a median period of 260 yrs.

ii) The short period binaries are circularized up to orbital periods of $\sim 11d$ due to the tidal evolution effects, a result compatible with the mean age of the galactic disk.

iii) The tight binaries not affected by tidal effects ($11 < P < 1000d$) may reflect the initial binary formation process, and they have a mean eccentricity $\bar{e} = 0.35 \pm 0.04$.

iv) The remaining binaries ($P > 1000d$) have an observed distribution, when corrected from detection biases, which tends smoothly toward $f(e) = 2e$.

v) The mass-ratio distribution shows no maximum for $q = 1$, but rather a regular increase toward small secondary masses, at least up to $q \simeq 0.3$. We find that this distribution is well fitted again by a Gaussian-like relation, with a maximum for $q \simeq 0.23$.

Concerning the systems with $M_2/M_1 \leq 0.1$, we have studied the possibility for the spectrometer CORAVEL to observe such systems. Using these simulations for the nearby G-dwarf sample, we tentatively derived the expected proportion of $(18\pm9)\%$ VLM secondaries mentioned above, a value in rather good agreement with the proportion of VLM detections among IAU standard stars. Among the latter sample and another sample of dM stars, we indeed detected and derived the orbits of 13 SB with probable VLM secondaries ($M_2 \sin i \leq 0.1 M_\odot$), around primary stars having roughly one solar mass. Added to 6 astrometric binaries with VLM secondaries candidates, they lead to the following results:

i) The brown dwarfs may not be as rare as quoted by Campbell (1987) around solar mass stars, or by Marcy (1989) around M-dwarf stars.

ii) The mean eccentricity for these VLM secondary binaries with $11 < P < 1000 d$ is $\bar{e} = 0.33 \pm 0.05$. We preliminary conclude that the binary formation process (fragmentation ?) seems to be the same for stars and for brown dwarfs. Contrarily, giant planets seem to form only with very small eccentricities.

Finally, we ask the question if companions in the mass range $0.001 - 0.010 M_\odot$ exist, and if yes, whether the eccentricity of their orbit around the primary is significantly non zero or not. The latter could be an information of their formation process and could become a test to distinguish if we deal with stars (or brown dwarfs) or real extra solar system planets.

Similar studies on duplicity remain to be done with K and M stars in the solar neighbourhood, to allow direct comparisons with the nearby G-dwarfs. This question is of great importance since K and M dwarfs represent respectively about 10 and 60% of the total number of stars in our galaxy. Important contributions (e.g. Tokovinin 1988, Marcy et al 1989) have been published recently but do not include yet a full description of the orbital elements among low-mass stars. We hope to be in position to discuss the K and M part of our survey in a near future.

REFERENCES

Abt, H.A., Levy, S.G. 1976, Astrophys. J. Suppl. Series **30**, 273
Ambartsumian, V.A. 1937, Soviet Astr. A. J. **14**, 207
Baranne, A., Mayor, M., Poncet, J.-L. 1979, Vistas Astron. **23**, 279
Boss, A.P. 1987, in 'Theory of Collapse and Protostar Formation', Interstellar Processes Symposium, Eds D. J. Hollenbach and H. A. Thronson, Jr., Reidel Publ.
Boss, A.P. 1988, Comments Astrophys. **12**, 169
Burki, G., Mayor, M. 1985, in 'Instrumentation and Research Programmes for Small Telescopes', 385-400, Eds. J.B. Hearnshaw and P.L. Cottrell, Reidel Publ.
Campbell, B., Walker, G.A.N., Yang, S. 1988, Astrophys. J. **331**, 902
Dommanget, J., Nys, O. 1982, Comm. Obs. Roy. Belgique, Ser.B no 124
Duquennoy, A. 1987, Astron. Astrophys. **178**, 114
Duquennoy, A., Mayor, M. 1988, Astron. Astrophys. **195**, 129

Duquennoy, A., Mayor, M. 1990a, Astron. Astrophys. Suppl. Ser. (in preparation)
Duquennoy, A., Mayor, M. 1990b, Astron. Astrophys. (in preparation)
Gliese, W. 1969, Veröff. Astron. Rechen Inst. Heidelberg, no 22
Griffin, R.F. 1967, Astrophys. J. **148**, 465
Halbwachs, J.-L. 1986, Astron. Astrophys. **168**, 161
Halbwachs, J.-L. 1987, Astron. Astrophys. **183**, 234
Harrington, R.S. 1975, Astron. J. **80**, 1081
Jasniewicz, G., Mayor, M. 1988, Astron. astrophys. **203**, 329
Kroupa, P., Tout, C.A. 1989 Mon. Not. Roy. Astr. Soc (preprint)
Latham, D.W., Mazeh, T., Carney, B.W., McCrosky, R.E., Stefanik, R.P., Davis R.J. 1988, Astron. J. **96**, 567
Latham, D.W., Mazeh, T., Stefanik, R., Mayor, M., Burki, G. 1989, Nature **339**, 38
Marcy, G.W., Benitz, K.J. 1989, Astrophys. J. **344**, 441
Mayor, M., Imbert, M., Andersen, J., Ardeberg, A., Baranne, A., Benz, W., Ischi., E., Lindgren, H., Martin, N., Maurice, E., Nordström, B., Prevot, L. 1983, Astron. Astrophys. Suppl. Ser. **54**, 495
Mayor, M., Maurice, E. 1985, in 'Stellar radial velocities', 299-310, Eds A. G. D. Philip and D. W. Latham, Davis Press
Mayor, M., Mermilliod, J.-C. 1983, in 'Les etoiles binaires dans le diagramme HR', Proceedings of the 5 th meeting of Strasbourg Observatory, p. 45
Mayor, M., Mermilliod, J.-C. 1984, in 'Observational Tests of the Stellar Evolution Theory', 411-414, Ed. A. Maeder and A. Renzini, Reidel Publ.
McClure, R.D., Griffin, R.G., Fletcher, J.M., Harris, H.C., Mayor, M. 1985, Publ. Astron. Soc. Pac. **97**, 740
Tassoul, J.-L. 1987, Astrophys. J. **322**, 856
Tassoul, J.-L. 1988, Astrophys. J. Letters **324**, L71
Tokovinin, A.A. 1988, Astrofizika **28**, 297, Engl. transl. in Astrophysics **28**, 173
Trimble, V. 1987, Astron. Nachr. **6**, 343
Van Albada, T.S. 1968, Bull. Astron. Inst. Neth. **19**, 479
Zahn, J.-P. 1977, Astron. Astrophys. **57**, 383
Zahn, J.-P. 1989, Astron. Astrophys. **220**, 112
Zahn, J.-P., Bouchet L. 1989, Astron. Astrophys. (preprint)

STELLAR STRUCTURE AND EVOLUTION

Scientific Organizer: E. Schatzman
Local Organizer: R. Rebolo

Activity of Young Low-Mass Stars

C. BERTOUT

Institut d'Astrophysique de Paris

1. INTRODUCTION

A review of observational properties and models of optical young stellar objects must include both the *classical* T Tauri stars (CTTSs) that match Herbig's (1962) definition and that were discovered from H_α survey; and the optical low-luminosity pre-main sequence stars first detected at some other wavelength, e.g., in X-rays (Walter et al. 1988) or in CaII (Herbig, Vrba, and Rydgren 1986). The latter category includes mostly weak emission-line pre-main sequence stars ($W_\alpha(H_\alpha) \lesssim 5\text{Å}$) which, following Herbig and Bell (1988), are called here *weak-line* T Tauri stars (WTTSs). It was realized only during the last decade that the WTTS population is several times larger than the CTTS population and that CTTSs and WTTSs form a continuum of objects with various degrees of activity.

In recent years, a picture has emerged in which the activity of WTTSs is quite similar in nature to that displayed by other late-type dwarfs; it is caused mainly by solar-type dynamo processes (cf. Feigelson, Giampapa, and Vrba 1989). Because rotation is the main parameter governing magnetic activity and because both WTTSs and CTTSs have similar rotation rates, one expects solar-type magnetic activity to be essentially the same in the two subclasses. Indeed, properties of dark spots found on the surfaces of WTTSs and CTTSs are comparable as are their X-ray flux levels (Bouvier and Bertout 1989; Bouvier 1989). The range and variety of CTTS activity are, however, much more dramatic, and it is now recognized that an energy source external to the star is needed to drive such intense activity.

Several lines of evidence discussed below strongly suggest that circumstellar disks surround CTTSs, and that both the presence of a disk and the interaction between disk and star may be responsible for some properties of CTTSs. Originally, this idea had been proposed independently by Walker (1972) to explain observed properties of a subgroup of CTTSs (the YY Orionis stars) and by Lynden-Bell and Pringle (1974) in their classical paper on accretion disks. In the current version of this scenario, the disk provides the energy reservoir needed to account for energetic mass and radiative losses observed in CTTSs; if the disk then dissipates on a time-scale smaller than or comparable to the duration of the T Tauri phase, a gradual transition from T Tauri star to main-sequence star can be expected. While several aspects of this tentative scenario are explored in several recent papers discussed below, it is far from being firmly established. In particular, the connections between disks and winds are not understood yet, and there is no widely accepted theoretical mechanism for driving T Tauri winds. Recent reviews summarizing the field of T Tauri disks and winds are those of Bertout (1989), and Bertout, Basri, and Cabrit (1989).

2. A BRIEF REVIEW OF OBSERVATIONS

2.1. Spectroscopic Properties: Optical Range

Walter et al. (1988) recently presented the first extensive study of WTTSs in the Taurus-Auriga region. They define a WTTS as an X-ray source with an optical counterpart showing pre-main sequence characteristics. Specifically, LiIλ6707 absorption is present with equivalent width larger than 100 mÅ, and stellar radial velocity indicates membership in the associated molecular cloud.

The main optical spectroscopic criterion that defines a CTTS according to Herbig (1962) is the presence of the hydrogen Balmer lines and CaII H and K lines in emission. LiIλ6707 absorption is also conspicuously strong in CTTSs. The strongest emission lines following the hydrogen and CaII lines are usually caused by FeII, TiII, and HeI. The emission line spectrum is superimposed on a continuous spectrum, which may range from a pure continuum (in *extreme* CTTSs) through a late-type absorption spectrum with anomalous line strengths (in *veiled* CTTSs) to an almost normal absorption spectrum of type F through M in *moderate* CTTSs. Figure 1 illustrates the various subgroups of T Tauri stars. There, medium-resolution spectrograms (2Å) covering the spectral range 3200 Å to 8800 Å (kindly communicated by G. Basri) are presented for 4 late K or early M stars. TAP 57 is a WTTS similar in many respects to a standard K7 dwarf, DN Tau a moderate M0 CTTS, DF Tau a veiled M0 CTTS, and DR Tau an extreme CTTS with probable K5 spectral type. Precise MK classification of CTTSs is difficult; the spectral type derived from blue spectrograms is usually earlier than that derived from red spectrograms. Note the strong Balmer emission present in the CTTS spectra.

2.2. Spectroscopic Diagnostics of T Tauri Winds

High-resolution profiles of emission and absorption lines provide evidence for complex gaseous flows in CTTS envelopes. Several studies of H_α show a variety of line profiles (Kuhi 1964; Herbig 1977; Kuhi 1978; Ulrich and Knapp 1979; Schneeberger, Worden, and Wilkerson 1979; Hartmann 1982; Mundt and Giampapa 1982; Mundt 1984). According to Kuhi (1978), a majority of CTTSs display Type III P Cygni profiles at H_α. In these profiles, the emission is broad (FWHM $\approx$ 200 km/s) and symmetric, and a blue-displaced absorption feature typically at 80 km/s is superimposed on the emission. While symmetric emission profiles are also common, only a

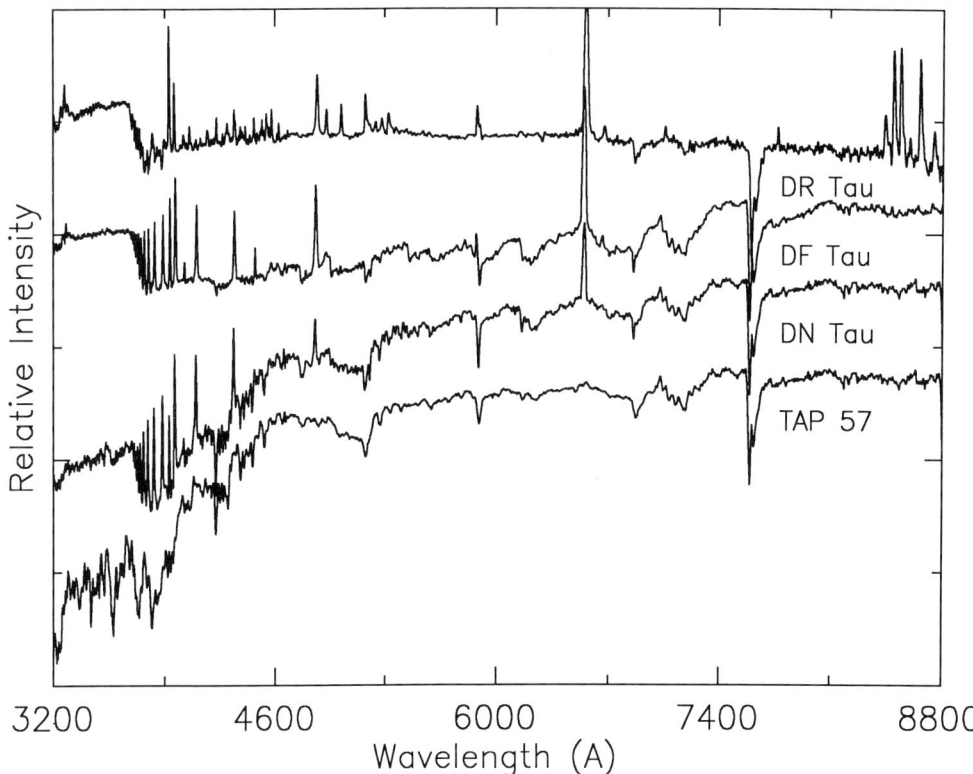

Figure 1 Medium-resolution spectrograms covering the spectral range 3200 Å to 8800 Å of four late K or early M T Tauri stars, shown in order of increasing emission levels. The relative intensity is displayed in wavelength units.

few CTTSs display Type I P Cygni profiles with blue-displaced absorption components reaching below the photospheric continuum. Unlike Type III profiles, Type I profiles unambiguously indicate that mass-loss is taking place. Inverse P Cygni profiles indicative of mass accretion are not observed at H_α (Herbig 1977), although a subclass of CTTSs named YY Orionis stars after their prototype by Walker (1972) displays such profiles at the higher members of the Balmer series. One must caution, however, that almost any statement about emission lines in CTTSs has many exceptions. The lines in any given star are always variable, sometimes drastically altering their shape in a few days or weeks, and sometimes being relatively stable for several years. In some extreme CTTSs, most notably DR Tau, Balmer line profiles can change from P Cygni to inverse P Cygni in a matter of days, and both P Cygni and inverse P Cygni profiles can occasionally be seen in the same spectrogram (Krautter and Bastian 1980).

Because of the faintness of these stars, it has only been in the last decade that reasonable numbers of high quality line profiles have been obtained. As an example

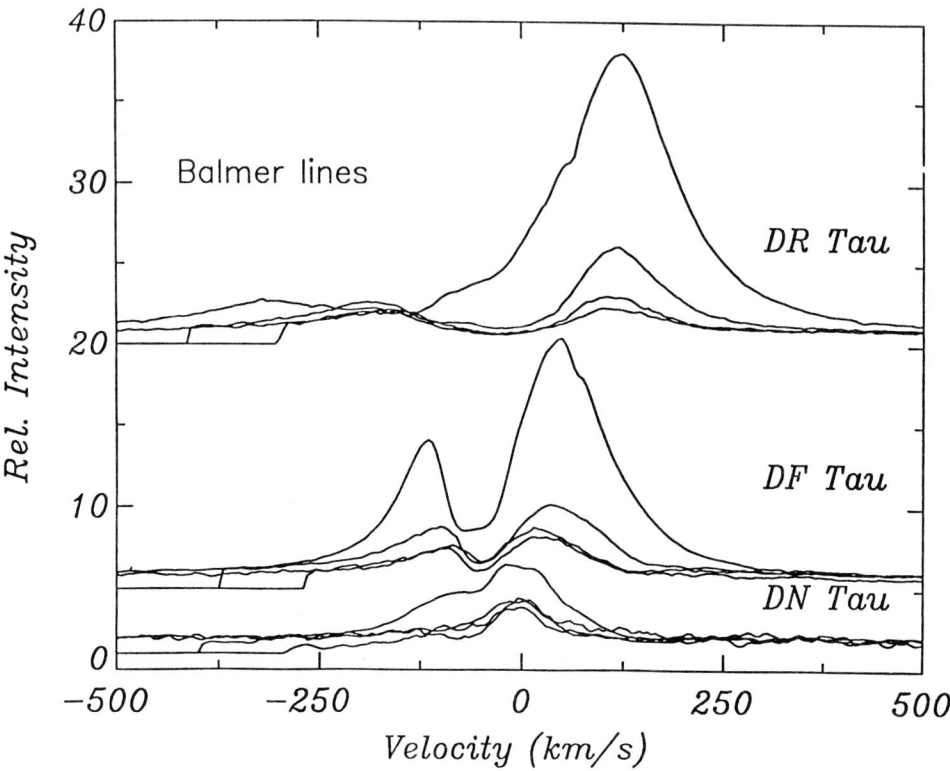

Figure 2 The first four members of the Balmer series for the stars in Figure 1. The line flux is normalized to the adjacent continuum and the velocity is in the stellar rest frame.

of data that is now becoming available, I show here a few line profiles from high-resolution spectrograms obtained by Gibor Basri at Lick Observatory using the Hamilton Spectrograph. These profiles are from three stars which span the range of line strengths found among CTTSs and whose overall spectra are shown in Fig. 1. The first four Balmer lines are shown for each star in Figure 2 on common continuum scales. Most striking there is the Balmer decrement; the H_α line is much stronger than the higher series lines. This cannot arise if the line forms on the stellar surface; one needs either low densities or a large emitting volume for H_α. On the other hand, if the higher Balmer lines are scaled up so they have the same peak heights, then they often look remarkably similar. The main difference is usually that the absorption component is less blue-shifted and often weaker as one goes up the series. It is also more common for the absorption component to go below the continuum in the higher lines. Certain lines, Na D in particular, also show evidence for cool sharp high-velocity components (Mundt 1984) that may be due to shells of cooler material far from the star. Figure 3 shows one of the Ca II triplet lines

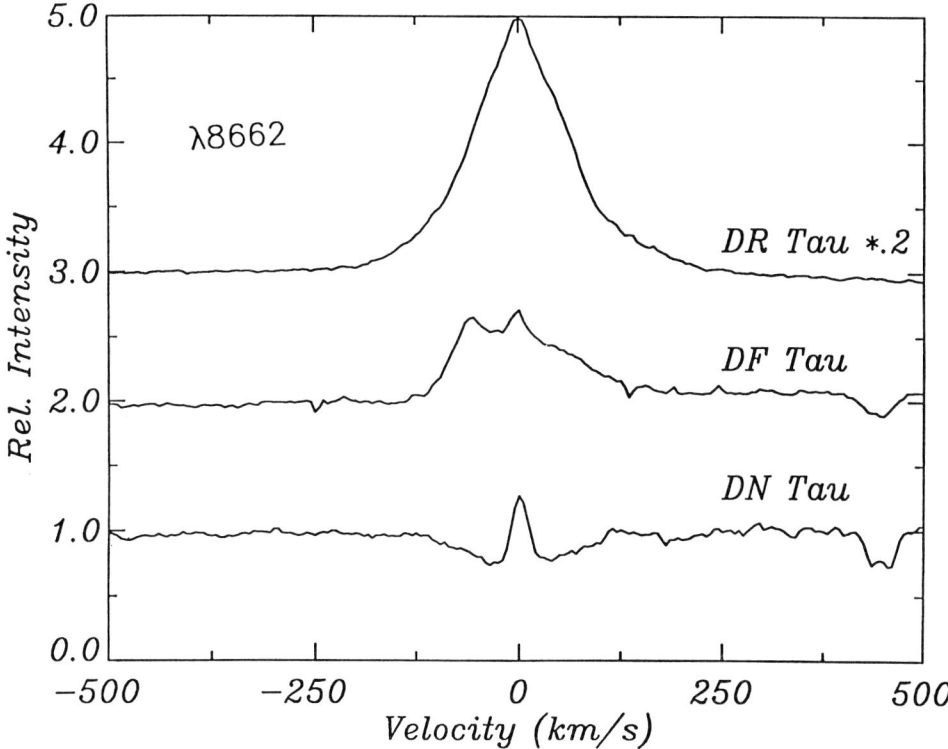

Figure 3 One of the CaII infrared triplet lines for each of the stars in Fig. 1, observed on the same night as the profiles of Fig. 2.

for the three stars discussed above. In DN Tau, the line could arise purely from a strong chromosphere, but there is clear high velocity material in the other cases. We note that on different occasions the line profiles can appear rather different, and in one instance the DF Tau profile showed no emission at all! The DR Tau profile is typical of the extreme stars, and unshifted triangular profiles such as this are fairly common among CTTS profiles. Interpretation of all these profiles is the subject of much current work.

2.3. Spectral Energy Distributions

When compared to standard stars, the spectral energy distributions of T Tauri stars display a wide range of both ultraviolet excesses (Kuhi 1974; Herbig and Goodrich 1986; Bertout, Basri, and Bouvier 1988) and infrared excesses (Mendoza 1966, 1968). Recently published databases of T Tauri optical and infrared photometry include those of Rydgren *et al.* (1984) and Bouvier, Bertout, and Bouchet (1988). Figure 4 displays the observed spectral energy distributions from 3600 Å to 100 μm of the stars whose spectra are shown in Figs. 1 through 3. The spectral energy

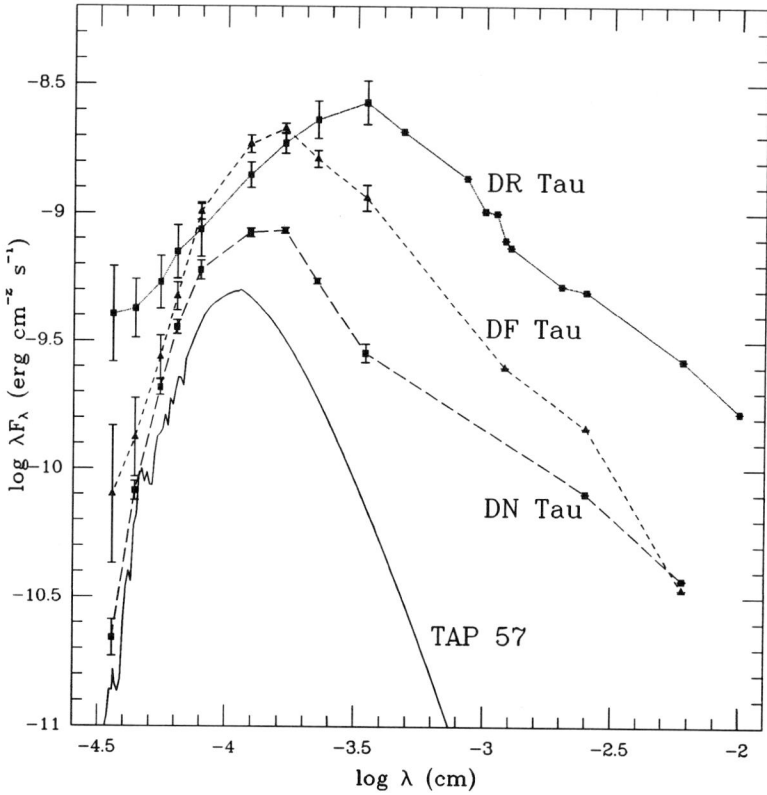

Figure 4 Observed spectral energy distributions from 3600Å to 100 μm of the stars whose spectra are shown in Fig. 1. When available, observed variability is indicated by error bars. When compared to WTTSs such as TAP 57, CTTSs display prominent ultraviolet and infrared excesses.

distribution of the WTTS TAP57, which is basically that of a normal K7V star, has been displaced downwards by 0.3 dex in order to facilitate comparison with CTTSs that are shown at their actual flux level. It is clear that the excess luminosity of these objects increases with activity as measured by optical line emission.

3. THE DISK HYPOTHESIS

Direct observations of solar system-sized disks surrounding CTTSs in close-by star-forming regions such as Taurus or Ophiuchus are not possible yet, but it will be a task for ESO's *Very Large Telescope*. There are however a number of indirect pieces of evidence suggesting the presence of optically thick, dusty circumstellar disks around CTTSs. That dust was present around CTTSs has been known for a long time (cf. Mendoza 1966; 1968), but it only recently became clear that it is not distributed isotropically around the star.

3.1 Geometry of Dust around CTTSs

A first line of evidence favoring the view that dust is located in a flat disk rather than in a more or less spherical shell surrounding the star is provided by observations of IRAS sources near dense cores by Myers *et al.* (1987), who show that the amount of dust necessary to explain the infrared spectrum is irreconcilable with the observed low optical extinction values. Polarization data provide an additional hint of disk geometry. Bastien and Ménard (1987) recently demonstrated that both the observed linear polarization maps and the circular polarization detected in some CTTSs result naturally from a model where a stellar photon is scattered first in an optically thin dusty envelope and then in an optically thick but geometrically thin disk. The profiles of the forbidden [OI] and [SII] lines seen in many CTTS spectra provide further indirect evidence for the presence of extended disks around CTTSs. Forbidden lines, formed in low density regions, primarily probe the outer parts of the warm stellar wind. High-resolution observations of forbidden line emission revealed that their intensity-weighted systemic velocity is blueshifted. Appenzeller, Jankovics, and Oestreicher (1985) and Edwards *et al.* (1987) show that an opaque screen (i.e., an optically thick circumstellar disk) must "hide" the redshifted emission, a conclusion which does not depend on details of disk or wind models.

3.2 Stellar Activity vs. Disk Activity

CTTSs lose large amounts of energy through both radiation and winds, which is a recognized difficulty in models that assume a late-type star is the only energy source (DeCampli 1981; Calvet and Albarrán 1984). While the disk hypothesis solves this problem by providing a huge reservoir of potential energy, it introduces another: which activity manifestations are due to the star and its chromosphere, and which to the disk?

Chromospheric activity

Joy (1945) first noted the similarity between the emission spectrum of active T Tauri stars and the chromospheric (flash) spectrum of the Sun. Later, Herbig (1970) proposed that the photospheric temperature minimum occurs at greater continuum optical depth in CTTSs than in the Sun. Following Herbig's suggestion, several detailed investigations of continuum and line formation in a chromosphere located at larger optical depths than in the Sun were performed (e.g., Calvet, Basri, and Kuhi 1984). These works assume a plane-parallel, homogeneous atmosphere in hydrostatic equilibrium and append a chromosphere to a late-type photosphere. Both the chromospheric temperature stratification and the optical depth of the chromosphere's anchorage in the atmosphere are free parameters that are adjusted until an acceptable fit to a set of spectral diagnostics is obtained. The results demonstrate that the metallic emission spectrum and the veiling can be reproduced by this model. They also show that an extended, optically thin region is needed to reproduce both the strong H_α flux and Balmer decrement of classical T Tauri stars.

While the Balmer continuum (in emission in many CTTSs; see Fig. 1) could, at least in principle, also be formed in such a chromospheric region, this would require a large optical depth at the bottom of the transition region, which also drives the Paschen continuum in emission and thus produces a strong veiling in the optical region. In contrast, many CTTSs (e.g., DN Tau) display strong Balmer continuous emission but little veiling. Others, like DF Tau display both Balmer emission and veiling which could conceivably be of chromospheric origin. In such strong chromospheres, the chromospheric luminosity then becomes comparable to the photospheric luminosity.

A drawback of these semi-empirical models is that they do not directly address the question of the origin of the non-radiative energy flux responsible for heating the deep-lying chromosphere. While solar-type magnetic activity is often mentioned as the most likely possibility, the amount of energy required to cover observed radiative losses in CTTSs such as DF Tau is a problem. Difficulty of quantitatively assessing the role of magnetic dynamo activity in the T Tauri phenomenon arises mainly because of the present lack of magnetic field measurements in these faint objects. It is therefore still possible to speculate that large convective regions, strong differential rotation, and fossil magnetic fields may all conspire to enhance surface magnetic field strengths to values much larger than found in other late-type stars, and large enough to drive the entire range of activity witnessed in these objects in the optical and ultraviolet ranges.

At least one major parameter governing solar-type magnetic activity, the rotation period, is now known for a large sample of T Tauri stars, which makes it possible to compare known activity diagnostics in CTTSs and other active late-type stars with similar periods, among them the RS CVn stars and the WTTSs. Walter and Bowyer (1981) first demonstrated a dependence between rotation rate and coronal luminosity in late-type stars. Since then, this connection has both been confirmed and shown to extend to chromospheric and transition region diagnostics as well (cf. Hartmann and Noyes 1987).

Figure 5 (from Bouvier 1989) displays relationships between the rotation periods of a sample of late-type stars (T Tauri stars and other active stars) and the observed flux in two major activity diagnostics: (a) the X-ray flux, and (b) H_α. In both plots, T Tauri stars are shown as black dots and their observed variability, when available, is indicated by a vertical bar. RS CVn stars are displayed as open squares, dMe stars as open circles, and late-type dwarfs as crosses.

The observed correlation between the X-ray flux of T Tauri stars and their rotation period (Fig. 5a) closely matches the relationship derived from other late-type stars; this is the strongest argument so far in favor of solar-type coronal heating in T Tauri stars (Bouvier et al. 1985; Bouvier 1987, 1989).

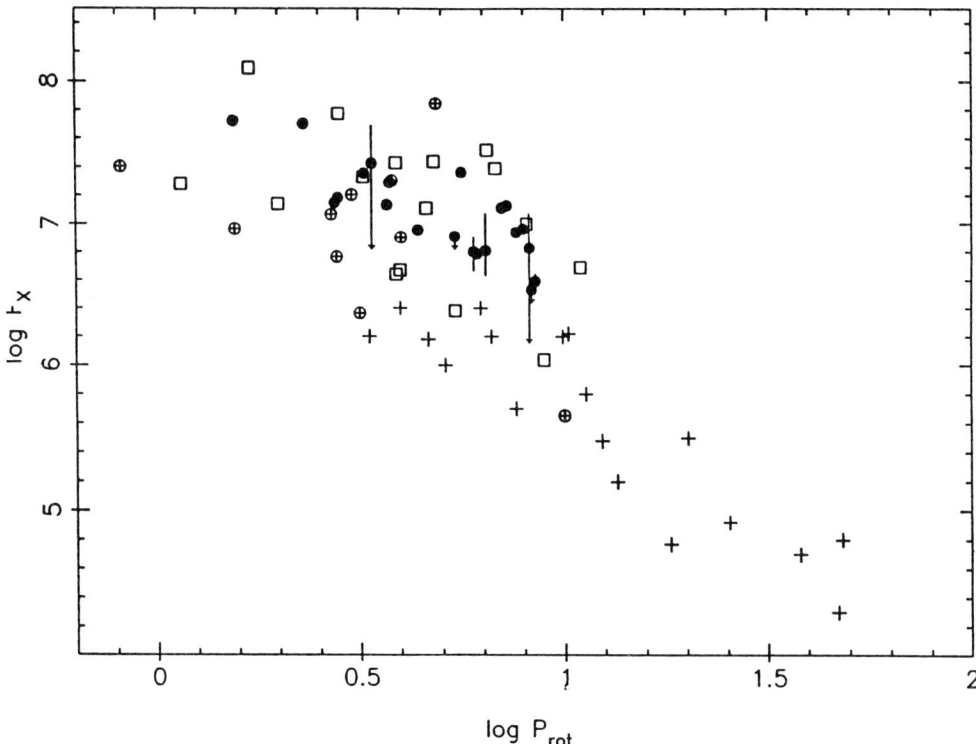

Figure 5a Relationship between rotation period and extinction-corrected X-ray flux for a sample of T Tauri stars and other late-type stars (from Bouvier 1989). See text for symbol meanings and explanations.

The relationship between H_α flux and rotation period (Fig. 5b) is, however, quite different. A weak correlation is observed for late-type dwarfs and RS CVn stars. The behavior of T Tauri stars in Fig. 5b is quite peculiar. While stars with low H_α flux (WTTSs) display emission levels comparable to dwarfs with similar rotation rates, the bulk of CTTSs is way above the correlation line. Thus, "chromospheric" non-radiative losses as measured by H_α appear quantitatively similar in WTTSs and in other late-type active stars, but they can be up to a factor 100 larger in CTTSs; and there is no connection in CTTSs between this activity indicator and rotation rates. It is therefore probable that H_α is formed not only in a solar-type chromosphere in CTTSs. In contrast, the H_α line of WTTSs could be entirely of chromospheric origin. Similar results are found for the CaII H and K, and for the MgII h and k lines. These empirical comparisons support Calvet and Albarrán's (1984) conclusion (based on theoretical estimates) that only low-level WTTS activity could possibly be driven by MHD wave dissipation.

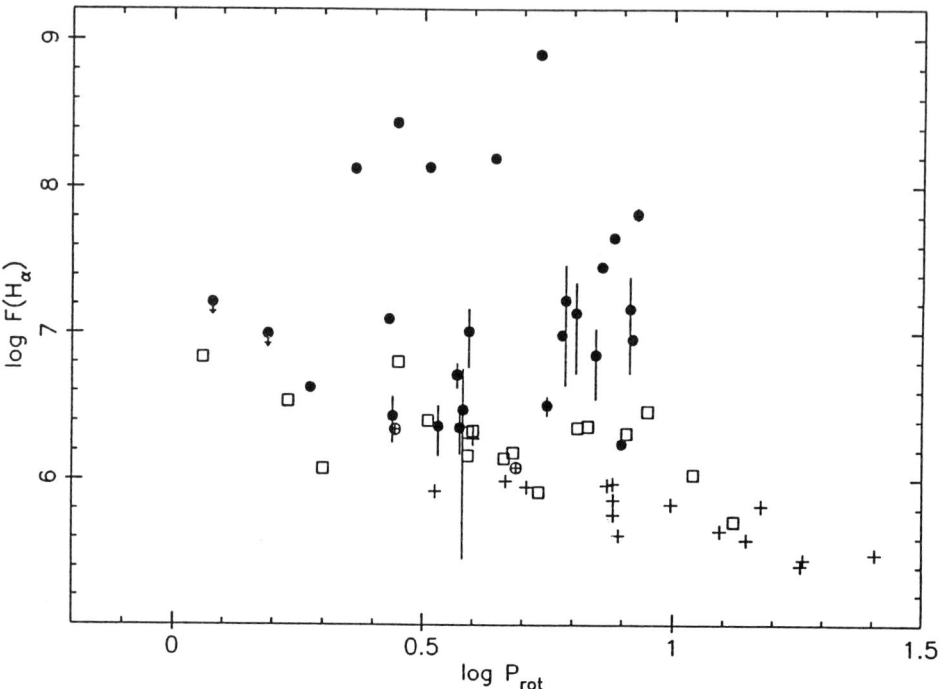

Figure 5b Relationship between rotation period and observed H_α flux for a sample of T Tauri stars and other late-type stars (from Bouvier 1989).

Disk activity

In their first disk models for CTTSs, Adams and Shu (1986) considered only passive disks that merely reprocess stellar light (i.e., stellar optical and UV photons are absorbed in the dusty disk and re-emitted thermally in the infrared). Although the star is the only photon source in this picture, the observed bolometric flux of the star/disk system can be up to 50% larger than the stellar photospheric flux because stellar photons that would not reach the observer if the disk were absent are redirected by the disk in the observer's direction. Adams and Shu (1986) showed that the resulting disk temperature distribution is proportional to $r^{-3/4}$ at large distances from the star and that resulting infrared spectra depend only on the disk viewing angle. These models, while sucessful in explaining the infrared excesses of several young stellar objects (cf. Adams, Lada, and Shu 1987), fail to account for several details of CTTS energy distributions. Specifically, they account neither for the strong blue and ultraviolet excesses typically observed in CTTSs nor for stars with infrared luminosities obviously larger than the limit mentioned above. UV and IR excesses are often correlated in CTTSs, and we have mentioned above the difficulties involved in trying to attribute large UV (and IR) excesses to solar-type chromospheric emission. With purely reprocessing disks, we are confronted with the very same problem. We can now interpret the main features of IR spectral energy distributions as the result of disk emission, but we must postulate still another physical process to produce an ultraviolet excess luminosity comparable, in some cases, to the photospheric luminosity.

This, together with the problem presented by stars with IR excess luminosity in excess of half the photospheric luminosity, can be overcome if the disk is not only reprocessing photons from its central star, but is also self-luminous. Since accretion of disk matter onto the star is a natural consequence of energy dissipation within the disk, it will occur in the active disks envisioned here. Lynden-Bell and Pringle (1974) were first to suggest that a viscous accretion disk might account for the continuous excesses of CTTSs, but this proposal was only given full attention recently (cf. Basri and Bertout 1989 and literature therein). Now, an assessment of this model's successes and shortcomings is possible.

Current work generally assumes that the viscosity is parametrized according to the so-called α–prescription as defined by Shakura and Sunyaev (1973). The theory of optically thick accretion disks predicts an asymptotic temperature distribution proportional to $r^{-3/4}$ in steady-state disks, i.e., identical to that of reprocessing disks. The emitted infrared spectra therefore have the same shape in these two classes of models ($\lambda F_\lambda \propto \lambda^{-4/3}$). But one half of the accretion luminosity $L_{acc} = GM_*\dot{M}_{acc}/R_*$ (where R_* is the stellar radius, M_* the stellar mass, and $\dot{M}_{acc}$ the constant mass-accretion rate within the disk) is now emitted from the disk together with the reprocessed luminosity. As long as the accretion luminosity is smaller than or comparable to the reprocessed luminosity, i.e., $\dot{M}_{acc} \lesssim 10^{-8} M_\odot \mathrm{yr}^{-1}$, the infrared spectrum alone does not allow one to distinguish between these two models. However, a quasi-keplerian disk extending down to the stellar photosphere joins the star in a hydrodynamically complex boundary layer in which disk matter (rotating at ≈ 200 km/s) is slowed down to photospheric rotational velocity (≈ 10–30 km/s) and where about half of the accretion luminosity is dissipated (Lynden-Bell and Pringle 1974). As it turns out, emission from the boundary layer peaks in the blue or ultraviolet range in CTTSs.

Bertout, Basri, and Bouvier (1988) compared quasi-simultaneous sets of data in the ultraviolet/optical and optical/near-infrared ranges to synthetic spectra emitted by models of a T Tauri system made up of a late-type active star, an accretion disk and its boundary layer. They found that typical T Tauri disks are optically thick over most of their surface as long as $\alpha \leq 1$ and that the spectral energy distribution of typical CTTSs can be reproduced from about 0.2 to 10 μm if emission from the boundary layer is confined to an equatorial region with width comparable to the local disk scale height ($\approx 2\%$ of the stellar radius). The isothermal boundary layer temperature, determined from the condition that half the accretion luminosity be emitted from its surface, is then in the 7000 – 12000 K range. The disk temperature varies from about 3000 K close to the star to interstellar temperatures in its outer parts.

Positive aspects of this simple model are its self-consistency and its small number of free parameters (essentially the disk mass-accretion rate and view angle), while a major drawback is the assumption of an optically thick boundary layer. Observed Balmer jumps (Fig. 1) indeed indicate that the Paschen continuum is at least partially optically thin. Basri and Bertout (1989) therefore used a classical atmosphere code to compute monochromatic gas opacities in the boundary layer, which is again assumed to be isothermal. They chose boundary layer width rather than α to be the free parameter needed to control the optical depth, and computed emergent spectral energy distributions that they compared to observations of the Balmer and Paschen continua region. While the head of the Balmer continuum is optically thick in these models, the Paschen continuum is partially optically thin and the Balmer jump consequently appears in emission. Line emission from the Balmer lines with high quantum number appears consistent with optically thick line emission from the boundary layer, the width of which is again comparable to the local disk height-scale; but there is obviously a more extended and more optically thin region of emission which contributes to the flux in the lowest members of the Balmer series.

3.3 Properties of Current Disk Models

Finding accurate mass-accretion rates from the disk onto the star is an important task because: (i) if sufficiently high, accretion is expected to affect the evolution of the star in the H-R diagram (Hartmann and Kenyon 1989); (ii) one also expects that accretion of disk matter (which possesses high specific angular momentum) will affect the evolution of rotation in CTTSs; and (iii) there is growing evidence of a relationship between mass-accretion and mass-loss (Cohen, Emerson, and Beichmann 1989; Cabrit et al. 1989). The efficiency of mass gain/loss conversion is important both for understanding the mass-loss mechanism and for issues (i) and (ii) above. Basri and Bertout (1989) demonstrated that the best mass-accretion diagnostic to date is the Balmer continuum of CTTSs. By assuming a classical Lynden-Bell and Pringle disk model, mass-accretion rates ranging from a few times 10^{-9} to a few times 10^{-7} $M_\odot yr^{-1}$ can be determined from models of individual stars. But crucial questions remain about the validity of these values: are they highly dependent on the assumed disk model, and are they more or less unique, or can we find several solutions leading to the same spectrum but with highly different mass-accretion rates? These questions are discussed in detail by Bertout and Bouvier (1989) and Bouvier, Bertout, and Basri (1989), who used newly obtained simultaneous ultraviolet, optical and near-infrared spectroscopic and photometric data (Basri et al. in preparation) to derive disk and stellar parameters from spectral energy distributions.

Bertout and Bouvier studied the uniqueness problem by constructing maps of the quantity $1/\chi^2$, which measures the goodness of the fit between observed and computed spectral energy distributions, for all parameter couples. There are 6 computational parameters: the stellar radius R_*, the visual extinction in front of the system A_V, the system's view angle i, the accretion rate $\dot{M}_{acc}$, the viscosity parameter α, and the width δ_{BL} of the emission region associated with the boundary layer. Figure 6 displays some examples of $1/\chi^2$ maps for DF Tau. In these figures, the highest contour indicates the best solution, and acceptable solutions are usually confined within the five highest contours. The four panels of Figure 6 show the topology of $1/\chi^2$ for 4 parameter couples. Each couple involves the mass-accretion rate (in $M_\odot yr^{-1}$) and one of the following parameters: R_* (in units of $R_\odot$), A_V (in mag.), $\log\alpha$, and $\log\delta_{BL}$ (in units of R_*). Although none of these four parameters is particularly well-constrained, the best solutions (with high $1/\chi^2$) span a small range of mass-accretion rates: $\dot{M}_{acc} \approx 1 - 2 \cdot 10^{-7} M_\odot yr^{-1}$. It thus appears that the mass-accretion rate can be estimated within a factor of two from these models. This result stems from the fact that the mass-accretion rate primarily reflects the relatively well-determined quantity of integrated excess flux in the near infrared.

The two parameters α and δ_{BL} contain most of the assumed physics for the disk and boundary layer. Given the large range of parameter values that our solution spans, we found it truly remarkable that best fits to the overall spectrum (also including the Balmer jump) were obtained for values $\alpha \approx 1$ and $\delta_{BL}/R_* \approx 0.02$. This appears physically reasonable and gives us some confidence both in the validity of the underlying disk physics and in the α parametrization. We also found similar values of α for most CTTSs that we investigated.

Figure 7 displays the best fits found for the star DF Tau at two epochs of observations. For computing these models, we took advantage of the knowledge of both the stellar rotation period and projected rotational velocity, so that i is not a computational parameter anymore. These models were computed using a χ^2 minimization procedure and involve little human prejudice. It is thus quite reassuring that the stellar parameters R_* and A_V are basically the same for the two solutions and that observed spectral variations are accounted for by changes in $\dot{M}_{acc}$ and δ. The model parameters are as follows.
Upper panel: $R_* = 4.1 R_\odot$; $i = 62°$; $A_V = 1.3$mag; $\alpha=1$; $\delta_{BL} = 0.03 R_*$; $\dot{M}_{acc} = 2.9 \cdot 10^{-7} M_\odot yr^{-1}$.
Lower panel: $R_* = 3.9 R_\odot$; $i = 69°$; $A_V = 1.4$mag; $\alpha=1$; $\delta_{BL} = 0.09 R_*$; $\dot{M}_{acc} = 4.9 \cdot 10^{-7} M_\odot/yr^{-1}$.

Similar computations were made for 10 CTTSs (20 simultaneous datasets) spanning the spectral-type range K1–M1. They yield an average mass-accretion rate of

$$<\dot{M}_{acc}> = (1.4 \pm 1.2) \cdot 10^{-7} M_\odot yr^{-1}.$$

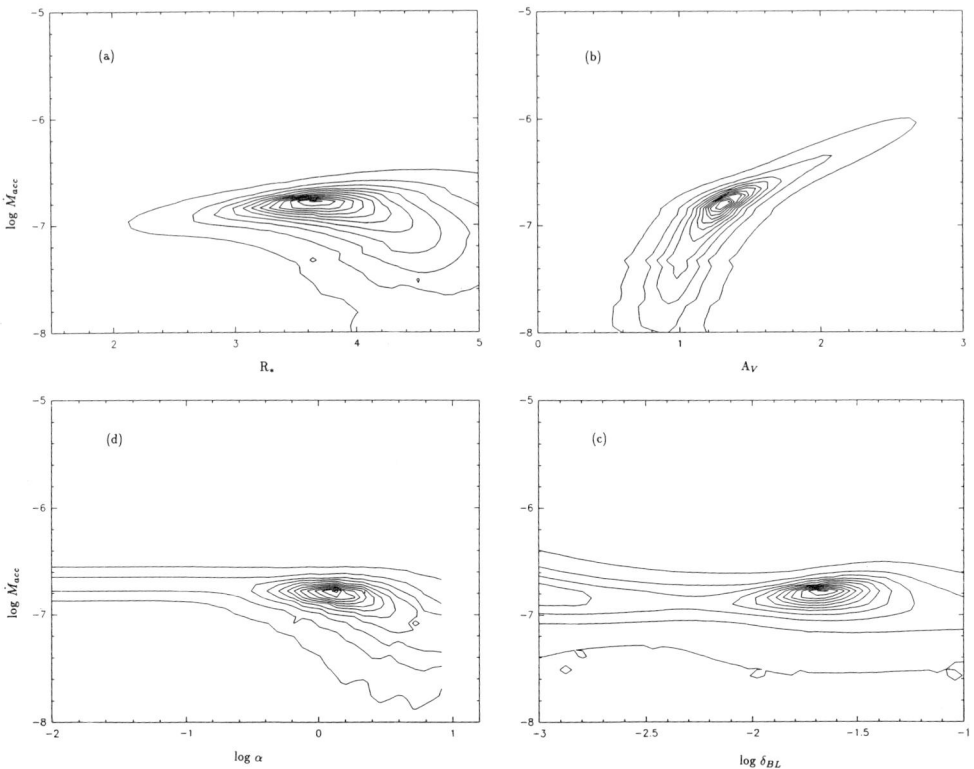

Figure 6 Contours of constant $1/\chi^2$ for models of DF Tau, as explained in the text. Each panel represents the topology of a 2D slice in the 6D parameter space. The value of $1/\chi^2$ decreases from the best fit value (highest contour) in steps of 8%. Acceptable models are usually found within the 5 highest contours.

If steady-state accretion proceeds during the disk lifetime of a few million years (Strom et al. 1989), disk accretion may therefore contribute a significant fraction of the stellar mass in CTTSs, and drastic consequences may be expected in the star's evolution. It is however difficult at this point to make a clear-cut statement about the actual amount of mass that is ultimately accreted by the star since (i) it is becoming increasingly clear that either the disk or the boundary layer is also driving the strong wind that characterizes the CTTS phase, and (ii) disk lifetime estimates depend crucially on the validity of the convective-radiative evolutionary tracks, which may not be all that relevant to accreting CTTSs.

Approximate estimates of the mass and maximum radius of CTTS disks can be made by modelling the spectral energy distribution in the infrared and millimeter range, using the sharp turnover of the spectrum that occurs in that spectral range

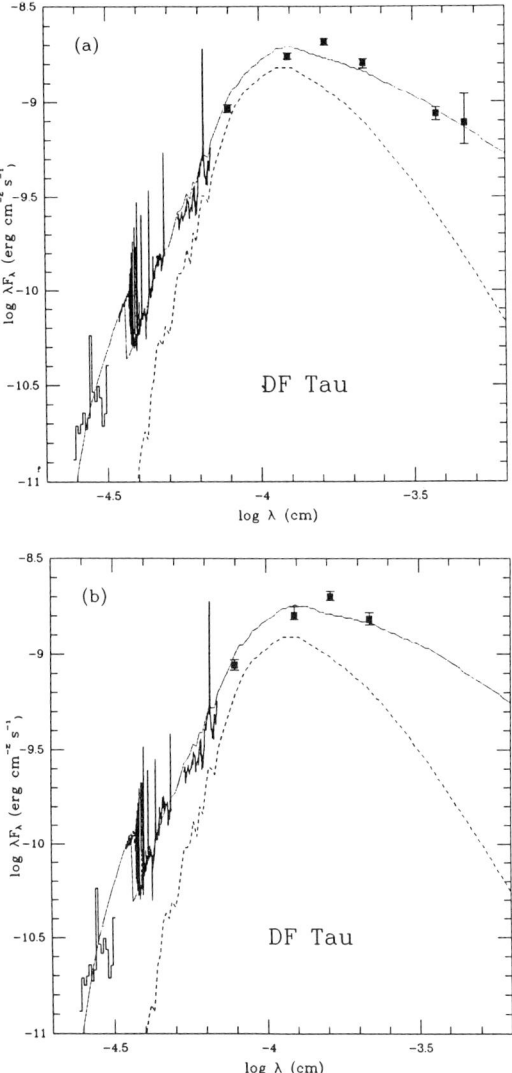

Figure 7 The observed and computed spectral energy distributions fo DF Tau on 2 dates. The dotted line shows the predicted total flux due to the star + disk + boundary layer system. The dashed line shows the photospheric and chromospheric stellar contribution. Low-resolution IUE and optical flux-calibrated spectra are shown as solid lines, while filled squares show IJHKLM broad-band photometric measurements with the error bars indicating observed flux variability during the observing period.

because the outer disk becomes optically thin; Adams, Emerson, and Fuller (1989) recently presented such computations. A difficulty that arises when making models of the far-infrared part of many CTTSs' spectral energy distributions is that their spectral slope in that range is often shallower than expected for a standard viscous accretion disk (Rydgren and Zak 1987). More precisely, the observed slope of λF_λ is typically $\propto \lambda^{-3/4}$ while the expected slope for a viscous disk is $\propto \lambda^{-4/3}$. Furthermore, about 10% of the most extreme CTTSs have flat far-infrared spectral energy distributions. These peculiarities of the far-infrared spectrum gave rise to two opinions. Some (Bertout, Basri and Bouvier 1988; Kenyon and Hartmann 1988) defend the standpoint that a T Tauri disk may be a viscous disk whose emitted spectrum is modified by radiative transfer effects in a dusty environment while others (Kenyon and Hartmann 1987; Adams, Lada, and Shu 1988; Adams, Emerson and Fuller 1989) conclude that the basic physics of T Tauri disks must be different from classical viscous disks.

Adams, Emerson, and Fuller thus feel justified in using the temperature distribution in the disk as a free parameter that they adjust to get an overall fit to a given spectral energy distribution. While this ad-hoc assumption introduces some uncertainty in the disk masses derived by Adams *et al.*, the main source of uncertainty stems from lack of knowledge of dust opacities in the millimeter range: depending on the assumed opacity law, mass estimates can vary by about one order of magnitude! Taking this problem into account, one can conclude from the millimetric data that masses of observed CTTS disks may range from less than 10^{-2} to perhaps as much as $1 M_\odot$. The same data also indicate that the disk maximum radius is typically 100 AU, in agreement with estimates based on the emission measure of forbidden lines (Edwards *et al.* 1987). Uncertainties concerning the mass of the disk make it difficult to say at this point whether derived age, mass-accretion rate, and mass are all compatible. They certainly are if the disk masses actually correspond to the upper limit given above, that is, if dust opacities are comparable to those assumed by Adams, Emerson, and Fuller. If, on the other hand, opacities prove higher, as would happen if, for instance, grain growth were important in T Tauri disks, one may have difficulty reconciling all quantities.

4. WINDS FROM DISKS?
Connections between the winds and disks of T Tauri stars being actively researched today, a major revision of past ideas about protostellar and T Tauri winds should result from current work on this topic.

4.1. Observational Constraints
Mass-loss rates can in principle be computed from the profiles of both permitted and forbidden optical lines. Since forbidden lines have low absorption probabilities, they are optically thin, which considerably simplifies the line formation problem. Thus, mass-loss rates computed from these lines should in principle be more reliable than those computed from H_α but they span the range from 10^{-9} to a few $10^{-7} M_\odot \text{yr}^{-1}$ (Edwards *et al.* 1987). These values are close to those found from H_α

(e.g., Kuhi 1964). As first noted by DeCampli (1981), it is extremely difficult to think of a theoretical stellar wind-driving mechanism that can produce mass-loss of $10^{-7} M_\odot \text{yr}^{-1}$ (see below). DeCampli proposed that H_α-broadening could be due in part to turbulent rather than organized motions. One may also try to revise down the high mass-loss rates indicated by the forbidden lines by assuming that the different components of the forbidden lines come from several atmospheric regions (cf. Hartmann and Raymond 1988; Kwan and Tademaru 1988; Cabrit et al. 1989).

At the low end of the mass-loss scale, WTTSs show little evidence of strong winds. With few exceptions, no forbidden line emission is observed in these objects (Walter et al. 1988; Strom et al. 1989). And yet, like moderately active CTTSs, WTTSs also display X-ray luminosities equal to about 10^{-3} times their bolometric luminosities; these radiative losses are comparable to those of a solar-type wind with mass-loss rate $\lesssim 10^{-9} M_\odot \text{ yr}^{-1}$. That massive outflows appear confined to CTTSs raises the question of the role of disks in driving their outflows.

Cohen, Emerson, and Beichmann (1989) and Cabrit et al. (1989) offer the first evidence that the disks and winds of CTTSs must indeed be somehow causally connected. Both groups find correlations between [OI]λ6300 emission line flux and infrared excess in a large sample of stars. Since the [OI] line flux is not correlated with photospheric luminosity even for stars along the same convective track, both groups conclude that the infrared excess is not age-dependent. This result, combined with the presence of WTTSs on convective tracks, suggests that the mass and extent of the circumstellar disks formed with the young stellar objects during the protostellar phase is largely determined by the initial conditions of star formation, that is, by physical conditions in the molecular core from which the star was born. This, in turn, means that the most prominent T Tauri characteristics, which are caused largely by the interaction between the star and its circumstellar disk, do not depend primarily on the age of the central star.

If the [OI] flux reflects the wind mass-loss rate, the correlation between [OI] flux and infrared excess indicates that T Tauri winds might be powered by mass-accretion. Estimated ratios of mass-loss to mass-accretion rates – computed by assuming that the entire [OI] flux originates from the wind – range from a few percent to almost one (with considerable uncertainty). These numbers imply that the conversion of mass accretion to mass outflow must be quite efficient, and Edwards and Strom (1988) find similar conversion ratios for embedded protostellar objects driving molecular outflows. By using the range of mass-loss rate values derived from the forbidden lines or from the H_α line $10^{-9} \lesssim \dot{M}_{wind} \lesssim 10^{-7} M_\odot \text{yr}^{-1}$, one can derive the force (momentum rate) and the luminosity of the stellar wind emanating from optical

T Tauri stars, two quantities which can then be compared to theoretical predictions. One finds

$$F_{wind} = \dot{M}_{wind}V_{wind} \approx 10 - 10^3 L_{bol}/c \tag{1}$$

and

$$L_{wind} = \frac{1}{2}\dot{M}_{wind}V_{wind}^2 \approx 0.003 - 0.3 L_{bol} \tag{2}$$

4.2 Mass-loss mechanisms

In pre-main sequence stars, there are three energy sources that can be used to drive a stellar wind: (i) photons – themselves resulting from gravitational contraction, (ii) rotation, and (iii) magnetic field. These energy sources are probably also present in the circumstellar disk associated with the CTTS, as follows. (i) If energy dissipation occurs within the disk because of viscosity or some other torques, then gravitational energy will be transformed into photons and the produced luminosity will be comparable to the accretion luminosity. (ii) In optical CTTSs, which have low rotational velocities (as measured from the stellar spectrum), the rotational energy is presumably much larger in the Keplerian disk than in the central star. (iii) Whether typical protostellar disks are magnetized is unknown at present, but is not unlikely, if, for example, the protostellar dynamical collapse is magnetically controlled. Mechanisms taking advantage of these energy sources have been proposed to drive outflows from both the stellar object and its circumstellar disk. Bertout, Basri, and Cabrit (1989) present a detailed assessment of current models, most of which prove unsuitable for driving T Tauri winds either because one of their basic requirements clearly contradicts observational facts or because their efficiency is too low. Each one of the three remaining models discussed below seems compatible with at least some aspects of the observations.

Coronal expansion

The solar wind is driven by thermal expansion, i.e., the thermal velocity in the corona is comparable to the escape velocity. In that case, however, radiative losses are proportional to $\dot{M}_{wind}^2$, and observed X-ray flux of T Tauri stars indicates that thermal coronal expansion cannot be responsible for mass-loss rates larger than about $10^{-9} M_\odot yr^{-1}$ in these objects (DeCampli 1981). As noted earlier, this is compatible with observations of WTTSs, so there is no reason at this point to assume another wind-driving mechanism for these solar-type stars.

Centrifugal ejection

Pudritz and Norman (1983, 1986) worked out an analytical description of centrifugal ejection from a magnetized disk which assumes that the field lines are globally parallel to the system's rotation axis but are pinched toward the center in the disk plane because of accretion in the disk. If the angle between field lines and rotation axis is larger than about 30°, the radiatively heated disk envelope is accelerated along magnetic field lines by the centrifugal force up to the Alfvén radius r_A (Blandford and Payne 1982). The ejected gas thus reaches the asymptotic speed $V_{wind} = \Omega r_A = (r_A/r_D)V_D$, where V_D is the rotational velocity of disk matter at its starting radius r_D and where $\Omega = V_D/r_D$. Pudritz and Norman assume $r_A/r_D = 10$, so that the specific angular momentum of the ejected matter is 100 times larger than its original specific angular momentum, which results in mass-accretion within the disk with rate $\dot{M}_{acc}$. In steady-state, mass-accretion and mass-loss rates are related by

$$\frac{\dot{M}_{wind}}{\dot{M}_{acc}} = \left(\frac{r_A}{r_D}\right)^2 \approx 0.01. \tag{4}$$

An interesting feature of Pudritz and Norman's model is its ability to produce the different observed wind components that are observed in protostellar infrared sources, as well as in CTTSs. The cool outer parts of the disks eject a massive molecular wind component that reaches end velocities ranging from 10 to 50 km s^{-1} at $r_A \approx 0.1$ pc, while the innermost parts of the disk, which are photoionized by UV radiation from the accretion zone, eject an ionized wind with mass-loss rate equal to only 2% of the molecular mass-loss rate but with velocity ranging from 200 to 600 km s^{-1}. In intermediate regions, atomic disk matter is ejected and reaches intermediate velocities. After denoting by $V_{esc} = \left(\frac{GM_*}{R_*}\right)^{1/2}$ the escape velocity from the central object and assuming again that $r_A = 10r_D$, the following relations are found:

$$\frac{L_{wind}}{L_{acc}} = \frac{1}{2}V_{wind}^2 \left(\frac{r_D}{r_A}\right)^2 \left(\frac{R_*}{GM_*}\right) = \frac{1}{2}\left(\frac{V_D}{V_{esc}}\right)^2 \approx 0.001 \left(\frac{V_{wind}}{V_{esc}}\right)^2 \tag{5}$$

$$\frac{cF_{wind}}{L_{acc}} = \left(\frac{r_D}{r_A}\right)\left(\frac{V_D}{V_{esc}}\right)\left(\frac{c}{V_{esc}}\right) \approx 10 \cdot \left(\frac{300\text{km}\cdot\text{s}^{-1}}{V_{wind}}\right)\left(\frac{V_{wind}}{V_{esc}}\right)^2 \tag{6}$$

This model is therefore able to account for the average correlations observed in T Tauri stars between L_{bol} and the wind parameters, as long as (i) $L_{bol} \approx L_{acc}$ and (ii) $V_{wind} \approx 1 - 5 \cdot V_{esc}$, i.e., $V_D \approx 0.1 - 0.5V_{esc}$. For T Tauri stars, which have $V_{wind} \sim V_{esc}$ and Keplerian disks, the wind must therefore originate within the innermost parts of the disk. One problem raised concerning this model is whether the poloidal field lines remain attached to their footpoints in the disk long enough for the centrifugal acceleration to take place. Pringle (1989) argues that the disk field is a dynamo field with too short a characteristic time-scale for this requirement to be met.

Magnetodynamical ejection

Pringle (1989) outlines a possible magnetodynamical ejection mechanism that rests on the transformation of shear energy into magnetic energy in the boundary layer between disk and star. He argues that the magnetic field of disk material flowing into the boundary layer is amplified by the strong shear forces there and finds that a large fraction of the boundary layer luminosity emerges from the boundary layer in the form of a strongly toroidal field. The magnetic flux could then be driven away in a magnetic wind. If the wind speed is comparable to the maximum disk rotational velocity $v_\phi = (GM_*/R_*)^{1/2}$, then mass-loss and mass-accretion rates are found to be related by

$$\frac{\dot{M}_{wind}}{\dot{M}_{acc}} \lesssim \frac{H}{R} \qquad (7)$$

where H is the disk scale height and R is the stellar radius. Equality in Eq. 7 would occur in case of complete efficiency of the mechanism converting magnetic energy into wind kinetic energy. Using the maximum temperature reached in the disk to compute H, Pringle finds from Eq. 7 that

$$\frac{L_{wind}}{L_{acc}} \lesssim 0.03 \left(\frac{M_*}{0.5\,M_\odot}\right)^{-3/8} \left(\frac{R_*}{3\,R_\odot}\right)^{1/4} \left(\frac{\dot{M}}{10^{-5}M_\odot\mathrm{yr}^{-1}}\right)^{1/8}; \qquad (8)$$

and one can then calculate

$$\frac{cF_{wind}}{L_{acc}} \lesssim 2 \cdot \frac{c}{v_\phi} \cdot \frac{H}{R} = 100 \left(\frac{M_*}{0.5\,M_\odot}\right)^{-7/8} \left(\frac{R_*}{3\,R_\odot}\right)^{3/4} \left(\frac{\dot{M}}{10^{-5}M_\odot\mathrm{yr}^{-1}}\right)^{1/8}. \qquad (9)$$

These upper limits demonstrate that the expected efficiency of this possible layer wind-driving mechanism, while perhaps not quite large enough to drive the most energetic of the observed winds, makes a detailed investigation of the hydromagnetic processes taking place in the boundary layer worthwhile. Only then will one be able to determine more precisely both the feasibility of the mechanism and its efficiency.

REFERENCES
Adams, F.C., Emerson, J.P., Fuller, G.A. 1989, preprint
Adams, F.C., Lada, C.J., Shu, F.H. 1987, *Ap. J.* **312**, 788
Adams, F.C., Lada, C.J., Shu, F.H. 1988, *Ap. J.* **326**, 865
Adams, F.C., Shu, F.H. 1986, *Ap. J.* **308**, 836
Appenzeller, I, Jankovics, I., Oestreicher, R. 1985, *Astron. Astrophys* **141**, 108
Basri, G., Bertout, C. 1989, *Ap. J.* **341**, 340
Bastien, P., Ménard, F. 1987, *Ap. J.* **326**, 334
Bertout, C. 1989, *Ann. Rev. Astron. Astrophys.* **27**, 351
Bertout, C., Basri, G., Bouvier, J. 1988, *Ap. J.* **330**, 350
Bertout, C., Basri, G., Cabrit, S. 1989, in *The Sun in Time*, (Tucson: Univ. Ariz. Press), in press
Bertout, C., Bouvier, J. 1989, preprint
Blandford, R.D., Payne, D.G. 1982, *Mon. Not. Roy. Astron. Soc.* **199**, 883

Bouvier, J. 1987. in *Protostars and Molecular Clouds* Eds. T. Montmerle, C. Bertout (Gif-sur-Yvette: CEN) p. 189
Bouvier, J. 1989, submitted to *A. J.*
Bouvier, J., Bertout, C. 1989, *Astron. Astrophys* **211**, 99
Bouvier, J., Bertout, C., Basri, G. 1989, preprint
Bouvier, J., Bertout, C., Benz, W., Mayor, M. 1985. dans *Nearby Molecular Clouds* Ed. G. Serra (Berlin:Springer-Verlag), p.222
Bouvier, J., Bertout, C., Bouchet, P. 1988, *Astron. Astrophys. Suppl.* **75**, 1
Cabrit, S., Edwards, S., Strom, S.E., Strom, K.M. 1989, *A. J.* , submitted
Calvet, N., Albarrán, J. 1984, *Rev. Mexicana Astron. Astrof.* **9**, 35
Calvet, N., Basri, G., Kuhi, L.V. 1984, *Ap. J.* **277**, 725
Cohen, M., Emerson, J.P., Beichmann, C.A. 1989, *Ap. J.* **339**, 445
DeCampli, W.M. 1981, *Ap. J.* **244**, 124
Edwards, S., Cabrit, S., Strom, S.E., Heyer, I., Strom, K.M., Anderson, E. 1987, *Ap. J.* **321**, 473
Edwards, S., Strom, S.E. 1988, dans 5th *Cambridge Cool Star Workshop* Eds. J.L. Linsky, R. Stencel (Berlin: Springer-Verlag), p. 443.
Feigelson, E.D., Giampapa, M.S., Vrba, F.J. 1989, , in *The Sun in Time*, (Tucson: Univ. Ariz. Press), in press
Hartmann, L. 1982, *Ap. J. Suppl.* **48**, 109
Hartmann, L., Kenyon, S.J. 1989, preprint
Hartmann, L., Noyes, R.W. 1987, *Ann. Rev. Astron. Astrophys.* **25**, 271
Hartmann, L., Raymond, J.C. 1989, *Ap. J.* , sous presse.
Herbig, G.H. 1962, *Advances Astron. Astrophys.*,**1**, 47
Herbig, G.H. 1970, *Mem. Soc. Roy. Sci. Liège* **19**, 13
Herbig, G.H. 1977a, *Ap. J.* **214**, 747
Herbig, G.H., Bell, K.R. 1988, *Lick Observatory Bulletin No. 1111*
Herbig, G.H., Goodrich, R.W. 1986, *Ap. J.* **309**, 294
Herbig, G.H., Vrba, F.J., Rydgren, A.E. 1986, *A. J.* **91**, 575
Joy, A.H. 1945, *Ap. J.* **102**, 168
Kenyon, S.J., Hartmann, L. 1987, *Ap. J.* **323**, 714
Kenyon, S.J., Hartmann, L. 1988, dans *Formation and Evolution of Low-Mass Stars*, Ed. A.K. Dupree (Dordrecht: Reidel) sous presse
Krautter, J., Bastian, U. 1980, *Astron. Astrophys* **88**, L6
Kuhi, L.V. 1964, *Ap. J.* **140**, 1409
Kuhi, L.V. 1974, *Astron. Astrophys. Suppl.* **15**, 47
Kuhi, L.V. 1978, dans *Protostars and Planets* Ed. T. Gehrels (Tucson: University of Arizona Press) p. 708
Kwan, J., Tademaru, E. 1988, *Ap. J.* **332**, L41
Lynden-Bell, D., Pringle, J.E. 1974, *Mon. Not. Roy. Astron. Soc.* **168**, 603
Mendoza, V.E.E. 1966, *Ap. J.* **143**, 1010
Mendoza, V.E.E. 1968, *Ap. J.* **151**, 977

Mundt, R. 1984, *Ap. J.* **280**, 749

Mundt, R., Giampapa, M.S. 1982, *Ap. J.* **256**, 156

Myers, P.C., Fuller, G.A., Mathieu, R.D., Beichman, C.A., Benson, P.J., Schild, R.E., Emerson, J.P. 1987, *Ap. J.* **319**, 340.

Pringle, J.E. 1989, *Mon. Not. Roy. Astron. Soc.* **236**, 107

Pudritz, R.E., Norman, C.A. 1983 *Ap. J.* **274**, 677.

Pudritz, R.E., Norman, C.A. 1986 *Ap. J.* **301**, 571.

Rydgren, A.E., Schmelz, J.T., Zak, D.S., Vrba, F.J. 1984. *Publ. US Nav. Obs.* Vol 25, Part 1

Rydgren, A.E., Zak, D.S. 1987, *Publ. Astron. Soc. Pacific* **99**, 141

Schneeberger, T.J., Worden, S.P., Wilkerson, M.S. 1979. *Ap. J. Suppl.* **41**, 369

Shakura, N.I., Sunyaev, R.A. 1973, *Astron. Astrophys* **24**, 337

Strom, K.M., Strom, S.E., Edwards, S., Cabrit, S., Strutskie, M.F. 1989, *A. J.*, in press

Ulrich, R.K., Knapp, G.R. 1979, *Ap. J.* **230**, L99

Walker, M.F. 1972, *Ap. J.* **175**, 89

Walter, F.M., Bowyer, C.S. 1981, *Ap. J.* **245**, 677

Walter F.M., Brown, A., Mathieu, R.D., Myers, P.C., Vrba, F.J. 1988, *A. J.* **96**, 297

Turbulent Shear Flow and Rotation

J.-P. ZAHN

Observatoire Midi-Pyrénées, Toulouse, France
and
Columbia University, New York, U.S.A.

ABSTRACT. Differential rotation shares most properties of plane parallel shear flow. In the absence of stabilizing effects due to vertical stratification, magnetic field, etc., it becomes unstable for sufficiently large Reynolds number. Depending on the profile of the flow, the instability is either linear, or of finite amplitude. In differentially rotating stars, those shear instabilities reach a turbulent regime. The properties of that turbulence are sketched out, and it is shown how to estimate the transport of chemicals and angular momentum.

1 INTRODUCTION

In recent years, much attention has been paid to the transport processes that are likey to occur in the radiative interior of stars. One reason is that some chemical abundances at the surface of the Sun and solar-like stars can only be interpreted by invoking a mild transport of matter below their convective envelope, in their stable radiative region (Schatzman 1969, 1977; Schatzman and Maeder 1981). Moreover, the internal rotation of the Sun, which we discover now through helioseismology, is also the signature of the transport of angular momentum within the radiation zone.

Such transport can be achieved through large-scale circulations, for which two causes have been identified. One has been thoroughly studied: a rotating star can no longer achieve radiative equilibrium, and this thermal imbalance induces a meridional advection of heat which is called the Eddington-Sweet circulation (Eddington 1925, Vogt 1925, Sweet 1950, Mestel 1953, Kippenhahn 1958, McDonald 1972, Tassoul and Tassoul 1982, etc.).

Much less attention has been paid so far by the astrophysicists to the so-called Ekman circulation, which is generated in the boundary layer connecting two regions that rotate at different speeds (Ekman 1905; see also Pedlosky 1979). Such a circulation is likely to occur at the bottom of the solar convection zone, as suggested already by Bretherton and Spiegel (1968): there, a strong differential rotation is maintained by the convective motions, whereas the radiation zone below appears to rotate much more uniformly (Brown et al. 1989).

Another means of transport probably operates in stellar radiation zones, namely *turbulent diffusion*. This process is very familiar to the astrophysicist, who knows well that it is responsible for the transport of heat in a convection zone: there, it is the powerful convective instability which causes the turbulent motions. In a radiation zone, on the other hand, it appears that the major cause of such turbulence is the differential rotation, which is liable to various instabilities (see Knobloch and Spruit 1982, 1983, Zahn 1983). Among those instabilities, and given the rotation laws that are likely to occur in stars, the most powerful are the shear instabilities due to differential rotation. It is the properties of those instabilities that we shall now examine.

2 SHEAR INSTABILITIES IN PARALLEL FLOWS

The simplest shear flow is a plane-parallel flow whose velocity U is constant, downstream, in the x-direction, but varies, across the stream, in the z-direction. The intensity of the shear is measured by the gradient dU/dz, which is also the magnitude of the vorticity. Locally, any flow may be approximated by such a "tangent", plane-parallel shear.

In the absence of other restoring forces, all shear flows are unstable, provided the Reynolds number which characterizes them is large enough. But the detailed properties of the instability depend on the profile $U(z)$. The instability is of the linear type when U has an inflexion point somewhere in the domain, as it has been shown by Rayleigh (1880) (see also Fjørtoft 1950); it is then a dynamical instability, with a growth-rate of order $|dU/dz|$. When the flow has no such inflexion point (or local extremum of vorticity), it generally requires a perturbation of finite amplitude to become unstable. The theoretical study of that case is much more intricate; it is found that the instability threshold depends not only on the strength of the perturbation, but also of its shape (Gill 1965, Lerner and Knobloch 1988).

But shear flows can be stabilized by a restoring force, and the best example of this is provided by the buoyancy. In a stratified medium, when a parcel of matter is perturbed from its equilibrium position, it oscillates with the buoyancy frequency N, which in a homogeneous star is given by

$$N^2 = \frac{g}{H_P}(\nabla_{ad} - \nabla_{rad}), \qquad (1)$$

with the usual notations for the gravity, the presssure scale height, and the logarithmic temperature gradients of stellar structure. If the perturbations associated with the motions were adiabatic, the shear instability would be prevented as soon as

$$\frac{N^2}{(dU/dz)^2} > 1/4, \qquad (2)$$

a condition known as the Richardson criterion. In a star, however, the buoyancy force is weakened through radiative damping, and the stability criterion above takes the form (Zahn 1974)

$$\frac{N^2}{(dU/dz)^2} \sigma R_c > \approx 1, \qquad (3)$$

where R_c is a critical Reynolds number, and σ the Prandtl number (ratio between the viscosity and the thermal diffusivity, a very small quantity in stellar interiors).

But if there is a stratification of molecular weight, due for instance to nuclear reactions, or to gravitational settling, then the buoyancy force is only partly reduced by radiative damping, and one recovers the original Richardson condition, the characteristic frequency now being

$$(N_\mu)^2 = \frac{g}{H_P} \frac{d\ln\mu}{d\ln P}. \qquad (4)$$

One may wonder whether the Coriolis force, in a rotating fluid, could not play a role similar to the buoyancy. This possibility has been examined by Johnson (1963), who found that the most unstable modes are not affected by the rotation. Another proof is provided by the very nice experiment performed by Rabaud and Couder (1983), which clearly demonstrates that the onset of the shear instability is insensitive to the Coriolis force.

Therefore, the results of plane-parallel shear flow can be transposed as such to *differential rotation*. For convenience, let us distinguish between vertical differential rotation, in which the angular velocity Ω varies with depth r, and horizontal differential rotation, where it varies with latitude. In the first case, the instability criterion is

$$(s\partial\Omega/\partial r)^2 > \approx N^2 \sigma R_c, \qquad (5)$$

(s being the distance to the rotation axis), which holds for a homogeneous star. When there is a gradient of molecular weight, one recovers the Richardson criterion

$$(s\partial\Omega/\partial r)^2 > \approx (N_\mu)^2. \qquad (6)$$

In the second case, that of horizontal differential rotation, the instability cannot be hindered by the stratification, since the buoyancy force acts only in the vertical direction. Therefore, due to the large Reynolds number characterizing this shear flow in stars, a differential rotation in latitude is always unstable (Zahn 1975). Since the velocity profile has then probably no inflexion point, it is liable to finite amplitude instabilities, as mentioned above.

Another restoring force may be provided by a magnetic field. Such a field will wind up in a differentially rotating star until its azimuthal component becomes strong enough to react back on the rotation (Mestel 1953, Mestel and Moss 1986, Mestel and Weiss 1987). That interplay between shear and magnetic field certainly occurs in the regions where the magnetic field is generated through a dynamo mechanism (such a site seems to be located just below the convection zone of a solar-type star, see Spiegel 1987). But if the field alternates its polarity in time, as in the Sun, it will not penetrate deep into the star, and it should not interfer much with the differential rotation (and with the shear instabilities mentioned above). The possibility of a fossil field pervading the star is not very likely either; it is hard to see how it would survive the pre-main-sequence phase and the various instabilities which have been identified so far (see Tayler 1982).

When a horizontal shear flow becomes unstable, it generates eddies which have the same vorticity as the mean flow, and which are therefore horizontal, and two-dimensional. Those billows, in turn, undergo a three-dimensional instability, unless the vertical motions are hindered by some restoring force. If that is the case, the turbulent motions are anisotropic, and one has to take this into account when estimating the transport efficiency of such turbulence, as will be done next.

3 TURBULENCE IN DIFFERENTIALLY ROTATING STARS

Before we examine in more detail the properties of the turbulence arising in differentially rotating stars, let us recall how one estimates the transport efficiency of such turbulence. To first approximation, the transport in a turbulent medium of a scalar quantity (such as temperature, the concentration of chemical species, etc.) can be described as a diffusion process; vector fields, in addition, experience stretching. For simplicity, we shall assume first that the velocity field is isotropic, in which case the turbulent diffusivity reduces to a scalar, D_t. This coefficient is given by

$$D_t = \frac{1}{3} u\ell, \tag{7}$$

where u is the r.m.s. velocity and ℓ the correlation length of the turbulent velocity field (see Knobloch 1978). In the phenomenogical approaches, one identifies these as the velocity and the size (or the mixing length) of the turbulent eddies.

For the mixing length ℓ, one can take the dimension of the turbulent region, since the largest, most efficient eddies are of often of that size. A better recipe, due to Prandtl, is to choose the distance to the nearest boundary. But in some instances the kinetic energy is injected at a scale which is smaller than that of the whole unstable region; the most vigorous eddies, which contribute most to the turbulent transport, are then of intermediate size. To take this in account when modelling stellar convection zones, one generally follows E. Vitense (1953), and one relates the mixing length to the local pressure scale height, $\ell = \alpha H_P$, the coefficient α being calibrated with the observations.

To estimate the velocity u, various prescriptions are available. There also, the largest turbulent velocity is often comparable with the variation ΔU of the mean flow speed over the unstable domain. But in most cases that amplitude ΔU is itself controlled by the strength of the turbulence, which in turn is determined by another condition, such as the flux of momentum or of heat that has to be transported. When it is possible to estimate the rate ε_t at which kinetic energy is injected into the turbulent motions at the scale ℓ, it follows from dimensional considerations based on physical arguments that

$$\varepsilon_t \approx u^3/\ell \tag{8}$$

(see Landau and Lifschitz 1987). An alternate approach is to estimate u through the growth rate $1/\tau$ of the considered instability :

$$u \approx \ell/\tau. \tag{9}$$

That growth rate is often derived from the linear perturbation theory. But it is more correct to calculate it in the non-linear regime, i.e. in the stationary state which is established by the instability. In the classical mixing-length theory for convection, for instance, the velocity is estimated by following the acceleration of an eddy in the *actual* superadiabatic density gradient; the expression for the convective flux is found to be

$$F_c \approx \rho u^3 \frac{H_P}{\ell}. \tag{10}$$

Notice that this expression again involves the ratio u^3/ℓ, as in eq. 8 above: the energy injection rate is thus closely related to the prescribed convective flux.

Let us now turn specifically to the turbulence which is generated in the shear flow of differential rotation. That turbulence is strongly anisotropic: the motions are more vigorous in the horizontal than in the vertical direction. The diffusion coefficient is thus no longer a scalar, but a tensor (which is diagonal if one coordinate axis is chosen vertical).

To determine the turbulent diffusivity in the vertical direction, we proceed as follows. We have seen that the horizontal shear is always unstable; it gives rise to horizontal vortices, which themselves are unstable and split into three-dimensional eddies. But this transition from 2 to 3 dimensions is hindered in the presence of a restoring force.

Among such forces, the *Coriolis force* plays a major role: it governs the dynamics of all eddies for which

$$u/\ell < \Omega, \tag{11}$$

a property which is well known in geophysical fluids, and which it is illustrated by various laboratory experiments (see Hopfinger *et al.* 1982). Those large-scale motions will remain horizontal and two-dimensional, and they may also take the form of inertial waves; in any case, they will not contribute to the turbulent transport in the vertical direction. But the smallest eddies, i.e. those which do not obey the inequality above, are not sensitive to the Coriolis force; they are three-dimensional, and they follow to first approximation the Kolmogorov law (see Landau and Lifshitz 1987). Their distribution begins at the scale which verifies both

$$u'/\ell' \approx \Omega \quad \text{and} \quad (u')^3/\ell' \approx \varepsilon_t,$$

and therefore the vertical turbulent diffusivity is given by

$$D_v \approx u'\ell' \approx \varepsilon_t/\Omega^2. \tag{12}$$

The other restoring force present in a star is the *buoyancy*, but it only operates when there is a vertical gradient of molecular weight, since the temperature differences are completely smoothed out by radiative damping. Such a μ-gradient will inhibit three-dimensional turbulence for the eddies whose turn-over rate is less than the residual buoyancy frequency (eq. 4)

$$u/\ell < N_\mu, \tag{13}$$

and it will suppress it entirely when (see Zahn 1983)

$$(N_\mu)^2 > \varepsilon_t/\nu, \tag{14}$$

where ν is the kinematic viscosity (which fixes the size of the smallest scales, where the kinetic energy is dissipated through viscous friction). Even a rather small gradient of molecular weight suffices to prevent turbulent diffusion in the vertical direction (for instance, that due to the varying composition of ^{3}He in the Sun). However, such a "μ-barrier" can still allow the transport of momentum, which will then be accomplished through isothermal gravity waves (Zahn 1989).

To proceed further, one needs an estimate of ε_t, the generation rate of turbulent kinetic energy. This quantity is determined by the very cause of the differential rotation: the contraction or expansion of the star while it evolves, angular momentum loss through a wind, the coupling with a differentially rotating convection zone, meridional circulation, tidal braking in a binary star. As an illustrative example, we shall estimate this rate in the case where the differential rotation is produced by a meridional circulation.

Such a circulation advects angular momentum, whose conservation requires

$$\frac{\partial}{\partial t}\left(s^2\Omega\right) + \mathbf{U} \cdot \nabla\left(s^2\Omega\right) = \Gamma, \qquad (15)$$

where $\mathbf{U}$ is the meridional velocity and Γ the torque exerted per unit mass by the turbulent motions (as above, s is the distance to the rotation axis). Likewise, we may express the rate of variation of the rotational kinetic energy:

$$\frac{\partial}{\partial t}\frac{1}{2}(s\Omega)^2 + \Omega\,\mathbf{U} \cdot \nabla\left(s^2\Omega\right) = \Omega\,\Gamma. \qquad (16)$$

In a stationary state, the advection term balances the right hand side, which is the work done by the turbulent torque, and therefore the injection rate of kinetic energy into the turbulence. We now split the angular velocity into its mean and fluctuating parts $\Omega(r) + \delta\Omega(r,\theta)$ (over a level surface, θ being the colatitude), and subtract the kinetic energy of the mean flow from that which is advected into the layer; we thus obtain the following expression for the turbulent energy input, averaged over a the whole horizontal layer (Zahn 1987):

$$\varepsilon_t(r) = -\int_0^1 \delta\Omega(r,\theta)\,\mathbf{U} \cdot \nabla\left(\Omega_0(r)\sin^2\theta\right)\,d(\cos\theta). \qquad (17)$$

Let us stress that this expression is valid for any type of meridional circulation, either the Eddington-Sweet or the Ekman circulation.

To proceed further, one must specificy which of those circulations one is considering, and one also needs an additional prescription to determine the strength of the horizontal differential rotation. This would take us beyond the scope of this paper, and we refer the reader to works where those questions have been discussed in more detail (Zahn 1987, 1989).

4 CONCLUSION
Turbulent diffusion appears to play a major role in mixing the chemical species within a star. It brings to the surface some elements which are produced in the

deep interior, such as ^{3}He in the Sun (Geiss et al. 1972), or ^{13}C in the giant stars of the first ascending branch (Lambert et al. 1980). At the same time, it carries fragile elements from the surface to depths where they are destroyed, and it is probably responsible for the depletion of Li observed in solar-like stars (Boesgard 1976, Duncan 1981, Cayrel et al. 1984). For the same reason, it may also affect the abundance of Li in the old halo stars, which is taken as a test for the validation of cosmological theories (Spite and Spite 1982). Moreover, such diffusion competes with other processes (transport by waves, magnetic torquing) in distributing angular momentum within a star.

The theory of that turbulence is still in infancy, and one may complain that it relies too much on phenomenological arguments. That is certainly true, but cannot the same be stated about the mixing-length approach to stellar convection, whose shortcomings are well known, but which has not yet been replaced by a better procedure? Progress will certainly be achieved by implementing the few recipes which are available in evolutionary models, and by comparing the theoretical predictions with the observations. This was done by Schatzman and Maeder (1981), Baglin et al. (1985), and more recently by Vauclair (1988), Charbonneau and Michaud (1989), and by Deliyannis et al. (1989). Some discrepancies which have been noticed may be ascribed to physical processes that have been overlooked so far, such as the stabilizing effect of a μ-gradient, or the momentum transport by waves. As usual, the complexity of the problem surpasses our imagination!

ACKNOWLEDGEMENTS. I wish to thank P. Charbonneau, E. Hopfinger, E. Knobloch, G. Michaud, E. Schatzman, E. Spiegel and S. Vauclair, with whom I discussed many of the questions raised above. Part of this work was supported by grant AFOSR 89-0012 of the U.S. Air Force.

REFERENCES

Baglin, A., Morel, P. and Schatzman, E.: 1985, *Astron. Astrophys.* **149**, 309
Boesgard, A.: 1976, *Publ. Astron. Soc. Pacific* **88**, 353
Bretherton, F.P. and Spiegel, E.A.: 1968, *Astrophys. J.* **153**, 277
Brown, T.M., Christensen-Dalsgaard, J., Dziembowski, W.A., Goode, P., Gough, D.O. and Morrow, C.A.: 1989, *Astrophys. J.* **343**, 526
Charbonneau, P. and Michaud, G.: 1989, *Astrophys. J.* (in press)
Cayrel, R., Cayrel de Strobel, G., Campbell, B. and Däppen, W.: 1984, *Astron. Astrophys.* **283**, 205
Deliyannis, C.P., Demarque, P. and Kawaler, S.D.: 1989, *Astrophys. J.* (in press)
Duncan, D.K.: 1981, *Astrophys. J.* **248**, 651
Ekman, V.W.: 1905, *Arkiv Matem. Astron. Fysik* **2**, 11
Eddington, A.S.: 1925, *Observatory* **48**, 78
Fjørtoft, R.: 1950, *Geofys. Publ.* **17**, 52
Geiss J., Buhler, F., Cerutti, H., Eberhardt, P. and Filleux, C.H.: 1972, Apollo 16 Prel. Sci. Rep. NASA (SP 315)
Gill, A.E.: 1965, *J. Fluid Mech.* **21**, 503
Hopfinger, E.J., Browand, F.K. and Gagne, Y.: 1982, *J. Fluid Mech.* **125**, 505
Johnson, J.A.: 1963, *J. Fluid Mech.* **17**, 337
Kippenhahn, R.: 1958, *Z. Astrophys.* **46**, 26
Knobloch, E.: 1978, *Astrophys. J.* **225**, 1050
Knobloch, E. and Spruit, H.C.: 1982, *Astron. Astrophys.* **113**, 261
Knobloch, E. and Spruit, H.C.: 1983, *Astron. Astrophys.* **125**, 59
Lambert, D.L., Dominy, J.F. and Sivertsen, S.: 1980, *Astron. Astrophys.* **235**, 114
Landau, L. and Lifschitz, E.: 1987, Fluid Mechanics (English translation, 2nd edition; Pergamon)
Lerner, J. and Knobloch, E.: 1988, *J. Fluid Mech.* **189**, 117
McDonald, B.E.: 1972, *Astrophys. Space Sci.* **19**, 309
Mestel, L.: 1953, *Montly. Not. Roy. Astron. Soc.* **113**, 716
Mestel, L. and Moss, D.L.: 1986, *Montly. Not. Roy. Astron. Soc.* **221**, 25
Mestel, L. and Weiss, N.O.: 1987, *Montly. Not. Roy. Astron. Soc.* **226**, 123
Pedlosky, J.: 1979, Geophysical Fluid Dynamics (Springer)
Rabaud, M. and Couder, Y.: 1983, *J. Fluid Mech.* **136**, 291
Rayleigh, Lord: 1880, *Scientific Papers*, **1**, 474 (Cambridge Univ. Press)
Schatzman, E.: 1969, *Astrophys. Lett.* **3**, 139
Schatzman, E.: 1977, *Astron. Astrophys.* **56**, 211
Schatzman, E. and A. Maeder: 1981, *Astron. Astrophys.* **96**, 1
Spiegel, E.A.: 1987, The Internal Solar Angular Velocity (eds. B.R. Durney and S. Sofia; Reidel), 321

Spite, F. and Spite, M.: 1982, *Astron. Astrophys.* **115**, 357
Sweet, P.A.: 1950, *Monthly. Not. Roy. Astron. Soc.* **110**, 548
Tassoul, J.-L. and Tassoul, M.: 1982, *Astrophys. J. Suppl.* **49**, 317
Tayler, R. J.: 1982, *Monthly. Not. Roy. Astron. Soc.* **198**, 811
Vauclair, S.: 1988, *Astron. Astrophys.* **335**, 971
Vitense, E.: 1953, *Z. Astrophys.* **32**, 135
Vogt, H.: 1925, *Astron. Nachr.* **223**, 229
Zahn, J.-P.: 1974, Stellar Instability and Evolution (eds. P. Ledoux, A. Noels and R.W. Rodgers; Reidel), 185
Zahn, J.-P.: 1975, *Mém. Soc. Roy. Sci. Liège, 6e série* **8**, 31
Zahn, J.-P.: 1983, Astrophys. Processes in Upper Main Sequence Stars (Publ. Observatoire Genève), 253
Zahn, J.-P.: 1987, The Internal Solar Angular Velocity (eds. B.R. Durney and S. Sofia; Reidel), 201
Zahn, J.-P.: 1989, Inside the Sun (eds. G. Berthomieu and M. Cribier; Kluwer) (in press)

Surface Abundances of Light Elements in Stars

R. Rebolo

Instituto de Astrofísica de Canarias
38200 La Laguna. Tenerife
Spain

SUMMARY

The surface contents of light elements provide sensitive probes to the internal structure of stars in different evolutionary stages. In particular, the precise determination of lithium and beryllium abundances in the atmospheres of stars has produced a challange to our knowledge of the hydrodynamical processes in stellar envelopes. This review discusses the more important observations and the transport mechanisms proposed to explain them. Emphasis is given on the role that rotational velocity, metallicity and mass have controlling the processes that cause lithium depletion during pre-Main and Main Sequence evolution. The galactic evolution of lithium, beryllium and boron and current problems in the determination of the primordial lithium abundance are briefly discussed.

1 INTRODUCTION

Observations of Deuterium (D), Helium (^{3,4}He), Lithium (^{6,7}Li), Beryllium (^{9}Be) and Boron (^{10,11}B) from different astrophysical sites provide key information to clarify fundamental problems in Cosmology, Stellar Structure and Evolution, Nucleosynthesis and Chemical evolution of the Galaxy. Since more than 20 years ago it is known (Peebles, 1966; Wagoner, Fowler and Hoyle, 1967) that the origin of several of these elements (D, ^{3,4}He, ^{7}Li) lies, totally or partially, in nucleosynthesis processes that took place at very early phases in the evolution of the Universe. Stars are unable to produce a significant contribution to the present-day observed abundances of D, ^{6}Li, ^{9}Be and ^{10,11}B. The case for ^{4}He, ^{7}Li and ^{3}He is different since the former is produced in quantity by H burning during normal evolution of nearly all types of stars and the latter two isotopes can also be produced at some stages of stellar evolution. The synthesis of light elements is not limited to the primordial fireball or to stellar sites; other mechanisms like the spallation reactions triggered by Galactic Cosmic Rays on the Interstellar Medium (Reeves, Fowler and Hoyle, 1970; Meneguzzi, Audouze and Reeves, 1971) have provided a reasonable explanation for the galactic abundances of ^{6}Li, ^{9}Be and ^{10,11}B. Only a small fraction, ~10%, of the observed ^{7}Li could have been produced by these spallation reactions, leaving the origin of its present-day galactic abundance as a very interesting astrophysical puzzle.

Understanding the evolution of the galactic abundance of light elements is further complicated by the fact that all of them, but ^{4}He, are fragile in stellar interiors, where they are destroyed via (p,α) reactions at relatively low temperatures, 0.5-7 10^6 K. These temperatures are easily reached in the interiors of nearly all types of stars during some stage of their evolution. Convective motions and other transport mechanisms can circulate matter from the external zones to deeper regions where the temperatures are high enough to make the (p,α) reactions efficient, thereby modifying the initial content of stars at the time they were formed. Because of their lower binding energy, the D and ^{6}Li nuclei are the more easily destroyed, and are also very difficult to observe in stellar atmospheres. ^{7}Li and ^{9}Be, although more fragile than the ^{10,11}B isotopes, can be measured easily in stars via their resonance doublets at 6708 Å and 3130 Å respectively. The sensitivity of their abundances to superficial mixing is a very powerful tool that can be utilized to improve our knowledge of stellar structure and hydrodynamics at different evolutionary stages.

This review will discuss the present status of the observational work on the abundances of Li, Be and B in stars, and the main implications of these measurements for stellar structure and nucleosynthesis processes. For reviews of deuterium and helium abundances in stars and in other astrophysical contexts the reader is referred to Boesgaard and Steigman (1985), and Pagel (1988). Reeves (1974), and Boesgaard (1976a) have reviewed the earliest measurements of Li, Be and B; I will emphasize here the recent work on these elements. Much of the progress achieved is due to the development of more sensitive detectors over the last few years and the availability of high resolution spectrographs at intermediate and large telescopes. The solution to many of the problems concerning light elements will require the employment during the coming decades of the new generation of very large telescopes.

2 LITHIUM, BERYLLIUM AND BORON IN THE SUN, METEORITES AND THE INTERSTELLAR MEDIUM

2.1. The Sun

Greenstein and Richardson (1951) measured for the first time the solar Li abundance (in the following the ^{7}Li isotope will be referred to as Li). Stellmacher and Wier (1971) and Müller, Peytreman and De la Reza (1975) gave a more accurate measurement of the solar abundance, Li/H=1.1($\pm$) 10^{-11}, which was essentially confirmed by the NLTE analysis carried out by Steenbock and Holweger (1984). Comparison of this abundance with that of the meteorites (see below) clearly suggested that lithium was depleted in the Sun.

The first measurement of the solar ^{9}Be (hereafter Be) was obtained by Greenstein and Tandberg-Hanssen (1954). More recent determinations give Be/H=1.2-1.4 10^{-11} (Ross and Aller, 1974; Chmielewski, Müller and Brault, 1976). Since this abundance is similar to that found in meteorites, it is thought that beryllium has not been depleted (or just slightly) in the Sun. The solar boron abundance is quite uncertain, as opposed to Li and Be. Kohl et al. (1977) claimed a detection of B/H=4 10^{-10} (with an uncertainty of a factor 2) using the BI 2496 Å resonance doublet. Although this value would be compatible with measurements in carbonaceous chondrites and other meteorites (see references in Reeves and Meyer, 1978), measurements with reduced error bars are needed to shed light on a small, but possible, depletion of boron in the Sun.

2.2 Meteorites

Reeves and Meyer (1978) and, more recently, Anders and Grevesse (1989) present an extensive compilation of Li, Be, and B measurements in different types of meteorites. In general there is good agreement between the abundances derived from the different types of meteorites, boron having the largest dispersion. Li/H=$2.2(\pm 0.4)10^{-9}$, Be/H=$3.6(\pm 1.3)10^{-11}$ and B/H=1.4 10^{-10} (with an uncertainty of a factor 2). The ratios of the Li and B isotopes have been measured ^{7}Li/^{6}Li=12 ± 0.2 (Balsiger et al., 1968) and ^{11}B/^{10}B=4.05 ± 0.1 (Mason, 1971). They are of great interest to constrain the mechanism of Galactic Cosmic Ray spallation in the ISM.

2.3. The Interstellar Medium

Since the first detection of interstellar Li by Traub and Carleton (1973) along the line of sight to ζ Oph, considerable effort has been devoted to the measurement of Li abundances at different lines of sight (see e.g. Vanden Bout et al., 1978; Hobbs, 1984; Ferlet and Dennefeld, 1984). The measurement of the ratio of the two Li isotopes in diffuse clouds made by Ferlet and Dennefeld (1984) is remarkable. These authors reported a ^{7}Li/^{6}Li ratio in the range 25 to 180 in front of ζOph. The absolute abundance of each of the isotopes is uncertain because of the possible depletion in the gas phase by condensation onto grains, but corrections to this effect, based on the behaviour of the alkalides Na and K, indicate the best estimate of the ISM Li abundance to be Li/H$\sim$ 1 10^{-9} (log N(Li)=3, when adopting the standard scale log N(H)=12). Comparison with measurements in meteorites suggest a nearly constant Li abundance in the solar neighbourhood during the last 5 Gyr. We will come back to this point later when dealing with Li abundances in young and old open clusters.

Very recently Baade and Magain (1988), Sahu et al. (1988), and Vladilo and Molaro (1989, private communication) also obtained upper limits to the interstellar Li abundance towards SN 1987A. While the first two works set very restrictive upper limits of order Li/H$\sim$1 10^{-10} for the abundance, the latter suggests an upward correction by at least a factor 6 for this upper limit. To solve this controversy, precise observations of other alkalides along the same line of sight are required. Unfortunately, so far no positive detection of Li in the SMC ISM has been claimed.

Upper limits to the Be abundance in the ISM were obtained by York et al. (1982) and Boesgaard (1985a). These upper limits are of the same order as the measured solar abundance. Since boron is more volatile than Li and Be (less condensation into grains), the ISM data on B are easier to interpret than those on Li and Be. Interstellar singly-ionized boron was detected by Meneguzzi and York (1980) towards the line of sight of κ Ori; they measured an abundance B/H=1.5-1.7 10^{-10}. York et al. (1982) also established upper limits along two lines of sight.

3 LITHIUM IN PRE-MAIN SEQUENCE STARS

The first detection of the $\lambda 6708$ Å Li resonance doublet in stars was attained by Sandford (1947) in T Tauri. Ten years later, Hunger (1957) published the discovery of a very strong Li doublet in two T Tauri stars. Quantitative analyses in these stars were made by Bonsack and Greenstein (1960) who showed that Li was overabundant by a factor 100 with respect to the Sun. This result, together with the Li measurements in meteorites, gave rise to the idea that Li had been depleted in the Sun.

Bodenheimer (1965) presented the first calculations of the destruction of Li during the pre-main sequence evolution of solar type stars. He showed that in stars with B-V$\leq$0.63 (M/M$_\odot$ $\geq$1) this destruction was less than 50 %, while in K0 stars Li disappeared almost completely from the convective envelope. Zappala (1972) found Li/H$\sim 10^{-9}$ in a sample of very young stars belonging to the stellar association NGC 2264 and also in the T Tauri stars observed by Bonsack and Greenstein (1960). These were all relatively massive stars (M$\geq$2 M$_\odot$), with Li abundances very similar to those in F and early G stars of young open clusters. Thus, as expected from Bodenheimer's calculations, there was not Li depletion in these stars. The question of whether or not Li is depleted during pre-Main Sequence evolution of solar type and lower mass stars has remained open until very recently. The relevant observations have been carried out in T Tauri and low mass stars in very young open clusters.

3.1 T Tauri Stars

Magazzù and Rebolo (1989) and Strom et al. (1989) have recently investigated the Li abundance in T Tauri stars. The first authors derived the Li abundance in a sample of 30 T Tauri stars belonging to the stellar associations Taurus-Auriga, Ophiucus, Chamaleon and Lupus. Most of the stars are younger than 2 10^7 yr. and have masses in the range 0.4-2 M$_\odot$. In a preliminary LTE analysis, abundances log N(Li)=3.2-3.5, in the more massive stars, and a trend of decreasing Li abundance when stellar mass decreases below 1 M$_\odot$ were found. Stars with M$\leq$ 0.6 M$_\odot$ showed Li depletions higher than 1.0 dex in several cases. Most of the stars in this study were classical T Tauris (as defined by Herbig, 1962), having Hα equivalent widths larger than 10 Å. Their Li abundances are not free from serious uncertainties because pronounced spectral veiling (see e.g. Hartigan et al. 1989, Bertout 1989) can affect the estimation of stellar parameters and also the measurement of equivalent widths. Moreover, since the Li doublet is rather strong in T Tauris, the abundances may also be affected by NLTE. Our analysis of Li abundances in the naked T Tauris (stars that do not exhibit the extreme properties of classical T Tauris) observed by Walter et al. (1988) however has produced similar results to the sample of classical T Tauris (Magazzù and Rebolo, 1990) thus allowing us to feel more confident about the trends described above for PMS Li depletion.

Recently, Strom et al. (1989) have also published preliminary results of a extensive survey of Li in PMS stars. They observed 65 PMS stars in Taurus-Auriga and 65 X-ray PMS stars in the Lynds 1641 complex. Veiling in their stars was taken into account using the expression given by Hartigan et al. which relates Hα intensity with the veiling factor. After veiling correction, the Li equivalent widths produced an increase of the Li abundance which in several cases reached log N(Li)=4. Because such large overabundances would have important implications both, on Li nucleosynthesis mechanisms, and on our understanding of stellar structure at these very early stages of stellar evolution, much more detailed analysis of veiling, and Li (and other elements) abundances are needed. The trend of higher depletion in the less massive stars still persists after veiling corrections, again indicating that the trend is not an artifact. Explanations for the scatter in this trend have to be investigated. Differences in age, rotational velocities, chromospheric activity and the presence of accretion disks around some of the stars may be the cause.

3.2 Low Mass Stars in Very Young Open Clusters.

In this section I will deal with the Li abundances in stars of three open clusters: IC 2391, α Per and the Pleiades. According to the turn-off of the upper main sequence, these clusters have ages of $3 \ 10^7$ yr., $5 \ 10^7$ yr., and $1.5 \ 10^8$ yr., respectively (Stauffer et al., 1989; Balachandran, Lambert and Stauffer, 1988; Mazzei and Pigatto, 1989). The Pleiades was the first one to be studied for Li abundances; Duncan and Jones (1983) found a large spread of the Li abundances in the coolest stars they observed (early K type), and suggested a possible age spread several times larger than the nuclear age of the cluster which was thought to be 6-7 10^7 yr. New observations by Butler et al. (1987) of four rapidly rotating (vsin i$\geq$50 kms^{-1}) early K dwarfs, have shown that these contain an order of magnitude more Li than slow rotators of the same spectral type (see diamonds in Figure 1). It is argued by the authors that stars with low vsin i had enough time to decrease their velocity and Li abundances while rapid rotators could have just arrived to the main sequence. This argument assumes that the rates of Li destruction during the early main sequence evolution are essentially the same for all stars with the same mass, which implies that the differences in Li abundances are mainly due to differences in age.

Li observations by Balachandran et al. (1988) in α Per and by Stauffer et al. (1989) in IC 2391 have shown that Li depletion may depend more on stellar angular momentum loss (braking) than on stellar age. In Figure 1, the Li abundances for stars in these two clusters and the measurements made in the Pleiades by Ducan and Jones (1983) and Butler et al. (1987), are represented versus effective temperature. Relative errors in the abundances are not higher than 0.2 dex and typical errors in temperatures are ±150 K. For comparison, the Li depletion pattern of the older cluster, the Hyades, was also plotted. The information contained in the figure can be briefly summarized as follows: i) At any given temperature there is a large scatter in Li abundances for the three clusters. It is particularly interesting to see how in α Per there are stars with effective temperatures T_{eff} ~6000 K having lower Li abundances than similar stars in the Pleiades and the Hyades (much older clusters). ii) In α Per and the Pleiades, the Li poor stars are slow rotators. There are no stars with high rotational velocity (vsin i$\geq$50 km s^{-1}, open symbols in Figure 1) and low Li abundances. In IC 2391, Stauffer et al. (1989) were unable to detect Li in the coolest rapid rotators that they observed, and consequently we do not know if the same pattern is also valid in this cluster. More precise measurements in IC 2391 (and other very young open clusters) have the double interest of confirming the connection between rotation and Li destruction, and of settling stringent constraints to the PMS Li depletion. The cool stars in IC 2391 provide direct evidence for substantial (a factor 100) PMS Li destruction at effective temperatures below 4000 K (m$\leq$0.7 M$_\odot$). Such amount of Li depletion is consistent with those measured by Magazzú and Rebolo and also by Strom et al. in T Tauri stars of similar and lower masses.

3.3 Confronting theory to observations

Li burning during PMS and MS evolution was considered in detail by D'Antona and Mazzitelli (1984) for masses in the range 0.6-1.2 M$_\odot$. Their Li depletion curve without extra mixing at an age of 10^9yr. has been represented as a dot-dashed line in Figure 1. (This model is normalized to an initial Li abundance of log N(Li)=3.2). It is obvious that their calculation does not reproduce the observational data, yielding an upper envelope to them. Better agreement is obtained when a certain amount of extra mixing is allowed in the model, but still the predicted Li depletions cannot be reconciled with the data.

More recently, Proffitt and Michaud (1989) have produced, using new opacities (Huebner et al. 1977, Alexander 1975), a series of PMS models that follow the nuclear burning of ^{6}Li and ^{7}Li. Their predictions for an age of 7 10^7 yr. and metallicity Z=0.0169 have also been plotted in Figure 1 (dashed line). In this case, agreement with the observations is much better for stars cooler than 5000 K but there are hotter stars, with excessively high depletions, which are difficult to explain unless unjustified variations of model parameters like Z (the stellar metallicity) or α (the mixing length to pressure scale height) have to be adopted. VandenBerg and Poll (1989), also using the new opacities, essentially agree with the previous work on the amount of PMS Li depletion and with the dependence of this depletion on Z and α. In fact, both works show how the observed pattern of Li depletion in the Hyades (Figure 1) could be matched adopting adequate (Z,α) parameters (see also Swenson, Stringfellow and Faulkner, 1990) and a certain amount of extra mixing.

However, the observations in younger clusters have shown stars with 1-1.2 M$_\odot$, with much higher depletions than Hyades stars, which appear to demand a different explanation, possibly connected with the rotational history of the stars. This connection was already investigated by Dicke (1972), who quantitatively related the slowing of rotation in young solar type stars and the rate of of angular momentum loss to the loss of Li. More recently, Pinsonneault et al. (1989) have considered PMS and MS Li depletion using a new rotating stellar evolution code. They start with a hydrostatic fully convective PMS model and evolve it to the age of the Sun, accounting for angular momentum loss via a magnetic wind (Kawaler, 1988), and angular momentum redistribution by rotationally induced instabilities. In their model, the amount of rotationally induced mixing is proportional to the amount of angular momentum transport and as a result, is also proportional to the initial angular momentum. Figure 2 shows their predicted curves of Li depletion versus age for different values of initial angular momentum J_0 and the depletion curve obtained by Proffitt and Michaud (1989) for 1 $M_\odot$ (Z=0.024, α=1.5). Pinsonneault et al. predict higher depletions than Proffitt and Michaud at the end of the PMS evolution, suggesting that the scatter in Figure 1 could be a consequence of differences in the initial angular momentum of the stars. A spread in Li abundances of 0.5 dex can be accounted for in their model introducing differences in J_0 by a factor 10. Further computations for lower masses would be of great interest to test the validity of these models. Measurements of Be abundances in these very young low mass stars, and their correlation with v sin i, should also be pursued, because they may provide a very valuable test for these models of rotationally induced mixing and for other alternative descriptions of turbulent mixing (see Schatzman, 1989 and references therein). A search for ^{6}Li in very young stars would also help understand transport processes at these early stages of stellar evolution.

4. LITHIUM, BERYLLIUM AND BORON IN SOLAR METALLICITY MAIN SEQUENCE STARS

Lithium has been measured in solar metallicity stars since the early 60's. Herbig (1965) found an empirical correlation between Li abundance and age for field stars: younger MS stars had more Li. Herbig and Wolff (1966) showed that there was a Li abundance range at each spectral type, maximum abundances decreasing towards cooler stars, thus giving additional support to the idea that Li could be destroyed in stellar interiors. Further observations in open clusters like the Hyades, Pleiades and Praesepe (Wallerstein, Herbig and Conti, 1965; Danziger and Conti 1966, Danziger 1967) provided more evidence about the mass and age dependence of this element in solar type stars.

At the end of the 60's the data available were consistent with the picture that stars were formed with the same initial Li abundance (log N(Li)~3), and progressively destroyed Li during the Main Sequence (MS) lifetime, as convection transported it to inner regions where the temperatures are hot enough for (p,α) reactions to be efficient (T$\geq$2.4 10^6 K). More recent observational work in field stars (Duncan, 1981; Soderblom, 1983; Boesgaard and Tripicco ,1986a; Pallavicini, Cerruti-Sola and Duncan, 1987; Balachandran, 1990), and specially in open cluster stars (Cayrel et al., 1984; Boesgaard and Tripicco, 1986b; Rebolo and Beckman, 1988; Hobbs and Pilachowski, 1988; Garcia Lopez et al. 1988), has shown this picture is too simple. The pattern of Li depletion vs. effective temperature (mass) within a cluster (see Figures 1 and 3) has proved to be rather complicated. In general, open clusters provide stellar samples with homogeneous chemical composition in which the fundamental parameters of the stars can be determined with high precision. The determination of the Li depletion curve versus mass in clusters of different ages is the observational key to answer the question of how Li is depleted during MS evolution. Important evolutionary effects have been observed in mid F stars and also in late type G and K stars. They are discussed below separately.

A systematic study of Be in MS stars was carried out by Boesgaard (1976b) who found that the abundance of this element was not as sensitive as Li to the stellar effective temperature. A mean Be abundance of Be/H= $1.3(\pm 0.4)10^{-11}$ was found. Although she noted that F stars in the sample showed evidence of Be destruction. Boesgaard, Heacox and Conti (1977) measured a mean Be abundance of Be/H=1 10^{-11} in six Hyades stars and Be abundances in old disc stars were also published by Dravins and Hultqvist (1977) and Boesgaard and Chesley (1976). The B abundance has been investigated in A and B stars by Boesgaard and Heacox (1978), and these are the only measurements available in solar metallicity stars.

4.1 The F Stars

The Li abundance gap in mid F stars was discovered in the Hyades (0.76 Gyr.) by Boesgaard and Tripicco (1986b). The pronounced absence of Li in stars with effective temperatures in the range $6800 \geq T_{eff} \geq 6400$ K (see Figure 3) was subsequently confirmed in other open clusters: by Hobbs and Pilachowski (1986a) in NGC 752 (1.7 Gyr.); Boesgaard (1987) in Coma Berenices (0.5 Gyr.); Boesgaard, Budge and Burck (1988) in the Ursa Major Group (0.3 Gyr.) and Boesgaard and Budge (1988) in the Hyades and Praesepe (0.7 Gyr.). All the observations confirmed the dramatic absence of Li (a depletion by nearly 2 orders of magnitude) in the mentioned effective temperature range (see Figure 4). Unfortunately, the uncertainties in individual T_{eff} determinations do not allow us to distinguish significant differences between Li abundances of stars with similar parameters in different clusters. It is interesting to note that in the Ursa Major Group (the youngest of these associations) at least four stars have upper limits log $N(Li) \leq 1.3$, which implies a very rapid Li depletion. Again, observational limitations of the minimum equivalent widths which can be reliably achieved render the detection of Li at the base of the gap difficult.

Observations by Pilachowski, Booth and Hobbs (1987) and Boesgaard, Budge and Ramsay (1989) in F stars of the Pleiades and α Per clusters revealed a nearly constant Li abundance with a value log $N(Li) \sim 3.0$ for all of them. Only around 6600 K a slight dip, down to values of log $N(Li)= 2.85$, might be present. The strong conclusion is that Li depletion in the mid F stars did not occur during the PMS phase. These observations, taken together with those in Ursa Majoris, constrain the decline in surface Li by a factor larger than 50 to have taken place in less than 4 10^8 yr. (if the age for the Pleiades recently reported by Mazzei and Pigatto (1989) is adopted). The explanation for this depletion may come from diffusion processes, as first suggested by Michaud (1986) who showed that in the effective temperature range 7000 K - 6400 K the depth of the He II convection zone increases rapidly (several orders of magnitude) as T_{eff} is reduced, and that radiative acceleration is larger than gravity at $T_{eff} \geq 6900$ K, allowing Li atoms to be supported by the radiative pressure at the bottom of the convection zone in stars at the hot end of the Li gap, while they are not supported at lower temperatures. Below $\sim$ 6400 K the convection zone is so deep that diffusion does not have sufficient time to modify the initial Li abundance during the lifetime of the Hyades or Ursa Majoris stars.

There are two possible objections to this model:

i) It fails to explain underabundances by more than a factor 30. This is a serious drawback, since the first analysis by Boesgaard and Tripicco (1986b) produced very strong upper limits (log N(Li)$\leq$0.3 in some stars. But Boesgaard and Budge (1988) have adopted a more conservative position correcting these upper limits upwards to a value $\sim$1.6, and therefore observations and model are compatible.

ii) At T_{eff} $\geq$7000 K the predicted large Li overabundances are not observed. To solve this point, Michaud argued that the addition of a mass loss rate of 10^{-15} $M_\odot$ yr.$^{-1}$ (comparable to that in Am-Fm stars) could be adequate to reduce the predicted overabundances to the observed level.

An alternative to the microscopic diffusion model is meridional circulation, as suggested by Charbonneau and Michaud (1988). Their calculations show that meridional circulation can reduce considerably any overabundance of Li produced by radiative acceleration at the hot end of the Li dip, while at lower temperatures (below $\sim$6900 K) Li is no longer supported and meridional circulation carries Li-depleted matter into the convection zone. Considering only the competition between transport by meridional circulation and radiative acceleration, they are able to reproduce the low T_{eff} side of the Li gap in the Hyades, using a linear relationship between rotational velocities and effective temperatures that gives 70 Kms^{-1} and 25 Km s^{-1} for stars with 7000 K and 6300 K respectively. Charbonneau and Michaud's (1988) computations are rather sensitive to the adopted rotational velocities; changes by a factor 1.5 (which is reasonable) in these velocities may cause the models to fail to reproduce the observations. A major objection to this scenario is posed by the stars that do not show the expected Li underabundances if meridional circulation were effective. For instance, there are fast rotators in the Hyades and UMa with T_{eff} $\geq$ 6900 K having Li abundances near the cosmic value. Li observations in field F stars also show that some of the most rapid rotators are undepleted (see also Balachandran 1990), ruling out meridional circulation as the cause of Li depletion.

Following Zahn's ideas (see this volume), Vauclair (1988) proposed a different explanation for the Li gap in terms of turbulent mixing induced by rotation. The decrease of the Li abundance in the cool side of the Li dip is attributed to nuclear destruction related to the increase of stellar rotation when we move from G to F stars, while in the hot side of the gap, two separate mixed zones would prevent Li from being destroyed.

Beryllium observations in the Hyades mid-F stars (Boesgaard and Budge, 1989) have shown a remarkably constant abundance with a value close to the cosmic Be/H$\sim$1 10^{-11}. As noted by these authors, these observations do not represent a critical test of meridional circulation, but they can be of great interest to constrain the turbulent diffusion model. Theoretical predictions for Be are needed. Observations of other elements, like the CNO group and measurements of the ^{7}Li/^{6}Li ratios in F stars, will help to distinguish between the different mechanisms of Li depletion (microscopic diffusion, meridional circulation and turbulent diffusion) so far proposed. So far, only upper limits of order 10-20 to the ^{7}Li/^{6}Li isotopic ratio have been established in solar metallicity F and also G stars (Andersen, Gustafsson and

Lambert, 1984; Hobbs, 1985; Rebolo, et al. 1986). Possible detections have been claimed, but the difficulty of the measurement requires further confirmation.

4.2 The G and Late Type Stars

The behaviour of Li abundances in very young late type stars has already been described in section 3.2. I discuss here the Li abundances of older G and K MS stars. In fact, not many measurements are available for old stars later than K0, where Li is destroyed very quickly during PMS and/or MS evolution. Even with high S/N and high spectral resolution, Cayrel et al. (1984) could not detect the Li feature in early K Hyades stars and Li rich main sequence K stars are extremely rare (see Cayrel et al., 1989). On the contrary, many Li measurements are available for G field and open cluster stars. Observations in intermediate age and old open clusters confirm that Li depletion has occured during MS evolution of G stars. Figure 5 shows how the Hyades Li depletion curve provides a higher envelope to the diagram at $T_{eff} \leq 5500$ K. As can be appreciated Figure 3 the dispersion of the measurementes in the late Main Sequence of this cluster is very small, as it is the dispersion of Li versus rotational velocity (Rebolo and Beckman, 1988). Comparison of Figures 1 and 3 suggest that young G-K stars experience rotational braking and destroy their Li in such a way that at the end of their first Gyr. of life on the MS they have minimized possible starting differences in angular momentum, primordial magnetic fields or any other stellar parameters that can affect the evolution. Observations in older clusters like NGC 752 (Hobbs and Pilachowski, 1986a), M67 (Hobbs and Pilachowski, 1986b; Spite et al., García López, Rebolo and Beckman, 1988), NGC 188 (Hobbs and Pilachowski, 1988a) have shown that the picture is more complicated. Li abundances in NGC 752 have been represented versus T_{eff} in Figure 4, where it can be appreciated that: i) the Li abundance in the hot side of the Li gap is 3.1; ii) the Li abundances in the G stars are considerably lower and show high dispersion (more than 1.5 dex) between 5700 K and 5800 K (~ 1 $M_\odot$ stars). A similar situation is seen in M67 stars (Figure 5) where the scatter is of order 0.7 dex, larger than what is to be expected from relative errors in the abundance determinations (bottom left corner in the figure).

Li measurements in NGC 188 by Hobbs and Pilachowski (1988a) have also been plotted in Figure 5 where it is clear that their Li abundances at a given effective temperature are similar to those in M67 stars (in some cases even higher). The age of this cluster has recently been revised by Twarog and Anthony-Twarog (1989) who reduced to 6 10^9 yr. the previous, generally accepted, estimation by VandenBerg (1985) of 10^{10}yr.. Considering VandenBerg's age estimation it was surprising that such old solar metallicity stars could have higher Li abundances than similar stars in the much younger M67. Hobbs and Pilachowski consequently argued that the initial content of NGC 188 should have been log N(Li)$\sim$3 or even higher. After this revision of the cluster's age, comparison of Li abundances in the older open clusters shows a similar behaviour, both in absolute abundances and in the internal scatter within each cluster.

The observational data for intermediate age and old open clusters lead to the following conclusions:

i) Stars with masses in the range 1.1-0.9 $M_\odot$ suffer destruction of their original Li during their MS lifetime. This destruction is much higher in the less massive ones and much more efficient during the first Gyr. in the MS. The higher mass stars in this range seem to have preserved a large amount of Li.

ii) The Li content in the galactic material has not changed significantly during the last 6 Gyr. Stellar observations constrain the range of variation during this time interval between log N(Li)=2.6 and 3.2. Observations in meteorites and the Li depletion pattern in the various clusters suggest that the real variation may have been smaller.

These two results have important implications on transport processes in stellar superficial zones and on nucleosynthesis processes within the Galaxy; while the latter will be considered in the last section of this work, I discuss in the following a variety of processes that have been suggested to explain the MS depletion of Li in solar type stars: convective overshooting, mass loss, microscopic diffusion, turbulent diffusion mixing, turbulent mixing induced by rotational braking and meridional circulation. A more detailed description of many of them can be found in the review by Spruit (1987).

i) Mass loss. The first theoretical efforts made to explain the observed abundances of Li in stars took place in the mid 60's. Weymann and Sears (1965) calculated the mass loss rate needed to explain the depletion of Li observed in the Sun. The mass rates found were nearly 3 orders of magnitude higher than the present-day solar wind. This has been confirmed in a very recent study by Hobbs, Iben and Pilachowski (1989) on the Li removed through mass loss induced dilution of the surface material. Such high mass loss rates could have existed in earlier epochs of the life of the Sun, and in fact there is observational evidence (see Willson, Bowen and Struck-Marcell, 1987) for high mass loss rates in slightly more massive stars.

ii) Convective overshooting. The possibility exists that downward motions, which start in the convection zone, do not stop when they reach the region where the temperature gradient changes from super-adiabatic to sub-adiabatic, descending to unknown distances in their penetration of the radiative interiors. The effect of this overshooting on the depletion of Li was investigated by Weymann and Sears (1965); Strauss, Blake and Schramm (1976) and Cayrel et al. (1984) in their study of the Hyades. The Li destruction curves predicted via this mechanism (see Strauss et al.) are characterized by a steep fall of the abundance at a given mass which varies depending on the fraction of the scale height adopted for overshooting. This fall reflects the high efficiency of the reaction $^7Li(p,\alpha)^4He$ destroying Li when the convective elements reach deep regions with temperatures of order $2.5\ 10^6$K. To explain the observations, different amounts of overshooting for different stellar masses and also at a given mass (if the scatter of Li abundances in clusters were to be accounted for) would be necessary. There is no obvious reasons for that, and it might be that overshooting is not necessary at all as suggested by Swenson, Stringfellow and Faulkner (1990). They claim that the Li depletion curve in the Hyades is a natural result of convective mixing when the new Los Alamos opacities are used in the code of stellar evolution. Opacities play a critical role in the location of the base of the convection zone and consequently on the amount of Li destroyed at each stellar mass. Their computations are very sensitive to small variations (0.2 dex) in the global metallicity of the stars, thus allowing a direct observational test which may be carried out by comparing accurate measurements of the Li abundance in similar stars of the Hyades and Coma Berenices clusters (both have the same age).

iii) Microscopic diffusion. As a result of gravitational force, radiative pressure and also of temperature and concentration gradients, microscopic diffusion can be present in stellar atmospheres. Vauclair et al. (1978) concluded that these processes cannot explain the abundances of Li, Be and B in the Sun. The depletion of Li by a factor 100 via microscopic diffusion would have caused a similar deficiency of the solar Fe abundance, which is not observed. While in mid F stars (Michaud, 1986) these processes may be the primary cause of Li depletion, in late F and G stars, the time scale for gravitational settling increases with the depth of the convection zone, failling as a possible explanation of the Li observations.

iv) Meridional circulation. The effect of meridional circulation on the depletion of the superficial Li of G stars was also investigated by Vauclair et al. (1978). They explained the depletion of Li in the Sun by a factor 100, but the model also predicted a Be depletion by a factor 10 which is not compatible with the observations. More recently, Charbonneau and Michaud (1988) reconsidered meridional circulation and reported that no depletion should occur via this mechanism in F stars cooler than 6400 K at the age of the Hyades since the time required to transport matter from the Li-burning region to the convection zone is larger than $8\ 10^8$ yr., but they do not make predictions for G stars.

v) Turbulent mixing. It is generally believed that late type stars lose their angular momentum via stellar winds driven by the magnetic fields generated in the convection zone (Schatzman, 1962; Weber and Davis, 1967). In a pioneer work, Dicke (1972) concluded that the turbulence and transport of material invoked to slow the stars should have associated severe depletions of Li and Be. Several years later, Schatzman (1977) considered Li transport by turbulent diffusion (originated by some sort of marginal instability) from the bottom of the convection zone to the burning layer, during the MS lifetime of a star. The diffusion coefficient D_t was adopted proportionally by a constant factor R_e (a pseudo-Reynolds number) to the ordinary microscopic viscosity ν of the plasma and found that such a simple relation accounted for the destruction of Li in MS stars and for the rate of angular momentum loss in the Sun. Other important problems, like the observed width of the upper MS and the low value of the solar neutrino flux or the ^{3}He enrichment of the solar surface (Schatzman et al. 1981), could also find an explanation with the introduction of turbulent diffusion. Despite this success, using this coefficient Baglin, Morel and Schatzman (1985) were not able to explain the mass dependence of the Li depletion in the low MS of the Hyades. They also considered a new turbulent diffusion coefficient $D_t = \beta\Omega^2(\nabla_{ad} - \nabla_{rad})^{-1}$, proportional to rotational velocity and to the difference in adiabatic and radiative gradients, estimated according to Zahn's ideas (1987) on the production of meridional circulation currents and differential rotation as a result of the thermal imbalance due to stellar rotation. With the new coefficient a much more reasonable fit to the dependence of Li with mass is obtained for the Hyades lower

MS, even if no dependence of Ω with age was considered in the calculation, and, as already pointed out by Law et al. (1984), the mixing from differential rotation induced instabilities should be much stronger initially when the spindown rate is larger, i.e. a strong dependence of the turbulent diffusion coefficient on age is expected. Schatzman (1989) proposed a model of turbulent diffusion mixing and of electromagnetic braking which might solve this problem.

According to Pinsonneault et al. (1989), Li depletion in the Sun and solar type stars in old open clusters is explained in terms of rotationally induced mixing. The amount of mixing being different for models of different initial angular momentum J_0 (see Figure 2). The scatter of Li abundances in stars with similar mass, age and composition (see Figure 5) is explained as a consequence of a spread in J_0. Confirmation for this mixing mechanism may come from observations of CNO abundances in the subgiant phase and also from confrontation of observations with theoretical predictions of Li depletion at different stellar masses.

vi) Other mixing mechanisms. Given the space limitations it is impossible to describe here all the proposed mixing mechanisms. In the review by Spruit (1987) the reader can find details of nonmagnetic mixing mechanisms like gravity waves or unstable nuclear burning and also of those involving magnetic fields.

5. LITHIUM, BERYLLIUM AND BORON IN UNEVOLVED LOW METALLICITY STARS

5.1 Lithium in the Old Disk Population

By Old Disk Population I refer to those stars having metallicities ([Fe/H]) between -0.5 and -1.4 which satisfy the kinematical criteria for the thin or thick disc (Gilmore and Reid, 1983; Sandage and Fouts, 1987). Because of their high metal deficiency, these stars were probably formed at the first stages of the disc evolution. Thus, their Li abundances can provide very interesting clues of the value of the abundance in the ISM at that epoch. In Figure 6 the Li abundance in the old disk population is compared to that of the old open cluster M67. As noted by Rebolo, Molaro and Beckman (1988), stars cooler than 5700 K have in general higher Li abundances than the open cluster stars of the same effective temperature; for hotter stars the Li abundances are lower in the old disc stars than in the open cluster stars. Two explanations can be suggested: i) the nature of the dominant mechanism of Li depletion is different for metal deficient stars cooler and hotter than 5900 K; while Li depletion is inhibited in the cool metal-deficient stars, in the hotter, the less Li depleted appear to be the more metallic stars. ii) The metal deficient stars were formed with a lower initial Li abundance and their depletion rates are lower than in cluster stars.

There is no definitive evidence supporting either of these two options: to accept the first, new models for mixing are required since those already proposed do not explain such behaviour; the second option has the drawback of a number of hot metal deficient stars with very low Li abundances. It would be worthwhile to investigate the evolutionary state of these stars because they may have evolved out of the MS, and have undergone Li dilution because of progressively deeper convection zones. As discussed in the following section, Li abundances in halo dwarf stars seem to give support to the second option.

5.2 Lithium in Halo Dwarfs

Li observations in halo stars were pioneered by Spite and Spite (1982). They found that Li was present in stars of the halo with effective temperatures between 5500 and 6250 K with a remarkably constant value of log N(Li)=2.05 (±0.15). Additional studies in halo stars have been carried out by Spite, Maillard and Spite (1984), Boesgaard (1985b), Spite and Spite (1986), Hobbs and Duncan (1987) and Rebolo, Molaro and Beckman (1988). Special attention has been given to extreme halo dwarfs (Spite et al., 1987b; Rebolo et al., 1987; Hobbs and Pilachowski, 1988b). Each study has selected its "halo stars" according to different criteria: the Spites adopted [Fe/H]$\leq$-1 as the division line between halo and disc stars; Hobbs and Duncan required a [Fe/H]$\leq$-0.6 and the magnitude of the stellar space motion with respect to the local standard of rest, $V_{LSR} \geq$100 Km s^{-1} for a star to be considered as belonging to the halo. Rebolo, Molaro and Beckman (1988) defined "extremely metal deficient" (EMD) stars as those having [Fe/H]$\leq$-1.4. All of them satisfied the kinematic halo population membership criteria of Sandage and Fouts (1987). The reason for definition of EMD comes from Figure 3 in Rebolo et al. (1988), where it is appreciated how the scatter of the Li abundances increases when [Fe/H] is higher than -1.4.

In Figure 7, Li versus effective temperature has been plotted for all the stars with [Fe/H]$\leq$-1.4 that satisfy at least one of the halo criteria mentioned before. While stars with $T_{eff} \geq$5500 K have a nearly constant Li abundance, at lower temperatures the Li abundance decreases rapidly. Rebolo et al. (1988) have pointed out the existence of a slight slope in the plateau: stars at the hotter end of the *plateau* appear to have higher abundances than stars at the cooler side. The least square fit to the data suggests a difference of 0.2 dex between both sides of the *plateau*. The mean value of the Li abundance for all the EMD stars in the *plateau* is 2.08 ±0.09 while the mean value of the Li abundance in the hot stars (T$\geq$6000 K), calculated giving equal weight to the measurements by different authors, is 2.16±0.04. This value sets a lower bound to the Li abundance in the pregalactic matter. Whether these stars have altered (reduced) their initial Li content during their long life in the MS is a matter of debate. So far three different theoretical approximations to this problem have been presented:

i) Michaud et al. (1984) studied the effect of microscopic diffusion and nuclear destruction of Li in stars with masses between 0.7 and 0.9 M$_\odot$ and [Fe/H]=-2. They concluded that a reduction of the Li abundance in the halo dwarfs by more than a factor 2.5 is unavoidable at all masses, microscopic diffusion being the dominant depletion mechanism at $\sim$ 0.9 M$_\odot$ and convective transport dominating at 0.7 M$_\odot$. The combination of both mechanisms (without considering overshooting) gives a poor fit to the observations because higher depletions are found at the hot end of the *plateau* than at the cool one, in contradiction with the observations. Microscopic diffusion can be avoided if turbulence below the base of the convection zone, induced by hydrodynamical instabilities, had taken place. As Vauclair et al. (1978; see also Vauclair 1988) already showed, a turbulent diffusion coefficient larger than 10 times the gravitational, would prevent microscopic diffusion.

ii) Vauclair (1988) claims that a Li *plateau* in the hotter halo stars appears naturally in the framework of Zahn's theory, because of the rapid increase of the turbulent diffusion coefficient, D_t with radius. According to the value of D_t at the bottom of the convection zone the initial Li may have been preserved (if $10 \leq D_t \leq 1000$ cm^{2}s^{-1}) or reduced by a factor 10 (if D_t is of order 10000 cm^{2}s^{-1}). Rotational velocity is the critical parameter governing this coefficient. To preserve Li in the halo dwarfs, average rotational velocities in the range 0.3 to 3 Km s^{-1} have to be adopted. Although the rotation braking law of halo stars is unknown at present, and it is a quantity of intrinsically difficult measurement, it can be argued that for a solar type star, it becomes easy to get a very low rotational velocity since they brake very soon after their arrival to the MS (see Figure 1). As an example, the Sun has a v_{rot} ~2 Km s^{-1} and its age is lower by a factor ~3 than the age of an extreme halo dwarf. It seems therefore plausible that halo stars could have preserved their original Li in this scenario.

iii) D'Antona and Mazzitelli (1984) already noted that the Li depletion curve in extreme halo dwarfs can be explained in terms of less deep convection zones in these low metallicity stars. Deliyannis, Demarque and Kawaler (1990) have examined in detail several mechanisms of Li depletion during the evolution from PMS to the giant branch of low metallicity stars. They consider nuclear burning at the bottom of the convection zone, diffusion by gravitational settling, convective dredge-up and dilution. Different physical processes dominate Li depletion when a star evolves. Their calculations show that nuclear burning occurs in the less massive halo stars (0.5-0.6 M$_\odot$) during PMS and that some nuclear destruction also takes place during MS. Diffusion dominates the late MS and just beyond the MS turn-off, low mass stars may suffer convective dredge-up bringing hidden Li to the surface. During the subgiant phase dilution becomes the dominant mechanism.

Using standard evolution models, Deliyannis et al. (1990) are able to explain not only the halo Li *plateau* but also the depletion curve of the coolest halo stars. They find that for stellar masses in the range 0.75-0.65 M$_\odot$ no substantial burning (less than 1-4%) occurs during the PMS and MS evolution. When diffusion (gravitational settling) is also taken into account in the models, the highest T_{eff} halo dwarfs suffer a severe depletion of Li, giving a significant curvature to the theoretical depletion curve. Current observations seem to indicate that this is not the case, but further Li measurements in halo dwarfs with $T_{eff} \geq 6200$ K would provide a definitive test. Also, some theoretical improvements can be produced if He diffusion were considered in the models. Rotationally induced mixing near the base of the convection zone may also be of growing importance when increasing the metallicity, since the deeper base is located nearer to the Li burning region. Predictions for other light elements (Be and B) and the CNO isotopes during post MS evolution will help when comparing this theory with observations.

The question of whether Li has been preserved or not in halo dwarfs still deserves much investigation. A definitive answer may be provided by the detection of the isotope ^{6}Li in halo dwarfs. Since this isotope is much more fragile than ^{7}Li in stellar interiors, its detection would be a strong case for the preservation of ^{7}Li. Brown and Schramm (1988) calculated which is the expected ratio between both isotopes when a ^{7}Li depletion curve is assumed to be similar to that represented in Figure 7. They found that for the hotter halo stars the ^{6}Li isotope should be depleted by a factor 1.45, while cool stars should not have any measurable amount of ^{6}Li.

Previous measurements in cool halo dwarfs (Maurice, Spite and Spite, 1984) had shown the ^{7}Li/^{6}Li to be higher than 10. In a more recent work by Pilachowski, Hobbs and De Young (1989) an upper limit of 10 to this ratio has been established in the hot low metallicity star HD 84937 ([Fe/H]=-2.1). An additional difficulty for detecting ^{6}Li is that, assuming that the standard model of Big Bang nucleosynthesis is correct, no measurable abundance for this isotope would be expected in the pregalactic matter, (but see Dimopoulos et al. (1988) for alternative models which overproduce ^{6}Li). As a consequence, stars formed before the interaction between Cosmic Rays and CNO nuclei in the ISM of the Galaxy has produced a measurable ^{6}Li abundance (say ^{6}Li/H$\geq 10^{-11}$) are not suitable candidates for the detection of this isotope. Following Rebolo et al. (1988b) halo stars with measured Be abundances can give us an indication of this spallogenetic production of ^{6}Li and consequently Be abundances should be pursued before measurements of Li isotopic ratios.

5.3 Beryllium and Boron in unevolved halo stars

Using IUE spectra of the metal deficient star HD 76932, Molaro and Beckman (1984) established an upper limit to the Be abundance of 25 % of the solar Be abundance. The first detection of Be in highly metal deficient stars was claimed by Rebolo et al. (1988b) who measured Be/H values between 10^{-12} and $2.5\ 10^{-12}$ in stars with metallicity in the range $-1 \geq$[Fe/H]≥ -1.3. Upper limits of $2.5\ 10^{-12}$ in lower metallicity stars were also obtained. Since all these stars have "normal" Li abundances, no Be depletion is expected. These measurements are represented in Figure 8 with others in more metallic stars taken from the literature, and with new measurements by Beckman, Abia and Rebolo (1989) and Ryan et al. (1990).

The observations indicate that:
i) Be has never had a higher abundance than the solar value at any time in the past history of the Galaxy.
ii) The maximum contribution to the ^{7}Li abundance in halo stars which can be due to spallation of Cosmic Rays in the ISM is 0.1 dex (see Rebolo et al., 1988), in agreement with the constraints coming from ^{6}Li (see Pilachowski, Hobbs and De Young, 1989).

So far, only the upper limit to the B abundance in a very low metallicity star (HD 140283) by Molaro (1987) has been published. He used a high resolution IUE spectrum to search for the BI 2496Å resonance doublet and derived an upper limit to the B abundance of B/H$\leq 10^{-11}$

(more than one order of magnitude less than the solar value). B depletion is ruled out by the presence of Li in the star, thus this upper limit provides support, as derived, to the galactic enrichment of B.

6 LITHIUM, BERYLLIUM AND BORON IN POST-MAIN SEQUENCE STARS

6.1 The Subgiants

During Post-Main Sequence evolution, dilution becomes the dominant process for changes in surface Li abundance of solar type stars. The deeper convection zones favour the mixing with Li depleted layers and increasingly dilute the Li content. Iben (1967) predicted post-MS Li dilution for 1.0, 1.25 and 1.5 $M_\odot$ stars, firstly confronted with observational data by Herbig and Wolff (1966). Boesgaard and Chisley (1976b) made similar predictions for the post-MS dilution of Be (which occurs later than Li dilution) and also determined Be abundances in 14 of the stars observed by Herbig and Wolff. These measurements essentially confirmed the ideas about convective dilution.

A definitive test for dilution can be provided by the subgiants in open clusters. Pilachowski (1986) measured Li in two subgiants of the open cluster NGC 7789 (age 1.6 10^9 yr.; stellar mass at the turn-off 1.5 $M_\odot$) and found a reduction by a factor 4 with respect to the cosmic value. Beckman and Rebolo (1988) have reported cosmic Li abundances for a few slightly evolved stars in the cluster NGC 752. Another much cooler subgiant in the open cluster NGC 188 was observed by Hobbs and Pilachowski (1988a). They set an upper limit log N(Li) ≤ 0.4 which, when compared with the Li abundance in the turn-off stars (log N(Li)$\sim$2.5) suggest a strong dilution effect (higher than expected from theory) or higher depletion during the MS lifetime. Very recently, Balachandran (1990) has measured or established upper limits to the Li abundance in a sample of subgiants in M67, whose temperatures are between 6200 and 4800 K. She finds the Li abundance to decrease with effective temperature. The subgiants in M67 are stars that were placed in the Li "dip" during their MS lives. Assuming as starting point an abundance log N(Li)=2.5 (the turn-off Li abundance of the cluster), Balachandran calculated the expected Li abundances for the stellar mass at turn-off, 1.25 $M_\odot$, using Iben's ideas on dilution. From the comparison with the observations, she points out that the strong depletion in the M 67 subgiants can be explained if MS depletion occured. She reports an upper limit for her coolest subgiant in agreement with the subgiant observed in NGC 188. These observations appear to indicate that nuclear destruction of Li has taken place in the Li "dip" stars. If Li had been "hidden" by microscopic diffusion in them we could expect that convective dredge-up, as the star evolves to subgiant, had temporaly restored hidden Li back to the surface and this is not seen. Be observations in these stars are worth obtaining.

Li in subgiants may also help to test the links between different types of evolved stars. In particular, Li measurements in subgiant CH stars (Smith and Lambert, 1984) led to the conclusion that these subgiants are not the progenitors of the mild and classical Ba stars.

6.2 The Giants

Lithium dilution is predicted as a star evolves up the giant branch, a deepening convective envelope dilutes the remaining Li from the MS and subgiant evolution. Predicted Li-dilution for the first dredge up by Iben (1967) range from a factor of 60 for a 3.0 $M_\odot$ model to 2.8 for a 1 $M_\odot$ model. Therefore G and K giants are expected to have low Li abundances, and this is indeed the case for most of them. Bonsack (1959) provided the first comprehensive study of Li abundances in field giants. Results for a large sample of M giants by Lambert, Dominy and Sivertsen (1980) and by Luck and Lambert (1982), combined with those in Alschuler (1975) confirmed that a majority of giants have log $N(Li) \leq 1.5$. However, rare Li rich G-K giants have also been discovered (Wallerstein and Sneden, 1982; Hänni 1984; Gratton and D'Antona, 1988) showing Li abundances near to the cosmic value. To assess the frequency of these Li rich giants, Brown et al. (1989) carried out a spectroscopic survey on 644 giant stars. They found that all stars but a few have log $N(Li) \leq 1.5$ and just one star (HD 9746) was found with log $N(Li) = 3$. This star has also a low carbon isotope ratio which suggests that mixing has occured in a fully convective envelope which should have caused the destruction of Li. Possible explanations for the observed Li can be given in terms of the Cameron-Fowler mechanism for beryllium transport at the helium shell-flash stage; large scale diffusion of Li during the MS lifetime followed by homogenization via the giant's deep convective envelope (Lambert and Sawyer, 1984). High chromospheric activity and the presence of large atmospheric spots may account for the enhanced Li abundance.

Li abundances in open cluster giants are of particular value for stellar evolution and provide the advantage of stellar samples with known and homogeneous mass and age. Li measurements in NGC 7789, NGC 752 and M67 have been reported in Pilachowski (1986) Pilachowski, Saha and Hobbs (1988), García López, Rebolo and Beckman (1988). Pilachowski noted that Li is more abundant than normal in NGC 7789 giants, perhaps because of a higher mass (smaller amount of MS depletion) than typically observed field giants. Abundances in the other two clusters appear to be normal.

Beryllium abundances were studied by Boesgaard, Heacox and Conti (1977) in a sample of giants in the Hyades. A Be deficiency by a factor 40 with respect to the cosmic value was reported. They argued that this value is in agreement with the depletion predicted by dilution if a stellar wind of 0.02 $M_\odot$ had removed the outermost layers of the star. Be observation in three weak G band giants were reported by Parthasarathy, Sneden and Bohm Vitense (1984).

6.3 Highly evolved cool stars

AGB stars are perphaps one of the more interesting sites to search for Li. The discovery by Smith and Lambert (1989) that massive AGB stars in the SMC are Li rich may be a confirmation of the Hot Bottom Convective Envelope predicted by Scalo et al. (1975). The observed stars have $M_{bol} \sim -6$ to -7 and their Li abundances are 10^2 to 10^4 higher than those found in normal giants. As suggested by Scalo et al., mass loss from these stars may return to the ISM large amounts of Li that could have produced a significant enrichment of our Galaxy.

Other super Li-rich cool evolved stars have been known (Boesgaard, 1976a and references therein) to show anomalously high Li abundances (orders of magnitude higher than the cosmic value). Cameron and Fowler (1971) suggested the ^{7}Be transport mechanism to explain the range of Li content in S and C stars. They found that the super Li-rich stars should be the most massive (as apparently confirmed by Smith and Lambert) but a large number of flashes were needed to produce the Li overabundances. Sackman, Smith and Despain (1974) found that just a single deep mixing of $\sim 10^3$yr. could be enough. A systematic search for Li in AGB stars must be carried out to determine the precise evolutionary stage at which Li is produced. More observations in the SMC and also in galactic clusters are called for.

7 CONSTRAINTS ON NUCLEOSYNTHESIS

The abundances of the light elements provide a unique test for very different nucleosynthesis scenarios which are reviewed in many works (e.g. Reeves, 1974; Boesgaard and Steigman, 1985; Arnould, 1986). Here I will briefly summarize the constraints that recent observations of Li, Be and B may have added to classical nucleosynthesis mechanisms or to those more recently developed.

7.1 Primordial Nucleosynthesis

Many new developments in the field of Primordial nucleosynthesis have been recently achieved. The standard scenario (Wagoner et al., 1967; Yang et al., 1984) predicts measurable abundances of D, ^{3,4}He and ^{7}Li, which after comparison with astrophysical observations allow to set stringent constraints on Ω_b, the fraction of the density of the Universe contributed by baryons. The ^{4}He primordial abundance, derived from extragalactic HII regions, seems to be the more solid determination (Pagel, 1988). The current D abundance in the ISM may be seriously affected by astration during galactic evolution and consequently it can only impose upper limits on Ω_b. The evolution of the galactic abundance of ^{3}He is not well understood, competition between production and destruction processes makes very difficult a reliable estimation of its primordial abundance.

The primordial abundance of ^{7}Li is also a subject of debate. The observations described in previous sections suggest two possible values: a) the low value of the abundance observed in hot halo dwarfs, log N(Li)=2.2±0.2 or; b) the value of the abundance observed in very young early F stars of open clusters and massive pre-MS stars, log N(Li)=3.2±0.2. I will refer to them as the Pop II and Pop I values for the primordial ^{7}Li abundance. The dilemma is whether the primordial value is the Pop II and Li has been further enriched in the galactic matter or, it is the Pop I and halo dwarfs have depleted Li during their long lifetimes. Empirical arguments for considering the Pop II value as the primordial have been given by Spite and Spite (1982) and Rebolo, Molaro and Beckman (1988). Deliyannis, Kawaler and Demarque (1989) have shown that standard codes of stellar evolution do not predict Li depletion in the hot halo dwarfs and can reproduce the halo Li *plateau*, providing a strong theoretical argument in favour, but other "non-standard" models like those considering turbulent diffusion can also reproduce the *plateau* starting with a factor 10 higher Li abundance. The question will remain open until a fully consistent theory is developed for explaining the

Li observations described above, or until detection of ^{6}Li in low metallicity stars is carried out.

Recent studies of primordial nucleosynthesis give an increasingly important role to the Li and Be primordial abundances. These elements can be used to discriminate among the standard and non-standard cosmological models. Different groups have investigated the effects of density perturbations, arising in the phase transition from the QCD phase to the hadron phase, on the primordial production of light elements and whether the primordial abundances can be reconciled with $\Omega_b=1$ (Applegate, Hogan and Scherrer, 1987; Alcock, Fuller and Mathews, 1987). Although first computations showed a large overabundance of ^{7}Li (other light elements abundances were not different than in the standard model), Malaney and Fowler (1988), taking into account the diffusion of neutrons during the nucleosynthesis epoch, were able to find a parameter space of their model where the observed primordial abundances including ^{7}Li could be reconciled with $\Omega_b=1$. Neutron diffusion during the nucleosynthesis epoch has been studied by other groups (Kurki-Sounio and Matzner 1989; Teresawa and Sato, 1989) and obtained controversial results that makes it difficult to set constraints on the model from Li abundances alone. Fortunately, recent calculations of the ^{9}Be production in inhomogeneous models (Boyd and Kajino, 1989; Malaney and Fowler, 1989) show that this element is overproduced with respect to the standard model by more than 3 orders of magnitude. A Be abundance of order Be/H=1 10^{-13} is predicted, very near to present observational capabilities. In fact, Ryan et al. (1990) have set upper limits to the Be abundance in several halo dwarfs (see Fig. 8) that set strong constraints to these predictions, ruling out certain values of the parameter space in the model. It is exciting to see how accurate spectroscopic observations are setting tight constraints to new cosmological scenarios.

7.2 Galactic nucleosynthesis

Assuming that the Li abundance presently observed in halo stars is primordial, we have to explain how the Galaxy have been enriched in Li by a factor 10. Several models of galactic evolution, taking into account the different production mechanisms of light elements (spallation by Cosmic Rays in the ISM, thermonuclear reactions in stellar interiors) and astration can certainly explain it (Audouze et al., 1983; Audouze and Silk, 1989). But they can also explain, within the uncertainties of the model parameters, that the present galactic abundance could effectively be the primordial one, having remained nearly unchanged during the life of the Galaxy (see Abia and Canal, 1988). The new observations in very old disc stars and in AGB stars may provide the constraints to distinguish between them. It seems that most of the required enrichment must take place within a few Gyr. after the halo collapose, and more precisely, before the origin of NGC 188, thus restricting the possible galactic mechanisms of Li production to those able to produce it efficiently at an early epoch.

We already know from Be observations (see Figure 8) that there is a rapid increase in the galactic Be abundance during this epoch. The Be enrichment is mostly attributed to synthesis in the ISM by spallation reactions. According to Walker, Mathews and Viola (1985) no more than 10% of the present Li may have been produced in that period.

The most probable mechanism for Li production at early epochs is mass loss from massive peculiar red giants (carbon stars, AGB and so on). These stars, are more massive than novae progenitors and hence the Li enrichment that they produce could occur earlier in the evolution of the Galaxy. Scalo (1976) already concluded that steady mass loss in such stars could give an important contribution to the amount of Li in the Galaxy. The solution to the problem of Li enrichment can be provided by these stars. A firm calibration of the Li yield in them, and better determinations of mass loss rates should be pursued. If the enrichment is confirmed we have a strong case for the Pop II value of the primordial Li abundance.

ACKNOWLEDGEMENTS

I would like to thank J. Beckman, R.J. García López and A. Magazzù for many valuable discussions.

REFERENCES

Abia, C., Canal, R.: 1988, *Astron. Astrophys.* **189**, 55.
Alcock, C., Fuller, G.M., Mathews, G.J.: 1987, *Astrophys. J.* **320**, 439.
Alexander, D.: 1975, *Astrophys. J. Suppl. Ser.* **29**, 363.
Alschuler, W.R.: 1975, *Astrophys. J.* **195**, 649.
Anders, E., Grevesse, N.: 1989, *Geochim. Cosmochim. Acta* **53**, 197.
Andersen, J., Gustafsson, B., Lambert, D.L.: 1984, *Astron. Astrophys.* **136**, 65.
Applegate, J.H., Hogan, C.J., Scherrer, R.J.: 1988, *Astrophys. J.* **329**, 572.
Arnould, M.: 1986, *Prog. Part. Nucl. Phys.* **17**, 305.
Audouze, J., Bouladé, O., Malinie, G., Poilane, Y.: 1983, *Astron. Astrophys.* **127**, 164.
Audouze, J., Silk, J.: 1989, *Astrophys. J. Lett.* **342**, L5.
Baade, D., Magain, P.: 1988, *Astron. Astrophys.* **194**, 237.
Baglin, A., Morel, P.J., Schatzman, E.: 1985, *Astron. Astrophys.* **149**, 309.
Balachandran, S.: 1989, in Proc. of the 5th Cambridge Workshop *Cool stars, Stellar Systems and the Sun*, Seattle, Washington.
Balachandran, S.: 1990, *Astrophys. J.* in press.
Balachandran, S., Lambert, D.L., Stauffer, J.R.: 1988, *Astrophys. J.* **333**, 267.
Balsiger, H., Geiss, J., Groegler, N., Wyttenbach, A.: 1968, *Earth Planet. Sci. Letters* **5**, 17.
Beckman, J.E., Rebolo, R.: 1988, in IAU Symp. 132, *The Impact of Very High S/N Spectroscopy in Stellar Physics*, G. Cayrel de Strobel and M. Spite (eds.), p. 473.
Beckman, J.E., Abia, C., Rebolo, R.: 1989, *Astrophys. Sp. Sci.* **157**, 41.
Bertout, C.: 1989, *Ann. Rev. Astron. Astrophys.* **27**, 351.
Bodenheimer, P.: 1965, *Astrophys. J.* **142**, 451.
Boesgaard, A.M.: 1976a, *Pub. Astron. Soc. Pacific.* **88**, 353.
Boesgaard, A.M.: 1976b, *Astrophys. J.* **210**, 466.
Boesgaard, A.M.: 1985a, *Pub. Astron. Soc. Pacific.* **97**, 37.
Boesgaard, A.M.: 1985b, *Pub. Astron. Soc. Pacific.* **97**, 784.
Boesgaard, A.M.: 1987, *Astrophys. J.* **321**, 967.
Boesgaard, A.M., Budge, K.G.: 1988, *Astrophys. J.* **332**, 410.
Boesgaard, A.M., Budge, K.G.: 1989, *Astrophys. J.* **338**, 875.

Boesgaard, A.M., Budge, K.G., Burck, E.E.: 1988, *Astrophys. J.* **325**, 749.
Boesgaard, A.M., Budge, K.G., Ramsay, M.E.: 1988, *Astrophys. J.* **327**, 389.
Boesgaard, A.M., Chesley, S.E.: 1976, *Astrophys. J.* **210**, 475.
Boesgaard, A.M., Heacox, W.D.: 1978, *Astrophys. J.* **226**, 888.
Boesgaard, A.M., Heacox, W.D., Conti, P.S.: 1977 *Astrophys. J.* **214**, 124.
Boesgaard, A.M., Steigman, G.: 1985, *Ann. Rev. Astron. Astrophys.* **23**, 319.
Boesgaard, A.M., Tripicco, M.J.: 1986a, *Astrophys. J.* **303**, 724.
Boesgaard, A.M., Tripicco, M.J.: 1986b, *Astrophys. J. Lett.* **302**, L49.
Bonsack, W.K.: 1959, *Astrophys. J.* **130**, 843.
Bonsack, W.K., Greenstein, J.L.: 1960, *Astrophys. J.* **131**, 83.
Boyd, R.N., Kajino, T.: 1989, *Astrophys. J.* **336**, L55.
Brown, L., Schramm, D.N.: 1988, *Astrophys. J. Lett.* **329**, L103.
Brown, J.A., Sneden, C., Lambert, D.L., Dutchover, E.Jr.: 1989, *Astrophys. J. Supp. Ser.* **71**, 293.
Butler, R.P., Cohen, R.D., Duncan, D.K., Marcy, G.W.: 1987, *Astrophys. J. Lett.* **319**, L19.
Cameron, A.G.W., Fowler, W.A.: 1971, *Astrophys. J.* **164**, 111.
Cayrel, R., Cayrel de Strobel, G., Campbell, B., Däppen, W.: 1984, *Astrophys. J.* **283**, 205.
Cayrel de Strobel, G., Cayrel, R.: 1989 *Astron. Astrophys. Lett.* **218**, L9.
Charbonneau, P., Michaud, G.: 1988, *Astrophys. J.* **334**, 746.
Chmielewski, Y., Müller, E.A., Brault, J.W.: 1975, *Astron. Astrophys.* **42**, 37.
D'Antona, F., Mazzitelli,I.: 1984, *Astron. Astrophys.* **138**, 431.
Danziger, I.J., Conti, P.S.: 1965, *Astrophys. J.* **146**, 392.
Danziger, I.J.: 1967, *Astrophys. J.* **150**, 733.
Deliyannis, C.P., Demarque, P., Kawaler, S.D.: 1990, *Astrophys. J.* in press
Dicke, R.H.: 1972, *Astrophys. J.* **171**, 331.
Dimopoulos, S., Esmailzadeh, R., Hall, L.J., Starkman, G.D.: 1988, *Phys. Rev. Lett.* **60**, 7.
Dravins, D., Hultqvist, L, : 1977, *Astron. Astrophys.* **55**, 463.
Duncan, D.K.: 1981, *Astrophys. J.* **248**, 651.
Duncan, D.K., Jones, B.F.: 1983, *Astrophys. J.* **271**, 663.
Ferlet, R., Dennefeld, M.: 1984, *Astron. Astrophys.* **138**, 303.
García López, R.J., Rebolo, R., Beckman, J.: 1988, *Pub. Astron. Soc. Pacific.* **100**, 1489.
Gilmore, G., Reid, I.N.: 1983, *Monthly Notices Roy. Astron. Soc.* **202**, 1025.
Gratton, R.G., D'Antona, F.: 1989, *Astron. Astrophys.* **215**, 66.
Greenstein, J.L., Richardson, R.S.: 1951, *Astrophys. J.* **113**, 536.
Greenstein, J.L., Tandberg-Hanssen, E.: 1954, *Astrophys. J.* **119**, 113.
Hänni, L.: 1984, *Soviet Astr. Lett.* **10**, 51.
Hartigan, P., Hartmann, L., Kenyon, S.J., Hewett, R. ,Stauffer, J.: 1989, *Astrophys. J. Suppl. Ser.* **70**, 899.
Herbig, G.H.: 1962, *Advances Astron. Astrophys.* **1**, 47.
Herbig, G.H.: 1965, *Astrophys. J.* **141**, 588.
Herbig, G.H., Wolff, R.J.: 1966, *Ann. Astrophys.* **29**, 593.
Hobbs, L.M.: 1984, *Astrophys. J.* **286**, 252.
Hobbs, L.M.: 1985, *Astrophys. J.* **290**, 284.

Hobbs, L.M., Duncan, D.K.: 1987, *Astrophys. J.* **317**, 796.
Hobbs, L.M., Iben, I.Jr., Pilachowski, C.A.: 1989, *Astrophys. J.* **347**, 817.
Hobbs, L.M., Pilachowski, C.A.: 1986a, *Astrophys. J. Lett.* **309**, L17.
Hobbs, L.M., Pilachowski, C.A.: 1986b, *Astrophys. J. Lett.* **311**, L37.
Hobbs, L.M., Pilachowski, C.A.: 1988a, *Astrophys. J.* **334**, 734.
Hobbs, L.M., Pilachowski, C.A.: 1988b, *Astrophys. J. Lett.* **326**, L23.
Huebner, W.F., Merts, A.L., Magee, N.H.Jr., Argo, M.F.: 1977, *Los Alamos Report*, LA-6760-M.
Hunger, K.: 1957, *Astron. J.* **62**, 294.
Iben, I.Jr.: 1967, *Astrophys. J.* **147**, 624.
Kawaler, S.D.: 1988, *Astrophys. J.* **333**, 236.
Kurki-Sounio, H., Matzner, R.A.: 1989, *Phys. Rev. D.* **37**, 1380.
Kohl, J.L., Parkinson, W.H., Withbroe, G.L.: 1977 *Astrophys. J. Lett.* **212**, L101.
Lambert, D.L., Dominy, J.F., Sivertsen, S.: 1980, *Astrophys. J.* **235**, 114.
Lambert, D.L., Sawyer, S.R.: 1984, *Astrophys. J.* **283**, 192.
Law, W.Y., Knobloch, E., Spruit, H.C.: 1984, in *Observational tests of Stellar Evolution Theory*, A. Maeder and A. Renzini (eds.), Reidel, Dordrecht p. 523.
Luck, R.E., Lambert, D.L.: 1982, *Astrophys. J.* **256**, 189.
Magazzù, A., Rebolo, R.: 1989, *Mem. Soc. Astr. It.* **60**, 105.
Malaney, R.A., Fowler, W.A.: 1988, *Astrophys. J.* **333**, 14.
Malaney, R.A., Fowler, W.A.: 1989, *Astrophys. J. Lett.* **345**, L5.
Mason, B.: 1971, *Handbook of Elemental Abundances in Meteorites*, New York, Gordon-Breach.
Maurice, E., Spite, F., Spite, M.: 1984, *Astron. Astrophys.* **132**, 278.
Mazzei, P., Pigatto, L.: 1989, *Astron. Astrophys. Lett.* **213**, L1.
Meneguzzi, M., Audouze, J., Reeves, H.: 1971, *Astron. Astrophys.* **15**, 337.
Meneguzzi, M., York, D.G.: 1980, *Astrophys. J. Lett.* **235**, L111.
Michaud, G.: 1986, *Astrophys. J.* **302**, 650.
Michaud, G., Fontaine, F., Baudet,G.: 1984, *Astrophys. J.* **282**, 206.
Molaro, P.: 1987, *Astron. Astrophys.* **183**, 241.
Molaro, P., Beckman, J.E.: 1984, *Astron. Astrophys.* **139**, 394.
Müller, E.A., Peytremann, E., de la Reza, R.: 1975, *Solar Phys.* **41**, 53.
Pagel, B.E.J.: 1988, in *Origin and Distribution of the Elements*, ed. G.J. Mathews, World Scientific, Singapore, p. 253.
Pallavicini, R., Cerruti-Sola, M., Duncan, D.K.: 1987, *Astron. Astrophys.* **174**, 116.
Parthasarathy, M., Sneden, C., Böhm-Vitense, E.: 1984, *Pub. Astron. Soc. Pacific.* **96**, 44.
Peebles, P.J.E.: 1966, *Phys. Rev. Lett.* **16**, 410.
Pilachowski, C.A.: 1986, *Astrophys. J.* **300**, 289.
Pilachowski, C.A., Booth, J., Hobbs, L.M.: 1987, *Pub. Astron. Soc. Pacific.* **99**, 1288.
Pilachowski, C.A., Hobbs, L.M., De Young, D.S.: 1989, *Astrophys. J. Lett.* **345**, L39.
Pilachowski, C.A., Saha, A., Hobbs, L.M.: 1988, *Pub. Astron. Soc. Pacific.* **100**, 474.
Pinsonneault, M.H., Kawaler, S.D., Sofia, S., Demarque, P.: 1989, *Astrophys. J.* **338**, 424.
Proffitt, C.R., Michaud, G.: 1989, *Astrophys. J.* **346**, 976.

Rebolo, R., Beckman, J.E.: 1988, *Astron. Astrophys.* **201**, 267.
Rebolo, R., Beckman, J.E., Molaro, P.: 1987, *Astron. Astrophys.* **172**, L17.
Rebolo, R., Crivellari, L., Castelli, F., Foing, B., Beckman, J.E.: 1986, *Astron. Astrophys.* **166**, 195.
Rebolo, R., Molaro, P., Beckman, J.E.: 1988, . *Astron. Astrophys.* **192**, 192.
Rebolo, R., Molaro, P., Abia, C., Beckman, J.E.: 1988b, *Astron. Astrophys.* **193**, 193.
Reeves, H.: 1974, *Ann. Rev. Astron. Astrophys.* **12**, 437.
Reeves, H., Fowler, W.A., Hoyle, F.: 1970, *Nature* **226**, 727.
Reeves, H., Meyer, J.P.: 1978, *Astrophys. J.* **226**, 613.
Ross, J., Aller, L.H.: 1971, *Solar Phys.* **36**, 11.
Ryan, S.G., Bessell, M.S., Sutherland, R.S., Norris, J.E.: 1990, *Astrophys. J. Lett.* **348**, L57.
Sackmann, I.J., Smith, R.L., Despain, K.H.: 1974, *Astrophys. J.* **187**, 555.
Sahu, K.C., Sahu, M., Pottasch, S.R.: 1988, *Astron. Astrophys. Lett.* **207**, L1. parSandage, A., Fouts, G.: 1987, *Astron. J.* **92**, 74.
Sandford, R.F.: 1947, *Pub. Astron. Soc. Pacific.* **59**, 134.
Scalo, J.M.: 1976, *Astrophys. J.* **206**, 795.
Scalo, J.M., Despain, K.H., Ulrich, R.K.: 1975, *Astrophys. J.* **196** 809.
Schatzman, E.: 1962, *Ann. Astrophys.* **25**, 18.
Schatzman, E.: 1969, *Astron. Astrophys.* **3**, 331.
Schatzman, E.: 1977, *Astron. Astrophys.* **56**, 211.
Schatzman, E.: 1989, in *Turbulence and non-linear dynamics in magnetohydrodynamic flows*, ed. P.L. Sulem, July 4-9, 1988, Cargèse (Corsica).
Schatzman, E., Maeder, A., Angrand, F., Glowinski, R.: 1981, *Astron. Astrophys.* **96**, 1.
Smith, V.V., Lambert, D.L.: 1986, *Astrophys. J.* **311**, 843.
Smith V., Lambert, D.: 1989, *Astrophys. J. Lett.* **345**, L75.
Soderblom, D.R.: 1983, *Astrophys. J. Suppl. Ser.* **53**, 1.
Spiegel, E.A.: 1968, in *Highlights of Astronomy*, ed. L. Perek, Dordrecht: Reidel), p. 261.
Spite, F., Spite, M.: 1982, *Astron. Astrophys.* **115**, 357.
Spite, F., Spite, M.: 1986, *Astron. Astrophys.* **163**, 140.
Spite, M., Maillard, J.P., Spite, F.: 1984, *Astron. Astrophys.* **141**, 56.
Spite, F., Spite, M., Peterson, R.C., Chaffee, F.M.Jr.: 1987a, *Astron. Astrophys. Lett.* **171**, L8.
Spite, M., Spite, F., Peterson, R.C., Chaffee, F.M.Jr.: 1987b, *Astron. Astrophys. Lett.* **172**, L9.
Spruit, H.: 1987, in *The Internal Solar angular Velocity* eds. B.R. Durney and S. Sofia, D. Reidel, Dordrecht.
Stauffer, J., Hartmann, L.W., Jones, B.F., McNamara, B.R.: 1989, *Astrophys. J.* **342**, 285.
Steenbock, W., Holweger, H.: 1984, *Astron. Astrophys.* **130**, 319.
Stellmacher, G., Wiehr, E.: 1971, *Solar Phys.* **21**, 96.
Strauss, J.M., Blake, J.B., Schramm, D.N.: 1976, *Astrophys. J.* **204**, 481.
Strom, K.M., Wilkin, F.P., Strom, S.E., Seaman, R.L.: 1989, *Astron. J.* **98**, 1449.
Swenson, F.J., Stringfellow, G.S., Faulkner, J.: 1990, *Astrophys. J. Lett.* **348**, L33.

Teresawa, N., Sato, K.: 1989, University of Tokyo, preprint. parTraub, W.A., Carleton, N.P.: 1973, *Astrophys. J. Lett.* 184, L11.
Twarog, B.A., Anthony- Twarog, B.J.: 1989, *Astron. J.* **97**, 759.
VandenBerg, D.A.: 1985, *Astrophys. J. Suppl. Ser.* **58**, 711.
VandenBerg, D.A., Poll, H.E.: 1989, *Astron. J.* **98**, 1451.
Vanden Bout, P.A., Snell, R.L., Vogt, S.S., Tull, R.G.: 1978, *Astrophys. J.* **221**, 598.
Vauclair, S.: 1988, *Astrophys. J.* **335**, 971.
Vauclair, S., Vauclair, G., Schatzman, E., Michaud, G.: 1978, *Astrophys. J.* **223**, 567.
Walker, T.P., Mathews, G.J., Viola, V.E.: 1985, *Astrophys. J.* **299**, 745.
Wagoner, R.V., Fowler, W.A., Hoyle, F.: 1967, *Astrophys. J.* **148**, 3.
Wallerstein, G., Sneden, C.: 1982, *Astrophys. J.* **255**, 577.
Wallerstein, G., Herbig, G.H., Conti, P.S.: 1965, *Astrophys. J.* **141**, 610.
Walter, F.M., Brown, A., Mathieu, R.D., Myers, P.C., Vrba, F.J.: 1988, *Astron. J.* **96**, 297.
Weber, E.J., Davis, L.Jr.: 1967, *Astrophys. J.* **148**, 217.
Weymann, R., Sears, R.L.: 1965, *Astrophys. J.* **142**, 174.
Willson, L.M., Bowen, G.H., Struck-Marcell, C.: 1987, *Comments Astrophys.* **12**, 17.
Yang, J., Turner, M.S., Steigman, G., Schramm, D.N., Olive, K.A.: 1984, *Astrophys. J.* **281**, 493.
York, D.G., Meneguzzi, M., Snow, T.P.: 1982, *Astrophys. J.* **255**, 524.
Zahn, J.P.: 1987, in *The Internal Solar Angular Velocity*, eds. B.R. Durney and S. Sofia (Dordrecht: Reidel) p. 201.
Zappala, R.R.: 1972, *Astrophys. J.* **172**, 57.

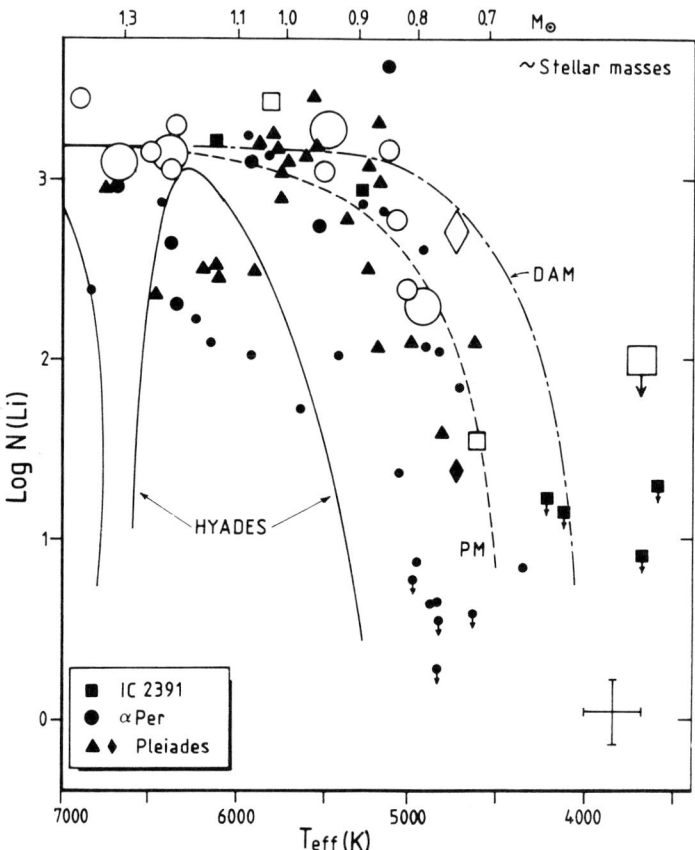

Figure 1.-

Lithium abundances versus effective temperature for stars in the young open clusters: IC 2391 (squares), α Per (circles) and the Pleiades (triangles and diamonds). The data were taken from Stauffer et al. (1989), Balachandran, Lambert and Stauffer (1988), Duncan and Jones (1983) and Butler et al. (1987). The diamonds are averages of the Li measurements in four rapid and four slow rotators in the Pleiades (Butler et al. 1987). Open and filled symbols represent projected rotational velocities (v sini) higher and lower than 50 Km s^{-1}, respectively. Approximate stellar masses are given in the top of the diagram. Two curves of predicted Li pre-main sequence depletion are shown: *dashed* line, D'Antona and Mazzitelli's (1984; DAM) Li isochrones for an age 10^{9}yr. without extra mixing; *dot-dashed* line, Li depletion curve at 7 10^7 yr by Proffitt and Michaud (1989). The solid line represents a fit to Li data in the Hyades (see Figure 3). A representative error bar for Log N(Li) and T$_{eff}$ is shown in the lower right-hand corner.

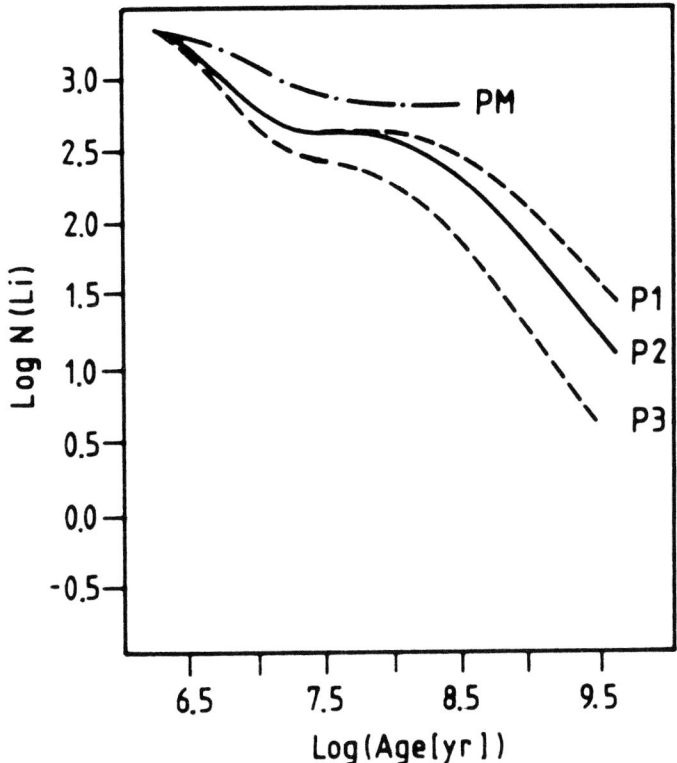

Figure 2.-

Li depletion as a function of time for 1 $M_\odot$ stars with solar metallicity. Rotating models computed by Pinnsoneault et al. (1989) are labelled with P1, P2 and P3. The solid line (P2) is their "best" solar model (initial angular momentum $J_0=1.63\ 10^{50}$ gcm^{2}s^{-1}). P1 and P3 are models with J_0 values of $5\ 10^{49}$ and $5\ 10^{50}$ gcm^{2}s^{-1}, respectively. The *dot-dashed* line is the pre-Main Sequence Li destruction curve computed by Proffitt and Michaud (1989) for 1 $M_\odot$ ($Z=0.024$, $\alpha=1.5$).

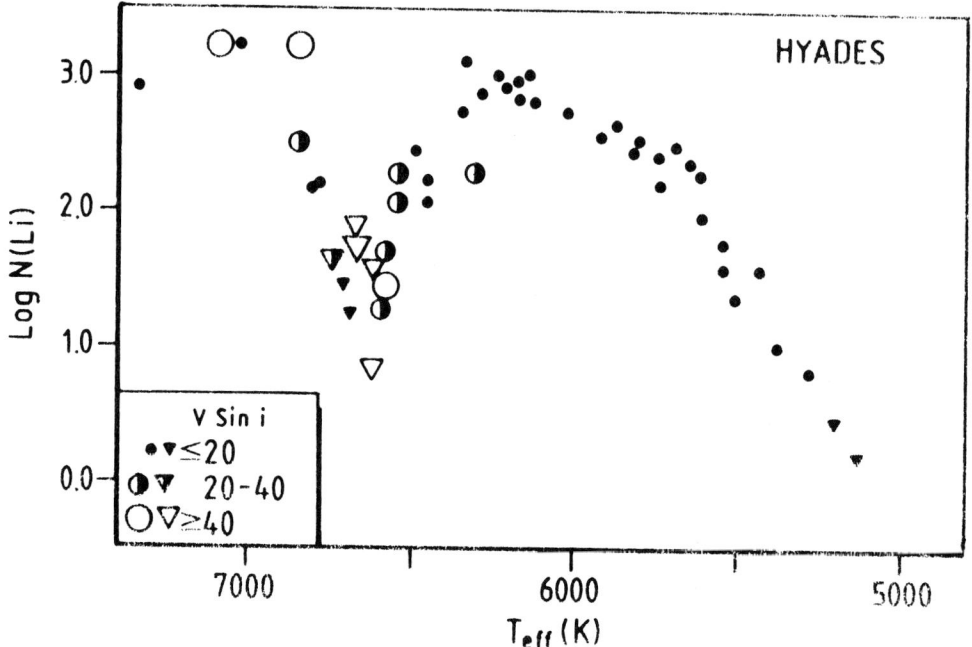

Figure 3.-

Lithium abundances versus effective temperatures and rotational velocities for Hyades stars. Circles are measurements and triangles upper limits. Small filled, semi-filled and large open symbols refer to rotational velocities vsin i as described in the legend at the lower left-hand corner. The data were taken from Boesgaard and Tripicco (1986b), Boesgaard and Budge (1988), Cayrel et al. (1984) and Rebolo and Beckman (1988). Measurements in the Li gap include the upward corrections reported by Boesgaard and Budge (1988).

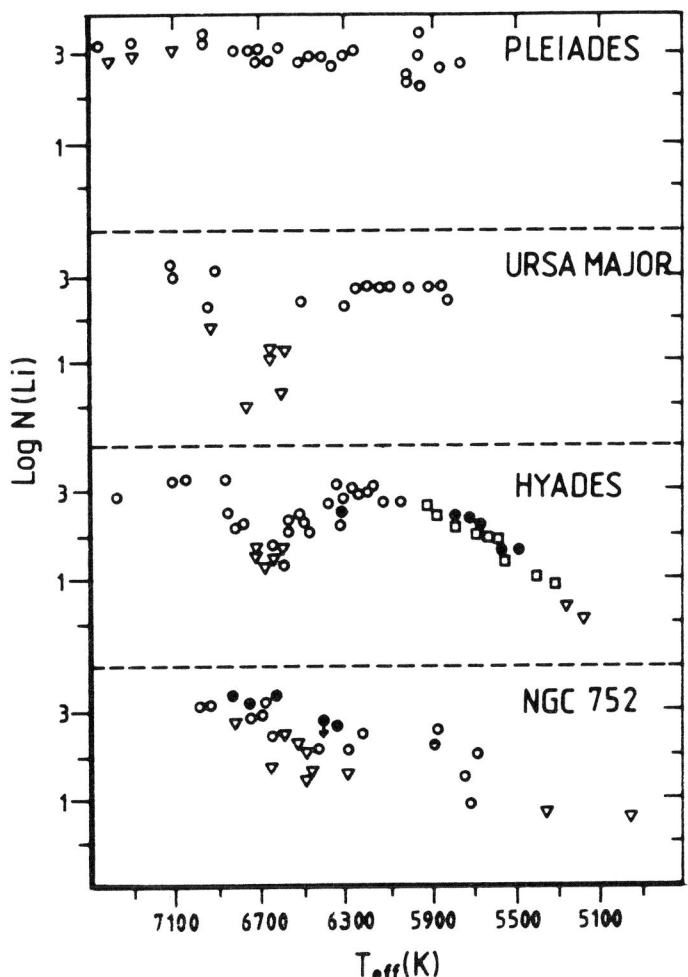

Figure 4.-
Lithium abundances as a function of effective temperature for MS stars in young and intermediate old open clusters. Dots, circles and squares represent measured values; triangles are upper limits. Pleiades data are taken from Pilachowski, Booth and Hobbs (1987) and Boesgaard, Budge and Ramsay (1988). Ursa Major data are taken from Boesgaard, Budge and Burck (1988). Hyades data are as in Figure 3. NGC 752 data are taken from Hobbs and Pilachowski (1986a), Pilachowski and Hobbs (1988) and Beckman and Rebolo (1988).

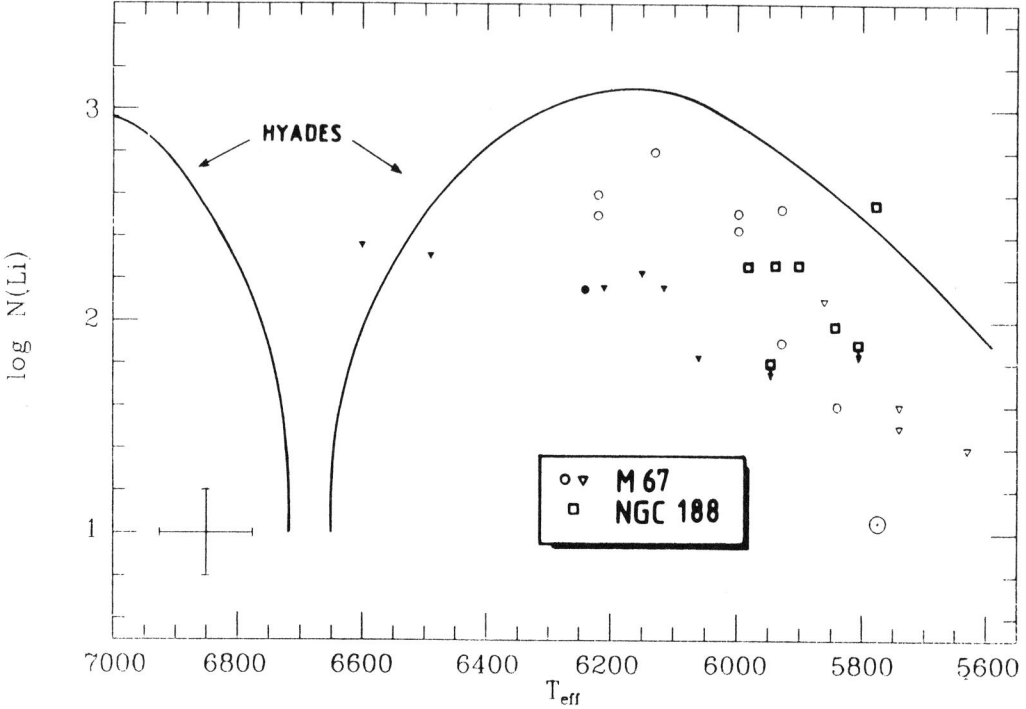

Figure 5.-

Lithium abundances against effective temperatures for all MS stars so far observed in M 67 and NGC 188. Circles (M 67 data) and squares (NGC 188) are measured values. Triangles and arrows attached to symbols indicate upper limits. The solar value is represented by the conventional symbol. The sources of data are Hobbs and Pilachowski (1986b), Spite et al. (1987) and Garcia López, Rebolo and Beckman (1988a). The solid line represents the observed Li depletion curve in the Hyades.

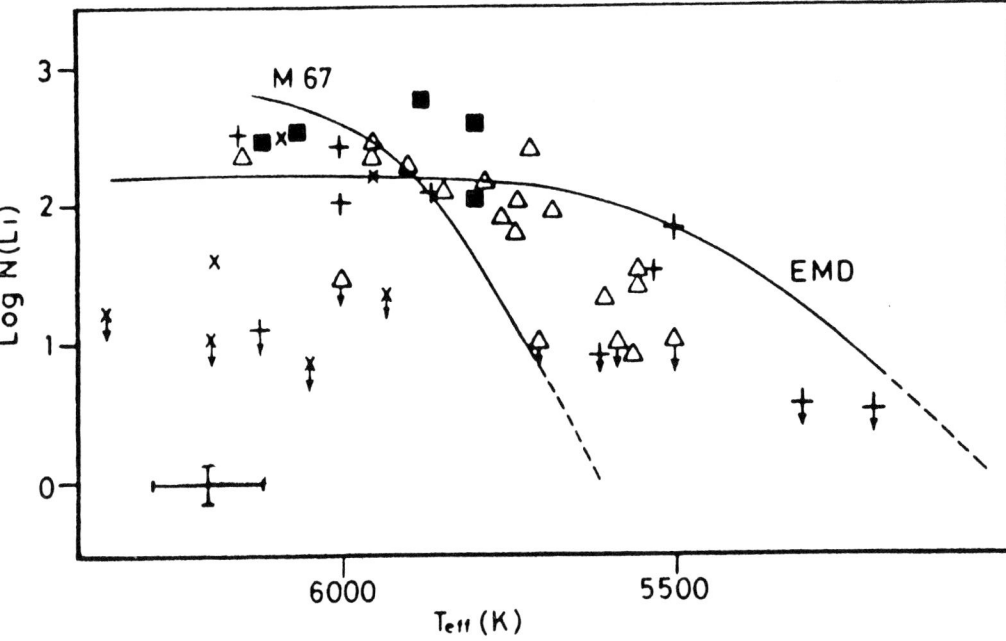

Figure 6.-
Lithium abundances in stars with metallicities in the range $-0.3 \geq [Fe/H] \geq -1.3$ in the effective temperature range 5000 - 6300 K. Sources of data: filled squares, Duncan (1981); pluses, Spite and Spite (1982); crosses, Boesgaard and Tripicco (1986a); Triangles, Rebolo, Molaro and Beckman (1988). Also included for comparison are fits to the data for extreme metal deficient stars (EMD, see text for definition) and for M 67 (previous Figure).

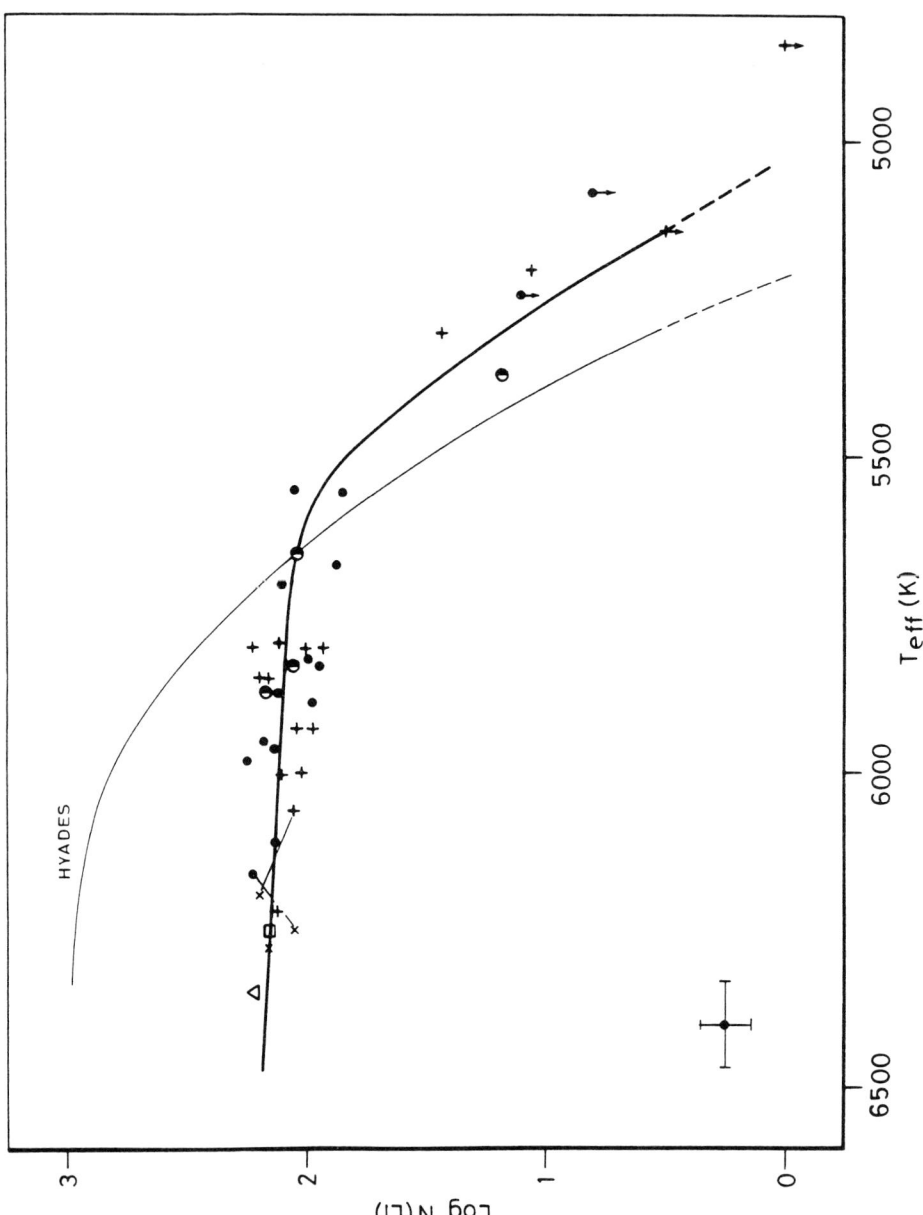

Figure 7.-
Lithium abundances versus effective temperatures for stars with [Fe/H]≤-1.4. All of them satisfy the kinematics requirements for halo population. Sources of data: dots, Rebolo, Molaro and Beckman (1988) (semifilled circles indicate stars with measurements also reported by Spite et al. (1984)); pluses, Spite and Spite (1982, 1986), Spite et al. (1984, 1987b); open triangle, measurement in G64-12 ([Fe/H]=-3.5) Rebolo et al. (1987); open square, Boesgaard (1985); crosses, measurement in LP 608-62 (Hobbs and Duncan ,1987) and the two stars reported in Hobbs and Pilachowski (1988b) which are connected by thin solid lines to measurements by other authors in the same stars.

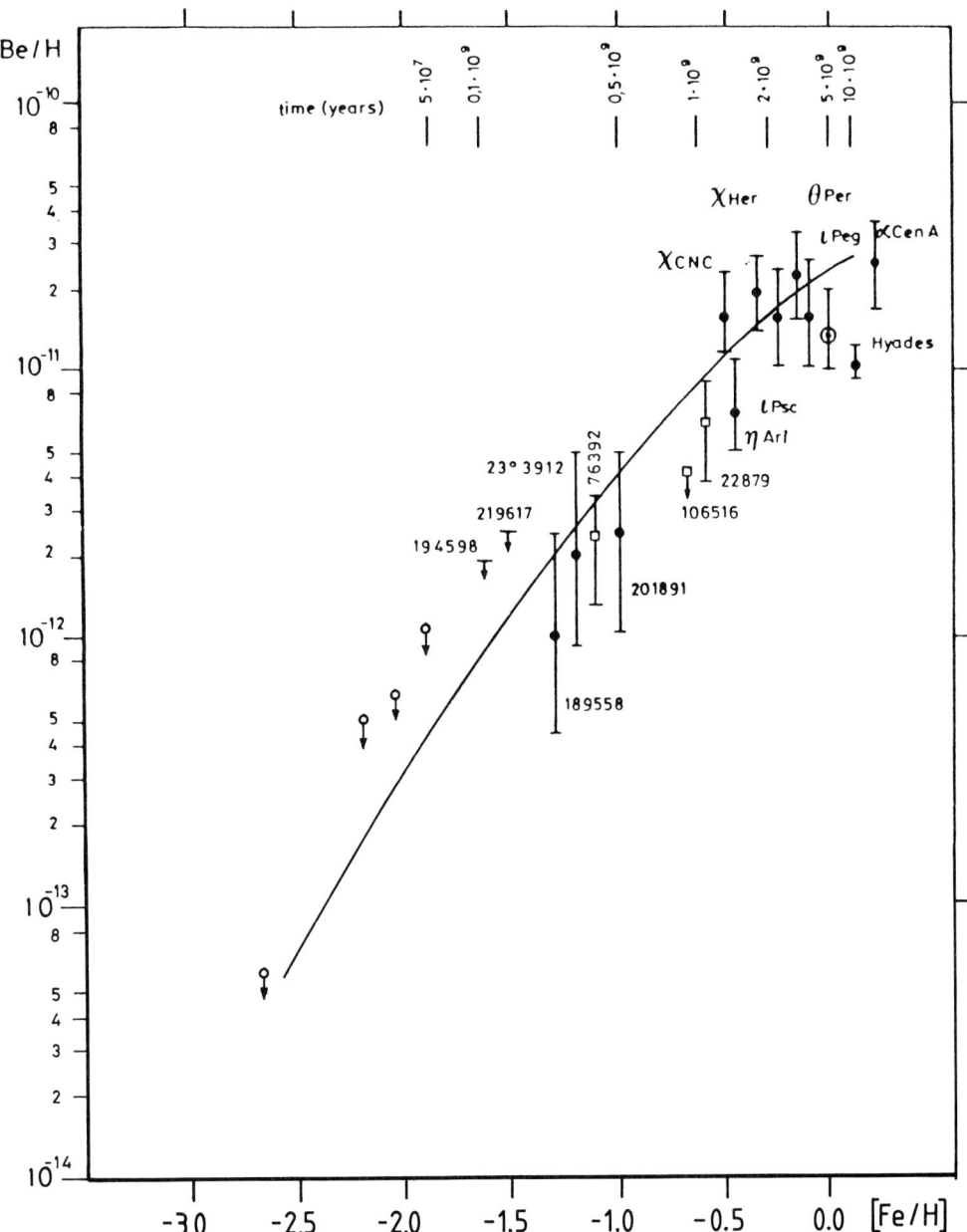

Figure 8.-
Beryllium abundance (Be/H), versus metallicity for stars over a wide metallicity range. Dots and squares are measured values, and arrows attached to symbols indicate upper limits. At metallicities below -0.5, the sources of data are: dots, Rebolo et al. (1988b); open squares, Beckman, Abia and Rebolo (1989) and open circles (all are upper limits), Ryan et al. (1990). A typical error bar in [Fe/H] is ±0.2 dex. The continuous line represents a theoretical model for the galactic evolution of Be (Abia and Canal, 1988), which reproduces well the observations. Age, shown as an alternative abscissa, was also derived from this model.

On the Ages of Galactic Globular Clusters

F. Fusi Pecci[1] and C. Cacciari[2]

[1] Dipartimento di Astronomia, Bologna, Italy
[2] Osservatorio Astronomico, Bologna, Italy

1. INTRODUCTION

Only a few types of objects, a part from the universe itself, are suitable to estimate cosmological ages. We have to look in fact at constituents of the universe whose *time constants* are neither too short — as they would give information only on recent and specific events, nor too long — as no detectable change might have occurred yet. In this respect, globular clusters are among the best subjects to study, as there are strong observational indications suggesting that a *complete and populated* main sequence has existed since the beginning of the cluster lifetime, whose characteristics have *significantly changed* with stellar evolution.

Very schematically, this means that we use the stars themselves as *clocks* to gauge the age of the whole cluster (Renzini 1985). Actually, what we currently do is to determine "the time it takes for a supposed coeval main sequence to change into the complex array of stars in a color magnitude diagram that we observe today" (Edwards 1982). In other words, we have to *evolve a model* until it best reproduces all the *present* observed characteristics of the cluster. We can thus say that *stellar clocks* are intrinsically based on theoretical models; hence, the *verification* of the validity of these models is *complementary* but, at the same time, *strictly necessary* to drawing firm conclusions from any procedure of clusters' age determination.

The main goal of much of the research on globular clusters is to find reliable answers to two basic questions: *how old is the oldest globular cluster* and *are all the globular clusters coeval?* Although globular clusters (GCs) do not contain the very first stars formed in the Galaxy as they are *somewhat* "metal enriched", the oldest cluster sets a strong lower limit to the age of the universe, while the spread in clusters' age constrains the models of formation and early evolution of the Galaxy.

Several papers have recently analysed and discussed in detail the problems related to these two items, for example Bolte (1989), Buonanno *et al.* (1989a,b), Chieffi and Straniero (1989), Fusi Pecci *et al.* (1989), Lee *et al.* (1989), Rood (1989), Sandage and Cacciari (1989), Sarajedini and King (1989), VandenBerg (1989), VandenBerg and Durrell (1989).

2. AGE DETERMINATION: DEFINITION OF THE PROBLEM

Adopting an extremely simplified general approach, one can say that any measure of the age of a cluster rests on two specific items:

• the **CLOCK READING** in the observational plane (CMD) (for instance the (V,B−V)-plane) where one has:

CLOCK: Main Sequence Turnoff (TO) (∗)

OBSERVABLES: V_{TO}, $(B-V)_{TO}$

SPECIFIC PROBLEMS:

○ Measure the values of the observables and of any other "reference" parameter (like for instance the apparent luminosity of the unevolved horizontal branch stars at the instability strip $-V_{HB}-$) with the highest possible accuracy.

○ Determine the cluster distance.

(∗) NOTE: the cluster luminosity function has been occasionally proposed as *alternative optimal* clock (Paczyński 1984, Ratcliff 1987, Demarque 1988). However, for several reasons (cf. Renzini 1986,1988), its use seems to be inadequate to achieve datations to better than ±5 Gyr.

• the **CLOCK RUNNING** in the theoretical plane (HRD), the (log L, log T_e) plane where one has:

CLOCK: Main Sequence Turnoff (TO)

OBSERVABLES: log L_{TO}, log $T_{e,TO}$

SPECIFIC PROBLEMS:

○ Both L_{TO} and $T_{e,TO}$ depend on other parameters beside age (t):
$$L_{TO} = L_{TO}(t, Y, Z)$$
$$T_{e,TO} = T_{e,TO}(t, Y, Z, \alpha)$$

where Y is the helium abundance, Z the metal content, and α the mixing length parameter involved in the treatment of convection. This implies that *a)* the correct determination of the age requires the precise knowledge of the chemical abundances, and *b)* ages derived using luminosities are generally more reliable than those based on temperatures, as the relationship linking the latter to age depends also on α, and the dependence on Z is stronger.

o The models must be *calibrated* and, in particular, α has to be fixed using "known" reference objects (mostly, the Sun).

o The validity of the models and of the transformation to the observational plane must be carefully checked.

In synthesis, besides the photometric observables, *in order to get "absolute" ages one has necessarily to know the cluster distance (and reddening), the helium and metal abundances, and the value of α.* The determination of "relative" ages is somewhat easier, but not so much, as we will see in the following.

3. UNCERTAINTIES AND ERRORS ADOPTING THE "STANDARD AND CANONICAL" MODELS

Many ingredients are necessary to build up stellar evolutionary models; some of them are reasonably well known, others are "somehow" parametrized, many are simply neglected. We speak about "standard and canonical" models when we assume the validity of the "standard" input physics (*i.e.* opacities, reaction rates, neutrino losses, equation of state, and the like) and the adequacy of the "canonical" assumptions which neglect phenomena such as rotation, mixing of nonconvective origin, and exotic particles effects. Although they differ in many "details" (which, as a matter of fact, might be crucial), most of the available stellar evolutionary models can be considered to be "standard and canonical". In particular, such a label is usually attributed to those computed by Iben and Rood, VandenBerg, the people at Yale and Frascati, and by the Danish group. The detailed comparison of the different sets is beyond the purpose of the present paper; latest useful analyses, discussions and references can be found for instance in the Proceedings of the meeting on "Calibration of Stellar Ages" held in Middletown (Philip 1988, editor) and in Chieffi and Straniero (1989).

A caveat: Each set of evolutionary models has been computed using a *tool* based on different input physics, assumptions, and codes. As recalled for instance recently by Renzini and Fusi Pecci (1988) and Fusi Pecci *et al.* (1989), this implies that such tool *must* first be *tested*, and then *calibrated and tuned* by adjusting basic parameters like Y and α to fit (usually) the observed properties of the Sun and a few other specific targets. As a consequence, each model yields only *differential* results with respect to the adopted "calibrators", and variations of the input parameters are only allowed at *constant Sun*.

Assuming however for the moment that no significant discrepancy between model predictions and observations exists, one can quite easily figure out the size of the errors induced on the derived age by the uncertainties affecting the various parameters involved in the procedure:

A) *TO-Luminosity*

$$\frac{dLogt_9}{d\log L_{TO}} \sim -1 \quad \Rightarrow \quad \frac{dLogt_9}{dM_{bol,TO}} \sim 0.4$$

This means that an error of 0.1 mag in the TO bolometric magnitude translates into a corresponding error of about 10% on the derived age, which is equivalent to about 1.5 Gyr. Since a "conservative" estimate of the present uncertainties in the measures of the distance moduli of most Galactic globular clusters alone (not to speak of the errors due to reddening and BC estimates) is hardly less than ~ 0.2 mag, it is immediate to say that errors up to $3 - 4$ Gyr are "normal".

B) *Helium Abundance*

$$\frac{dLogt_9}{dY} \sim 0.5$$

Most estimates of the helium abundance in Galactic globular clusters agree on the conclusion that the helium abundance of main sequence stars $-Y_{MS}-$ is practically coincident with the primordial one $-Y_p-$, at least for the most metal poor objects (see for instance Buzzoni et al. 1983), and is $Y_p = 0.23 - 0.24$ (Boesgaard and Steigman, 1985). However, the possibility that more metal rich clusters have slightly higher values of Y_{MS} cannot be totally excluded on the basis of the available data. Adopting for example $\delta Y \sim 0.03$, one gets an uncertainty on the age of about 3% (~ 0.5 Gyr).

C) *Metal Abundance* (with solar abundance ratios)

$$\frac{dLogt_9}{dLogZ} \sim 0.13$$

Even without worrying about the major systematic problems which affected the metallicity scale a few years ago (see Kraft 1985 and references therein), the *rms* errors affecting the mean metallicities of most globular clusters are typically $\delta Log Z \sim 0.1 - 0.2$. An uncertainty of $\delta Log Z = 0.1$ translates into an error on the age of about 3% (~ 0.5 Gyr).

D) *Color* (what follows is just indicative to show the possible influence of the errors in the reddening, the calibration of α, and the color transformations on age determinations based on colors).

From:
$$\frac{dLogt_9}{dM_{V,TO}} \sim 0.4 \quad \text{and} \quad \frac{dM_{V,MS}}{d(B-V)} \sim 7 \Rightarrow$$
$$\Rightarrow \frac{dLogt_9}{d(B-V)} \sim 2.8$$

one has that, when datation is based on colors, an error in $(B-V) = 0.01$ mag leads to $\delta t_9 \sim 1$ Gyr.

4. CLOCK READING

Since the TO luminosity is intrinsically a better age indicator than the TO temperature (see Sect. 2), it is crucial to reduce as much as possible the errors in the measure of the TO-absolute luminosity. A first "direct" improvement may be achieved by getting more precise measures of the *apparent* luminosity of the TO in the observed CMD. In the $V, B-V$-plane the TO-region of the Main Sequence (MS) is nearly vertical, which makes the TO proper location very hard. The use of a more extended baseline color (for instance $V-I$), or of different photometric bands (like the Gunn and/or the Strömgren systems) changes the shape of the TO region, and may help decrease the errors. Unfortunately, a significant improvement upon the present situation requires the reduction of the error in the measure of the TO apparent luminosity to less than $0.03 - 0.05$ mag; this aim seems beyond the reach of the current photometric techniques. On the other hand, even assuming a very precise knowledge of V_{TO}, the central problem remains the measure of the cluster distance using the best available *standard candle*.

4.1 Possible Standard Candles

The requirements the *optimal* standard candle must fulfil in a globular cluster are essentially those of being: *a)* sufficiently bright and numerous to be easily detected and measured, *b)* "well known", in the sense that the intrinsic properties and evolutionary status are properly understood, and *c)* as much as possible independent of parameters whose measure is uncertain or whose relationships with the absolute luminosity are dubious. Since what is essentially needed is to use *marked* objects or branches having a "well defined, known" luminosity, there are in principle many candidate candles spread out allover the CMD, for instance:

i) The brightest stars in the red giant branch (RGB) close to experience the helium flash. Stochastic effects on very poor samples have up to now prevented a precise check of the RGB-tip models which predict a very narrow luminosity range (∼ 0.1-0.2 mag) over the full metallicity interval covered by Galactic GCs (see Frogel 1983, Rood and Crocker 1985, Renzini and Fusi Pecci 1988, VandenBerg and Durrell 1989). However, the advent of "panoramic, high resolution" IR-detectors, will offer quite soon the required samples and the necessary calibration of this candle.

ii) The stars populating the horizontal branch (HB). Due to its "horizontal" placing and the presence of RR Lyrae variables, the HB has been the natural *reference branch* since the construction of the first CMDs. We will discuss in the following the many problems still affecting HB stars as distance indicators.

iii) The faint unevolved stars ($M_V \geq +5.5$) belonging to the Main Sequence. Although they are intrinsically faint, they can be detected with good accuracy in most globular clusters using large telescopes and the most advanced CCD detectors. The crucial uncertainties are here due to the lack of an appropriate sample of stars with known trigonometric parallaxes and covering a sufficiently wide range in metallicity (*i.e.* from [Fe/H] = 0 down to [Fe/H] ∼ -2.3), and to the difficulty of getting *unique* solutions from the fitting procedures (see for a discussion Buonanno *et al.* 1989a,b, hereafter referred to as BCF and BCCF).

iv) The white dwarfs' cooling line. Since the locus of the DA and $nonDA$ white dwarfs in a globular cluster CMD is expected to be very narrow (≤ 0.03 mag), Fusi Pecci and Renzini (1979) suggested that the ridge line of this locus could be used as a very precise "reference branch" for the *direct* determinations of relative distances, and also for getting absolute distances once a calibration based on "local" white dwarfs is available. This method, still hardly usable at present due to the very preliminary detection of white dwarfs in Galactic globulars (Richer and Fahlman 1987, 1988, Ortolani and Rosino 1988), might become much more effective as soon as new extensive data will be available, for instance with the *Hubble Space Telescope*.

In conclusion, although it is easy to foresee that in the very near future all the quoted candles will prove to be very fruitful, at present only HB and MS stars are sufficiently reliable distance indicators for Galactic globular clusters. We shall discuss them in more detail in the following Section.

4.2 The "basic" methods for the TO luminosity determination

Two different methods have been commonly used so far to derive the TO absolute luminosity and, in turn, the age of globular clusters: a *MS-fitting* procedure, and the so-called ΔV_{HB}^{TO}-*method*. We currently favour the second approach, which is however not free from significant problems.

4.2.1 MS-fitting

Main sequence fitting is a tricky procedure. It depends on so many parameters, assumptions, and consequently uncertainties (see Fig.11 of Sandage 1989), that one should not be surprised when rather different results are obtained following only slightly different procedures, and/or adopting different sets of observational data and theoretical isochrones. A detailed discussion of this method is given by BCF and BCCF. For the sake of example, we comment here on two results, which are also listed in Table 1. The first was found by BCF using a sample of 19 galactic globular clusters, and a main sequence fitting procedure based on VandenBerg and Bell's (1985) isochrones, calibrated on the few available metal-poor subdwarfs and solar abundance ratios ($[CNO/Fe] = 0$) for all the clusters. The second was obtained by King et al. (1988) via a similar procedure based on the Revised Yale Isochrones (Green et al. 1987). As stressed by BCCF and Sandage and Cacciari (1989), the significant difference of the above two results seems to be due mainly to the different fitting procedures.

From the inspection for instance of Fig. 9 in Hesser et al. (1987), one can see that the use of CCD cameras and of powerful data reduction packages (such as DAOPHOT, ROMAFOT, etc.) have dramatically improved the quality of globular cluster CMDs. Nevertheless, a *totally free* MS-fitting procedure (*i.e.* a procedure where the basic parameters like chemical compositions, distance modulus and age are determined at the same time) does not yield "unique" solutions yet. In fact, limiting the discussion to the MS and TO regions only, it is clearly impossible to choose the "best" solution among the various fits presented in Figs.18−19 of Hesser et al., or in Fig. 4 of Green (1988). The *simultaneous* fit of the whole CMD (*i.e.* using also the giant branch and, specially, the HB) limits more effectively the range of possible parameters, but a significant *liberty* still remains. A statement quite frequently put forward to support this method is that "overall fits" make use of the "global" available information, rather than resting on the comparison of *just a few* (maybe one, the TO) points. However, as shown for instance by Vanden-Berg (1983) and discussed by Renzini and Fusi Pecci (1988), the *shapes* of both theoretical isochrones and globular cluster loci are so sensitive to parameters such as mixing length, transformations of the isochrones into the observational plane, chemical abundances and "non solar" abundance ratios, age, distance modulus, and reddening, that it might be unwise to put much weight on the local agreement of the morphology of the various branches. This opinion is further strengthened if one adds the possibility of preliminary color-shifts of the isochrones, to allow for errors in their computation or transformation into the observational plane. In other words, if for the sake of discussion, one fixed an age for a cluster (say, within 14 and 18 Gyr), it might be not so difficult to get a "good" overall fit playing a bit with permitted variations of the above parameters. Therefore, one should perhaps better combine *local* star counts and complete stellar luminosity functions to get a contemporaneous check of the validity of the adopted models.

Does this mean that the age determinations carried out using this method are totally "unreliable"? Not at all! We simply believe that they cannot yield ages to ± 1 Gyr as sometimes claimed, but they can surely be accepted as good estimates to ± 3-4 Gyr. Note however that such a difference has *important* consequences on the general astrophysical scenario.

4.2.2 The ΔV_{HB}^{TO}-method.

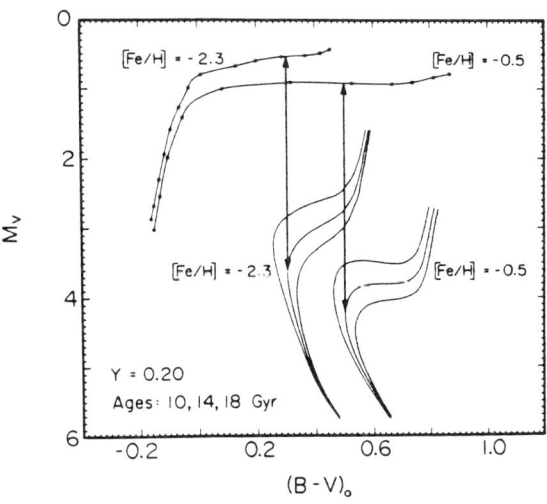

Fig. 1: Plot of VandenBerg and Bell (1985) isochrones for the noted metallicities, along with fully consistent Zero Age HB's (computed by VandenBerg). Arrows illustrate the definition of the age-dependent quantity ΔV_{TO}^{HB}. (Taken from Fig.1 of VandenBerg 1988).

Fig. 1 explains the essence of this method. The quantity ΔV_{HB}^{TO}, defined as the difference in luminosity between stars at the cluster TO and the Zero Age HB stars at the RR Lyrae gap, is quite a strong age indicator. The problem is to define an "appropriate" calibration of the ΔV_{HB}^{TO} vs age -relationship. Many authors have discussed in detail the pros and cons of this procedure in comparison with other methods (see for references BCF, BCCF, and Sandage and Cacciari 1989). Briefly, the advantages are:
i) it is not affected by uncertainties in convection theory, since convection does not play a significant role in TO and HB star luminosity;
ii) it is independent of the interstellar reddening, since it uses differences in luminosity;
iii) it is independent of color-temperature transformations and bolometric correc-

tions;
iv) it does not require the independent knowledge of the cluster distance modulus;
v) the derived age is less sensitive to the adopted metallicity than in most other methods, in fact:

$$\frac{dLogt_9(\Delta V_{HB}^{TO})}{dLogZ} \sim \frac{1}{2} \frac{dLogt_9(MS_{fit})}{dLogZ}$$

The main disadvantages are:
a) many HB's are populated only in the very red or very blue tails, and locating the Zero Age HB (ZAHB) at the mean color of the instability strip requires some careful extrapolation;
b) the distribution of TO stars in the CMD is almost vertical, and the definition of V_{TO} can be quite uncertain;
c) the dependence of the HB luminosity on metallicity and helium content must be known very accurately (i.e. with an error < 0.05 mag), since this affects the TO luminosity determinations accordingly;
d) the validity of the input physics and model computation throughout core He-burning phases should be verified and "guaranteed";
e) an accurate knowledge of the helium abundance is crucial, since:

$$\frac{dLogt_9(\Delta V_{HB}^{TO})}{dY} \sim 4 \frac{dLogt_9(MS_{fit})}{dLogZ}$$

Hence, a full exploitation of the capabilities of this method requires, besides the very accurate knowledge of the observed ΔV_{HB}^{TO}-values, the *correct* calibration of the adopted standard candle which is in this case "the ZAHB luminosity level at the color of the instability strip". In particular, it is crucial to know the dependence of the absolute magnitude of the ZAHB $-M_V^{HB}-$ on metallicity, as Galactic globular clusters span a very wide range in metallicity.

At this point, it may also be worthwhile adding a relation useful to estimate the error in the age determination via this procedure, and an important *procedural* comment. The dependence of the derived age on ΔV_{HB}^{TO} (of course via M_V^{TO}) is:

$$\frac{dLogt_9}{d(\Delta V_{HB}^{TO})} \sim 0.4$$

This implies that an uncertainty of $\delta(\Delta V_{HB}^{TO}) \sim 0.1$ mag leads to $\delta t_9 \sim 10\%$ corresponding to about 1.5 Gyr.

The problem of determining the absolute age of a cluster can therefore be split into two *separate* steps:
a) the determination of the *slope* of the M_V^{HB} vs $[Fe/H]$ relation, which yields relative ages with respect to a reference cluster (see for instance BCF);
b) the determination of the *zero-point* of the $M_V^{HB} vs [Fe/H]$ relation, in order to obtain the absolute age of the adopted *reference* cluster.

5. THE SLOPE AND THE ZERO-POINT OF THE $M_V^{HB} vs [Fe/H]$ RELATION

Five methods have been used so far to study the dependence of the absolute magnitude of HB stars on chemical abundance. They can be substantially divided into two main groups, those based on *variable* HB stars (*i.e. RR Lyrae variables*) and those using *non variable* stars. The first group includes:
a) Statistical parallaxes of field RR Lyrae stars.
b) Baade-Wesselink method (BW), applied only to field RR Lyrae variables, so far.
c) Study of the pulsational properties of RR Lyrae variables in clusters and in the field under the assumption that $M_V^{RR} = M_V^{HB}$.

Table 1. The M_V^{HB} vs metallicity relation: $M_V^{HB} = a\,[Fe/H] + b$

Method	a	b	Label
Statistical parallax	...	0.68	SP
Baade-Wesselink	0.19	1.13	BW
Pulsation Properties of the RR Lyrae, i.e., the Sandage Period Shift	0.39	1.27	pulsation
Main sequence fitting— Buonanno et al. 1989a	0.37	1.29	MS1
Main sequence fitting— King et al. 1988	0.20	0.84	MS2
RGB "bump"——— Fusi Pecci et al. 1989	0.15	0.44	RGB-bump

The second group includes:
d) Globular cluster MS-fitting to a fiducial MS, which can be either observational (*i.e.* the unfortunately small sample of suitable metal-poor subdwarfs) or theoretical (*i.e.* the "calibrated" isochrones).
e) The use of the observational and theoretical properties of the so-called "RGB-bump" in the luminosity function of the Red Giant Branch

These different approaches have been recently analyzed and discussed in detail by us in BCF, Sandage and Cacciari (1989) and Fusi Pecci et al. (1989). The current results are summarized in Table 1.

The five procedures require some comment:

a) The statistical parallax determination was taken from Barnes and Hawley (1986). This approach is still affected by large uncertainties due to the small size of the available samples for different metallicities. A precise determination of the metallicity dependence of M_V^{RR} therefore is not currently possible with this method, and only the zero point can be obtained.

b) The latest estimates of the absolute magnitudes of field RR Lyraes based upon updated versions of the BW method led to similar, consistent results (Longmore et al. 1986, Cacciari et al. 1988, Jones et al. 1988, Liu and Janes 1988), but a conservative estimate of the errors affecting each determination of M_V^{RR} is hardly less than ± 0.2 mag. The adopted result is the least squares regression over the 25 variables analyzed by the above 4 groups.

c) It is well known that globular clusters are divided into two distinct groups based on the observed mean periods of the ab-type RR Lyrae variables, i.e. $< P_{ab} > \sim 0.^d55$ and $\sim 0.^d65$, for group I and II respectively (Oosterhoff 1939,1944); the Oosterhoff separation into period groups is a separation by cluster metallicity, and the *period-shift* effect exists for each RR Lyrae star at a given color (temperature) (Sandage 1982a). Field RR Lyraes seem to share a period-shift effect similar to cluster variables (Sandage 1982b). Using the stellar pulsation theory, these charactestics may be explained by a significant dependence of luminosity on metallicity – i.e. $\frac{\Delta M_V^{HB}}{\Delta [Fe/H]} \sim 0.35$ (Sandage 1982a) – which however does not quite agree with the predictions of theoretical models, unless an anticorrelation of helium and metal abundance is assumed, contrarily to intuition. Also the zero point turns out to be slightly fainter than the theoretical predictions, and the results of other methods. For illustration purposes, in Table 1 we report the calibration currently adopted by Sandage and Cacciari (1989) using this method.

d) This procedure has been discussed in 4.2.1. We only remind here that from this method we cannot obtain determinations, neither of the slope nor of the zero point of the HB luminosity-metallicity relation, more accurate than ± 0.15 mag and ± 0.20 mag respectively.

e) In standard stellar models, at some point on the lower part of the RGB, the convective envelope penetrates deep enough into the star to reach the region of varying hydrogen abundance established during core hydrogen-burning. When the convective envelope retreats from the advancing H-burning shell, a discontinuity in the H-profile is left. Eventually the H-burning shell passes through this discontinuity leading to a potentially observable "bump" in the differential luminosity function. The physical origin of the "bump" was first pointed out by Iben (1968), and the possibility of finding it in an observed luminosity function has been the subject of many papers since that time, most recently Rood and Crocker (1985) and Castellani et al. (1989). Fusi Pecci et al. (1989) have been able to identify the RGB-bump in 11 clusters having metallicities in the range $-2.15 <$[Fe/H]< -0.71. Since the absolute luminosity of the RGB-bump is "known" from the theoretical models, a measurement of V_{bump} and V_{HB} gives a value for M_V^{HB} which can be compared with other calibrations. Moreover, as we will stress in Sect. 7, the overall general agreement of position, size, and "sharpness" of the detected RGB-bumps with the predictions of theoretical computations represents an important direct check of the validity of current models. Using this approach, Fusi Pecci et al. (1989) have found that the level of the HB varies with metallicity as $\frac{dM_V^{HB}}{d[Fe/H]}$ in the range $0.15 - 0.20$, a value completely consistent with "standard" HB theoretical models (Sweigart et al. 1987, and references therein).

From the data in Table 1 we can therefore see that the slope estimates of the $M_V^{HB} vs [Fe/H]$ relation group around two values, i.e. 0.20 and 0.38, and different results can be obtained even using the same method and data, but different procedures and/or assumptions.

Turning to the zero-points, the definition of the various zero-points reported in Table 1 have different weight and usefulness. They have been derived via the RR Lyrae variables themselves or alternative methods (see BCF), except the zero-point listed for the RGB-bump method, which is "mathematic" (Fusi Pecci et al. 1989).

The basic importance of the correct determination of this zero-point can easily be understood by properly combining the different values for M_V^{HB} vs $[Fe/H]$ relations listed in Table 1 and the observed values of $[Fe/H]$ and ΔV_{HB}^{TO}, with the "age equation":

$$Log t_9 = -0.13[Fe/H] + 0.37 M_V^{TO} - 0.51$$

derived by BCF interpolating on the VandenBerg and Bell (1985) models for $t_9 = 15$ and $Y = 0.23$. Using the expressions listed in Table 1, one gets respectively:

$$Log t_9 = -0.130[Fe/H] + 0.37 \Delta V_{HB}^{TO} - 0.26$$
$$Log t_9 = -0.060[Fe/H] + 0.37 \Delta V_{HB}^{TO} - 0.09$$
$$Log t_9 = 0.014[Fe/H] + 0.37 \Delta V_{HB}^{TO} - 0.04$$
$$Log t_9 = 0.007[Fe/H] + 0.37 \Delta V_{HB}^{TO} - 0.03$$
$$Log t_9 = -0.056[Fe/H] + 0.37 \Delta V_{HB}^{TO} - 0.20$$
$$Log t_9 = -0.056[Fe/H] + 0.37 \Delta V_{HB}^{TO} - 0.27$$

It is thus immediately clear that even inserting the same values for the observed

ΔV_{HB}^{TO} and [Fe/H] (taken for instance from Tables 3 and 5 of BCF), the various ΔV_{HB}^{TO} vs age calibrations may lead to differences in age up to 7-8 Gyr, with the *oldest* cluster being from ~ 13 to ~ 20 Gyr old. Therefore, we must conclude that our *current* knowledge is *insufficient* to *properly answer* the two basic questions posed in the introduction (*i.e.* how old is the oldest cluster? and are all the clusters coeval?).

Before trying to "converge somewhow" towards a more *optimistic* outlook on the current situation, we have to briefly introduce a further aspect which plays an important rôle in the cluster age determination, *i.e.* the observational evidence for nonsolar chemical abundance ratios, and, in particular, the suggested existence of CNO-enhancements.

6. EFFECTS OF "REVISED" ASSUMPTIONS: [CNO/Fe] > 0

As we will briefly comment in Sect. 7, there are reasons to believe that "nonstandard or non-canonical" assumptions might play a rôle in Pop II stellar evolution and, in turn, on age-determinations. One of these standard assumptions, that may have to be released, is that CNO-abundances scale with Fe like in the Sun, independent of the [Fe/H] values.

Only a few, non systematic theoretical studies have produced evolutionary tracks computed adopting different levels of CNO overabundances (see BCF for references). Very briefly, all these models show that *a)* CNO enhancements decrease the age of the cluster for a given TO luminosity, *b)* O overabundances are most important, and *c)* $[O/Fe] = +0.3$ implies a decrease in age of about 1 Gyr, independent of the adopted method.

There are now observational measures of the O-abundance for a sufficient number of stars, reviewed by Sneden *et al.* (1989), suggesting the existence of systematic trends in the data. In particular, there is a quite general consensus on the evidence that O is enhanced in metal poor stars (*i.e.* $[Fe/H] < -0.7$) and that at $[Fe/H] \sim -1$, $[O/Fe]$ is $\sim +0.4 \pm 0.2$. This "automatically" implies a decrease of the absolute ages by *at least* 1 Gyr. With respect to the problem of relative ages, it is also very important to know whether the O enhancement reaches some approximately constant value at $[Fe/H] \leq -1$, as found by most of the observers, or whether it increases linearly as metallicity decreases, as claimed very recently by Abia and Rebolo (1989) who observed the O I IR-triplet in 30 unevolved field stars covering a wide metallicity range $-0.2 \geq [Fe/H] \geq -3.5$.

The study of the $[O/Fe]$ vs $[Fe/H]$ relation is crucial because its *slope* and *zero-point* combine with the slope and the zero-point of the $M_V^{HB}vs[Fe/H]$ relation to offer a very complex scenario, where the four quantities may lead to hardly predictable "amplifying" or "compensating" effects. For example, even if one adopts for simplicity $\Delta V_{HB}^{TO} = const$ for all the clusters (which may well not be the case for a few objects at a level of accuracy of $\sim \pm 0.2$ mag, or for most of them at a level of ± 0.05 mag), one can get various "scenarios" covering almost every proposed model for the formation and early evolution of the Galaxy. Table 2 presents a schematic overview of the possible alternatives one may have from the different combinations of the two slopes and the two zero-points. "Intermediate, mixed" solutions are moreover possible.

Table 2 : Schematic overview of "possible scenarios" with varying the $M_V^{HB} vs [Fe/H]$ dependence

---

with [CNO/Fe]=0.0

$\Delta M_V^{HB}/\Delta[Fe/H] \simeq 0.38$	$\Delta M_V^{HB}/\Delta[Fe/H] \simeq 0.20$
★ All the clusters having ΔV_{HB}^{TO} = const. are coeval.	★ All the clusters having ΔV_{HB}^{TO} = constant have ages decreasing with increasing [Fe/H].

---

with [CNO/Fe]>0

$\Delta M_V^{HB}/\Delta[Fe/H] \simeq 0.38$	$\Delta M_V^{HB}/\Delta[Fe/H] \simeq 0.20$
1. [CNO/Fe]>0,constant with varying [Fe/H]	1. [CNO/Fe] > 0,constant with varying [Fe/H].
a) All the clusters with ΔV_{HB}^{TO}=constant are coeval , but the absolute ages decrease with respect to the case with [CNO/Fe] = 0 , by a quantity depending on the degree of CNO enhancement (with [CNO/Fe] = 0.3 , $\Delta t \simeq -1$ Gyr).	a) All the clusters with ΔV_{HB}^{TO} = constant have ages decreasing with increasing [Fe/H] : metal rich objects are younger than metal poor ones. The absolute ages decrease by a quantity depending on the degree of CNO enhancement (with [CNO/Fe] = 0.3 , $\Delta t \simeq -1$ Gyr).
2. [CNO/Fe]>0,increasing with decreasing [Fe/H]	2. [CNO/Fe]>0,increasing with decreasing [Fe/H]
a) The condition ΔV_{HB}^{TO} = constant for all clusters does not implies anymore coevality. Due to the differential CNO enhancement , metal poor clusters are younger, while the age of metal rich clusters is only marginally decreased. This combination of slope and abundances may lead to the following two indications against intuition : • Y anticorrelated with [Fe/H] • t of metal poor clusters < t of metal rich clusters	a) All the clusters having ΔV_{HB}^{TO} = const. may be coeval for a "proper" difference in the CNO enhancement between metal rich and metal poor clusters. The absolute age may be decreased if also the metal rich clusters are slightly O-overabudant.

The general effect of O enhancement on globular cluster age determinations is, therefore, to decrease both the absolute ages and the age spread among clusters, except for a few combinations of parameters which produce, however, side results against intuition.

7. CLOCK RUNNING

Since, independent of the procedure used for the "clock reading and calibration", the *core* of the clock is based on theoretical models, we must ask ourselves how much confidence we may put in the available models. In particular we must check whether:
1) *classical input physics* (*e.g.* reaction rates, opacities,...) is correct;
2) *left-out classical physics* (*e.g.* rotation, diffusion, non-convective mixing,...) is actually important;
3) *left-out non-classical physics* (*e.g.* massive neutrinos, WIMPS,...) plays a rôle.

Significant revisions of the clock-running cannot be totally excluded yet, as a *systematic and meticulous* check of the models is still far from beeing achieved (Renzini and Fusi Pecci 1988). It is however possible to estimate the effects of the theoretical parameters and assumptions. A discussion of the uncertainties and errors related to the observable parameters in the age determination procedure was given in Sect. 3. Rood (1989) has presented an excellent and complete overview of how an *expert and cunning model-maker* can get quantitative estimates of the errors on the three above mentioned items. Using the results of tests he carried out with Adler, Rood has reached the conclusion that errors in *standard* input physics (item 1) should not affect the derived age to more than ± 1 Gyr.

Concerning non-convective mixings and diffusion, the present status of the knowledge is quite meager. As reviewed by Renzini and Fusi Pecci (1988), these phenomena might *rejuvenate* the stars, decreasing the cluster age by $\sim 20\%$ (at least according to some mechanisms and treatments). Unfortunately, it is hard to prove either their existence or the validity of their neglection. The best way to show their *insidious* effects is to study the luminosity function and the spectral peculiarities of sub-giant and giant branch stars. In our opinion, an important new result in this respect is the quoted detection of the RGB-bump in a number of globular clusters by Fusi Pecci *et al.* (1989), and the general agreement of its observed characteristics with the theoretical predictions. This implies that mixing processes (either episodic or diffusive) do not extend inside the region where hydrogen is depleted on the MS. Since the predicted star lifetimes depend on the "actual fuel burnt" (Renzini 1981), the available data on the RGB-bumps tend to exclude that a significant quantity of "extra-fuel" might be brought into the central regions of MS stars. Hence, this decreases the probability of a "big mistake" in the lifetime-scale of globular cluster stars.

Turning finally to the possible influence of "exotic" particles like WIMPS, investigations are at a very preliminary stage. Rood and Renzini (1989) have specifically attempted to set quantitative limits on WIMPS in globular cluster stars. Their conclusion is essentially that the presence (if any) of WIMPS in the core of GC stars should not alter significantly the star evolution and, consequently, the cluster ages.

In conclusion, there is at present no obvious *breakdown* in the "standard and canonical" assumptions, which might imply significant age-variations with respect to the estimates currently available.

8. CONCLUSIONS AND FUTURE PERSPECTIVES

The main conclusions of this review on the ages of galactic globular clusters are the following:

1. The observational quantities that represent the CLOCK READING in the age determination procedure need to be better defined and/or measured. The aspects that need to be particularly improved are the definition and measure of the TO luminosity, and the distance modulus determination. Strictly related to both items is the determination of the HB luminosity and its dependence on metallicity, which has a direct consequence on the age determination. The *rms* error in the age estimates due to the uncertainties in the CLOCK READING is realistically not less than 3-4 Gyr.

2. The theoretical models, input physics and assumptions used to construct the isochrones, *i.e.* the CLOCK RUNNING in the age determination procedure, seem to be basically correct. Although a number of specific items still need to be verified, improved and calibrated, the "internal" accuracy of the isochrones yields *rms* errors probably $\sim$ 1-2 Gyr in the age determination.

3. Relative ages: the main uncertainty here is the slope of the $M_V^{HB} vs [Fe/H]$ relation. On the assumption that ΔV_{HB}^{TO}=constant for all clusters, and that [CNO/Fe] = solar, if the above slope is $\sim$ 0.38 then all clusters are nearly coeval ($\Delta t \leq 1$ Gyr), while if the slope is $\sim$ 0.20 then the most metal-rich clusters would be younger by $\sim$ 3-5 Gyr. CNO enhancement tends to decrease the age spread among the clusters (see Table 2 for details).

4. Absolute ages: the main uncertainty here is the zero-point of the M_V^{HB} vs

$[Fe/H]$ relation, or more generally the correct determination and calibration of the distance modulus. Most determinations are currently leading to an age of $\sim 16^{+4}_{-2}$ Gyr for the oldest galactic globular cluster. The reason for the different errors is that, with reasonable variations of the involved parameters and assumptions, it is much easier to make older than younger clusters. As a matter of fact, if one derived an age of, say, ~ 10 Gyr for the oldest globular cluster, this would imply a major flaw in theoretical models or distance determinations. CNO enhancement decreases the absolute value of the age, by an amount that depends on the degree of the enhancement.

The perspectives for future improvements look quite promising.

On the observational side, a large amount of ground-based work is devoted to better abundance determinations (in particular [O/Fe]), accurate determinations of V_{TO} and V_{HB}, and the study of the RGB characteristics for distance determinations.
On the other hand, space observations (in particular with the *Hubble Space Telescope*) will soon provide highly accurate information on topics such as the subdwarfs, the WD cooling line, the $M_V^{HB} vs [Fe/H]$ relation in a number of extra-galactic globular clusters (*e.g.* in M31 and Fornax), and the luminosity functions of huge samples of post-MS stars in globular clusters.
On the theoretical side, a better knowledge about phenomena such as rotation, non-convective mixing, mass loss, magnetic fields, and influence of non-solar abundance ratios can be expected soon, as well as their proper treatment in new stellar models.

Acknowledgements: It is a pleasure to thank all the colleagues with whom we discussed the problem of GC age determination over the years. Specially we wish to thank Alvio Renzini, Bob Rood, and Don VandenBerg for some illuminating seminars on this subject.

REFERENCES

Abia, C., and Rebolo, R. 1989, preprint
Barnes, T.G.III, and Hawley, S.L. 1986, *Astrophys. J. Lett.* **307**, L9
Bolte, M. 1989, preprint
Boesgaard, A., and Steigman, G. 1985, *Ann. Rev. Astron. Astrophys.* **23**, 319
Buonanno, R., Corsi, C.E., and Fusi Pecci, F. 1989a, *Astron. Astrophys.* **216**, 80 (BCF)
Buonanno, R., Cacciari, C., Corsi, C.E., and Fusi Pecci, F. 1989b, *Astron. Astrophys.* in press (BCCF)
Buzzoni, A., Fusi Pecci, F., Buonanno, R., and Corsi,C.E. 1983, *Astron. Astrophys.* **128**, 94
Cacciari, C., Clementini, G., and Buser, R. 1988, *Astron. Astrophys.* **209**, 154
Castellani, V., Chieffi, A., and Norci, L., 1989, *Astron. Astrophys.* **216**, 62
Chieffi, A., and Straniero, O. 1989, *Astrophys. J. Suppl.* in press
Demarque, P. 1988, in *Globular Cluster Systems in Galaxies*, eds. J.A. Grindlay and A.G. Davis Philip, Kluwer Academic Pub., p. 121
Edwards, A.C. 1982, in *Progress in Cosmology* ed. A.W. Wolfendale, Reidel, Dordrecht, p.291
Frogel, J.A. 1983, *Astrophys. J.* **272**, 167
Fusi Pecci, F., and Renzini, A. 1979, in *Astronomical Uses of the Space Telescope*, eds. F. Macchetto, F. Pacini and M. Tarenghi, ESO,Garching, p. 181
Fusi Pecci, F., Ferraro, F.R., Crocker, D.A., Rood, R.T., and Buonanno, R. 1989, *Astron. Astrophys.*, submitted
Green, E.M. 1988, in *Calibration of Stellar Ages*, Workshop at Van Vleck Observatory, ed. A.G.D. Philip, L. Davis Press, Schenectady, N.Y., p. 81
Green, E.M., Demarque, P., and King, C.R. 1987, *The Revised Yale Isochrones and Luminosity Functions,* Yale Univ. Obs., New Haven
Hesser, J.E., Harris, W.E., VandenBerg, D.A., Allwright, J.W.B., Shott, P., and Stetson, P.B. 1987, *Pub. A.S.P.* **87**, 739
Iben, I., Jr. 1968, *Nature* **220**, 143
Jones, R.V., Carney, B.W., and Latham, D.W. 1988, *Astrophys. J.* **332**, 206
King, C.R., Demarque, P., and Green. E.M. 1988, in *Calibration of Stellar Ages*, Workshop at Van Vleck Observatory, ed. A.G.D. Philip, L. Davis Press, Schenectady, N.Y., p. 211
Kraft, R.P. 1985, in *Production and Distribution of CNO Elements*, eds. I.J. Danziger, F. Matteucci, and K. Kjar, ESO, Garching, p.21
Lee, Y.-W, Demarque, P., and Zinn, R.J. 1989, preprint
Liu, T., and Janes, K.A. 1988, in *Calibration of Stellar Ages*, Workshop at Van Vleck Observatory, ed. A.G.D. Philip, L. Davis Press, Schenectady, N.Y., p. 141
Longmore, A.J., Fernley, J.A., and Jameson, R.F. 1988, *Mon. N.R.A.S.* **220**, 279
Oosterhoff, P.Th. 1939, *Observatory* **39**, 104

Oosterhoff, P.Th. 1944, *Bull. Astron. Inst. Neth.* **10**, 55
Ortolani, S., and Rosino, L. 1987, *Astron. Astrophys.* **185**, 102
Paczýnski, B. 1984, *Astrophys. J.* **284**, 670
Philip, A.G.D. 1988, ed. *Calibration of Stellar Ages*, Workshop at Van Vleck Observatory, L. Davis Press, Schenectady, N.Y.
Ratcliff, S.J. 1987, *Astrophys. J.* **318**, 196
Renzini, A. 1981, *Ann. Phys. Fr.* **6**, 87
Renzini, A. 1985, in *Galaxy Distances and Deviation from Universal Expansion*, eds. B.F. Madore, R.B. Tully, Reidel, Dordrecht, p.177
Renzini, A. 1986, in *Stellar Populations*, eds. C.A. Norman, A. Renzini and M. Tosi, Cambridge Univ. Press, p.73
Renzini, A. 1988, in *Globular Cluster Systems in Galaxies*, eds. J.A. Grindlay and A.G. Davis Philip, Kluwer Academic Pub., p. 443
Renzini, A., and Fusi Pecci, F. 1988, *Ann. Rev. Astron. Astrophys.* **26**, 199
Richer, H.B., and Fahlman, G.G., 1987, in *Stellar Evolution and Dynamics in the Outer Halo of the Galaxy*, eds. M. Azzopardi and F. Matteucci, ESO, Garching, p. 299
Richer, H.B., and Fahlman, G.G., 1988, *Astrophys. J.* **325**, 218
Rood, R.T., 1989, in *Astrophysical Ages and Datation Methods*, 5th IAP Astrophysics Meeting, ed. E. Vangioni-Flam, in press
Rood, R.T., and Crocker, D.A. 1985, in *Production and Distribution of CNO Elements*, eds. I.J. Danziger, F. Matteucci, and K. Kjar, ESO, Garching, p.61
Rood, R.T., and Renzini, A. 1989, in *Astronomy, Cosmology, and Fundamental Physics*, eds. M. Caffo, R. Fanti, G. Giacomelli, and A. Renzini, Kluwer Academic Pub., p.287
Sandage, A.R. 1982a, *Astrophys. J.* **252**, 553
Sandage, A.R. 1982b, *Astrophys. J.* **252**, 574
Sandage, A.R. 1989, in *The Use of Pulsating Stars in Fundamental Problems of Astronomy*, ed. E.G. Schmidt, Cambridge Univ. Press, p. 000
Sandage, A.R. and Cacciari, C. 1989, *Astrophys. J.* in press
Sarajedini, A. and King, C.R. 1989, preprint
Sneden, C., Wheeler, J.C., and Truran, J. 1989, *Ann. Rev. Astron. Astrophys.* **27**, p. 000
Sweigart, A.V., Renzini, A., and Tornambè 1987, *Astrophys. J.* **312**, 762
VandenBerg, D.A. 1983, *Astrophys. J. Suppl.* **51**, 29
VandenBerg, D.A. 1988, in *Globular Cluster Systems in Galaxies*, eds. J.A. Grindlay and A.G. Davis Philip, Kluwer Academic Pub., p.107
VandenBerg, D.A. 1989, in *Astrophysical Ages and Datation Methods*, 5th IAP Astrophysics Meeting, ed. E. Vangioni-Flam, in press
VandenBerg, D.A., and Bell, R.A. 1985, *Astrophys. J. Supp.* **58**, 561
VandenBerg, D.A., and Durrell, P.R. 1989, preprint

COSMOCHRONOLOGY: AN INTRODUCTORY OVERVIEW

Marcel Arnould and Kohji Takahashi
Institut d'Astronomie, d'Astrophysique et de Géophysique
Université Libre de Bruxelles, B-1050 Bruxelles, Belgium

quot homines, tot sententiae
(So many men, so many opinions)

ABSTRACT: We briefly review the various techniques that have been developed in order to evaluate the age of the Universe, and more particularly of the galactic globular clusters and disk. They involve recourse to cosmological models, fitting of (globular and open) cluster color-magnitude diagrams or luminosity functions, theoretical modelling of white dwarf luminosity functions, or to nucleo-cosmochronological methods. From this overview, we are forced to conclude that the age of the Universe is not yet known to better than a factor of about two.

1. INTRODUCTION

Anybody who could afford such a luxurious thought must have once in a while wondered when, where, and how the Universe was born. Indeed, understanding the birth of the Universe and its subsequent evolution remains one of the most basic and tantalizing tasks in modern science. While the very physical mechanisms relevant to that question are reserved for <u>Cosmology</u>, the determination of the age (or of the lower bounds of it) of the Universe is specifically referred to as <u>Cosmochronology</u>.

Cosmochronology deals with different ages, each of which corresponds to an epoch-making event in the past. They are of the Universe, T_U, the globular clusters, T_{GC}, the Galaxy [as (a typical ?) one of many galaxies], T_G, the galactic disk, T_{disk}, and of the non-primordial chemical elements in the disk, T_{el}, where $T_U \geq T_{GC} \approx (\geq?) T_G \geq T_{disk} \approx T_{el}$. Consequently, cosmochronology involves not only cosmological models and observations, but also various other astronomical and astrophysical studies, and even invokes some nuclear physics information.

Notwithstanding much efforts, T_U is not yet well known: at the most, $9 \leq T_U(Gyr) \leq 20$ is somewhat assured, while values outside this range cannot yet be categorically excluded. In the following, we do not dare attempt to narrow that range, but rather present a brief overview of the various age determination techniques which have often left the results in controversy. In due course, we cannot avoid some overlaps with the reviews in the same vein, such as van den Bergh (1984), Fowler and Meisl (1986) and Fowler (1987). Other related review articles are, to name a few, Tayler (1986) on cosmology, Tammann (1987) and Tully (1987) on the cosmological constants, VandenBerg (1988) and Rood (1990) on globular clusters, and Arnould and Takahashi (1990) on nucleo-cosmochronology.

2. THE COSMOLOGICAL MODELS

These models can help determining T_U, as well as T_{GC} and T_{disk}, at least to a certain extent.

2.1 Isotropic and Homogeneous Universe

If the Universe has been, by and large, isotropic in motion and homogeneous in distribution, as is neatly the case at present, its evolution can be described by Friedmann's equation (e.g. Zel'dovich and Novikov, 1983)

$$H^2 a^2 \equiv (da/dt)^2 = H^2 a^2 (\Omega+\lambda) - kc^2, \qquad (1)$$

k=0, +1, or -1 providing the ordinary Euclidean "flat", the "closed", or the "open" metric, respectively. In eq. (1), a is a spatial scale length, H is the Hubble parameter, $\Omega=\rho/\rho_c$ is the ratio of the matter and radiation mass density ρ to the "critical" density $\rho_c=3H^2/(8\pi G)$. The parameter λ is related to Einstein's cosmological parameter Λ (or $c^2\Lambda$) by $\lambda=c^2\Lambda/(3H^2)$.

2.2 Matter Dominated Universe

The matter mass density is thought to have dominated over the radiation mass density since less than one million years after the birth of the Universe. For cosmochronological purposes, one can therefore consider just the matter dominated Universe, such that ρ scales with a^{-3}. Then, eq. (1) has analytical solutions in

(I) the Standard $\Lambda=0$ Model:
$$H_0 t = (\Omega_0/2)(1-\Omega_0)^{-3/2} [x - \sinh^{-1}(x)], \qquad (2a)$$
with $x=2[(1-\Omega_0)(1+\Omega_0 z)]^{1/2}/[\Omega_0(1+z)]$;
and in the
(II) $\Lambda \neq 0$, $k=0$ Model,
$$H_0 t = (2/3)(1-\Omega_0)^{-1/2} \sinh^{-1}(x), \qquad (2b)$$
with $x=[(1-\Omega_0)/\Omega_0]^{1/2}/(1+z)^{3/2}$,

where $z=(a_0/a)-1$ is the redshift, and the suffix 0 indicates the value at the present time $t=t_0$. If $\Omega_0=1$, $H_0 t=(2/3)(1+z)^{-3/2}$ in both cases. For $\Omega_0>1$, remember that $\sinh^{-1}(ix) \equiv i \sin^{-1}(x)$. Also note that i) $\Omega_0+\lambda_0=1$ if $k=0$, ii) in case (I) $\Omega_0<1$ if $k=-1$ and $\Omega_0>1$ if $k=+1$, and iii) $\Omega=1-(1-\Omega_0)/(1+\Omega_0 z)$ in Model (I), and $\Omega=1-(1-\Omega_0)/[1-\Omega_0+\Omega_0(1+z)^3]$ in Model (II). This suggests that $\Omega \sim 1$ at early epochs (very large z), and $\Omega=1$ if $\Omega_0=1$. For later discussions, we define $f(\Omega_0, z) \equiv [\text{RHS of eq.(2)}]$.

3. THE AGE OF THE UNIVERSE

For given H_0 and Ω_0, eq. (2) can provide the age of the Universe $T_U=t_0$ once it is noted that $z_0=0$ and $f(\Omega_0, z=\infty)=0$. If we introduce the "Hubble time" $T_0 \equiv 1/H_0$, T_U/T_0 becomes a formal function of Ω_0.

3.1. The Determination of Ω_0

Although it depends on H_0 (through $\rho_{c,0}$), Ω_0 can be determined independently. One way is to assume that clusters of galaxies obey the Virial Theorem $V+2T=0$ between the potential (V) and kinetic (T) energies. (Observations usually lead to $T>>|V|$, suggesting the presence of some "hidden mass".) The total mass of a cluster is approximately given by $<v^2>R/G$, where $<v^2>$ is the average random velocity squared, and R is the effective radius of the cluster. Since R scales with H_0^{-1}, ρ is proportional to H_0^2, as is $\rho_{c,0}$. Redshift and angular diameter measurements generally lead to the Virial mass that corresponds to $\Omega_0 \approx$ 0.1-0.2. A few other methods (see Tully, 1987) also lead to rather low values $\Omega_0 \approx 0.1-0.4$. However, higher values ($\Omega_0=0.85\pm0.16$ by Yahil et al., 1986; $\Omega_0 \approx 0.5$ by Meiksin and Davis, 1986) have been proposed following the finding of the apparent proximity of the sources of the dipole anisotropies in the microwave background and in the galaxies of the Infrared Astronomical Satellite (IRAS) catalog. On the theoretical side, the standard big bang nucleosynthesis of light elements (He, D, Li) is known to favor $\Omega_{0,baryon} \sim 0.1$, whereas the inflationary Universe scenario concludes $\Omega_0 = 1$ in an inappeasable manner.

A graphical presentation of the $T_U/T_0 = f(\Omega_0, z=0)$ values from eq. (2) is found in Fowler (1987). Within the Standard Model, for instance, $1 \geq T_U/T_0 \geq 2/3$ for $0 \leq \Omega_0 \leq 1$. (If $\Omega_0 = 1$, $T_U/T_0 = 2/3$ in both Models.) Considering that those limiting cases are for an empty space and a closed space, it may be said that the uncertainty of $\sim 1/3$ in the ratio is modest. For $\Omega_0 > 1$, the universe turns out to be "uncomfortably" young.

3.2 The Determination of H_0

It is no secret that there have been incessant disputes over the two distinct ranges of H_0, i.e. ~ 50 and ~ 100 km/s/Mpc. An apparent good news is that some intermediate values appear more frequently in the recent literature. Unfortunately, this seems to have fomented a three-way battle (the enemy of an enemy is still an enemy !). Let us briefly review some of the recently claimed H_0 values.

The Hubble constant H_0 can be determined by examining Hubble's ratios $H_0 = v/r = cz/r$ (for low z) first in nearby galaxies, and then in more and more distant objects (e.g. the Virgo cluster). Here, v is the recession velocity in a fixed (i.e. co-moving) frame such that the observed (i.e. heliocentric) values be corrected for the "non-Hubble" flows (notably the component of the Local Group velocity in the direction of the Virgo cluster). The distance r is related to the apparent and absolute magnitudes in the usual way.

The accurate determinations of H_0 are severely hampered by the difficulty of setting the absolute distance scale. The "classical" approach is to assume that v and r are linearly related, and to calibrate the resulting constant H_0 with respect to some nearby galaxies with known distances. The alternative route to H_0 is to find a "distance indicator" that is well correlated with the absolute magnitudes. An extensive list of proposed distance indicators is found in Tammann (1987).

In any event, a major uncertainty arises from the fact that the total number of calibrating galaxies with reliable data is quite limited. Naturally, different choices of calibrators introduce some uncertainties. Moreover, when reinforced by the apparent more or less large intrinsic dispersions of an assumed relationship, the question is raised of how to deal with the potential bias in selecting a sample. In fact, a closer look at the literature leads us to conclude that much of the differences in the estimated H_0 values arises from differences in <u>interpreting</u> the possible selection bias, and has hardly anything to do with observational data themselves. In order to demonstrate this, let us compare results derived from basically the same method.

The Tully-Fisher Relations   Many of the recent attempts to estimate H_0 rely on the distance indicator referred to as the Tully-Fisher relations between absolute magnitudes M and 21cm HI line widths, Δv_{21}. Table I lists some of the results from locally calibrated Tully-Fisher relations (in blue and/or infrared regions) applied to distant clusters. The controversies over the absolute distance scale (and particularly those concerning the bias), and consequently over the H_0 values, manifest themselves in the discrepant results for the distance r_{Vir} to the Virgo cluster. In contrast, only a minor (5-10 %) uncertainty remains in v_{Vir}.

Other Methods   Other distance indicators rely on (i) the "corrected" luminosity index Λ_c, and the correlation between the apparent diameters (D_n) and the central velocity dispersions (σ) in elliptical galaxies, (ii) the assumption that a galaxy has the same M as that of its look-alike ("sosie") with regard to various observed quantities, (iii) the assumption that the maximum luminosity occurs at the same M(max) in the Local and distant globular clusters, or (iv) the resemblance in nova or in supernova light curves. This can be taken as evidence for the similarity of progenitors, so that the same M(max) can be assigned. Type Ia supernovae are of particular interest. Carbon-deflagration models can indeed account for the decline of the light curves in terms of the decay of ^{56}Ni, and can provide a fair estimate of M(max).

Table I Some of the H_0 determinations from locally calibrated Tully-Fisher relations applied to distant clusters. The numbers of calibrators adopted range from 3 to 13. The entries are selected in such a way that authors just appear once. For comparison, the entry at the bottom refers to a study of field galaxies.

author	cluster	T - F relation	Virgo r_{Vir} (Mpc)	v_{Vir} (km/s)	H_0 (km/s/Mpc)
de Vaucouleurs (1982)	Virgo	blue	15.0±0.6[a]	~1415(±133)	~ 95[a]
Richter and Huchtmeier (1984)	Virgo	blue	24.0±3.3	(~ 1300)[b]	~ 55[b]
van den Bergh (1984)	Virgo	blue infrared	19.5±5.1 16.1±3.0	1322±150	68±17 82±18
Aaronson et al. (1986)	Virgo +10	infrared[c]	14.6±0.8[d]	1350±75	~ 90
Bottinelli et al. (1987)	Virgo +10	blue	16.6±0.9[e]	1200	72±5
Pierce and Tully (1988)	Virgo +UMa	blue + infrared	15.6±1.5	1316±120	85±10
Kraan-Korteweg et al. (1988)	Virgo +UMa+10	blue + infrared	20.9±1.3	1196	56.6±0.9[f]
Sandage (1988b)	field galaxies	blue + infrared	-	-	56±13

[a] including the values obtained via another (Λ_c) distance indicator
[b] values apparently preferred by the authors
[c] M quadratic rather than linear in $\log_{10}(\Delta v_{21})$
[d] value derived from H_0 for 10 clusters and v_{Vir}
[e] errors estimated from the original work by the present reviewers
[f] <u>sic</u> (internal errors only)

Some recent H_0 values derived from these different method are shown in Table II, along with the one based on the classical method that makes use of the linear velocity (redshift) - distance relation.

Table II Some of the recent H_0 determinations based on various methods but the Tully-Fisher relations.

method	H_0 (km/s/Mpc)	author
classical	42±11	Sandage (1988a)[a]
Λ_c	59±20	Tammann (1987)
D_n/σ correlations (Virgo)	67±10	Dressler (1987)
look-alike[b] galaxies	99±15	de Vaucouleurs and Corwin (1986)
globular clusters (M87)	68±10	van den Bergh et al. (1985)
novae (Virgo)	69±14	Pritchet and van den Bergh (1987)[c]
Type Ia supernovae	58±5	Fowler (1987)[d]
	53±11	Tammann (1987)

[a] see also Tammann and Sandage (1985) for distant clusters: $H_0=50\pm7$
[b] in particular with regard to Λ_c and Δv_{21} (or rotation velocity)
[c] c.f. Sandage and Tammann (1988)
[d] analysis based on C-deflagration models

It has also been proposed to determine H_0 (along with the cosmological "deceleration parameter" $q_0 = \Omega_0/2 - \lambda_0$) from the analysis of the ultra-relativistic jets exhibited by certain radio galaxies. A preliminary value of $H_0=100\pm30$ km/s/Mpc is derived from this interesting technique (Roland and Pelletier, 1990).

3.3 The Determination of T_U from Cosmological Models

The age of the Universe $T_U = f(\Omega_0, z=0) \times [978/H_0 (km/s/Mpc)]$ Gyr. If $\Omega_0=1$, $T_U \approx 13.0$-6.5 Gyr for $50 \leq H_0 (km/s/Mpc) \leq 100$. The corresponding values for $\Omega_0=0.2$ are $T_U \approx 16.6$-8.3 Gyr in the Standard Model, and $T_U \approx 21.0$-10.5 Gyr in the k=0, $\Lambda \neq 0$ Model. Those for $\Omega_0= 0.1$ are $T_U \approx 17.6$-8.8 Gyr, and $T_U \approx 25.0$-12.5 Gyr.

4. THE AGE OF THE GLOBULAR CLUSTERS

A rigorous lower bound for T_U can be set by the age T_G of the Galaxy. An apparent measure for the latter is the age T_{GC} of the old metal-poor globular clusters, at least if the globular clusters do not predate the formation of the Galaxy (see below). We now summarize different T_{GC} determination techniques.

4.1 The Determination of T_{GC} from Cosmological Models

Once H_0 and Ω_0 are given, the cosmological models [eq.(2)] provide T_{GC} from the redshift z_{GC}. From $T_{GC} = t_0 - t_{GC}$,

$$T_{GC}/T_0 = f(\Omega_0, z=0) - f(\Omega_0, z=z_{GC}) \equiv \Delta f(\Omega_0, z_{GC}),$$

or
$$T_{GC} = \Delta f(\Omega_0, z_{GC}) \times [978/H_0 \,(\text{km/s/Mpc})] \text{ Gyr}.$$

From the largest redshift values observed <u>so far</u> for quasars and radio galaxies, it seems that galaxies formed at z ~ 4-5. This conclusion gets some support from the cold dark matter cosmology (e.g. Couchman and Rees, 1986). If our Galaxy is indeed typical, that z range would be a good measure of z_G. It would then be natural to assume that $z_{GC} \sim z_G$. However, Peebles (1984) has suggested from the study of a cold dark matter cosmological model that z_{GC} could be as high as 50. The possibility of pregalactic globular clusters has been studied further by Fall and Rees (1985), who are skeptical, and by Rosenblatt et al. (1988), who are cautious about such a high z_{GC} value.

If $z_{GC} = 5$ and the 50-100 km/s/Mpc range for H_0 are adopted, the Standard Model leads to T_{GC} = 15.5-7.8 Gyr for $\Omega_0 = 0.1$, T_{GC} = 14.9-7.4 Gyr for $\Omega_0 = 0.2$, and $T_{GC} = 12.2-6.1$ Gyr for $\Omega_0 = 1$. The corresponding values from the $\Lambda \neq 0$, k=0, Model are T_{GC} = 22.2-11.1 Gyr for $\Omega_0 = 0.1$, and T_{GC} = 19.1-9.5 Gyr for $\Omega_0 = 0.2$. Of course, higher (lower) z_{GC} lead to increased (decreased) T_{GC} values. If, for instance, $\Omega_0 = 0.1$ and H_0 = 50-100 km/s/Mpc, the Standard Model gives T_{GC} = 14.1-7.0 Gyr if $z_{GC} = 3$, and T_{GC} = 16.7-8.3 Gyr if $z_{GC} = 10$. For a tabulation of $f(\Omega_0, z)$, see Tayler (1986).

4.2 The Determination of T_{GC} from Color-Magnitude Diagrams

This question has been discussed in detail by Fusi-Pecci (1989), so that we limit ourselves to some brief and general comments.

From the basic assumption that all stars in a globular cluster formed with much the same composition and within a time interval much shorter than the cluster age T_{GC}, many attempts have been made to determine T_{GC} by fitting theoretically the characteristic features (see Fig. 1) of observed color-magnitude diagrams.

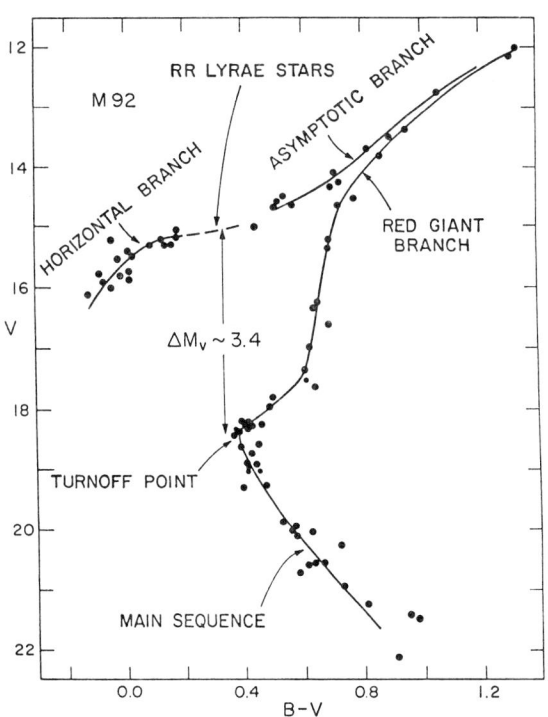

Fig. 1 An example of color-magnitude diagram taken from Iben (1972)

First, T_{GC} can be derived by fitting the main sequence and lower part of the red-giant branch (including the turnoff point) of globular cluster color-magnitude diagrams. This "isochrone fitting" procedure requires the calculation of series of evolution models for stars with different (constant) masses and initial compositions. The isochrones that can be constructed from these grids of models are then mapped into a color-magnitude diagram for the sake of comparison with observation. This mapping inevitably introduces some uncertainties. Through an iterative procedure, one can select the isochrone that best fits the data for a given globular cluster. Recent calculations of that type have particularly emphasized the importance of not specifying the metallicity by the single parameter Z, but of using the individual C, N, and O abundances instead. These can indeed evolve differently with time, as suggested by observations in the galactic disk (e.g. Lambert, 1987). For example, an O enhancement by a factor of 3 relative to C and N reduces by ≈15 % the T_{GC} values derived from isochrones entirely specified by Z (VandenBerg, 1985a). Also note that the difficulties in precise distance determinations are another importance source of uncertainties in age predictions (e.g. Vandenberg, 1988).

Another procedure is to fit just the luminosity at the turnoff, where the stars are <u>currently</u> leaving the main sequence (Fig. 1). Calculated T_{GC} values can be parametrized in terms of the turnoff luminosity, and of the initial Z value and helium mass fraction Y. This method also faces the problem of accurate distance determinations. For instance, the knowledge of T_{GC} within 20 % errors requires distances to be known to better than ≈10 %. The distance to a globular cluster can be calibrated by the RR Lyrae variable stars on the horizontal branch (see Fig. 1). Based on their determination of the absolute magnitudes M_V(RR Lyrae) of RR Lyrae, Jones et al. (1988) have derived T_{GC}(M5)=18±3 Gyr from the turnoff luminosity fitting. On the other hand, Cacciari et al. (1989) claim a considerable Z-dependence of M(RR Lyrae), which implies a hardship for accurate distance determinations.

Being <u>in principle</u> independent of distances, the difference ΔM_V between M_V(RR Lyrae) and M_V(turnoff) has also been proposed as a quantity to fit (Iben, 1972). From observation, ΔM_V = 3.40 ± 0.12 seems to be representative for the galactic globular clusters. Various calculations imply that the M(RR Lyrae) values do not depend strongly on T_{GC}, but that the M_V(turnoff) values are roughly proportional to

$\log(T_{GC})$. This implies that ΔM_V can be a good indicator for cluster ages. For instance, adopting Iben and Renzini's (1984) parametrization and the primordial helium mass fraction $Y=0.245\pm0.003$ of Kunth and Sargent (1983), Fowler and Meisl (1986) obtain $T_{GC}(M92)=14.9\pm 2.4$ Gyr for M92. The red-giant branch of the color-magnitude diagram (Fig. 1) is also used in order to improve T_{GC} determinations (e.g. Fusi-Pecci, 1989).

The consistent fitting of as many features as possible of the observed globular cluster color-magnitude diagrams with detailed stellar evolution models has led to many recent T_{GC} determinations [see e.g. Rood (1990) for a summary]. Among others, let us mention $T_{GC}(M68) \sim 14$ Gyr (McClure et al., 1987), $T_{GC}(47$ Tuc$) \sim 13$-14 Gyr (Hesser et al., 1987), and $T_{GC}(M92) \sim 14$-17 Gyr (VandenBerg, 1988).

It should be warned, however, that predictions from stellar evolution models are still very uncertain. We have already mentioned that due consideration of individual abundances of the CNO elements reduces the estimated ages considerably. Similar reductions are also expected if He diffusion in Population II stars is taken into account (Stringfellow et al., 1983). Furthermore, mass loss effects would possibly join the conspiracy of reducing the estimated T_{GC} values, whereas the inclusion of some "overshooting" in convective cores would increase the ages (e.g. Maeder, 1990).

On top of that, the physics of subphotospheric convection is poorly understood, and there is no guarantee that the standard use of the mixing length theory with a constant parameter α (= mixing length/pressure scale height), which affects T_{eff} in particular, is sufficiently accurate.

Let us also emphasize that the ability of stellar models to reproduce the horizontal branch remains to be questioned. In particular, large uncertainties still remain in the modelling of the helium flash. The fitting of the red-giant branch has also problems of its own.

4.3 The Determination of T_{GC} from Luminosity Functions

Paczyński (1984) has emphasized the interest to evaluate T_{GC} by fitting luminosity functions, which display the total number of stars per absolute magnitude interval as a function of absolute magnitudes. Shapes of luminosity functions indeed vary with T_{GC}, but are also sensitive to Z, Y, and to the initial mass function. An apparent advantage of the use of luminosity functions lies in particular in the weakened dependence of the T_{GC} predictions on the details of model envelopes. A soft spot of the method lies in the requirement of completeness of the observational data, as well as of very accurate distance determinations. Based on this method, for instance, Ratcliff (1987) can merely conclude that $T_{GC}(M13) \geq 14$ Gyr (and $Z \leq 0.001$), mainly because of the 15-20 % uncertainty in the distance to M13, and partly because Z is unknown.

5. THE AGE OF THE GALACTIC DISK

A stringent lower bound for the age of the Galaxy is given by the age T_{disk} of its disk. Obviously, T_{disk} can be derived from age determinations of the oldest disk members. As briefly explained below, a few techniques have been developed for that purpose.

5.1 The Determination of T_{disk} from Cosmological Models

The idea is exactly the same as the one discussed for globular clusters. Redshift values of up to $z_{disk} \sim 3$ have been reported (see Wolfe et al., 1986). We do not repeat the exercise of calculating the corresponding ages. It is easy to see that the galactic disk has to form about 1 Gyr after the birth of the Galaxy if indeed $z_{disk} \sim 3$ and $z_G \sim 4$-5 (see Sect. 4.1).

5.2 The Determination of T_{disk} from Open Clusters

The other two age determination techniques used for globular clusters (Sects. 4.2 and 4.3) have also been applied to open clusters. The age of NGC 188, one of the few old disk clusters, has been estimated to be 8-10 Gyr by Vandenberg (1985b), or ~6.5 Gyr by Twarog and Anthony-Twarog

(1989). Such values can be considered as lower bounds for T_{disk}.

5.3 The Determination of T_{disk} from White Dwarf Luminosity Functions

This is particularly attractive an idea because the coolest white dwarfs are supposedly the oldest stars in the disk. In addition, the view is sometimes expressed that white dwarf evolutionary models are relatively "clean" when compared, for instance, with those of red giant stars. The basic idea of using the luminosity function of the coolest white dwarfs for determining T_{disk} is as follows.

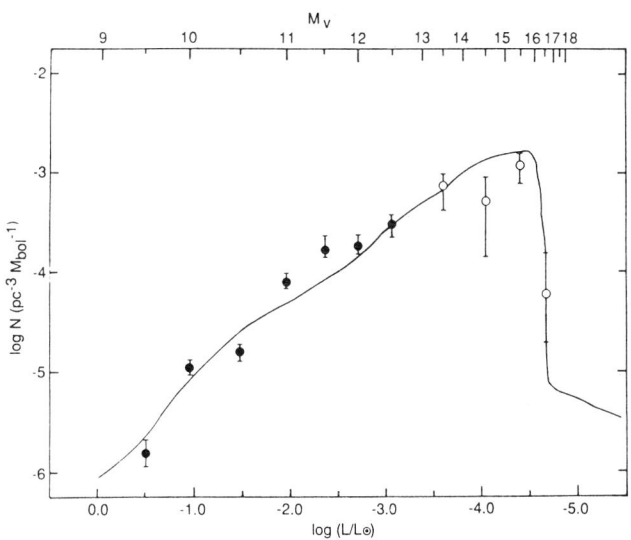

Fig. 2 White dwarf luminosity function taken from Winget et al. (1987). Observed values (dots) are compared with theoretical predictions (curve). The observed data are a preliminary version of those of Liebert et al. (1988). Additional white dwarfs with luminosities near the fall-off have been observed recently (Hintzen et al. 1989).

As seen in Fig. 2, the observed luminosity function exhibits a sharp fall-off at the lower end of the luminosities. This effect is explained by the finite time allotted for the white dwarf cooling. (The monotonous increase of the luminosity function relates to the fact that the hotter objects cool faster.) The fit of theoretical luminosity functions to observation then provides an estimate of the age of the oldest white dwarfs, which equals to time T_{WD} needed for cooling them down to that fall-off luminosity plus the time (<1 Gyr) they spent in the preceding (mostly main sequence) evolutionary phases.

Pre-white dwarf evolutionary models predict that white dwarf cores are probably composed of a mixture of C and O. Winget et al. (1987) have calculated luminosity functions by assuming a constant white dwarf birth rate, and a pure carbon white dwarf core [they argue that such a core model may well simulate white dwarfs with C-O cores and He (and H)- rich surface layers]. From a comparison with observation (Fig. 2), they derive T_{WD}=9.0±1.8 Gyr, or T_{disk}=9.3±2.0 Gyr. Some other calculations have led to quite different T_{WD} values, however. This results namely from the fact that white dwarf evolutionary models still suffer from various uncertainties. Some of them relate to the pre-white dwarf evolution, and more specifically to the exact C and O contents and abundance profiles in the white dwarf cores (e.g. Mazzitelli and d'Antona, 1986). Other uncertainties relate more directly to the physics of the white dwarf cooling itself. In particular, it has been suggested (e.g. Mochkovitch, 1983) that C and O may not be able to co-exist in the core after solidification. Such a redistribution, if it happens, would release a large gravitational energy, so that the cooling would be slower than in the case of a mixed C-O core. In absence of any such total C-O separation, Mazzitelli and d'Antona (1986) conclude that $5 \leq T_{WD}(Gyr) \leq 7$, which is much shorter than the predictions by Winget et al. (1987). This would mean either that the galactic disk is indeed young, or that the fall-off of the luminosity function (Fig. 2) does not provide a good measure of the age of the disk.

On the other hand, scenarios relying on the C-O separation predict a bump in the calculated luminosity functions. It usually appears just before the fall-off luminosity, and has no observed counterpart. According to Garcia-Berro et al. (1988), however, realistic calculations of white dwarf birth rates (taking account of the initial mass function, and of age-dependent scale heights) can lead to such a reduction of that bump that it can hide itself behind the uncertainties in the currently available data. From this, they argue that T_{WD} as high as ~ 15 Gyr cannot be excluded.

5.4 The Determination of T_{disk} from Nucleo-Cosmochronology

Nucleo-cosmochronology aims at determining the age T_{el} of the chemical elements in the galactic disk through the use of the observed abundances of long-lived heavy radionuclides that are produced by the s- and/or r-processes of nucleosynthesis. Consequently, it is hoped to provide at least a lower limit to T_{disk}.

The main difficulty with nucleo-cosmochronology is that it requires a clear picture of the history of those processes and of their yields during the evolution of the Galaxy. This is quite demanding. While everybody would agree this far, there are different ways to look at the problem.

A pessimistic view is that nucleo-cosmochronology can at best set limits on the age of the elements through the use of the so-called "model-independent" approach devised by Schramm and Wasserburg (1970). This formalism has been adopted and slightly extended by Meyer and Schramm (1986), who conclude that $9 \leq T_{el}(Gyr) \leq 27$. (The lower limit is model-independent, while the upper limit depends on certain model assumptions.) This is not a drastically constraining range !

An optimistic view is that, given the presumed complexity of the chemical evolution of the galactic disk, it is by far preferable to describe the nucleosynthesis history by a simple function with a few adjustable parameters. This view has been strongly advocated by Fowler over the years with the use of the so-called "exponential model".

A practitioner's view is that it is really worth studying nucleo-cosmochronology in the framework of chemical evolution models constrained by as many observational data as possible. This view was first expressed vividly by Tinsley (1977, 1980), and has been followed by Yokoi et al. (1983), Clayton (1985, 1988), and Pagel (1989).

Whichever stand is chosen, the search is to be conducted with nuclides that have decay lifetimes commensurate with galactic ages. Since Fowler and Hoyle (1960), several such nuclear chronometers have been proposed. The virtues and demerits of them have been thoroughly discussed by Clayton (1988) and Arnould and Takahashi (1990). Therefore, we will be very brief.

T_{disk} *from solar abundances* Nucleo-cosmochronology classically utilizes the solar abundance ratios of chronometric nuclides in order to estimate T_{el}, which has to be close to T_{disk}. The most widely studied chronometric ratios include $^{232}Th/^{238}U$, $^{235}U/^{238}U$, and $^{187}Re/^{187}Os$.

The chronology based on the former two ratios (Fowler and Hoyle, 1960) has the disadvantage of being dependent of the r-process, the study of which still suffers from many uncertainties. First of all, that process involves hundreds and thousands of unknown neutron-rich nuclei. On top of that, no astrophysical site has been found for the very r-process that was responsible for the solar system r-process composition. Even if we closed our eyes on these problems, the chronology is severely hampered by the still very uncertain observed abundance ratios. All in all, and despite much progress made recently in the nuclear physics involved, it is our opinion that any strong claims based on the Th and U chronologies are to be considered premature, to say the least.

The ^{187}Re-^{187}Os chronology (Clayton, 1964) has the advantage of being free from the r-process models, because the r-process can produce ^{187}Os only via ^{187}Re decays. On the other hand, one has to evaluate the s-process contribution to the solar ^{187}Os (and perhaps ^{187}Re) abundance so as to obtain the cosmoradiogenic abundance ratio of those two nuclides. Again referring the reader to Arnould and Takahashi (1990), we flatly mention here that this requires much nuclear physics information and a detailed knowledge of the s-process sites. The development of the Re-Os chronology is made still more difficult by the enormous enhancement of ^{187}Re decay rates that is expected at high temperatures in stellar interiors. In particular, this effect urges the use of a refined chemical evolution model [see Yokoi et al. (1983) for details].

T_{disk} *from G-dwarfs* The recent observation (Butcher, 1987) of Th/Nd line strength ratios in some twenty G-dwarf stars has paved a way for extending nucleo-cosmochronology to non-solar system material. Armed with his data, which show little change in the Th/Nd abundance ratios between the oldest and youngest stars, and adopting a simple exponential model of chemical evolution, Butcher (1987) argues that the Galaxy is as young as 10 Gyr, and that the current stellar age determination techniques largely and systematically overestimate the ages. (The oldest star in his sample is estimated to be 19 Gyr old !)

To make a long story short, we simply mention here that these conclusions (and similar ones based on Th/Eu) are premature, for many uncertainties are involved in the analysis of the observations and in the adopted chemical evolution models. We refer the reader to Arnould and Takahashi (1990) for details.

6. CONCLUSIONS

In this review, we have attempted to summarize very briefly the basic ideas underlying different age determination techniques, and some of the accompanying controversial results. Whether or not we have been successful in so doing, the following fact stands solid: the age of the Universe is not yet known to better than a factor of two or so. This view is substantiated by Fig. 3, which we consider as a quite unbiased summary of the different ages "determined" with different techniques AND by different people. Clearly, we come to "the happy conclusion that there is more work for all involved in cosmological chronology" (Fowler and Meisl, 1986).

As emphasized by Tayler (1986), accurate age determinations of the oldest globular clusters would have a tremendous impact on cosmological models. For the sake of simplicity, let us take the $12 \leq T_{GC}$ (Gyr) ≤ 17 and $50 \leq H_0$ (km/s/Mpc) ≤ 100 ranges at face value. It is seen in Fig. 3 that $T_{GC}=12$ Gyr is hardly compatible with the predictions of the $\Omega_0=1$ model as long as $z_{GC} \leq 5$. In fact, this model gives 12.2 Gyr for $z_{GC}=5$ and $H_0=50$ km/s/Mpc. If $T_{GC}=17$ Gyr instead, even $\Omega_0=0.1$ does not work without invoking a non-zero cosmological constant. Of course, a similar game can be played with regard to T_{disk}.

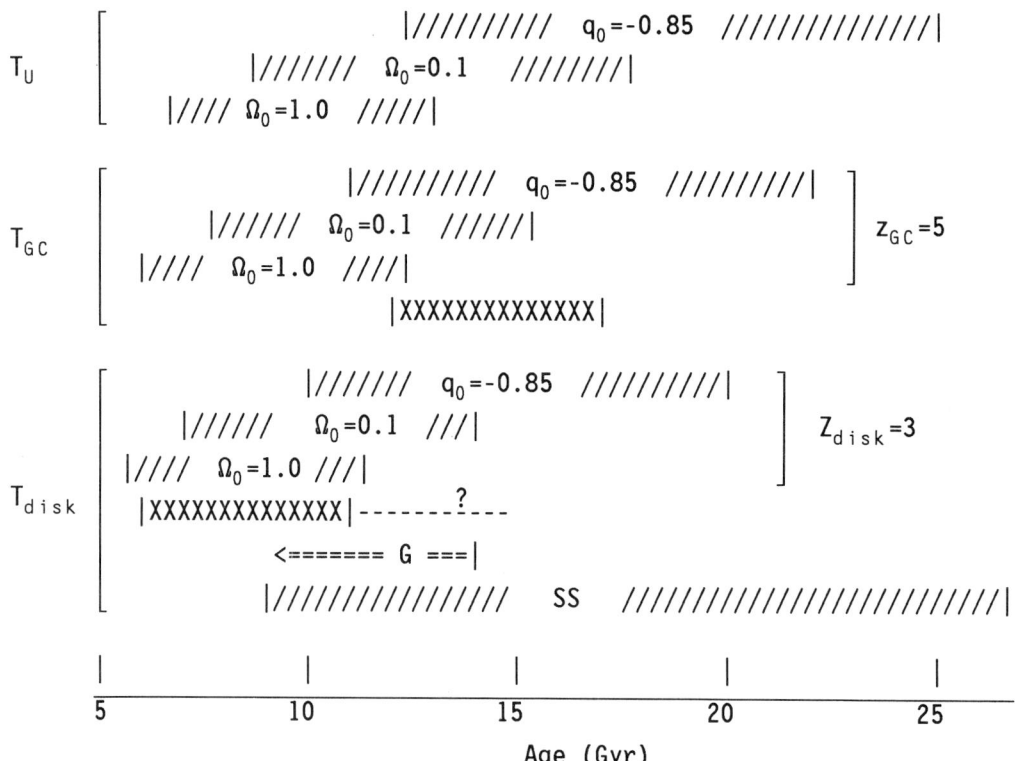

Fig. 3 Ages of the Universe, T_U, of galactic globular clusters, T_{GC} and of the galactic disk, T_{disk}. Estimates from cosmological models are specified by Ω_0 values for the Standard Model, and by $q_0 = \Omega_0/2 - \lambda_0 = 3\Omega_0/2 - 1$ in the $\Lambda \neq 0$, k=0 Model. In both cases, the displayed ranges correspond to $100 \geq H_0 (km/s/Mpc) \geq 50$. The adopted z_{GC} and z_{disk} are discussed in Sects. 4.1 and 5.1. The ages derived from color-magnitude diagram and luminosity function fits are in the ranges marked with crosses. Predictions from nucleo-cosmochronology are shown for G-dwarfs (G) and for the solar-system elements (SS).

Acknowledgements We thank M. Forestini for discussions on white dwarfs. This work has been supported in part by the Programme International de Collaboration Scientifique (PICS) No. 18, and by the Science Program SC1-0065 of the European Economic Community. One of us (M.A.) is Chercheur Qualifié F.N.R.S. (Belgium).

REFERENCES

Aaronson, M., Bothun, G., Mould, J., Huchra, J., Schommer, R.A., Cornell, M.E.: 1986, Astrophys. J. **302**, 536
Arnould, M., Takahashi, K.: 1990, in *Astrophysical Ages and Dating Methods* (Editions Frontières, Gif-sur-Yvette), to appear
Bottinelli, L., Fouqué, P., Gouguenheim, L., Paturel, G. and Teerikorpi, P.: 1987, Astron. Astrophys. **181**, 1
Butcher, H.R.: 1987, Nature **328**, 127
Cacciari, C., Clementini, G., Prevot, L., Buser. R.: 1989, Astron. Astrophys. **209**, 141
Clayton, D.D.: 1964, Astrophys. J. **139**, 637
Clayton, D.D.: 1985, in *Challenges and New Developments in Nucleo-Synthesis*, eds. W.D. Arnett, J.W. Truran (University of Chicago Press, Chicago), p. 65
_____ : 1988, Mon. Not. R. astr. Soc. **234**, 1
Couchman, H.M.P., Rees, M.J.: 1986, Mon. Not. R. astr. Soc. **221**, 53
de Vaucouleurs, G.: 1982, Astrophys. J. **253**, 520
de Vaucouleurs, G., Corwin, H.G.Jr.: 1986, Astrophys. J. **308**, 487
Dressler, A.: 1987, Astrophys. J. **317**, 1
Fall, S.M., Rees, M.J.: 1985, Astrophys. J. **298**, 18
Fowler, W.A.: 1987, Q. Jl. R. astr. Soc. **28**, 87
Fowler, W.A., Hoyle, F.: 1960, Ann. Phys. **10**, 280
Fowler, W.A., Meisl, C.C.: 1986, in *Cosmogonical Processes*, eds. W.D. Arnett, C.J. Hansen, J.W. Truran, S. Tsuruta (VNU Science Press, Utrecht), p.83
Fusi-Pecci, F.: 1989, this conference
Garcìa-Berro, E., Hernanz, M., Mochkovitch, R., Isern, J.: 1988, Astron. Astrophys. **193**, 141
Hesser, J.E., Harris, W.E., VandenBerg, D.A., Allwright, J.W.B., Shott, P., Stetson, P.B.: 1987, Pub. Astron. Soc. Pac. **99**, 739
Hintzen, P., Oswalt, T.D., Liebert, J., Sion, E.M.: 1989, Astrophys. J. **346**, 454
Iben, I.Jr.: 1972, in *IAU Coll. #17, Stellar Ages*, eds. G.C. de Strobel, A.M. Delplace (Paris-Meudon Observatory, Paris), p. XI-1
Iben, I.Jr., Renzini, A.: 1984, Phys. Rep. **105**, 329
Jones, R.V., Carney, B.W., Latham, D.W.: 1988, Astrophys. J. **332**, 206
Kraan-Korteweg, R.C., Cameron, L.M., Tammann, G.A.: 1988, Astrophys. J. **331**, 620
Kunth, D., Sargent, W.L.W.: 1983, Astrophys. J. **273**, 81
Lambert, D.L.: 1987, J. Astrophys. Astr. **8**, 103
Liebert, J., Dahn, C.C., Monet, D.G.: 1988, Astrophys. J. **332**, 891
Maeder, A.: 1990, in *Astrophysical Ages and Dating Methods* (Editions Frontières, Gif-sur-Yvette), to appear
Mazzitelli, I., d'Antona, F.: 1986, Astrophys. J. **308**, 706
McClure, R.D., VandenBerg, D.A., Bell, R.A., Hesser, J.E., Stetson, P.B.: 1987, Astron. J. **93**, 1144
Meiksin, A., Davis, M.: 1986, Astron. J. **91**, 191
Meyer, B.S., Schramm, D.N.: 1986, Astrophys. J. **311**, 406
Mochkovitch, R.: 1983, Astron. Astrophys. **122**, 212
Paczyński, B.: 1984, Astrophys. J. **284**, 670

Pagel, B.E.J.: 1989, in *Evolution Phenomena in Galaxies*, eds. J.E. Beckman, B.E.J. Pagel (Cambridge University Press, Cambridge), to appear
Pierce, M.J., Tully, R.B.: 1988, Astrophys. J. **330**, 579
Peebles, P.J.E.: 1984, Astrophys. J. **277**, 470
Pritchet, C.J., van den Bergh, S.: 1987, Astrophys. J. **318**, 507
Ratcliff, S.J.: 1987 Astrophys. J. **318**, 196
Richter, O.-G., Huchtmeier, W.K.: 1984, Astron. Astrophys. **132**, 253
Roland, J., Pelletier, G.: 1990, in *Astrophysical Ages and Dating Methods* (Editions Frontières, Gif-sur-Yvette), to appear
Rood, R.T.: 1990, in *Astrophysical Ages and Dating Methods* (Editors Frontières, Gif-sur-Yvette), to appear
Rosenblatt, E.I., Faber, S.M., Blumenthal, G.R.: 1988, Astrophys. J. **330**, 191
Sandage, A.: 1988a, Astrophys. J. **331**, 583
_____: 1988b, Astrophys. J. **331**, 605
Sandage, A., Tammann, G.A.: 1988, Astrophys. J. **328**, 1
Schramm, D.N., Wasserburg, G.J.: 1970, Astrophys. J. **162**, 57
Stringfellow, G.S., Bodenheimer, P., Noerdlinger, P.D., Arigo, R.J.: 1983, Astrophys. J. **264**, 228
Tammann, G.A.: 1987, in *IAU Symp. #124, Observational Cosmology*, eds. A. Hewitt, G. Burbidge, L.Z. Fang (Reidel, Dordrecht), p. 151
Tammann, G.A., Sandage, A.: 1985, Astrophys. J. **294**, 81
Tayler, R.J.: 1986, Q. Jl. R. astr. Soc. **27**, 367
Tinsley, B.M.: 1977, Astrophys. J. **216**, 548
_____: 1980, Fund. Cosmic Phys. **5**, 287
Twarog, B.A., Anthony Twarog, B.J.: 1989, Astron. J. **97**, 759
Tully, R.B.: 1987, in *IAU Symp. #124, Observational Cosmology*, eds. A. Hewitt, G. Burbidge and L.Z. Fang (Reidel, Dordrecht), p. 207
VandenBerg, D.A.: 1985a, in *Production and Distribution of C, N, O Elements*, eds. I.J. Danziger, F. Matteucci, K. Kjär (ESO, Garching bei München), p. 73
_____: 1985b, Astrophys. J. Suppl. **58**, 711
_____: 1988, in *Astron. Soc. Pac. Conf. Ser., vol. 4, The Extragalactic Distance Scale*, eds. S. van den Bergh, C.J. Pritchet (Brigham Young University Print Service, Provo, Utah), p. 187
van den Bergh, S.: 1984, Q. Jl. R. astr. Soc. **25**, 137
van den Bergh, S., Pritchet, C., Grillmair, C.: 1985, Astron. J. **90**, 595
Winget, D.E., Hansen, C.J., Liebert, J., Van Horn, H.M., Fontaine, G., Nather, R.E., Kepler, S.O., Lamb, D.Q.: 1987, Astrophys. J. **315**, L77
Wolfe, A.M., Turnshek, D.A., Smith, H.E., Cohen, R.D.: 1986, Astrophys. J. Suppl. **61**, 249
Yahil, A., Walker, D., Rowan-Robinson, M.: 1986, Astrophys. J. **301**, L1
Yokoi, K., Takahashi, K., Arnould, M.: 1983, Astron. Astrophys. **117**, 65
Zel'dovich, Ya.B., Novikov, I.D.: 1983, in *Relativistic Astrophysics*, vol. 2 (University of Chicago Press, Chicago)

Rotation, age and lithium*

by E. Schatzman
Observatoire de Meudon
92195 Meudon-Principal Cedex, France

Abstract. After a brief sketch of the lithium problem and a statement concerning the principles of its solution, the author presents the basic ideas :(1) the non-linear dynamo . The loss of magnetic energy is supposed to be due to buoyancy taking place in the neignborhood of the bottom of the convective zone. It is then asumed that the motion upwards of the flux tubes is taking place in a fluid having a high turbulent viscosity; (2) spin down. When carried into the equations describing the loss of angular momentum, with the assumption that the asymptotic value of the stellar wind is equal to the velocity of escape, it is found that the period of rotation increases like $(1+t/t_0)^{3/4}$. Proofs of the law of spin-down are looked for :(i) in the differential rotation in the convective zone, (ii) in the statistical distribution of v sin i for F stars, and (iii) in the distribution of the periods of rotation as a function of mass in the Hyades. It is shown that the available sample of F stars is not homogeneous and is probably made of at least two sub-groups with a very different characteristic time t_0 of spin down. (3) pre-main sequence stars. It is suggested that electromagnetic braking begins during contraction along the Hayashi track when both a high dynamo number and a high magnetic Reynolds number caracterise the properties of the outer layers. This leads to a relatively small velocity of rotation when reaching the main sequence, in contradiction with the observations in young clusters. (4) a consistent picture of lithium deficiency in late dwarfs is given, with lithium being carried to the level of nuclear processing by time dependent turbulent diffusion mixing. . The high abundance of lithium in fast rotators in α Per is explained by the presence of the remnants of an accretion disk.

---

*Paper presented at the colloquium held in Les Houches, June 19-23, 1989, "Frontiers in Stellar Structure Theory", M.J. Goupil and J.P. Zahn Eds, Springer Verlag (in press).

ASTEROSEISMOLOGY AS A PROBE OF STELLAR STRUCTURE AND EVOLUTION

Teodoro Roca Cortés and Juan A. Belmonte

Instituto de Astrofísica de Canarias 38200 La Laguna, Tenerife, Spain.

Summary: Asteroseismology, by studying the spectrum of global oscillations in stars, is probably the only way to probe their interiors, providing key information on the stellar structure and dynamics. Furthermore, by comparing spectra from different stellar objects, information on stellar evolution can be obtained. For the Sun, Helioseismology has already yielded very interesting results, both observationally and theoretically. In the case of stars, it is a new and exciting field. In this review, observational techniques to measure the oscillations in stars all over the H-R diagram and theoretical efforts being made will be discussed.

1. INTRODUCTION

It is not easy to make a review of an area in Astrophysics which, in fact, has just born. Although it is very difficult to tell when something has started, it is even more difficult to do so in this case. The reason is that variable stars are known for ages and the theoretical basis of how stars pulsate were set longtime ago (Ledoux and Walraven, 1958). However, the seismologycal techniques applied to the Sun started just a decade ago providing a very rich oscillation spectrum; therefore, this name should not be applied to stars from which we know only one mode of pulsation or, at best, two.

The word Asteroseismology will only be applied in those cases where at least more than two modes are known. In the last few years, hundreds of them have been observed in our Sun; nevertheless, the measured amplitudes are very small, a few micromagnitudes (μmag, in photometry) or a few cms^{-1} (in velocity). Furthermore, some stars seem to oscillate with several modes such as the roAp (rapid oscillating Ap; Kurtz, 1982 and 1986), δSct (Breger et al, 1987; Belmonte et al, 1989c), and some compact stars (mainly WD; Winget, 1988). Moreover, as the Sun oscillates, then other low mass main sequence (MS) stars have no reason for not doing the same. Therefore, provided we have the technical capability to measure such small oscillations, most of the stars in any stage of evolution will oscillate. The measurement of the oscillation spectra will provide the basis to know their structure and, by comparing similar mass stars, we will have a new tool to measure, and therefore better understand, their evolution.

The techniques to be used for such an adventure are, by any means, not yet well stablished. The photometry and spectrometry have been known for ages already, but the problem starts when we ask them to measure such small quantities as the ones observed in our Sun (see Table 1). New spectrometric techniques have emerged in the past years to study helioseismology such as resonant scattering techniques (Brookes et al, 1978; Fossat and Roddier , 1971), magneto-optical filters (Cacciani and Fofi,1978), Fourier tachometry (Brown, 1981) and heterodyne spectroscopy (Glenar et al, 1986) that have already been succesful there. Their application to measure star pulsations is just in its begginings. On the other hand, photometry , a technique which we believe is known, is deterrently limited by atmospheric noise (Fossat, 1984).

In this review we shall try to present the actual status of this new and exciting field, by showing where we are, how are we trying to do so and what perspectives do we have for the near future. In some phases we will refer to other reviews, which are so recent that are still valid and we will only update those sections which might be slightly old.

2. THEORY

There has been many theoretical review papers on this subject, e.g. see Stein and Leibacher (1974), Deubner and Gough (1984), Christensen-Dalsgaard (1989). Still, we will only say a few words in order to introduce the special wording and a few results that we will need to understand the observations or, at least, they can guide us in this sense.

When a linear, adiabatic theory is applied to a spherical symmetry star looking for its pulsations, it leads to a differential equation of the eigenvalue type problem. The equation, very similar to the hidrogen atom energy levels problem, has eigenfunctions which have separated variables (radial and angular) and whose eigenvalues are the angular frequencies of such pulsations. The solutions are of the type,

$$\Psi(r,\theta,\varphi,t) = \mathcal{R}_{n,\ell}(r,t) Y_\ell^m(\theta,\varphi,t) \qquad (2.1)$$

where n, is the order of the mode, ℓ is called the degree and m, degenerate up to now, is the azimutal index. If a perturbation destroys the spherical symmetry of the problem then the m-degeneracy breaks up and $(2\ell+1)$ new possible values for the frequencies appear. For instance, stellar rotation can play such a role, producing degeneracy that, to first approximation, can be expressed as:

$$\nu_{n,\ell,m} = \nu_{n,\ell} \pm m\,(1 - C_{n,\ell}) \frac{\bar{\Omega}_{n,\ell}}{2\pi} \qquad m = 0, 1, ..., \ell \qquad (2.2)$$

where $C_{n,\ell}$ is a positive constant (~ 0) and $\bar{\Omega}_{n,\ell}$, the average rotational velocity in the cavity where the standing wave is propagating.

Following Cowling(1942), the modes can be classified into p-modes and g-modes depending on the restauring force that produces them: if it is the pressure gradient then acoustic modes are produced, also called p-modes; if it is the bouyancy force then gravity modes are produced, also called g-modes. The p-modes , are confined mainly in the exterior part of the star, have frequencies between that of the acoustical cutoff frequency (ν_c) and the Brunt Vaisala frequency (ν_N) and they increase as n increases propagating almost vertically. On the other hand, the g-modes penetrate into the stellar core, they propagate almost horizontally and the frequencies are lower than the Brunt-Vaisala frequency, decreasing as the order n increases.

2.1. Asymptotic Theories

The equations governing the oscillations in stars admit approximate solutions that are valid when $n \gg \ell$; these solutions are usually called asymptotic solutions and are due mainly to Vandakurov (1976) and Tassoul(1980).
In the case of p-modes, the frequencies are approximated by:

$$\nu_{n,\ell} = \Delta\nu_0 \{n + \frac{\ell}{2} + \epsilon - \delta_{n,\ell}\} \quad \text{with} \quad \delta_{n,\ell} = \frac{\ell(\ell+1)\alpha + \beta}{n + \frac{\ell}{2} + \epsilon} \quad (2.3)$$

where $\Delta\nu_0 = \{2 \int_0^R \frac{dr}{c}\}^{-1}$, 2ϵ the politropic index of the model, and α and β are constants that depend on the internal structure of the star.
As it can be noticed the frequencies are almost equally spaced for modes of equal ℓ and consecutive n, $\Delta\nu_\ell = \nu_{n+1,\ell} - \nu_{n,\ell} \simeq \Delta\nu_0$. Moreover, if $\delta_{n,\ell} \simeq 0$, then modes of consecutive n and differing in ℓ by 2 have the same frequencies $\Delta\nu_{\ell,\ell+2} = \nu_{n+1,\ell} - \nu_{n,\ell+2} \simeq 0$. Provided enough frequency resolution is achieved, this parameter will be of extreme importance since it reflects conditions at the centre of the star.
As far as the g-modes is concerned, there exists also an asymptotic theory for them. This theory, valid as before when $n \gg \ell$, predicts the values of the periods of the modes, rather than their frequencies, in the form:

$$\frac{1}{\nu_{n,\ell}} = P_{n,\ell} \approx \frac{P_0}{\{\ell(\ell+1)\}^{1/2}} (n + \psi + \frac{\ell}{2}) \quad (2.4a)$$

$$\text{with} \quad P_0 = \frac{\pi}{\int_0^{r_c} \nu_N \frac{dr}{r}} \quad (2.4b)$$

where ψ is a constant depending on the core structure.
As before, for modes of given ℓ the periods for different n are equispaced as:

$$\Delta P_\ell = P_{n,\ell} - P_{n+1,\ell} = \frac{P_0}{(\ell(\ell+1))^{1/2}}. \quad (2.5)$$

These properties provide therefore clear signatures for both kind of modes.
So far, the linear theories provide only the frequencies of the modes. However Christensen-Dalsgaard and Frandsen(1983) have also provided an estimation of the amplitudes to be expected from different kind of stellar models assuming that the excitation mechanism of this modes is similar to the one that is believed works on the Sun, that is random excitation by turbulent convection. Their results for ZAMS models yield a higher amplitude for 1.5 $M_\odot$ (spectral type F0V) of 1.5 ms^{-1} and 13

μmag for periods around 18 minutes and p-mode spacings $\Delta \nu_0 \sim 90$ μHz. Moreover, the evolution of a 1M$_\odot$ yields an increase of the amplitudes and frequencies with age. They also calculate the characteristics of the oscillations in envelope models yielding periods in the range of days and amplitudes of tens of ms^{-1} for masses higher than 5 M$_\odot$.

Christensen-Dalsgaard (1988) has proposed an evolutionary diagram for Asteroseismology, similar to the H-R diagram (see Figure 1) where, provided we can determine $\Delta\nu_0$ and D_0, a parameter related to the small spacings in the form,

$$\Delta\nu_{\ell,\ell+2} \approx D_0\,(6 + 4\ell), \tag{2.6}$$

we will be able to stablish the stellar mass and central hydrogem abundance (hence, the age) to high precision (see Figure 1). such diagram is calibrated using the observed values for the Sun.

Obviously, the Sun is the best studied case and therefore we should aim to a similar knowledge for other stars. The solar oscillations spectrum has been measured to a high degree of accuracy, mainly the p-mode spectrum and, in Table 1, we present a certain number of parameters already found for the Sun, which should be the observables to be found in other stars.

Frequency Range (A$\geq$3 cms^{-1}):	2-4 mHz
Corresponding n range:	15$\leq n \leq$30
Maximum Amplitude Frequency:	3.1 mHz
Amplitudes:	Δm$\leq$4 μmag; ΔV$\leq$20 cms^{-1}
Observable ℓ:	0-2 (lum.); 0-3 (veloc.)
ΔV/Δm:	80-100 kms^{-1}
Average ($\forall\,\ell$) $\Delta\phi$:	140$\pm$4.9°
p-modes lifetime:	40 days at 2 mHz
$\Delta\nu_0$:	135.4 μHz
D_0:	1.54 μHz
ϵ:	1.35
ψ (g-modes):	1/4
P_0 (g-modes):	35.0 to 42.6 min

Table 1: Solar oscillations parameters (mainly for p-modes), when the Sun is observed in disk-integrated light (the Sun as a star); hence, these should be the observables to be found in other stars (from Belmonte, 1989 and references therein).

3. OBSERVATIONAL TECHNIQUES

The measurement of the solar-like oscillations in other stars is a difficult one due to the extremely low amplitudes involved. As we will see later, in only very special cases (δSct, roAp, WD) the amplitudes already quoted appear to be amplified by one or two orders of magnitude due to some forcing mechanism inside these stars (κ and/or γ mechanisms, magnetic overstability, etc...).

Moreover, the information that we can obtain from the modes is extremely important and, obviously, the more modes we can identify, more information we will obtain. The frequencies of the modes and the spacings between them will teach us about the global structure of the star. The amplitudes will provide information about the physical processes involved in the pulsation, especially the mechanisms responsables of the excitation and damping of the modes. Finally, the phases or, even better, the phase-lag between modes observed in different ways (e.g. velocity and luminosity simultaneously) will tell us about the conditions present in the cavity where the mode is resonantly propagating (e.g. adiabaticity conditions).

Up to now, two basic ways to look at the problem have been used. The first one, already sucessful in the Sun, is by means of spectrometric techniques, i.e. trying to measure the radial stellar velocity variations (ΔV) due to the presence of cyclic pulsational instability. Photometry is the other one, and specially it has been used in its version of high-speed (differential or not) to look at the Sun and other stars. In this later case, the luminosity variations ($\Delta L/L \sim \Delta m$) are studied. Excellent review papers on this matter have already been published (Campbell and Walker, 1985; Harvey,1988).

3.1. Spectrometric Techniques.

Succesive expansions and compressions on the external layers of a star supporting oscillations, will produce radial velocity variations that, eventually, might be detectable as wavelength shifts of the stellar spectral lines due to the Doppler effect. Asteroseismology requires an extremely high precission and stability, since to determine a $\Delta V < 1$ ms^{-1}, such us the one we need for solar-like stars, it is necessary to find $\Delta\lambda/\lambda$ with a precission better than 3×10^{-9}.

Nowadays, there are several methods to measure the spectral line shift $\Delta\lambda$, most of them, originated in the mask correlation procedure (Griffin, 1967). However, the standard method (using calibrating stars, autocorrelation, etc...) has the disadvantage of not being able to correct the instrumental drifts present in almost all spectrometers. For this reason, one should use a stable reference to compensate for such drifts. Some of the most common procedures, specially those that have yielded to some asteroseismologycal result, can be summarized as follows:

i) To impose a calibration arc to the stellar spectrum. This is the most common method in use and it has yielded the first possitive result in velocity observations for roAp (Matthews et al, 1988).

ii) To pass the stellar light through either an atomic or molecular vapour that adds its own absorption lines to the stellar spectrum (Campbell and Walker, 1979;). This system has been sucessfully used by Libbrecht (1988) in roAp stars, achieving a noise level of nearly 15 ms^{-1}.

iii) To pass the light of an stellar absorption line, previously isolated with an interference-filter, through cells, containing atomic vapour of the same element that produces the stellar line, where an atomic resonance effect is produced yielding very narrow bandpasses (magneto-optical filters) that permits to do differential spectrometry (Isaak and Jones,1988) . This technique has given some preliminary results on MS stars (Gelly, Grec and Fossat, 1986) and has been used also to detect stellar oscillations in Arcturus (Belmonte et al, 1989b), providing the lowest noise levels ever achieved in stellar spectrometry (<1 ms^{-1}).

iv) To transmit or reflect the stellar light in an interferometer (either a Fabry-Perot or a Michelson) that will provide a stable spectral feature, vapour gas, spectral lamp or laser, to be compared with the stellar spectrum. Such a device was used for the pioneering work of Traub, Mariska and Carleton (1979) where it yielded non-possitive results. However, succesful results have been obtained in the last years, especially in red giants (Smith, McMillan and Merline, 1987) and Ap stars (Belmonte et al, 1989a).

v) To use the telluric O_2 lines as a reference (Smith, 1983). However, it has the problem of the atmospheric instability (e.g. strong winds).

The major problem of spectrometric techniques, specially for those described in iii and iv, is the lack of photons. Increasing the number of stellar lines observed, the brightness of the star and/or the size of the telescope, being, so far, the easiest ways to overcome such a problem (Harvey, 1988).

3.2. Photometric Techniques.

Asteroseismology by means of photometric techniques measures the stellar brightness fluctuations as a function of time due to variations of the surface layers temperature and/or the stellar flux in a special wavelength range. So far, they have been the most commonly used techniques in stellar variability mainly due to their easiness and, consequently, their low cost. However, on the Sun, they only had good S/N when using space observations.

Photometry is severely limited by the effect of the Earth atmosphere contributing with two major sources of noise: the sky transparency fluctuations (correlated to small angular distances and dominant at low frequencies) and scintillation (not correlated at all and contributing to almost all frequency range). Standard differential photometry has been used for ages to avoid the effect of transparency fluctuations. However, Asteroseismology is mostly interested in variations with periods as short as minutes, so that continuous monitoring of the star is necessary. This is why high-speed (also called rapid) photometry has been mainly used to perform stellar oscillations studies photometrically (Deubner and Isserstedt, 1983; Kurtz, 1984). Of course, if a two or three channel photometer is available, the use of rapid differential photometry is, by far, the best way (Belmonte,1989).

However, to achieve the low noise levels needed (≤ 10 μmag), we still have to deal with the scintillation contribution to noise which, desperately, in the best cases, produces a flat noise level all over the frequency spectrum of nearly 100 μmag. Very recently (see Harvey, 1988), it has been proposed a spectrophotometric technique that could avoid this contribution, by using very narrow bandpass filters ($\Delta\lambda \leq 1$ Å) centred in the core of certain intense spectral lines. However, this technique has not been fully tested yet, although some results have been claimed on ϵEri by Noyes et al (1984).

Anyway, rapid photometry has yielded some spectacular results on roAp stars (Kurtz, 1986) and it has been proven useful in the detection on solar-like oscillations in MS stars (Belmonte, Pérez Hernández and Roca Cortés, 1989). Indeed, in its variant of rapid differential photometry, it has been found to be the best way to tackle the study of oscillations in the extremely faint compact pulsators (Winget, 1988) and in the detection of multimode pulsation in δSct stars (Mangeney et al, 1988).

Whatever the technique one can use, a continuous monitoring of the target for as long as possible (weather conditions permitting), has been found absolutely necessary for Asteroseismology. This is why two, three or even multisite observational campaigns have been performed in the last few years, trying to obtain the possible highest resolution, lowest noise level and minimum gaps in the data (Breger et al, 1987; Solheim, 1988; Kurtz et al, 1989; Kreidl et al, 1989; etc...).

4. TARGETS TO OBSERVE

As already quoted, there are several type of stars that are known to oscillate in various modes and hence we can learn a lot about the physics governing the oscillation and, consequently, about their internal structure and evolution. In this section we will deal with a few characteristics of these groups of stars.

4.1. δScuti stars.

The already known Cepheid instability strip crosses the MS in a region occupied by stars of spectral type F0V to A5V. In this region, pulsational instability has been known for more than 30 years (Eggen, 1956) with the name of δ Scuti like stars (see Baglin et al, 1973 and Breger, 1979 for excellent reviews). This type of stars is formed by a compact group of pulsating stars which show low amplitude variations ($\Delta m \sim 10$ mmag; $\Delta V \sim 5$ kms^{-1}) in periods from 30 min to several hours. Also, the driving mechanism of the pulsation (κ-mechanism) has been understood for several years.

However, their position, near the MS, goes together with a complexity of the observed frequency spectrum. Evolutionary effects have a strong influence on the ν_N distribution with depth in the core of these stars and one expects that some observed frequencies should carry information on the stellar core. By the way, non-linear coupling between modes of pulsation is not yet as severe as it is expected to be in earlier type variables. However, until very recently, only a few modes (less than 3) had been identified in the best studied stars of this type.

In the last few years, multisite differential photometric campaigns of observation have shown that this was merely a selection effect due to the low resolution and lower quality of previous data. Recent studies on, for example, θ^2 Tau (Breger et al, 1987) and 63Her (Belmonte et al, 1989c) have shown, at least, seven nearby frequencies modes, with amplitudes as low as 1 mmag, for the first of these two stars and six modes extended for a wide range of frequencies for 63Her. In this case, even some evidence of the presence of an $\ell=1$ g-mode split by rotation has been found (Mangeney et al, 1989). So that, it becomes extremely important to perform deep asteroseismologycal studies, with high quality observations, on δSct stars. These can provide us with a powerful tool to study the internal structure of a very interesting, and not well understood, region of the MS.

4.2. Rapid Oscillating Ap Stars.

The cooler (highest color indexes) sample of stars of that group known as A peculiar (Ap) is located inside the instability strip. In several of these stars, Kurtz (1982) found photometric rapid variations in the range form 4 to 12 min, with typical amplitudes of the order of 1 mmag. Nowadays, the photometric observations on these stars have quickly evolved and 13 Ap stars have been included in the roAp type so far. All them, are of spectral type Ap or Fp with strong lines of Sr, Eu and/or Cr in thir spectra. They have negative δC_1 indexes (very young stars) and color index b-y between 0.09 and 0.22 (Kurtz, 1986).

The periods of the oscillations have led to an interpretation of the frequencies as high order ($n \geq 15$), low degree ($\ell < 3$) p-modes. However, the amplitudes, 50 times bigger than expected, suggests a different mechanism than the one driving the solar-like oscillations. Several theoretical works have been undertaken and a good review of all them can be found in Shibahashi (1987). The excitation mechanism is not yet clear (magnetic overstability?, some sort of κ-mechanism?,...) but it should be probably related to the strong magnetic field present in all the roAp.

However, only one year ago (see Weiss and Schneider, 1989) the oscillations have been detected spectrometrically and the results are not conclusive. Therefore, the roAp stars form a group of very interesting objects to be studied since a lot of questions remain unclear about their oscillations. A correct answer to some of them can largely help us to interpret the true nature of these strange, but fascinating, type of stars.

4.3. Compact Oscillators.

Several years ago, Landolt (1968) discovered rapid variations (periods of~min) in the white dwarf (WD), : HL Tau-76. Several new variables of this kind have been found by means of rapid differential photometry since, forming what has been called as the ZZ Ceti type oscillating stars. This group is formed by WD of spectroscopic

type DA, however, oscillations have been found also in the hotter DB and DO stars and even in the extremely hot planetary nebula nuclei (Winget, 1988).
All these compact pulsators (oscillations are even expected in neutron stars but no one has been discovered pulsating so far) oscillate with period between 100 and 1000 seconds, with amplitudes lower than 0.1 mag. Something like 4 to 10 different frequencies of oscillation are identified; however, some of them can present as much as 30. Theoretically, they are interpreted as low degree g-modes driven and excited by the κ and γ mechanisms in the H and He ionization zones of the envelopes of these stars; therefore, they are related to other classes of pulsating stars sited inside the instability strip. An extensive review paper on this pulsators has been done by Winget (1988).
Regarding the information to be obtained from seismologycal studies of compact oscillators, it is specially important what we can learn about the behaviour of matter under extreme conditions of density and temperature. Indeed, the determination of the evolution time-scale of these objects ($\dot{P}$) can be used to calibrate the WD cooling sequence and, hence, yield a precise estimation of the galactic disk age (Winget et al, 1987). Finally, we should not forget what we can learn about the internal structure of WD itself. In conclusion, the study compact pulsators is also very interesting and it is hoped that new advances will take place, in this field, in the near future.

4.4. Solar-like Stars.

Claverie et al (1979), one decade ago, first identified the signature of the p-mode frequency spectrum for the Sun. Since then, several attempts have been conducted trying to find the frequency spectrum in other MS solar-like stars. Although some important and preliminary results have been obtained (see Table 2), no one of these attempts has been completely sucessful in finding the actual oscillation frequencies in any target star.
Nevertheless, several analytical processes (spectrum of the spectrum, folding spectrum, average spectrum, echelle diagram, etc... -see Gelly, Grec and Fossat, 1986; Belmonte, 1989 and Belmonte, Pérez Hernández and Roca Cortés, 1989 for a description of these procedures applied to spectrometric and photometric data, respectively-) of high quality data have been able to yield important outcomings (sometimes marginal) on some MS stars as for example, the range of frequencies with higher amplitude p-modes, relative amplitudes of these modes, mean spacings ($\Delta\nu_0$, D_0) between them, etc... Theoretical interpretation of some of these results become problematic, specially in the case of ϵEri (Däppen and Soderblom, 1988 and references therein) and αCen. On the other hand, the values obtained on αCMi (Gelly, Grec and Fossat,1986) and HD155543 seem to agree quite well to what one would expect from theoretical predictions. However, new results on αCMi (Schmider, 1989) does not seem to confirm previously claimed results.

In Figure 1, the H-R diagram for Asteroseismology already cited in Section 2 is presented. As it can be seen, a precise determination of $\Delta\nu_0$ and D_0 can give us good estimations of M and X_c. However, still very few observational points can be drawn to such a diagram up to now. Observational techniques are evolving very quickly in the last years and these promising initial results gives us the hope that, in a short, the whole oscillation frequency spectrum of a MS star might be discovered. Its theoretical interpretation will provide us with the most powerful and efficient tool to probe the stellar structure and evolution theories.

Finally, we would like to mention that, as a logical extension of solar-like stars Asteroseismology, new works have been started on the study of oscillation in the next evolutionary stage (normal red giants) and, even, in very massive, gaseous planets (e.g. Jupiter). First and promising results have been obtained on the K2III star Arcturus (Belmonte et al, 1989b; Schmider, 1989); also, both theoretical (Mosser, Gautier and Delache, 1988) and observational (Schmider, 1989) works have been undertaken on Jupiter, opening new viewpoints to the study of these two types of objects.

5. CONCLUSIONS AND FUTURE PERSPECTIVES

To draw conclusions on such a new subject can be premature although work has already started with very good results already. Nevertheless, one can foresee an explendid future for this field in the view of the potential information that can be extracted from observational seismological data.

In such small number of years of intensive observational work, old spectrometric and photometric techniques are being pushed to their limits. In parallel the development of new techniques, already succesfully applied to the sun, beguin to grow and they start showing their first results. Photometric thecniques do not seem to be the way to measure slar-like stars oscillations at least from ground. For these the new spectrometric techniques seem to be the ones to be used in the near future, although the lack of photons severely constrain its usefulness. However, it is within a foreseeable future that these techniques will improve allowing medium to low size telescopes (1 to 2 meters in diameter) to become usefull for these studies. Moreover, new thechniques, such as heterodyne spectroscopy, (Deeming et al., 1988) can also offer more posibilities.

The strategy of observation can be another handicap for future observations in this subject. Indeed very long (weeks) continuous runs on only one target are needed in order to obtain good S/N ratios. The possibilities of looking at several targets at once (CCD photometry of clusters, Gilliland (1988)) offers an atractive way of exploring this problem, provided enough precision is achieved. All these forces international cooperative ground based programs or space borne experiments to be launched to achieve the goals of continuous long term observation of any target.

Star	ST	P (min)	A_{max} (cms^{-1})	A_{max} (μmag)	$\Delta\nu_0$ (μHz)	D_0 (μHz)	n	ℓ	Ref.
ϵEri	K2V	10-12		1000	172.0	-	8-9	0-1	a
αCen	G2V	3.5-8.5	150		165.3	2.6	15-30	0-2	b
αCMi	F5IV	10-25	70		79.4	<.6	10-20	0-2	b
HD155543	F2V	5-16		20	97.3	1.4;1.8	14-33	0-2	c
Sun	G4V	3.5-8.5	15	6	135.4	1.54	14-30	0-3	d,e

Table 2: Oscillation parameters obtained so far in solar-like stars. For each star, the spectral type, period range of the p-modes, maximum amplitude either in velocity or luminosity, the mean spacings $\Delta\nu_0$ and D_0, and the range of suspected orders n and degrees ℓ are listed. The Sun is given for comparison. *References: a) Noyes et al, 1984; b) Gelly, Grec and Fossat, 1986; c) Belmonte, Pérez Hernández and Roca Cortés, 1989; d) Pallé, 1989.; e) Jiménez et al, 1987.*

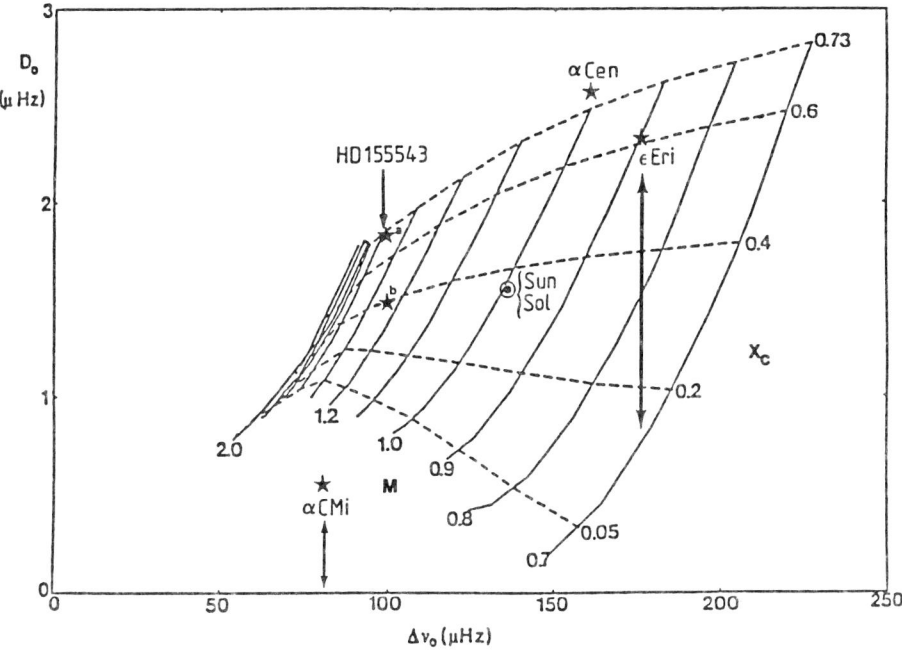

Figure 1: JCD diagram (for Jorgen Christensen-Dalsgaard) of asteroseismologycal data. Models with mass (in $M_\odot$) and central hydrogem abundance (X_c) , taken as free parameters, are used to calculate their p mode spectrum and the mean spacings $\Delta\nu_0$ and D_0. On it, the symbol $\star$ represents values of these two parameters given in Table 2 for all solar-like stars where some of them has been found. The Sun ($\odot$) is also plotted for calibration purposes. For HD155543 a and b stand for two possible values of D_0 consistant with actual observations.

The information to be gained is so rewarding that these efforts should be pursued and encouraged.

On the theoretical side, new and very important work has to be pursued. the efforts already made on the sun just show a way to initiate a deep study on the structure and physics of the stellar interiors. Inversion techniques are just being tried and applied to the sun but no doubt they will be of great importance and interest to the future of Asteroseismology and stellar structure and evolution.

6. REFERENCES

Baglin A., Breger M., Chevalier C., Hauck B., Le Contel J.M., Sareyan J.P. and Valtier J.C.; 1973. *Astron. Astrophys.* **23**, 221.

Belmonte J.A.; 1989. *Ph.D. Thesis. Universidad de La Laguna (Spain)*.

Belmonte J.A., Bell C.R., Leeper M., Pallé P.L., Pietraszewski K.A.R.B., Renton R.E., and Roca Cortés T.; 1989a. *Astron. Astrophys.* **221**, 41.

Belmonte J.A., Jones A.R., Pallé P.L. and Roca Cortés T.; 1989b. *Astrophys. Space Sci.* (In press).

Belmonte J.A., Mangeney A., Chevreton M., Praderie F., Saint-Pé O., Puget P., Alvarez M. and Roca Cortés T.; 1989c. *Astron. Astrophys.* (Submitted).

Belmonte J.A., Pérez Hernández F. and Roca Cortés T.; 1989. *Astron. Astrophys.* (In press).

Breger M.; 1979. *Pub. Astron. Soc. Pacific.* **91**, 5.

Breger M., Lin H., Shi-Yang J., Zihe G., Antonello E. and Mantegazza L.; 1987; *Astron. Astrophys.* **115**, 117.

Brookes J.R., Isaak G.R. and van der Raay H.B. ; 1978. *Mon. Not. Roy. Astron. Soc.* **185**, 1.

Brown T.M.; 1979; *in Solar instrumentation: What's next?*. ed. R.B. Dunn, 150.

Cacciani A. and Fofi M.; 1978. *Solar Phys.* **59**, 179.

Campbell B. and Walker G.A.H.; 1979. *Pub. Astron. Soc. Pacific* **91**, 540.

Campbell B. and Walker G.A.H.; 1985. *IAU Colloquium 88*, 5.

Christensen-Dalsgaard J.; 1982. *Mon. Not. R. Astron. Soc.* **199**, 735.

Christensen-Dalsgaard J.; 1988. *In Proc. IAU Sym. n° 123: Advances in Helio- and asteroseismology, J. Christensen-Dalsgaard and S. Frandsen (eds)*, 295.

Christensen-Dalsgaard J.; 1989. *Lecture Notes on Stellar Oscillations. Aarhus Universitet.*

Christensen-Dalsgaard J. and Frandsen S.; 1983. *Solar Phys.* **82**, 469.

Claverie A., Isaak G.R., McLeod C.P., van der Raay H.B. and Roca Cortés T.; 1979. *Nature.* **282**, 591.

Cowling T.G.; 1942. *Mon. Not. R. Astron. Soc.* **101**, 367.

Däppen W. and Soderblom D. 1988. *In Proc. of the ESA-IAC Sym. Seismology of the Sun ans sun-like stars. E.Rolfe (ed)*. SP-286, 653.

Deubner F.L. and Isserstedt J.; 1983. *Astron. Astrophys.* **126**, 216.

Deubner F.L. and Gough D.E.; 1984. *Ann. Rev. Astron. Astrophys.* **22**, 593.

Eggen O.J.; 1956. *P.A.S.P.* **68**, 238 and 541.

Fossat E. and Roddier F.; 1971. *Solar Physics* **18**, 204.

Fossat E.; 1984. *Proc. on Improvements to Photometry, Borucki and Young (eds.)*. NASA Conf. Pub. **2350**, 68.

Gelly B., Grec G. and Fossat E.; 1986. *Astron. Astrophys.* **164**, 383.

Gilliland R.L. and Brown T.M.; 1988. *Pub. Ast. Soc. Pacific* **100**, 754.

Glenar D.A., Deming D., Espenak F., Kostiuk T. and Mumma M.J. ;1986. *Applied Opt.* **25**, 58.

Griffin R.F.; 1967. *Ap. J.* **148**, 465.

Harvey J.W.; 1988. *In Proc. IAU Sym. n° 123: Advances in Helio- and asteroseismology*, S. Frandsen and J. Christensen-Dalsgaard (eds), 497.

Isaak G.R., Jones A.R.; 1988. *In Proc. IAU Sym. n° 123: Advances in Helio- and Asteroseismology*, S. Frandsen and J. Christensen-Dalsgaard (eds), 255.

Jiménez A., Pallé P.L., Roca Cortés T., Domingo V. and Korzennik S.; 1987. *Astron. Astrophys.* **172**, 323.

Kreidl T.J., Garrido R., Lin H., Zihe G., Belmonte J.A., Fernie D.J., Zwerko J. and Ziznovský J. 1989. *Mon. Not. R. Astron. Soc.* (Submitted).

Kurtz D.W.; 1982. *Mon. Not. R. Astron. Soc.* **200**, 807.

Kurtz D.W.; 1984. *In Proc. NASA/SDSU Workshop on improvements in photometry*, Borucki and Young (eds). NASA Conf. Pub. **2350**, 56.

Kurtz D.W.; 1986. *In Proc. NATO Workshop, Seismology of the Sun and the distant stars*, D. Gough (ed). NATO ASI series **169**, 417.

Kurtz D.W., Matthews J.M., Martinez P., Seeman J., Cropper M., Christopher Clemens J., Kreidl T.J., Sterken C., Schneider H., Weiss W.W., Kawaler S.D., Kepler S.O., van der Peet A., Sullivan D.J. and Wood H.J.; 1989. *Mon. Not. R. Astron. Soc.* (In press).

Landolt A.U.; 1968. *Ap. J.* **153**, 151.

Ledoux P. and Walraven Th.; 1958. *Handbuch der Physics.* **51**, 353.

Libbrecht K.G.; 1988. *Ap. J. Letters.* **330**, L51.

Mangeney A., Chevreton M., Belmonte J.A., Däppen W., Saint-Pé O., Praderie F., Roca Cortés T., Fuensalida J.J. and Alvarez M.; 1988. *In Proc. of the ESA-IAC Sym. Seismology of the Sun ans sun-like stars*. E.Rolfe (ed). SP-286, 551.

Mangeney A., Däppen W., Praderie F. and Belmonte J.A.; 1989. (In preparation).

Matthews J.M., Wehlau W.H., Walker G.A.H. and Yang S.; 1988. *Ap. J.* **324**, 1099.

Mosser B., Gutier D. and Delache Ph.; 1988. *In Proc. of the ESA-IAC Sym. Seismology of the Sun ans sun-like stars*. E.Rolfe (ed). SP-286, 593.

Noyes R.W., Baliunas S.L., Belserene E., Duncan D.K., Horne J. and Widrow L.; 1984. *Ap. J. Letters.* **285**, L23.

Pallé P.L.; 1989. *In Solar interior and atmosphere*. Cox A.N., Livingston W.C. and Matthews M.S. (eds). Chap. 4, (In press).

Pérez Hernández F.; 1989. *Ph.D. Thesis. Universidad de La Laguna (Spain)*.

Schmider F.X.; 1989. *Ph.D. Thesis. Universite de Paris VII (France)*.

Shibahashi H.; 1987. *Lec. Notes Phys.* **274**, 112.

Smith M.H.; 1983. *Ap. J.* **265**, 325.

Smith P.H., McMillan R.G. and Merline N.J.; 1987. *Ap. J. Letters.* **317**, L79.

Solheim J.E. 1988. *In Proc. of the ESA-IAC Sym. Seismology of the Sun ans sun-like stars.* *E.Rolfe (ed)*. SP-286, 637.

Stein R.F. and Leibacher J.W.; 1974. *Ann. Rev. Astron. Astrophys.* **12**, 407.

Tassoul M.; 1980. *Ap. J. Sup. Ser.* **43**, 469.

Traub W.A., Mariska J.T. and Carleton N.P.; 1978. *Ap.J.* **223**, 583.

Vandakurov Y.V.; 1967. *Astro Zh.* **44**, 786.

Weiss W.W. and Schneider H.; 1989. *Astron. Astrophys.* **224**, 101.

Winget D.E.; 1988. *In Proc. IAU Sym. n° 123: Advances in Helio- and asteroseismology*, S. Frandsen and J. Christensen-Dalsgaard (eds), 305.

Winget D.E., Hansen C.J., Liebert L., Van Horn H.M., Fontaine G., Nather R.E., Kepler S.O. and Lamb D.Q.; 1987. *Ap. J. Letters.* **315**, L77.

WHITE DWARFS

J. ISERN[1,4], E. GARCIA-BERRO[2,4], M. HERNANZ[3,4] and
R. MOCHKOVITCH[5]

[1] Centre d'Estudis Avançats de Blanes, CSIC, 17300 Blanes (Girona), Spain.
[2] Departament de Física Aplicada, ETSECCPB, Universitat Politècnica de Catalunya, Jordi Girona Salgado 31, 08034 Barcelona, Spain.
[3] Departament de Física i Enginyeria Nuclear, ETSEIB, Universitat Politècnica de Catalunya, Diagonal 647, 08028 Barcelona, Spain.
[4] Laboratori d'Astrofísica, Societat Catalana de Física, I.E.C.
[5] Institut d'Astrophysique du CNRS, 98 bis Boulevard Arago, 75014 Paris, France.

1 INTRODUCTION

In elementary books, white dwarfs are defined as stars with a mass of the order of the mass of the Sun, a radius of the order of the radius of the Earth and an average density of the order of 10^6 g/cm^3. The overall structure is supported by the pressure of degenerate electrons and, as they cannot longer burn their nuclear fuel, they can only cool forever. Beneath this simple picture, however, there is a fascinating astronomical object containing a lot of information on the behavior of matter at extreme densities and temperatures, on the late stages of stellar evolution and on the processes that have shaped the galaxy.

The first attempt to describe the cooling process of white dwarfs was due to Mestel (1952). He assumed a very thin outer envelope, an isothermal core, an opacity of the outer layers given by the Kramer's law, and ions behaving like an ideal gas. The outcome was a simple relationship linking the luminosity of the star to the cooling time: $\log t = -5/7 \log(L/L_\odot) + C$

This behavior can be directly tested by means of the luminosity function. This function is defined as the number of white dwarfs per unit of volume and per unit of magnitude interval versus the luminosity (or magnitude). Because the rate of cooling is slower for cooler stars, the simple theory described above predicts a monotonic function that increases as the magnitude increases, as well as a cutoff due to the finite age of the galaxy (Winget et al 1987).

The first attempts to construct the white dwarf luminosity function were due to Weidemann (1967), Kovetz and Shaviv (1976), Sion and Liebert (1977), but only after the completion of the Luyten Half Second (LHS) catalogue it has been possible to construct it from a reasonably complete sample of white dwarfs (Liebert et al 1988). The hot part of the function has been obtained from the Palomar-Green (PG) catalogue and the cool one from the LHS catalogue. The sources of error and incompleteness (white dwarfs hidden in binary systems, selection effects) have a contribution smaller than a factor of 2 in the computations (Liebert et al 1989). The luminosity function constructed in this way increases monotonically with the magnitude as expected, and displays a well defined cutoff at $M_v \approx 16$. In order to compare it with theory, visual magnitudes have to be converted into bolometric magnitudes (or, equivalently, luminosities). The bolometric corrections (BC) for hot white dwarfs are known, but for the coolest ones they are not available. In principle, it is possible to bound the BC assuming that the black body BC for a given effective temperature applies or that the bolometric correction is zero. This uncertainty implies that the cutoff is located within the interval $-4.2 \geq \log (L/L_\odot) \geq -4.6$ (Liebert et al 1988).

2 FUNDAMENTAL PROPERTIES OF WHITE DWARFS

Single stars with masses smaller than 10 $M_\odot$ develop a degenerate core and obtain their energy from the burning of H and He in shells. As the H burning shell consumes fuel faster than the He one, He accumulates and eventually flashes when it reaches a critical mass. This stage lasts for 10^5-10^6 years. At the end of this epoch, the star develops an instability not yet completely understood, the "superwind", that results in the ejection of the external layers. The bare core left becomes the central star of a planetary nebula and, finally, a white dwarf.

The newly formed object is characterized by its stratified structure. It is composed by an inner core made of by ONeMg or CO (or in some cases by He), that amounts to almost all the mass of the star, surrounded by an envelope composed by an inner He layer and an outer H layer that can be absent in a number of cases ($\approx 25\%$). The presence and the size of the H envelope is not well understood and depends on the details of the final evolutionary stages of AGB stars (Schönberner 1983, Iben 1984, Mazzitelli and D'Antonna 1986, Wood and Faulkner 1986).

According to their spectrophotometric properties, white dwarfs are classified as (see Sion 1986 for a complete phenomenological description): DA: They have pure H layers, with temperatures in the range of $9 \cdot 10^4$-$7 \cdot 10^4 K \geq T_e \geq 6 \cdot 10^3 K$. DO: They are the hottest white dwarfs known, with temperatures in the range $10^5 K \geq T_e \geq 4.5 \cdot 10^4 K$. DB: They have nearly pure He atmospheres. Their temperatures are in the range $3 \cdot 10^4 K \geq T_e \geq 1.2 \cdot 10^4 K$. Notice the gap between DO and DB. DQ: Their atmospheres are dominated by helium and they have C abundances in the range of 10^{-7} to 10^{-2}. Their temperatures are in the range $1.2 \cdot 10^4 K \geq T_e \geq 6 \cdot 10^3 K$. DZ: They only display metallic features (CaII H-K). The effective temperature is too small to show the dominant constituent. DC: They are so cool that the dominant component is not seen. They have no lines deeper than 5%.

The chemical composition of the inner core depends on the mass of the progenitor star. Stars in the range 0.8 to 8 $M_\odot$ (these figures are rather uncertain) produce CO white dwarfs with masses in the interval 0.5 to 1.15 $M_\odot$. Stars with masses higher than $8M_\odot$ produce ONeMg white dwarfs with masses in the range of 1.2 to 1.4 $M_\odot$ (Iben and Tutukov 1985). Stars with masses smaller than 0.8 $M_\odot$ have had no time to produce white dwarfs. Therefore, taking into account the shape of the initial mass function, it is clear that CO white dwarfs are representative of field white dwarfs.

Concerning the mass distribution of white dwarfs, it strongly peaks around 0.6 $M_\odot$ with a dispersion ± 0.1 $M_\odot$, irrespectively of their DA non-DA character (Weidemann and Yuan 1989).

Seismological data, however, provides an independent way to measure not only the total mass of the star (Kawaler 1987, Kawaler and Hansen 1989) and to probe the external layers (Winget et al 1981, 1982, Bradley et al 1989), but also to directly measure the evolutionary time scales (Kawaler and Hansen 1989).

3 THE INNER CORE

The inner core is also stratified because it is the result of the 3α reactions to give carbon and the $C\alpha$ reaction to give oxygen. It can be divided into three parts: A inner one produced during the central He burning, an intermediate one produced during the thick He shell burning and an outer one produced during the flashing phase (Mazzitelli and D'Antonna 1986, 1987). As T and ρ are different during such epochs, the final abundances are also different. As far as He is abundant, the 3α reaction is dominant and C accumulates. As soon as the He abundance is smaller than some critical value (about 0.1 in mass fraction) the $C\alpha$ reaction dominates and C is converted into oxygen. Of course the final abundances depend on the adopted reaction rates. If the rates of Harris et al (1983) are adopted for the $C\alpha$ reaction, the final abundance of oxygen is $X_0=0.5$. If the rates of Fowler et al 1975 are adopted, $X_0=0.75$. Furthermore, as $C\alpha$ reactions are favored by low temperatures, oxygen has the tendency to concentrate in the central layers. This effect is enhanced by the convective overshooting. Anyway, as the degree of stratification diminishes with the mass of the white dwarf and as the true ratio of the $C\alpha$ reaction is not known, it is reasonable to assume, at least provisionally, that CO white dwarfs are composed by a fifty fifty mixture of such elements.

As the core cannot ignite its nuclear fuel, its only source of energy is its thermal energy. Due to the high pressures and temperatures, elements are completely ionized and electrons, which have a high Fermi energy act like a neutralizing background. This plasma, called Coulomb plasma, can be characterized by the plasma coupling constant that is a measure of the Coulomb energy of the interacting ions and their kinetic energy: $\Gamma = (Z^2 e^2/a)/kT$, where a is the average distance between ions. If $\Gamma \simeq 0$, the Coulomb plasma behaves like an ideal gas. If $\Gamma \leq 1$ the Coulomb corrections are more and more important and in the interval $1 \leq \Gamma \leq 180$ the plasma behaves like a liquid while for higher values this plasma crystallizes into a BCC lattice. During the solidification process there is a release of latent heat of the order of $1 \approx (R\,T_{sol})/\mu$. Detailed descriptions of the thermodynamic properties of a Coulomb plasma composed by only one chemical species can be found in Ichimaru et al (1987).

White dwarf interiors, however, are not made of only one chemical element but by a mixture of elements, and as the solidification starts, there is always some degree of chemical separation, even in the case where the components are miscible in all proportions. This effect is not negligible. For instance, if carbon and oxygen were not miscible in the solid phase, pure carbon flakes would form and rise towards the surface, melting there. Meanwhile, pure oxygen flakes would sink and settle at the center. If this process would continue all the star would become completely differentiated and $\approx 10^{47}$ erg of gravitational energy would be released. As this would happen at $L = 10^{-4} L_\odot$, the cooling process would be delayed ≈ 8 Gyr.

To determine the miscibility of carbon and oxygen in the solid phase is a difficult task because it is necessary to know the free energy in both the fluid and solid phases and in the last case this depends on the configuration adopted by the mixture. For instance, Stevenson (1980), adopting the Wigner-Seitz approximation for the electrostatic energy ($u \propto \langle Z^{5/3} \rangle$), obtained a phase diagram of the spindle form thus, implying that the solid phase would be slightly enriched in oxygen, but adopting the electrostatic energy of a random alloy ($u \propto \langle Z \rangle^{5/3}$) he obtained an eutectic and total immiscibility of carbon and oxygen nuclei in the solid phase, thus implying that they would separate at the onset of crystallization. In this case, depending on the details of the cooling process, the final outcome was the formation of an oxygen core surrounded by a carbon-oxygen mantle with the eutectic composition or a pure oxygen core surrounded by a pure carbon mantle. The total energy released in the last case was $\approx 6 \cdot 10^{46}$ erg.

Barrat et al (1988), using a density functional approach to the free energy, found a phase diagram of the spindle form. This phase diagram leads to a smaller change than the former case in the distribution of the chemical elements and the energy release is 1/3 of the energy released in the previous case. This circumstance, together with the fact that the freezing temperature is higher, translates into a shorter cooling delay.

More recently, Ichimaru et al (1988) have shown that the electrostatic energy of a solid mixture can be better approached by using a linear interpolation of the thermodynamic functions of the single components than by using the random alloy approximation. They obtain, in this case, an azeotropic diagram. The final oxygen distribution is similar to that obtained by Barrat et al (1988) but, as solidification happens at lower temperatures the influence is slightly larger.

Nevertheless, an additional problem comes from the fact that the difference between the free energies of the liquid and the solid phase is of the same order as the uncertainties of the free energy in the solid phase (Gordon et al 1989).

4 THE OUTER LAYERS

The outer layers of white dwarfs are very important because they control the cooling process and they provide constraints to the previous evolution. To understand their behavior we have to remember that: i) Their initial chemical composition is not solar but that resulting from H and He burning in AGB stars. ii) Gravitational settling is very efficient to produce chemically pure atmospheres and white dwarfs only display at their surface the lightest component (Schatzman 1958). iii) Homogeneity can be partially restored by convective mixing, radiative levitation, thermal diffusion, accretion winds and nuclear burning (Koester 1989, Vauclair 1989).

5 THE HELIUM LAYER

The role of the helium layer is fundamental. Not only because of its energy release, but also because of its control of the diffusion of the outer H (if present) inwards and of the inner C outwards. Its effects can be summarized in the following way: i) If carbon encounters hydrogen, protons are burned and the mass of the H layer can be reduced by a factor two. ii) If the mass of the helium layer is smaller than some critical value, the encounter between hydrogen and carbon can even produce a thermal runaway (Iben and McDonald 1985). iii) If H is not present and the mass of the helium layer is in the range 10^{-3}-10^{-4} $M_\odot$, convection at the bottom of the helium layer can dredge up carbon and by diffusion expose it, converting DB types into DQ types (Shipman 1989). iv) Furthermore, as radiative opacities of C are higher than the He ones, thinner helium layers produce longer cooling times (Wood and Winget 1989).

6 THE HYDROGEN LAYER

The hydrogen layer also plays a very important role because: i) it is a source of opacity (the evolution of the non-DA stars is always faster) and ii) depending on the mass of the layer and on the presence of CNO elements H-burning can be an important source of energy.

The size of the hydrogen layer is an open question. There are two opposite points of view. On one hand, theoretical evolutionary models (Iben and Tutukov 1985; Koester and Schönberner 1986; Mazzitelli and D'Antonna 1986) predict a mass of the hydrogen layer, if present, of the order of 10^{-4} $M_\odot$, but the pulsating properties of ZZ Ceti stars (the DAVs) indicate that the H layer must have a mass smaller than 10^{-7} $M_\odot$. Thus, it is necessary a mechanism to reduce the mass of the layer below this quantity. In their models, Iben and Tutukov (op.cit.) have shown that H-burning via CNO-cycle is an important source of energy in the interval 10 to $10^{-2} L_\odot$. These results, however, depend on the total amount of CNO elements present in the surface layers (Koester, op.cit; D'Antonna and Mazzitelli, op.cit.) and the abundance of such elements depends not only on the initial abundances but also on the diffusion processes induced by gravitation, convective mixing and mass losses. For instance, if diffusion through the helium layer is taken into account, the mass of the hydrogen layer is never reduced by a factor bigger than four (Iben and MacDonald 1985). The change of the ratio of DA/non-DA white dwarfs as the cooling process goes on (Greenstein 1986) also demands small hydrogen layers (Shipman 1989). This change begins at 9000 K and is very clear at 7000 K (the number of DAs represents more than the 70% at T_e > 7000 K and only the 20% at T_e < 7000 K). This phenomenon is interpreted as due to the mixing of the H and He layers because of convection. A moderate wind could do the job, but this possibility has to be already explored.

On the other hand, it is assumed (Fontaine and Wesemael 1987) that the hydrogen layer is very thin ($\approx 10^{-13}$ $M_\odot$) in such a way that at very high T_e, all white dwarfs look like PG1159 and H is uniformly mixed with He. As the star cools down, H diffuses upwards and some white dwarfs begin to look like DAs. When T_e reaches 45000 K, all white dwarfs have become DAs (this would explain the gap between DBs and DQs, but see D'Antonna 1989). When T_e arrives to 30000 K, the He zone becomes convective at the interface and H mixes once more becoming gradually DBs. When T_e reaches a value of 5000 to 10000 K, a deep convection zone appears and most DAs turn into non-DAs.

7 THE COOLING PROCESS

The cooling process of a white dwarf can be divided into four stages (Iben and Tutukov, 1984; Koester and Schönberner 1986; D'Antonna and Mazzitelli 1989; D'Antonna, 1989). During the first one ($\log L/L_\odot \geq -1.5$; $\log T_e \geq 4.3$) the star is hot enough to support H burning (via CNO cycle and pp chains) as well as He burning and to lose copious amounts of energy because of thermal neutrinos. The importance of the nuclear contribution to the total luminosity and to the duration of this stage strongly depends on the input parameters (total mass of the white dwarf, initial mass of the H and He layers) and on the treatment of the diffusion in the outer layers. Although the time that a white dwarf lasts at this epoch has a small contribution to the total cooling time ($\leq 10^8$ years), it is fundamental to interpret the meaning of the relative populations of DO, DA and DB white dwarfs (D'Antonna 1989).

During the second phase ($-1.5 \geq \log (L/L_\odot) \geq -3$; $4.3 \geq \log T_e > 4.0$) all the energy sources continue to decline and the inner structure is dominated by the cooling process. Depending on the mass of the hydrogen layer, pp-burning could have a residual contribution to the total luminosity, but even if it were dominant, the slope of the cooling curve would not be significantly different from the Mestel's one (Iben and Tutukov, 1984).

At the third stage ($-3 \geq \log (L/L_\odot) \geq -4.5$; $4.0 \geq \log T_e \geq 3.6$) the central layers begin to crystallize and the cooling process is slowed down due to the latent heat release (Van Horn, 1971 and Shaviv 1979) and the gravitational settling due to phase transitions (Mochkovitch, 1983; Garcia-Berro et al 1988a,b; Isern et al 1989). At the same time, the densities reached by the external layers are so high and the temperature is so low that they enter into the pressure ionization region and the relationship between the central and the effective temperatures changes in a way not yet completely understood (D'Antonna 1989; D'Antonna and Mazzitelli 1989) but that can be very important in shaping the luminosity function.

Finally, the last stage ($\log L/L_\odot < -4.5$; $\log T_e < 3.6$) corresponds to the Debye cooling. When the temperature becomes smaller than the Debye temperature θ, the specific heat scales as $(T/\theta)^3$ and the star cools down very quickly. As in the previous phase, the cooling rate depends not only on the behavior of the inner layers but also on that of the outer ones which can prevent the acceleration of the process (D'Antonna 1989; D'Antonna and Mazzitelli 1989).

The total cooling time strongly depends on the physical and evolutionary input parameters. Fortunately, it is possible to bound the effect of each factor and reduce the results of the different models to a unique evolutionary track. Thus it seems natural to define a fiducial model and to refer all the calculations to it (Winget and Van Horn 1987). The parameters that define this model are: $M_{WD} = 0.6\ M_\odot$, $X_C = X_O = 0.5$, surface envelope parameters $X_H = 0.0$, $X_{He} = 1-Z$, $Z = 10^{-5}$, $M_{He} = 10^{-3} M_\odot$ and conductive opacities and neutrino emissivities from Itoh (1989). See figure 1 to compare the characteristic cooling times for the fiducial model, case a, the phase diagram of Barrat et al (1988), case b, that of Ichimaru et al (1989), case c, and that of Stevenson (1980) for partial, case d or total separation, case e).

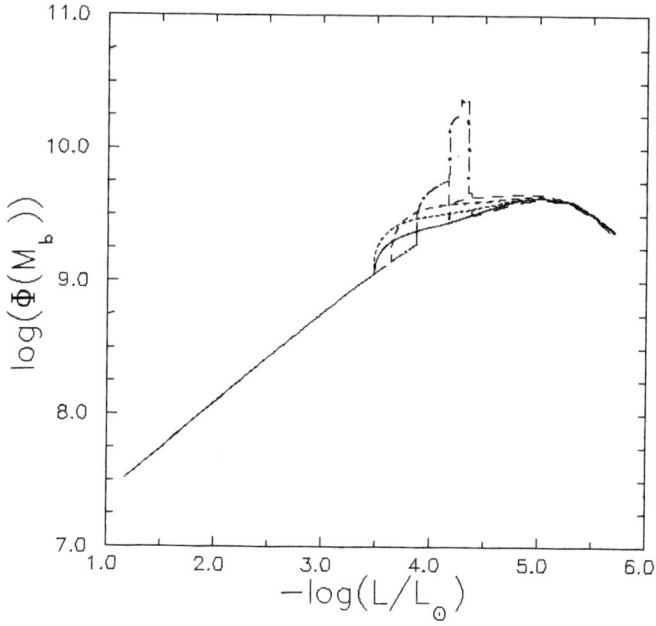

Figure 1: Characteristic cooling times (in years) for cases (a) solid line, (b) short dashed line, (c) dashed line, (d) long dashed line and (e) dashed-dotted line.

Notice that the settling of oxygen rich matter at the center of the star always introduces a delay in the cooling process. Table 1 displays the time necessary to reach the luminosity $\log L/L_{\odot}=-4.5$ and the corresponding delays respect to the fiducial model due to the different phase diagrams. Notice also, that minor species can play a very important role in the cooling process. For instance, ^{22}Ne amounts to 1 or 2% of the total mass. Due to its small number of electrons per nucleon, its sedimentation would induce a delay of 2 Gyrs.

To conclude we want to point out that the luminosity function contains fundamental information on the evolution of the solar neighborhood. Nevertheless, before extracting this information it is necessary to obtain a deep knowledge of the structure and evolution of white dwarfs.

Table 1

Ages (Gyr)	Delay (Gyr)	Phase diagram	
8.0	-		(a)
9.0	1.0	Barrat et al (1988)	(b)
9.5	1.5	Ichimaru et al (1989)	(c)
9.8	1.8	Stevenson (1980)	(d)
16.2	8.2	Stevenson (1980)	(e)

REFERENCES

Barrat J.L., Hansen J.P., Mochkovitch R.: 1988, Astron. Astrophys. **199**, L15.

Bradley P.A., Winget D.A., Wood M.A.: 1989, in IAU Col No 114 "White Dwarfs", ed G.Wegner, Springer-Verlag p.286.

D'Antonna F.: 1989, in IAU Col No 114 p.44.

D'Antonna F., Mazzitelli I.: 1989, preprint.

Fontaine G., Wesemael F.: 1987, in IAU Col No 95 "The Second Conference on Faint Blue Stars", ed A.G.D. Philip, D.S. Hayes and J.W. Liebert (Schenectady: L.Davis Press), p.319.

Fowler W., Caughlan G.R., Zimmermann B.A.: 1975, Ann. Rev. Astr. Astroph. **13**, 69.

Garcia-Berro E., Hernanz M. Mochkovitch R., Isern J.: 1988a, Astron. Astrophys. **193**, 141.

Garcia-Berro E., Hernanz M., Isern J. Mochkovitch R.: 1988b, Nature **333**, 644.

Gordon P., Shaviv G., Ashkenazi J., Kovetz A.: 1989, in IAU Col No 114, p.85.

Greenstein J.L.: 1986, Astrophys. J. **304**, 334.

Harris M.J., Fowler W.A., Caughlan G.R. and Zimmermann B.A.: 1983, Ann. Rev. Astr. Astroph. **21**, 165.

Iben I.: 1984, Astrophys. J. **277**, 333.

Iben I., McDonald J.: 1985, Astrophys. J. **296**, 540.

Iben I., Tutukov V.: 1985, Astrophys. J. Suppl. **58**, 661.

Ichimaru S., Iyetomi H., Tanaka, S.: 1987, Phys. Rep. **149**, 91.

Ichimaru S., Iyetomi H., Ogata S.: 1988, Astrophys. J. **334**, L17.

Isern J., Garcia-Berro E., Hernanz M., Mochkovitch R.: 1989, in IAU Col **114**, p.278.

Itoh N.: 1989, in IAU Col **114**, p.66.

Kawaler S.D.: 1987, in IAU Col No 95, p.297.

Kawaler S.D., Winget D.E., Iben I., Hansen C.J.: 1986, Astrophys. J. **302**, 530.

Kawaler S.D., Hansen C.J.: 1989, in IAU Sym No 114 ,p.97.

Koester D.: 1989, in IAU Col 114, p.206.

Koester D. and Schönberner D.: 1986, *Astron. Astroph.* 154, 125.

Koester D., Schulz H. Weidemann V.: 1979, *Astron. Astrophys.* 76, 262.

Kovetz A. and Shaviv G.: 1976, *Astron. Astrophys.* 52, 403.

Liebert J., Dahn C.C. and Monet D.G.: 1988, *Astrophys. J.* 332, 891.

------------------------------------: 1989, in IAU Symp. 114, p.15.

Mazzitelli and D'Antonna F.: 1986, *Astrophys. J.* 308, 706.

-------------------------: 1987, in IAU Col No 95 , p.351.

Mestel L.: 1952, *Monthly Notices Roy. Astron. Soc.* 112, 583.

Mochkovitch R.: 1983, *Astron. Astrophys.* 122, 212.

Oke J.B., Weidemann V., Koester D.: 1984, *Astrophys. J.* 281, 276.

Schatzman E.: 1958, in "White Dwarfs", North Holland.

Schönberner D.: 1983, *Astrophys. J.* 272, 708.

Shaviv G.: 1979, in IAU Col 53 "White Dwarfs and Variable Degenerate Stars", ed V.Weidemann and H.M. Van Horn, U. of Rochester Press, p.11.

Shipman H.: 1989, in IAU Col 114, p. 220.

Sion E.M.: 1986, *Publ. Astron. Soc. Pac.* 98, 821.

Sion E.M. and Liebert J.: 1977, *Astrophys. J.* 213, 468.

Stevenson D.J.: 1980, *J.Phys. Suppl.* 49, 375.

Van Horn H.M.: 1971, in IAU Col. 42, p.97.

Vauclair V.: 1989, in IAU Col. 114, p.176.

Weidemann V.:1967, *Zs. f. Astrophys.* 67, 286.

Weidemann V.: 1970, in IAU Symp. No42 "White Dwarfs", ed. W.J. Luyten. Reidel, p.81.

Weidemann V., Yuan J.W.: 1989, in IAU Symp No114, p.1.

Winget D.E., Hansen C.J., Liebert J., Van Horn H.M., Fontaine G., Nather R.E., Kepler S.O., Lamb D.Q.: 1987, *Astrophys. J.* 315, L77.

Winget D.E., Kepler S.O., Robinson E.L., Nather R.E., and O'Donoghue D.: 1985, *Astrophys. J.* 292, 606.

Winget D.E., Van Horn H.M.:1987, in IAU Col. 95, p 363.

Winget D.E., Van Horn H.M., and Hansen C.J.: 1981, *Astrophys. J.* 245, L33.

Winget D.E., Van Horn H.M., Tassoul M., Hansen C.J., Fontaine G., and Carroll B.W.: 1982, *Astrophys. J.* 252, L65.

Wood P.R. and Faulkner D.J.: 1986, *Astrophys. J.* 307, 659.

Wood M.A. and Winget D.E.: 1989, in IAU Col 114, p.282.

The Supernova 1987A

W. HILLEBRANDT and P. HÖFLICH

Max-Planck-Institut für Physik und Astrophysik, Institut für Astrophysik

Karl-Schwarzschild-Straße 1, 8046 Garching, FRG

1. INTRODUCTION

When on February 23, 1987, a supernova explosion was observed in the Large Magellanic Cloud (LMC) at a distance of roughly 50 kpc only, this event caused a lot of excitement in the astronomical community, because never since Kepler's supernova in 1604 has a supernova been so close. There was some hope, therefore, that most of the open questions in supernova theory would be answered. However, it now has become clear that this is not the case. Moreover, supernova 1987A raised new problems, which will be discussed in some detail in this article.

Spectra taken during the second night showed Balmer lines of hydrogen, indicating that the supernova was of type II. Its position coincided with that of a blue supergiant, Sanduleak -69° 202. When the UV-radiation from the supernova weakened a few weeks after the explosion it became obvious that this star had indeed disappeared. From its spectral class (B3) and its luminosity class (Ia) one could conclude that the main sequence mass of the progenitor had been close to 20 solar masses. The first big surprise was that, in contrast to theoretical expectations, it was a blue rather than a red supergiant. Possible explanations will be discussed in section 2, together with further properties of the progenitor inferred from the light curve and the spectra.

Section 3 is devoted to the explosion mechanism of massive stars. It had always been believed that the central cores of stars as massive as Sk-69° 202 should collapse to neutron star densities. Most of the binding energy of a newly born neutron star was thought to be radiated away in form of neutrinos during the formation process. Therefore, it did not come as a surprise that neutrinos were detected a few hours before the optical outburst because the theoretical models predicted that a few neutrinos should be detected by present neutrino detectors for a supernova at a distance of 50 kpc. Although, therefore, the arguments in favor of a core collapse are very convincing, many questions remain open. In particular, we have not learnt anything about the explosion mechanism itself from the neutrino observations. These questions will also be addressed in section 3. Finally, in section 4 some future prospects will be discussed, including the detectability of a possible neutron star remnant of SN 1987A, and its impact on stellar evolution theory, nucleosynthesis, and supernova models.

2. PROPERTIES OF SN 1987A INFERRED FROM THE LIGHT CURVE AND THE SPECTRA

2.1 The progenitor star

Astrometric determinations of the position of SN 1987A and comparisons with pre-outburst observations allowed to identify the progenitor to be one of the components of the multiple system Sanduleak -69°202 (Sk -69° 202) (Walborn et al., 1987; West et al., 1987). Its position coincided well with that of the main component which was surrounded by two companions at projected distances of about $1.2\ 10^{18}$ cm and $2.1\ 10^{18}$ cm, respectively.

Surprisingly, it was a blue supergiant of spectral class B3 and luminosity class Ia (Rousseau et al., 1978) at V=12.24^m instead of a red supergiant which was thought to be the progenitor of a type II supernova. The observed color indices (Isserstedt, 1975) indicated an effective temperature between 14000 and 17000 K. On the basis of the observed colors and brightness a total luminosity of the order of 3 to 6 10^{38} erg/sec could be estimated, assuming a distance of 49 kpc for the LMC (Graham et al., 1984; Walker, 1987). Spectral observations ($3700 \leq \lambda \leq 4840 \text{\AA}$) using objective prism plates (resolution 460 Å /mm) were obtained by Wamsteker at the ESO Schmidt telescope in 1977. Additional spectrograms were taken by the ESO astrograph in 1972-73 in the wavelength range between 3850 and 4150 Å at a resolution of 110 Å/mm (Rousseau et al., 1978). No outstanding chemical anomaly was observed, although a blend seen at about 4026 Å, which may be attributed to HeI or NII, is stronger than in comparison stars of about the same spectral class (Gonzalez et al., 1987).

IUE observations performed by Gilmozzi et al. (1987) and Walborn et al. (1987) definitely showed that the B3Ia star had dissappeared and, consequently, that it had been the progenitor of SN 1987A. The core mass luminosity relation derived from stellar models (Arnett, 1987; Hillebrandt et al., 1987; Nomoto et al., 1987; Truran and Weiss; 1987), together with the spectral type derived from the observed stellar spectrum and the UBV colors allowed to determine its He-core mass and 5 to 7 $M_\odot$ were found, indicating that the main sequence mass of the progenitor has been around 20 $M_\odot$ with an uncertainty of about $\pm 4\ M_\odot$.

2.2 Presupernova evolution of massive stars and Sk -69°202

Different evolutionary scenarios have been suggested which, starting from the main sequence with masses of about 16 -24 $M_\odot$, all have a blue supergiant as their final result. The differences of various models can be attributed to different assumptions concerning mass loss, convection theory and metallicity.

Brunish and Truran (1982) were the first who demonstrated that stars of 15 and 30 $M_\odot$ and a very low metallicity never evolve to the red supergiant phase, provided the mass loss rate was very small. In fact, the metallicity of the LMC is smaller than solar by a factor of 2 to 4 (Dufour, 1984). Therefore, a number of different groups reinvestigated and extended the calculations for various metallicities and confirmed the earlier results (e.g., Arnett, 1987; Hillebrandt et al., 1987; Truran and Weiss, 1988). In particular, it was shown that a somewhat lower metallicity than observed as a mean value in the LMC is necessary to explain that the progenitor was blue, in agreement with spectral analyses of the supernova envelope. All these models, however, have difficulties in explaining the existence of a high density slowly expanding circumstellar envelope which strongly argues for an earlier quite recent red giant phase (Cassatella, 1987; Fransson, 1987; Wampler and Richichi, 1989). Moreover, most models evolved to the red supergiant stage if moderate mass loss was allowed for and ended their lifes there (Weiss, 1988; Saio et al., 1988).

Mixing of different stellar layers seems by now to be the best way to obtain blue supergiant tpye II progenitors because of its effect on the opacities and subsequently on the energy transport and on the density structure of the envelope. Saio et al. (1988) have published models in which an ad hoc assumed mixing of helium into the hydrogen-rich layers during the red supergiant phase resulted in a return to the blue supergiant stage, i.e. forced a blue-red-blue evolution. Their model is very encouraging, because a mixing of helium into the outer envelope of post-main sequence stars is in good agreement with the high He/H ratios observed in some blue supergiants in the LMC (Kudritzki et al., 1987). Furthermore, such a mixing would also be consistent with the high N/C and C/O ratios in the shell of Sk -69°202.

So in conclusion, the evolution of Sk -69°202 is at present not well understood. However, it now seems to be very likely that various physical effects are responsible for the blue appearance of this star, namely the low metallicity of the LMC, the low mass-loss rate of the progenitor, and also mixing processes.

2.3 The light curve

Beginning on Febr 24.454, 1987, photometric measurements using different filter systems and spectral observations were available from the UV to the radio wavelength range. The complete sample of observations allowed for the first time to construct the bolometric light curve (BLC) of a type II supernova (Catchpole et al., 1987, 1988, 1989; Menzies et al., 1987; Whitelock et al., 1988; Hamuy et al., 1988; Suntzeff et al., 1988) (see also fig. 1).

The BLC of SN 1987A is quite unusual with respect to other type II supernovae. After a rapid decline during the first week, it increased very slowly until about day 84, then reached a broad maximum which was fainter by about a factor of 10 to 100. Finally, it showed a long linear decline in magnitude with mean half life of about 77 days until day 300, and then a somewhat faster decay.

Despite of these peculiarities the BLC can be well understood from the same general picture as other type II supernovae. In particular, its evolution is characterized by five main parts:

1) When the shock breaks through the photosphere, the outer layers are heated up.

The optical depths of this region is small. Therefore, internal energy stored from the shock wave can be radiated away immediately. At this time, most of the radiation is emitted in the far UV, i.e. a UV-light flash is to be expected with a duration of the order of a few minutes to one hour.

Although this UV radiation was not observed directly, Fransson and Lundquist (1989) have drawn attention to the fact, that the ionization stages seen in the circumstellar envelope constrain the color temperature at shock break out to values in the range of 400000 to 800000 K, in agreement with the effective temperatures predicted from hydrodynamical models (Arnett, 1987; Shigeyama et al., 1988; Utrobin, 1988; Woosley et al., 1988).

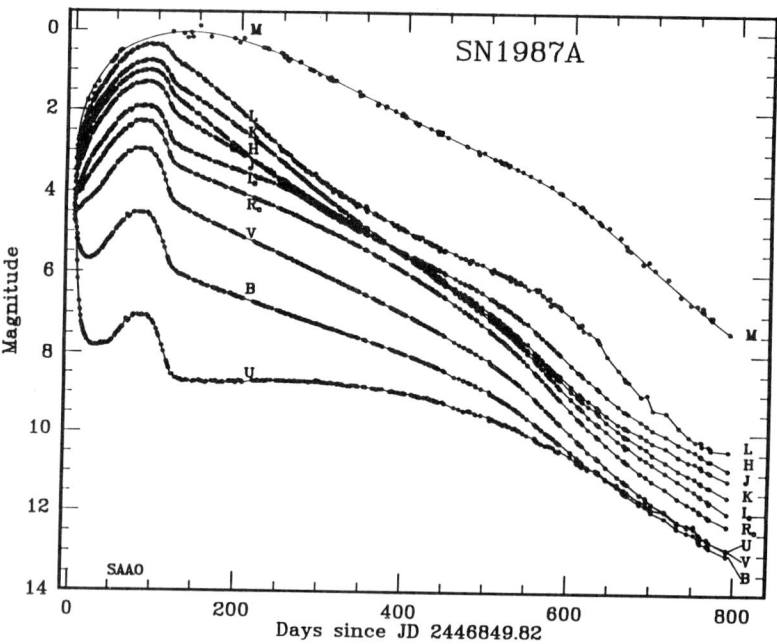

Figure 1 Light curves for SN 1987A in different filters of Johnson's (1966) system observed by Catchpole et al. (1989) (from Catchpole et al.(1989)).

2) Beginning on February 25, 1987, the UV flux was measured by the IUE satellite (Panagia et al., 1987). The UV flux of SN 1987A decreased during the first three days by about a factor of 1000 whereas the optical flux increased. All together, the bolometric luminosity showed only small changes in comparison to "normal" type II supernovae. This second phase, lasting for about 25 days, can be characterized by a free, nearly homologous expansion and adiabatic cooling of the envelope.

The low bolometric luminosity of SN 1987A relative to "normal" type II supernovae is a consequence of the more compact structure of progenitor (Arnett, 1987; Hillebrandt et al., 1987). In fact, fits to the early visual light curve confirm this conclusion, provided the explosion energy was slightly larger than 10^{51} erg. Adiabatic expansion then leads to a fast decrease of the effective temperature and an increase of the radius which cause an increase of the visual luminosity. Because the energy released in the hydrogen-rich envelope is about constant, the resulting bolometric luminosity shows only small changes with time.

3) During the following stages, the diffusion time scale decreased because of the decreasing density. Therefore, recombination and energy input from radioactive decay in the expanding envelope can contribute to the luminosity and it increases up to maximum light. Obviously, the diffusion time scale now governs the structure and consequently the peak luminosity of the light curve. It depends, therefore, strongly on the total mass of the envelope, and also on the opacities, i.e. the chemical composition of the expanding envelope. Woosley (1988b) from light curve models was able to determine the total mass of the hydrogen-rich envelope to be of the order of about 10 solar masses, consistent with what is obtained from the spectra. These results were also confirmed by other groups (Arnett, 1987; Shigeyama et al., 1988; Utrobin, 1988).

4) In the following stage the evolution of the light curve was mainly governed by the release of radioactive decay energy. The first direct evidence (besides the optical light curve) for radioactive decay is due to observations of SMM. In the beginning of August 1987, the γ-ray spectrometer on the SMM mission had detected significant net line fluxes at $\approx$ 847 keV in the background-subtracted spectrum of SN 1987A (Matz et al., 1988). There was also some evidence for flux in the 1238 keV line. This first appearance of the γ-ray lines coincided well in time with the detections of hard X-rays by the Ginga and MIR satellites (Dotani et al., 1987; Sunyaev et al., 1987) which can be well understood as a consequence of comptonization of γ-rays inside the expanding envelope. The early emergence of X- and γ-rays can be modelled if (ad hoc) mixing of ^{56}Co into the hydrogen rich envelope is assumed (Ebisuzaki and Shibazaki, 1988; Gebenev and Sunyaev, 1988; Itoh et al., 1987; Sutherland et al., 1988). However, the nearly constant hard X-ray flux observed by the Ginga satellite up to day 570 is in conflict with these models. A possible solution may be that the absorption by newly synthesized heavy elements is effectively reduced owing to the formation of chemically inhomogeneous clumps as it is also implied form the observed IR-spectra.

Thus it is apparent that from about day 110 on the light curve was governed by the radioactive decay of ^{56}Co, in particular because its exponential decline with a mean half life of 77 days is identical to the decay time of ^{56}Co. From the distance of 49 kpc to the LMC a total inital mass of radioactive Co can be estimated, and one finds of the order of 0.07 M$_\odot$, more or less independent of the details of the hydrodynamic models (Arnett, 1988; Shigeyama et al., 1988; Woosley et al., 1988).

5) Beginning at about day 300, the γ- photons were no longer completely comptonized, i.e. the envelope became partially optically transparent. Therefore, the bolometric light curve (which does not take the high energy radiation into account) decreased faster than expected from the ^{56}Co decay. Moreover, there is now strong

evidence for dust formation after day 500, leading to a more rapid decline of the bolometric light curve (Danziger et al., 1989).

No additional energy input has been confirmed up to now (September 1989). Most surprisingly, no evidence for a pulsar has been found so far in the light curve, i.e. the total luminosity of such a possible object could be limited to be less than a few times 10^{37} ergs, corresponding to a luminosity significantly less than that of the Crab pulsar. However, besides a pulsar, additional energy sources are to be expected in the future, e.g. radioactive heating due to ^{57}Co which was produced in smaller amounts (2 to 3 $10^{-3} M_\odot$) than ^{56}Co but has a much longer mean half life (227 days) and therefore is expected to dominate the energy input from radioactive decay after about 3 to 4 years (Nomoto et al., 1989). Further deviations may also occur from the dissipation of kinetic energy once the expanding envelope starts to interact with accumulated interstellar matter.

2.4 The observed spectra

The observed spectra give direct information on the physical and chemical conditions and the geometry of the expanding envelope of the supernova at a given time. Deeper layers of the expanding envelope are observable at later times. Therefore, a detailed analysis of the observed spectra allows one to determine the velocity at the photosphere, the density structure and chemical composition as a function of depth, and the mass of the hydrogen-rich envelope.

In order to understand the early spectral behaviour of type II supernovae and SN 1987A in particular, detailed model atmospheres are needed. In principle, the same models as for "standard" stellar atmospheres may be applied. However, there are a number of difficulties due to the physical properties which distinguish supernova envelopes substantially from stellar photospheres: i) the density profile cannot be assumed to be given by hydrostatic equations; ii) typical particle densities are much lower than in typical stellar atmospheres; iii) very large velocities and velocity gradients have to be taken into account and iv) the radial extension may become important.

For the early phase Balmer lines of hydrogen are the most prominent features in the optical wavelength region. It has turned out, that these Balmer lines cannot be well reproduced by models which assume pure scattering for hydrogen (Branch et al., 1981) or in which the interlocking between different energy levels is neglected (Lucy, 1987). However, the hydrogen lines can be well represented in a full non-LTE treatment i.e. in calculations which take the collisional processes and the coupling between the energy levels and ionization stages of different elements into accout. This has been shown for type II supernovae (Hempe, 1984) and in particularly for SN 1987A (Höflich, 1987, 1988ab). Moreover, blanketing effects due to elements other than hydrogen are most important at UV, optical and the IR-wavelengths at least during the phase when hydrogen is no longer completely ionized (Lucy, 1987, 1988; Höflich, 1988ab). Models in which it is assumed that the occupation numbers and ionization equilibria are given by the local thermodynamical equilibrium are very useful if attempts are made to identify lines and to give rough estimates of element abundances. However, detailed atomic models have to be used for quantitative analyses, at least, if only a small number of strong lines are used.

For the late phases the use of the nebular theory seems to be very attractive (Fransson, 1987) because it can be assumed that all the lines and the continua are optically transparent. Therefore, the total mass of each element can be determined. One disadvantage of this method is the fact that the depth dependent information comes in only quite indirectly due to line broadening. Although in the nebular case an analysis of the spectra is simpler than for the earlier stages (see above) there are also some severe new problems. No (or only very indirect) information about the density and temperature structure can be derived from the models. Therefore, in general, these quantities are taken as being constant. In addition, assumptions concerning the ionization equilibria and consequently the excitation and ionization mechanisms enter and strongly effect the analysis. In this respect the use of IR-lines is less model dependent. The particle density in the line forming regions of SN 1987A is rather high compared to "classical" HII regions even at later stages, implying that the thin LTE assumption for the population numbers is relatively accurate for IR-lines, even if the lines are optically thin.

The early observed spectra of SN 1987A showed a number of peculiarities with respect to 'normal' supernovae of type II in the ultraviolett as well as in the optical wavelength region:

i) During the first few days the energy distribution of optical and IR - flux showed very rapid variations but it changed only slowly during the following few months. In particular, a UV-flux deficit with respect to the optical flux was seen.

ii) Already at very early stages strong lines were observed in the spectra which showed a very rapid time evolution.

iii) The UV flux decreased by about a factor of 1000 during the first few days (Cassatella, 1987; Panagia et al., 1987).

The relatively rapid evolution of the spectra followed by small changes later on can be understood as a consequence of the rapid increase of the photospheric radius and of the strong coupling of the expansion of the photosphere to the expanding matter. Simultaneously the effective temperature decreased rapidly. One to two weeks after the explosion, the geometrical dilution of the material and the recombination of hydrogen outside a certain radius resulted in a much slower change of R_{ph} and, consequently, in smaller changes with time of the spectral energy distribution, as long as the continuum was determined by electron scattering (Figure 2). Therefore, the difference between the 'normal' spectral behaviour of type II supernovae and SN 1987A can be understood as being due to the relatively steep density gradient in the progenitor, and to the smaller luminosity, which resulted in a very early hydrogen recombination phase, i.e. as a consequence of the compact structure of the progenitor.

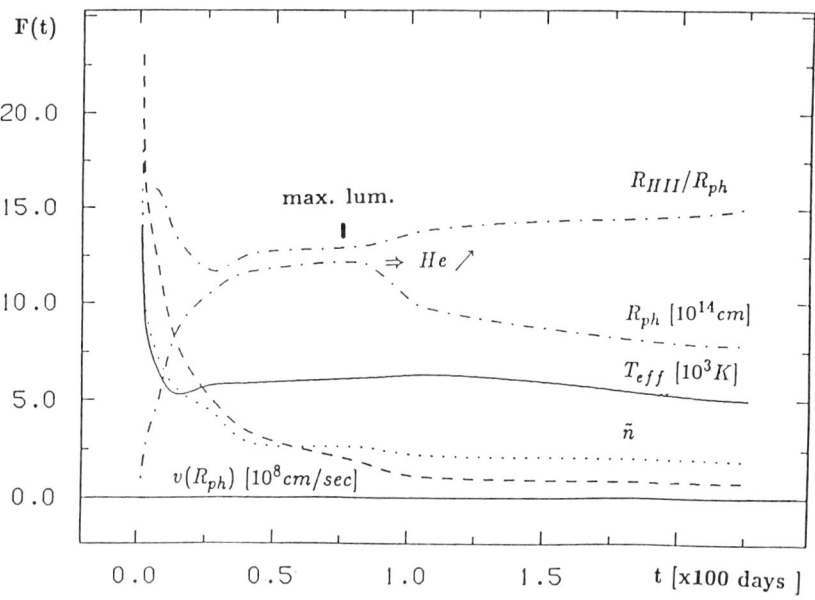

Figure 2 Radius R_{ph} at which the optical depth is 1 for true absorption at 5000 Å, effective temperature T_{eff}, velocity field v at R_{ph}, power law index $\tilde{n}$ of the corresponding density slope and distance R_{HII} up to which hydrogen is mainly ionized are given as a function of time (in days since the explosion). The data have been derived from the atmospheric diagnostics of SN 1987A by Höflich (1988b).

The optical spectra which have been observed during the first 7 months are well reproduced by NLTE-models (Höflich, 1987, 1988ab). In fact, one third of solar abundances is needed for all heavy elements except for some s-process elements. For Na, Sc and Ba which have been identified in the spectra overabundance factors of about 5, 2 and 10 relative to their solar values have to be chosen, with a possible uncertainty of the order of a factor of about 2, in order to fit the spectra. However, there is no evidence for abundance gradients in the hydrogen-rich layers. Starting around October 1987 helium-rich matter was seen at the photosphere. However, the presence of H with velocities of 800 km/s only requires significant mixing of the inner few solar masses of the hydrogen rich envelope with deeper layers during the explosion. This result is consistent with the light curve models and, in particular, with the observations of the hard radiation.

By integrating the continuity equation of the particle density and the photospheric velocities the total mass of the hydrogen-rich layers of the progenitor could be estimated to be of the order of 9–11 $M_\odot$ (Höflich, 1987, 1988ab). Note, that the value of the mass of the hydrogen-rich layers as determined from the spectral analysis is independent of the light curve models but gives essentially the same result (7–10$M_\odot$;

Arnett, 1988; Shigeyama et al., 1988; Woosley, 1988). Therefore, the value of 10 $M_\odot$ should be regarded as quite save, and, consequently, it is apparent that the progenitor has undergone only moderate mass loss prior to the explosion.

Spectroscopic IR-data have been obtained and published by most of the major southern observatories such as AAO (e.g. Bailey et al., 1988) CTIO (Elias et al., 1988), ESO (Danziger et al., 1988), MSSSO (McGregor, 1988) and SAAO (e.g. Catchpole et al., 1987, 1989) and from the Kuiper Airborn Observatory (Larson et al., 1988; Rank et al., 1988). The early spectra in the IR were characterized by a strong continuum on which hydrogen lines were superposed having a P-Cygni like structure. Beginning in June 1987, more emission features began to develop gradually. Although hydrogen emission lines from the Paschen-, Brackett and Pfund series still dominated the spectra, lines due to heavier species such as HeI, CI, OI, MgI, NaI, SiI, KI, CaI-II, FeII, CoII, NiI and SrII began to develop.

The temperatures derived from the IR-lines and from the continua (e.g. Danziger et al., 1989; Meikle et al., 1988) are consistent with the values derived from the optical spectra. Here, most of the observers have used the work of Branch (1987) as a guide for the ionization balance which can be expected. Analyses of the Co II lines using the nebulae theory led to an initial mass of ^{56}Co of about 0.07 $M_\odot$ which is consistent with the mass estimated from the light curve. This means that in spring 1988 all Co was seen in the IR spectra.

From about July to December 1987, the spectra changed significantly from continua superposed by lines to emission line spectra in a "nebular like" phase in the optical and in the IR-wavelength region. From then on, the observed spectra (Catchpole et al., 1989; Danziger et al., 1989; Whitelock et al., 1988) resemble closely those observed from other supernovae at similar stages of their evolution (Filippenko, 1989; Panagia, 1984). Features due to H_α and forbidden lines of the IR-duplet $[CaII]$ (7291 and 7323 Å), the $[OI]$ duplet (6300.23 and 6363.88 Å), and others, became stronger with time indicating a decreasing density.

It has been noted that in high resolution spectra accumulated since October 87 (Danziger et al., 1989) the profiles of the $[OI]$ duplet at about 6300 Å and 6363 Å contained significant structures which remain stable over several months. This clearly indicates the existence of some chemical inhomogeneities or geometrical clumpiness of the expanding inner regions of the envelope. Additional strong evidence for the fact that mixing processes must have occured comes from observations by the Kuiper Airborn Observatory during two flights in November 1987 and spring 1988 (Erickson et al., 1988; Rank et al., 1988) where strong emissions lines of NiI-II, ArII, CoII, FeI-II, Co and SiO were detected. These strong lines have also been seen in several ground based observations (see above). The full width half maximum of the roughly symmetric line cores correspond to about 3000 km/sec, showing that these elements move with about the velocities observed for the photospheric region in May 1987. This confirms that at least some Co has been mixed up to the inner layers of the hydrogen-rich region. However, at about May 1987 no enrichment of

Co was seen in the optical spectra (Höflich, 1988b), which means that the covering factor due to Co has been relatively small and that Co was mixed up in form of bubbles.

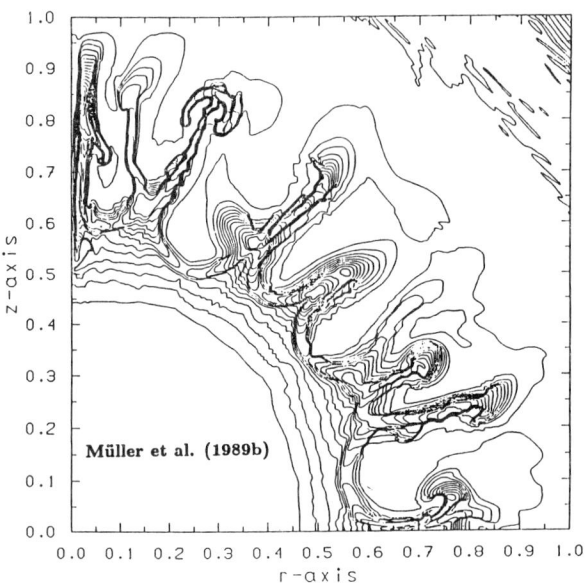

Figure 3 Density contours (5 percent spacing) showing Rayleigh-Taylor instabilities 9814 seconds after the explosion. An explosion energy of $2\ 10^{51}$ erg and a B3 supergiant structure were assumed. The "mushroom head" shapes which are characteristic for Rayleigh-Taylor instabilities are evident (from Müller et al.(1989b)).

2.5 Models for the mixing process

Strong inhomogeneous mixing processes of different layers are evident from the bolometric light curve, from the observation of X- and γ-rays and from the observed spectra. However, in the models mixing had to be introduced ad hoc in order to be able to fit the observations. Therefore, several groups have searched for possible mechanisms. Nagasawa et al. (1988) found Rayleigh-Taylor instabilities and fragmentation in supernova models based on polytropic density profiles and a 3-dimensional smooth particle hydrodynamic code. Müller et al. (1989a) have repeated these computations by means of both analytical methods in one dimension and numerical simulations using various codes and methods both in 2- and in 3-dimensions, but could not confirm the earlier results. Therefore, they concluded that the instabilities found by Nagasawa et al. (1988) were caused by their numerical method rather than by fluid dynamical effects.

However, Arnett et al. (1989) and Müller et al. (1989b) found that Rayleigh-Taylor instabilities (figure 3) might occur out to the hydrogen-rich envelope if a realistic stellar structure was used. These instabilities led to small scale structures in both the density and the chemical composition. Although so far these calculations have only been performed under the assumption of rotational symmetry, it is very likely that the instability persists in less symmetric situations.

Another possible mechanism for mixing may be the expansion of hot bubbles of radioactive nickel (e.g. Arnett, 1988; Woosley, 1988). In this picture, energy release by radioactive decay of ^{56}Ni during the early stages may cause instabilities and may result in mixing of different inner layers.

3. THE EXPLOSION MECHANISM

Prior to SN 1987A there was no direct evidence that type II supernova will lead to the formation of neutron stars, although several supernova remnants are known which do contain a neutron star (Crab, Vela, RCW 103, etc.). But in none of those cases do we know beyond doubt that the events were indeed of type II. Moreover, the remnant of an explosion of a rather massive star, CasA, apparently does not contain a neutron star. The assumption, therefore, that massive stars ($M \gtrsim 8M_\odot$) are the progenitors of type II supernovae and neutron stars rests mainly on theoretical models of stellar evolution (for recent reviews see, e.g., Hillebrandt (1987a) and Blinnikov et al. (1988)). On the other hand, a significant fraction of all galactic supernovae may escape visual detection, as did CasA some 350 year ago, and may only be discovered by their neutrino emission. It is interesting to note that, if SN 1987A had exploded near the center of our own galaxy, it probably would have been seen as a neutrino source only. In the following subsections we shall briefly discuss the theoretical models which have been developed over the last 20 years.

3.1 Core collapse models of type II supernovae

Several ways have been suggested which may, under favorable circumstances, convert a few percent or less of the gravitational binding energy of a forming neutron star into outward motion of the stellar envelope, among which at present two scenarios seem to be the most promising ones. Firstly, it has been proposed that a hydrodynamic shock wave created by the rebounding neutron star may release enough energy in the stellar envelope (Arnett, 1983; Bruenn, 1987; Hillebrandt, 1982; Hillebrandt, Nomoto and Wolff, 1984; Imshennik and Nadyozhin, 1983). However, as will be discussed later, this mechanism requires very special initial conditions and/or special properties of the nuclear equation of state. It may, therefore, work only for a narrow range of stellar masses, and in particular not for stars as massive as Sk-69° 202 (Hillebrandt, 1987b). Secondly, it has been found that neutrinos leaking out of a newly born neutron star may heat up certain mass zones of the stellar mantle sufficiently to revive a shock that otherwise would not reach the envelope (Wilson, 1985; Wilson et al., 1986). This effect, however, is extremely sensitive to the neutrino energies at very small "optical depth", and, therefore, numerical models may not be very reliable. Moreover, the explosion energy found in these simulations may be too low to explain the very energetic outburst of SN 1987A.

The outcome of computer simulations of stellar collapse will depend on the pre-collapse stellar models, the adopted microphysics input data and the numerical methods. Given the fact that the various computations use different initial models, different equations of state, and/or different numerical schemes it is not surprising that the results also differ considerably. There are, however, several features that are the same in all computations using the standard theory of weak interactions and non-rotating stellar models.

During the collapse phase the entropy stays low ($S \leq 2k_B$/nucleon) and, therefore, all models collapse to nuclear matter density. Moreover, about that fraction of the stellar core corresponding to the Chandrasekhar mass ($M_{CH} \simeq 5.72 M_\odot y_e^2$; $y_e =$ electron concentration) collapses homologously, $v = a(t)r$, where $a(t)$ is independent of r, in agreement with analytical considerations (Goldreich and Weber, 1980). Consequently, a sonic point must exist and the matter outside the sonic point has supersonic velocities close to free-fall velocity. Since M_{CH} is proportional to y_e^2 it will decrease during collapse. This decrease depends on the entropy of the initial model, on e^--capture rates, the ν-transport scheme, and the equation of state, and has turned out to be different in most computations.

At core-bounce the central density of the star is in general only slightly higher than nuclear matter density ($\rho_c \gtrsim 3 \times 10^{14}$g cm^{-3}), unless rather soft EOS are used (Baron et al., 1987). The inner core, defined by the condition that its velocity is subsonic, is stopped on a sound-crossing time ($\lesssim 1$ ms) once nuclei in the center of the star have dissolved into a homogeneous fluid of free nucleons and the EOS stiffens ($\gamma \simeq 2.5 - 3$). Because the outer material is still falling with supersonic velocity, a shock must form near the sonic point at M_{CH} (or a radius of about 20 km). The unshocked inner core does not expand against the ram-pressure of the supersonic outer layers and achieves a hydrostatic equilibrium soon after bounce. From energy conservation one can estimate the energy that is put into the shock is typically $E_{shock} \simeq E_B^{ic} \simeq (4-8) \times 10^{51}$erg, where E_B^{ic} is the binding energy of the unshocked inner core, in agreement with the results of numerical simulations. Here, the main source of uncertainties is the stiffness of the EOS near nuclear matter density, but also ν-transport can change these numbers considerably.

The outgoing shock wave is heavily damped by energy losses due to nuclear photo-dissociations, which cost about 8×10^{18} erg/g, and neutrino losses once the shock has passed the neutrino sphere. Neglecting neutrino losses for a moment we find that at most 0.5 $M_\odot$ of heavy nuclei can be dissociated by the shock, and a necessary condition for successful propagation is $M_{"Fe"} - M^{ic} \lesssim 0.5 M_\odot$, where $M_{"Fe"}$ is the mass of the original iron core. Numerical models predict $y_e \lesssim 0.38$ and thus $M^{ic} \lesssim 0.7 M_\odot$. It follows immediately that only stellar models with iron-core masses less than 1.2 $M_\odot$ can lead to prompt explosions. Stellar evolution calculations (Nomoto, 1984; Woosley, 1986) have shown that this condition limits the mass range of possible progenitor stars to values between 8 and 12 $M_\odot$ (or at most 15$M_\odot$) on the main sequence, and the only successful collapse computations have indeed been performed with stars in this mass range (Hillebrandt, 1987a; Baron et al., 1987). It should be noted, however, that in some simulations even the most favorable initial conditions do not lead to prompt explosions, probably because different equations of state and different ν-transport schemes were used. Therefore we reach the conclusion that at present we cannot answer the question whether or not stars can explode by the core-bounce mechanism and have to keep this unpleasant situation in mind when we try to interpret SN 1987A.

Because some supernova remnants (e.g., Cas A and Puppis A) show large oxygen overabundances indicating main sequence masses of at least 20 $M_\odot$, even prior to SN 1987A alternative explosion mechanisms had been searched for. A possibility which has been discussed extensively is the so-called delayed explosions model. The idea is that a few hundred milliseconds after core-bounce energy transport by neutrinos may revive a stalled shock. Wilson (1985) and Wilson et al. (1986) have computed the hydrodynamic evolution of several stellar models for approximately 1s after core bounce and found explosions in all cases considered. In most cases, however, the explosion energy ($\lesssim 4 \times 10^{50}$ erg) was too low to account for typical type II supernova light curves and in particular for the rather high kinetic energy seen in SN 1987A. Only for rather massive stars, $M \gtrsim 25 M_\odot$, did explosive oxygen burning add enough energy to the explosion to explain a typical outburst. This problem can be understood from a simple argument. Neutrino heating proceeds on a time scale much longer than the hydrodynamical time scale. Once, due to neutrino heating, a sufficient overpressure has been built up behind the accretion shock the heated zones will start to expand and further heating will be turned off. The thermal energy needed to build up such an overpressure will be of the order of the binding energy of the overlaying material, i.e., a few times 10^{50} erg. Consequently, one expects that the explosion energy is of the same order.

3.2 Neutrinos from supernova 1987A

As was already mentioned in the introduction and in section II, two neutrino pulses were detected prior to the optical outburst of SN 1987A, at February 23.12 (UT) (Aglietta et al., 1987a) and at February 23.32 (UT) (Hirata et al., 1987; Bionta et al., 1987), respectively. There is also a report of neutrino events close to the second burst in the Baksan detector (Alexeyev et al., 1987). Two photographs taken at about February 23.44 (UT) (Mc Naught, 1987) show the supernova at a visual magnitude of about $6^m.4$ only 2.8×10^4 s and 1.1×10^4 s, respectively, after the neutrino events were discovered. This close correlation indicates that the neutrinos were indeed emitted from the exploding star, although from energy arguments the Mont Blanc and Baksan events are considered as being instrumental.

The pulse seen in the Mont Blanc detector consisted of 5 events spread over $\Delta t \simeq 7s$ with measured positron energies between 5.8 and 7.8 MeV. From the second pulse, about 4.7 hours later, KAMIOKANDE detected 12 neutrinos spread over $\Delta t \simeq 13s$, whereas IMB saw 8 neutrinos with $\Delta t \simeq 6s$. The recorded electron (positron) energies in the second burst were significantly higher and ranged from 6 to 35 MeV (KAMIOKANDE) to 20 to 40 MeV (IMB). It is interesting to note that the angular distribution of electrons (positrons) was not isotropic, as one would expect from the reaction $\bar{\nu}_e + p \to e^+ + n$. In particular, the high-energy events ($E_e \gtrsim 20$ MeV) seem to be strongly forward peaked away from the LMC. In the case of IMB this effect was not caused by the fact that about 25% of the photomultipliers were inoperative when the neutrinos were detected, as recent Monte Carlo simulations have shown (Svoboda et al., 1987). Finally the Baksan group reported 5 events delayed by 30s relative to the IMB time, with energies from 12 to 24 MeV. The detections of neutrinos from a type II supernova are the first proof that the cores of massive star do collapse to neutron star densities.

An analysis of the energetics of the neutrino events has been performed by Sato and Suzuki (1987) and many others for the KAMIOKANDE and IMB data, assuming that all neutrinos detected were $\bar{\nu}$'s. It was found that the KAMIOKANDE data can be fitted by a thermal neutrino spectrum corresponding to a temperature of (2.8 ± 0.3) MeV, whereas the IMB data require a significantly higher temperature of (4.6 ± 0.7) MeV, again indicating that the data are only marginally consistent. The total energy in three neutrino flavours is then $(2.9 \pm 0.6) \times 10^{53}$ erg for KAMIOKANDE and $(1.5 +1.2/-0.6) \times 10^{53}$ erg for IMB. If some of the events are considered to be noise, this energy estimate is reduced by a factor of about 2. If, on the other hand, some of the events are assumed to be caused by (ν, e)-scattering, the energy in the burst would increase by about a factor of two. In any case, the data are consistent with the assumption that a neutron star of about 1.5 $M_\odot$ was born in SN 1987A and has radiated away a significant fraction of its binding energy in thermal neutrinos during the first few seconds of its life. It is also apparent, however, that this conclusion is still uncertain, because in the case of Mont Blanc we have to rely on the poor statistics of a few events and in the case of KAMIOKANDE and IMB we do not know for sure which (if any) of the detections were due to (ν, e)-scattering.

One may ask, moreover, whether or not the assumption of a thermal spectrum with zero chemical potential describes the neutrino energy distribution sufficiently well. In order to answer this question Janka and Hillebrandt (1989a) have performed Monte Carlo simulations of the neutrino transport in type II supernova explosions. Based on one particular model they could show that, in fact, the results of the Monte Carlo simulations are not well fitted by Fermi-Dirac distributions, even if one allows for a non-vanishing neutrino chemical potential. In particular they have demonstrated that the assumption of a thermal spectrum overestimates the number of neutrinos both in the high energy tail of the distribution function and at low energies. Much better fits could be obtained if an additiona parameter was introduced to generate the first three moments of the neutrino energy distribution, indicating the importance of geometrical effects. This result is not surprising because the thickness of the shell where neutrinos decouple from the matter is not small compared to the radius of the newly born neutron star, at least for the first few seconds after core-bounce.

Based on these results Janka and Hillebrandt (1989b) have reanalysed the KAMIOKANDE and IMB data. They performed maximum likelihood tests, varying the neutrino chemical potential, their temperature and the total energy emitted in form of electron antineutrinos from SN 1987A. As can be seen from the figures, the KAMIOKANDE (K2) and IMB data are only weakly consistent which is in agreement with earlier investigations. In particular, the IMB data indicate a significantly higher temperature and lower total energy. This may be due to the possibly non-thermal nature of the neutrino pulse or may indicate that the IMB detector, because of its higher threshold energy, has predominantly detected neutrinos from the very early cooling phase, whereas KAMIOKANDE has seen neutrinos from a different (and larger) time interval. Unfortunately, the detection time of KAMIOKANDE is uncertain by about 1 or 2 minutes and, therefore, the latter hypothesis can not be proven. In any case, the analysis of Janka and Hillebrandt (1989b) suggests that the neutrinos must have been emitted with higher average energies and that the total energy flow through the detectors might have been somewhat lower in comparison to estimates based on the assumption of blackbody emission.

3.3 Non standard scenarios

Given the apparent difficulties of the standard core collapse models one may search for alternative explosion mechanisms. Rotation, for example may alter the general picture outlined in section III. 1 considerably. A non-rotating stellar core will collapse until the adiabatic index γ exceeds the critical value of 4/3 around nuclear matter density. For a rigidly rotating stellar core the critical adiabatic index is given by $\gamma_c = \frac{4}{3} - \frac{2}{9}\frac{I\Omega^2}{|E_{pot}|} < \frac{4}{3}$, where I is the moment of inertia, ω is the angular velocity, and E_{pot} is the gravitational energy. So a rotating core may not collapse to nuclear matter density provided the equation of state is sufficiently stiff and/or the angular velocity is sufficiently high. Moreover, non-spherical velocity fields will lead to non-spherical shocks which may introduce large-scale circulations and thus a more efficient neutrino transport. Finally, rapidly rotating cores will be unstable to triaxial deformations or even fragmentation. Collapsing rotating cores have first been modelled by Le Blanc and Wilson (1970), Müller et al. (1980), Müller and Hillebrandt (1981), and Tholine (1984). More recently Symbalisty (1984) and Mönchmeyer and Müller (1989) have repeated those computations using more realistic initial stellar models and equations of state. So far, however, the conclusion was always that no prompt explosion triggered by rotation alone has been found. On the contrary, it seems that certain initial angular momentum distributions inside a collapsing stellar core rather weaken the chance of getting prompt explosions (Mönchmeyer and Müller, 1989).

Recently, Kristian et al. (1989) have reported the detection of sub-millisecond optical pulsations from SN 1987A which, if real, definitely require the presence of a neutron star. Moreover, the observed approximately 2 kHz frequency of the pulsations, if interpreted as being due to rotation, would require that the pulsar is spinning very rapidly close to break up. Implications from these exciting observations for neutron star equations of state will be discussed in section V. For our present discussion it is important to note that so far core collapse computations have only been performed for rather slowly rotating stellar models. One may, therefore, speculate that the results obtained do not apply to SN 1987A but have to be extended in order to include non-axisymmetric effects. We will come back to this question later in section 4. The observations of Kristian et al. (1989) together with the low bolometric luminosity of SN 1987A ($\lesssim$ a few times 10^{38} erg s^{-1}) in December 1988 can be used to set upper limits on the strength of the magnetic field of the pulsar and one obtains $B \lesssim 5 \; 10^9$ gauss. Therefore, it is obvious that magnetic fields were dynamically unimportant in SN 1987A and the explosion was certainly not driven by magnetic effects.

Fast rotation will also affect the neutrino signal emitted from a newly born neutron star. Janka and Mönchmeyer (1988) have demonstrated that an observer will detect more neutrinos if he sees the neutron star pole-on because he sees a larger surface area. Consequently, he will overestimate the total neutrino energy in that case. They therefore conclude that, if a rapidly spinning neutron star was born in SN 1987A, we should have seen it almost pole-on. Otherwise the energy detected by the KAMIOKANDE experiment would be in conflict with the maximum binding energies allowed by reasonable neutron star matter equations of state.

We conclude this section by noting that fast rotation will also change the dynamics of the mass zones outside the original iron-core considerably and may, in fact, lead to a thermo-nuclear explosion of the oxygen-shell as was shown by Bodenheimer and Woosley (1980). Their explosion mechanism may work if, supported by centrifugal forces, the oxygen-shell moves inwards with subsonic velocities and contains enough mass ($M_O \gtrsim 1.5$ to $2 \; M_\odot$). Both conditions may have been fulfilled for Sk - 69° 202. The prediction from this model is that less than approximately $0.5 \; M_\odot$ of oxygen should be observed in the supernova ejecta, but unfortunately the abundance determinations (Danziger et al., 1989) are still not sufficiently accurate to prove or disprove this scenario. Explosive oxygen-burning may, nonetheless, explain the observed early fragmentation of the supernova shell and the mixing of radioactive ^{56}Co with material of high velocity.

4. FUTURE PROSPECTS

Astronomers will continously monitor SN 1987A at all wavelength bands for decades to come. Optical and infra-red observations will reveal more information on the element abundances in the ejecta, on the formation of dust and grains, and, for the first time, on the dynamical evolution of a very young supernova remnant. In some 10 to 20 years the supernova shock will begin to interact with the circumstellar shell and then can be studied as a bright X-ray and radio source. SN 1987A will allow for an accurate determination of the distance to the LMC and thus will help to calibrate the cosmic distance scale.

The most important question, however, is: Did the supernova really leave behind an extremely rapidly spinning, weakly magnetized neutron star? Such a possibility had not been anticipated by theoretical models and, therefore, estimates of the time at which the pulsar should become visible may have been off by an order of magnitude. A confirmation of the observations of Kristian et al. (1989) would have far reaching consequences for theories of stellar evolution and supernova explosions, for nuclear equations of state, and, possibly, also for the interpretation of the observational data presented in this review, including the interpretation of the neutrino signals.

SN 1987A generally is being considered as a very unusual event in many respects such as the blue nature of its progenitor and its low visual peak luminosity. But this may be due to selection effects favoring the discovery of bright explosions. If the standard interpretation is correct and sub-luminous type II supernovae are likely if the progenitor stars have lower-than-solar metallicities, one would have to revise present day's estimates of supernova rates. Certainly, future automated supernova search programs will help to answer this question.

With respect to theoretical models SN 1987A has raised more new questions than it has answered. The apparent early fragmentation of the supernova envelope is not yet understood. If it was caused by a Rayleigh-Taylor instability in the expanding shell, as is suggested by recent numerical simulations, this would impose strong constraints on progenitor models. Moreover, even if we ignore the possibility of a rapidly rotating progenitor star, the observed (and unexpected) strong mixing calls for 2- or even 3-dimensional supernova models.

SN 1987A has also raised new interest in stellar evolution models. The presumably rather complicated history of Sk -69° 202 cannot be understood in terms of standard stellar evolution scenarios, the main uncertainties being the correct treatment of convection, mixing and mass loss. Further complications may arise if rotation has to be included.

With respect to nucleosynthesis detailed computations of synthetic spectra will, for the first time, allow us to determine rather accurately the element abundances ejected from a star which may be typical for the enrichment of galaxies with heavy elements, and to compare them with theoretical predictions.

Finally, the fact that the collapse of the core of a massive star did lead to a type II supernova explosion still lacks a reliable explanation. Although numerical simulations were successful in predicting the right neutrino luminosity and spectra, they failed so far in explaining the very energetic explosion. Whether this failure is due to our incomplete knowledge of the nuclear equation of state and of stellar evolution theory, or whether some important ingredients are still missing in our present supernova models is certainly an open question. We may hope, however, that SN 1987A will stimulate new activities which finally will lead to convincing answers.

References

Aglietta M. et al., 1987a *Europhys. Lett.* **3** 1315

Alexeyev E.N., Alexeyeva L.N., Krivosheina I.V., Volchenkov V.I., 1987 in *SN 1987A*, ed. by I.J. Danziger, ESO, Garching, p.237

Arnett W.D., 1982 *Astrophys.J.* **253** 541

Arnett W.D., 1983 *Astrophys.J.* **263** L55

Arnett W.D., 1987 *Astrophys. J.* **319** 136

Arnett W.D., 1988 *Proc. 4th George Mason Astrophysics Workshop: SN1987A in the LMC*, ed. by M. Kafatos and A.G. Michalitsianos, p. 301

Arnett W.D., Fryxell B.A., Müller E., 1989 *Astrophys.J.Lett.* **341** L63

Bailey J., Barton J.R., Conray P., Hiller D.J., Hyland A.R., Jones T.J., Shortridge K., Whittard D., 1988 *PASP* **100** 1178

Baron E., Bethe H.A., Brown G.E., Cooperstein, J., Kahana S., 1987 *Phys.Rev.Lett.* **59** 736

Bionta R.M. et al., 1987 *Phys.Rev.Lett* **58** 1494

Blinnikov S.I., Chugai N.N., Golenetskii S.V., 1988 *Astrophys. Space Phys. Rev.* **6** 197

Bodenheimer P., Woosley S.E., 1983 *Astrophys.J.* **269** 281

Branch D., 1987 *Astrophys.J.* **320** L23

Branch D., Falk S., McCall M., Rybski P., Uomoto A.K., Will B.J., 1981 *Astrophys.J.* **244** 780

Bruenn S.W., 1987 *Phys.Rev.Lett.* **59** 938

Brunish W.M., Truran J.W., 1982 *Astrophys.J.* **256** 247

Cassatella A., 1987 in *SN 1987A*, ed. by I.J. Danziger, ESO, Garching, p.101

Catchpole R.M. et al., D. 1987 *M.N.R.A.S.* **229** 15p

Catchpole R.M. et al., 1988 *M.N.R.A.S.* **231** 75p

Catchpole R.M. et al., 1989 *M.N.R.A.S.* **237** 55p

Danziger I.J., Bouchet P., Fosbury R.A.E., Gouiffes C., Lucy L.B., Moorwood A.F.M., Oliva E., Rufener F., 1988 *Proc. 4th George Mason Astrophysics Workshop: SN1987A in the LMC*, ed. by M. Kafatos and A.G. Michalitsianos, p. 37

Danziger I.J., Bouchet P., Gouiffes C., Rufener F. , 1989 *Big Bang, Active Galactic Nuclei an Supernovae*, ed. by S. Hayakawa and K. Sato, Universal Academic Press, p. 429

Dotani T. et al., 1987 *Nature* **330** 230

Dufour R.J., 1984 *IAU Symposium 108: Structure and Evolution of the Magellanic Clouds*, ed. by S. van den Bergh and K. de Boer, Reidel, p. 353

Ebisuzaki T., Shibazaki N., 1988 *Astrophys.J.* **328** 699

Elias J.H., Gregory B., Phillips M.M., Williams R.E., Graham J.R., Meikle W.P.S., Schwartz R.D. Wilking B., 1988 *Astrophys.J.* **331** L9

Erickson E.F., Haas M.R., Colgan S.W.J., Lord S.D., Burton M.G., Wolf J., Hollebach D.J., Werner M., 1988 *Astrophys.J.* **330** L39

Filippenko A.V., 1989 *Particle Astrophysics Workshop*, Berkely, ed. by C.Pennypacker, in press

Fransson C., 1987 in *SN 1987A*, ed. by I.J. Danziger, ESO, Garching, p. 467

Fransson C., Lundquist P., 1989 *Astron.Astrophys* in press

Gebenev S.A., Sunyaev R.A., 1988 *Soviet Astron. Let.* **14** 675

Gilmozzi R., Cassatella A., Clavel J., Fransson C., Gonzalez R., Gry C., Panagia N., Talavera A., Wamsteker W., 1987 *Nature* **328** 318

Goldreich P., Weber S.V., 1980 *Astrophys.J.* **238** 991

Gonzales R., Wamsteker W., Gilmozzi R., Walborn N., Lauberts A., 1987 in *SN 1987A*, ed. by I.J. Danziger, ESO, Garching, p. 33

Graham J.A., Nemec J.M., 1984 *IAU Symposium 108: Structure and Evolution of the Magellanic Clouds*, ed. by S. van den Bergh and K. de Boer, Reidel, p. 37

Hamuy M.I., Suntzeff N.B., Gonzalez R., Martin G., 1988 *Astron.J.* **95** 63

Hempe K., 1984 *Mitt.d.Astron.Ges.* **60** 107

Hillebrandt W., 1982 in *Supernovae: A Survey of Current Research*, ed. by M.J. Rees and R.J. Stoneham, NATO-ASI **C90**, Reidel, p. 123

Hillebrandt W., 1987a in *High Energy Phenomena around Collapsed Stars*, ed. by F. Pacini, NATO-ASI **C195**, Reidel, p. 73

Hillebrandt W., 1987b in *SN 1987A*, ed. by I.J. Danziger, ESO, Garching, p. 301

Hillebrandt W., Nomoto K., Wolff R.G., 1984 *Astron.Astrophys.* **133** 175

Hillebrandt W., Höflich P., Truran J.W., Weiss A., 1987 *Nature* **327** 597

Hirata K. et al., 1987 *Phys.Rev.Lett.* **58** 1490

Höflich P., 1987 *Proceedings of the 4th workshop on Nuclear Astrophysics*, ed. by W. Hillebrandt et al., Lect. Notes in Phys. **287**, Springer, p. 307

Höflich P., 1988a *Proc.Astron.Soc.Austrl.* **7** 434

Höflich P., 1988b *IAU Symposium 108: Atmospheric Diagnostic of Stellar Evolution*, ed. by K. Nomoto, Springer, p. 388

Imshennik V.S., Nadyozhin D.K., 1983 *Astrophys.Space Phys.Rev.* **2** 75

Isserstedt J., 1975 *Astron.Astrophys.Suppl.* **19** 259

Itoh H., Hayakawa S., Masai K., Nomoto K., 1987b *Nature* **330** 233

Janka H.-T., Hillebrandt W., 1989a *Astron.Astrophys.Suppl.*, **78** 375

Janka H.-T., Hillebrandt W., 1989b *Astron.Astrophys.*, in press

Janka H.-T., Mönchmeyer R., 1988 *Astron.Astrophys.* **209** L5

Kristian J.A. et al., 1989 *Nature* **338** 234

Kudritzki R.P., Groth H.G., Butler K., Husfeld D., Becker S., Eber F., Fitzpatrick E., 1987 in *SN 1987A*, ed. by I.J. Danziger, ESO, Garching, p. 39

Larson H.P., Drapatz S., Mumma M.J., Weaver H.A., 1988 *Proc. 4th George Mason Conference: SN 1987A in the LMC*, ed. by M. Kafatos and A.G. Michalitsianos, p. 74

Le Blanc J.M., Wilson J.R., 1970 *Astrophys.J.* **161** 541

Lucy L.B., 1987 *Astron.Astrophys.* **182** L31

Lucy L.B., 1988 *Proc. 4th George Mason Conference: SN 1987A in the LMC*, ed. by M. Kafatos, p. 323

Matz S.M., Share G.H., Leising M.D., Chupp E.L., Vestrand W.T., Purcell W.R., Strickman M.S., Reppin C., 1988 *Nature* **331** 416

McGregor P., 1988 *Proc.Astron.Soc.Austrl.* **7** 450

Meikle W.P.S., Allen D.A., Spyromilio J., Varani G.F., 1989 *M.N.R.A.S.* in press

Menzies J.W. et al., 1987 *M.N.R.A.S.* **227** 39p

Mönchmeyer R., Müller E., 1989 in *Timing Neutron Stars*, ed. by H. Ögelmann and E. van den Heuvel, NATO-ASI **C262**, Kluwer, p. 549

Müller E., Hillebrandt W., 1981 *Astron.Astrophys.* **103** 358

Müller E., Rozyczka M., Hillebrandt W., 1980 *Astron.Astrophys.* **81** 288

Müller E., Hillebrandt W., Orio M., Höflich P., Mönchmeyer R. Fryxell B.A., 1989a *Astron. Astrophys.* **220** 167

Müller E., Hillebrandt W., Orio M., Höflich P., Mönchmeyer R. Fryxell B.A., Arnett W.D., 1989b *Proceedings of the 5th workshop on Nuclear Astrophysics*, ed. by W. Hillebrandt and E. Müller, MPA-P1, Garching, p. 103

Nagasawa M., Nakamura T., Miyama S., 1988 *Publ.Astron.Soc.Japan* **40** 691

Nomoto K., 1984 *Astrophys.J.* **277** 791

Nomoto K. Shigeyama T., Hashimoto M., 1987 in *SN 1987A* ed. by I.J. Danziger, ESO, Garching, p. 325

Nomoto K., Hashimoto M., Shigeyama T., Kumagai S., Yamaoka H., Saio H., 1989 in *Big Bang, Active Galactic Nuclei and Supernovae*, ed. S.Hayakawa and K.Sato, Universal Academy Press, p. 495

Panagia N., 1984 *Supernovae as distance indicators* Lect. Notes in Physics **224**, Springer, p. 14

Panagia N., Gilmozzi R., Clavel J., Barylak M., Gonzales Riesta R., Lloyd C., Snas Fernandez de Cordoba L., Wamsteker W., 1987 *Astron.Astrophys.* **177** L25

Rank D.M., Gregman J., Witteborn F.C., Cohen M., Lynch D.K., Russell R.W., 1988a *Astrophys.J.* **325** L1

Rousseau J., Martin N., Prevout L., Reheirot E., Robin A., Brunet J.P., 1978 *Astron. Astrophys. Suppl.* **31** 243

Saio H., Kato M., Nomoto K., 1988 *Astrophys. J.* **331** 388

Sato K., Suzuki H., 1987 *Phys.Rev.Lett.* **58** 2722

Shigeyama T., Nomoto K., Hashimoto M., 1988 *Astron.Astrophys.* **196** 141

Sunyaev R. et al., 1987 *Nature* **330** 327

Suntzeff N.B., Heathocote S., Weller W.G., Caldwell N., Huchra J.P., Olowin R.P., Chambers K.C., 1988 *Nature* **334** 135

Sutherland P., Xu Y., McCray R., Ross R.P., 1988 *IAU Colloquium 108: Atmospheric Diagnostic of Stellar Evolution*, ed. by K.Nomoto, Springer, p. 394

Symbalisty E., 1984 *Astrophys.J.* **285** 729

Tholine J.E., 1984 *Astrophys.J.* **285** 721

Truran J.W., Weiss A., 1987 in *SN 1987A*, ed. by I.J. Danziger, ESO, Garching, p. 271

Utrobin V.P., 1988 *Atominform* p.20

Walborn N.R., Lasker B.M., Laidler V.G., Chu Y.-H., 1987 *Astrophys.J.* **321** L41

Walker A.R., 1987 *M.N.R.A.S.* **225** 627

Wampler E.J., Richichi A., 1989 *Astron.Astrophys.* **217** 31

Weiss A., 1988 *Astrophys.J.* **339** 365

West R.M., Lauberts A., Jorgensen H.E., Schuster H.-E., 1987 *Astron.Astrophys.* **177** L1

Whitelock P.A. et al., 1988 *M.N.R.A.S.* **234** 5p

Wilson J.R., 1985 in *Numerical Astrophysics*, ed. by J. Centrella, J. Le Blanc, and R. Bowers, Jones and Bartlett, p. 422

Wilson J.R. Mayle R.W., Woosley S.E., Weaver T.A., 1986 *Ann. NY Acad.Sci.* **479** 267

Woosley S.E., 1986 in *Nucleosynthesis and Chemical Evolution*, Saas Fee Lecture Notes, ed. by B. Hauck, A, Maeder, and G. Meynet, Geneva Observatory, p. 1

Woosley S.E., 1988a *Astrophys.J.* **330** 218

Woosley S.E., 1988b *Proc. 4th George Mason Astrophysics Workshop: SN1987A in the LMC*, ed. by M. Kafatos and A.G. Michalitsianos, p. 289

Woosley S.E., Pinto P.A., Weaver T.A., 1988 *Proc.Astron.Soc.Austrl.* **7** 355

Supernova Statistics

H.E. JORGENSEN

Copenhagen University Observatory, Oster Voldgade 3,
DK-1350 Copenhagen, Denmark

The frequencies of SNe in external galaxies as determined from optical searches are reviewed and discussed. In particular we comment on the fast SNe producing galaxies and relates the frequencies to the far infrared luminosity. Finally we discuss briefly the rates of SNe in distant clusters of galaxies.

1 OPTICAL SEARCHES FOR SNe
We will review the frequencies from optical searches as discussed by Tammann (1974, 1977, 1982), van den Bergh et al. (1987), Evans et al. (1989), Cappellaro and Turatto (1988) and Richter and Rosa (1989).

1.1 The Tammann Samples
Tammann (1974, 1977, 1982) discussed two samples. Sample A is a distance limited sample of 408 galaxies from the Revised Shapley-Ames (RSA) Catalogue by Sandage and Tammann (1981). The galaxies have $\delta > -36°$ and velocities corrected to the centroid of the Local Group $v_0 \leq 1200$ km/s but including Virgo Cluster members independent of velocity. The numbers of detected SNe as given by Tammann are N(SNI) = 31, N(SNII) = 24, N(unclassified) = 22 in total 77.

Sample B consists of 2955 galaxies from the Second Reference Catalogue of Bright Galaxies (RC2) by de Vaucouleurs et al. (1976) with N(SNI) = 44, N(SNII) = 28, N(unclassified) = 101 totaling 173 SNe. 75% of all classified SNe are common to sample A and B.

Sample A being much more complete than sample B would give absolute frequencies in SNu while sample B would give relative frequencies. (1 SNu = 1 SN per $10^{10} L_{B\odot}$ per 100 years). In practise however, sample A and B are given equal weight in the determination of relative frequencies while sample A is used to derive the absolute frequency of SNe in face-on Sbc-Sd galaxies ($i \leq 30°$). We shall therefore concentrate our discussion on the SN rates in Sbc-Sd galaxies. We refer the reader to Table 1 and 2 in Tammann (1982) for the actual numbers. The face-on Sbc-Sd galaxies have a SNe frequency of 1.38 SNu (for H_0=50 km/sec/Mpc) carrying a statistical error of only 20% because it is based on 30 SNe (Tammann, 1982).

The control time of 35.3 years for sample A galaxies was determined dividing the total number of detected SNe by the detection rate during the period 1960-1976 assuming that all SNe were caught during that period. This procedure allows for changing intensity of the search since 1885 where the first SN in an external galaxy was detected. However, it is not obvious that this control time is relevant for the sub-sample of face-on Sbc-Sd galaxies with less internal absorption than Sbc-Sd galaxies in general, which results in different detectability.

In Table 1 we give data from Tammann (1977) and updated information on SNe in face-on Sbc-Sd galaxies using the Asiago Supernova Catalogue by Barbon et al. (1988) adopting data on the galaxies from the RSA Catalogue. We have calculated the control time of 29.2 years for the sample of face-on Sbc-Sd galaxies using the detection rate of 12 SNe per 10 years for the period 1960-69, see Fig. 1.

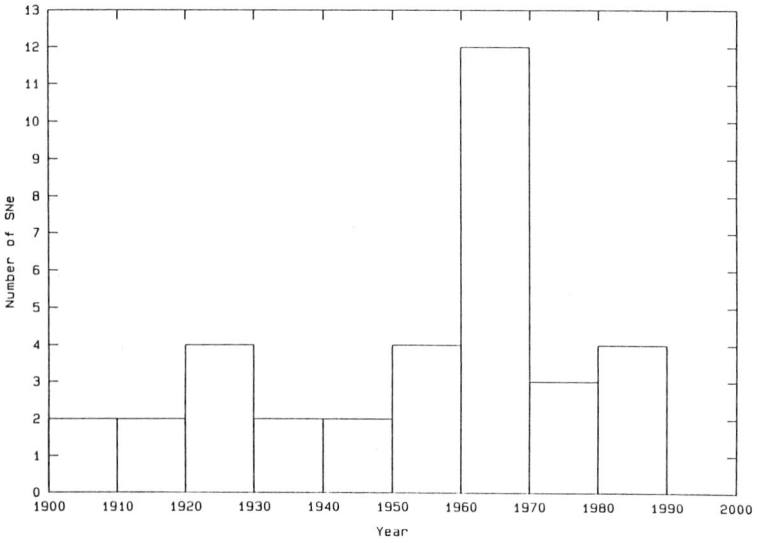

Figure 1. Number of detected SNe per 10 years in nearby ($v_o \leq 1200$km/s) face-on ($i \leq 30°$) Sbc - Sd galaxies.

Comparing to the Tammann (1977) data we notice minor differences in number and total luminosity of the face-on galaxies, which possibly is due to small differences between preliminary and final data on RSA galaxies. However it is clear from Table 1 that the overall rate of 1.44 SNu which we derive for face-on Sbc-Sd galaxies is very similar to the Tammann value.

Table 1. Nearby face-on and nearly face-one Sbc-Sd galaxies ($v_0 \leq 1200$ km/s). The luminosity is given in units of $10^{10} L_{B\odot}$.

	n_{gal}	L_B	contr. time	SNIa	SNIb	SNI	SNII[a)]	SN uncl.	SNu	i
Tamman (1977)	36	60.4	35.3	-	-	6	12	11.5	1.38	$\leq 30°$
This review	34	83.1	29.2	0	3	5[b)]	23[c)]	4[d)]	1.44	$\leq 30°$
This review	38	73.5	29.2	1	2	1	0	1[e)]		30°-45°

Remarks:
a) SNII-IV including SNII:
b) 1901B, 1921C, 1939C, 1959E, 1964F
c) Including 1909A Pec, which is a type II according to Young and Branch (1989)
d) 1914A, 1950B, 1957D, 1969P
e) 1974C

It is remarkable from Table 1 that not a single supernova in our sample of face-on Sbc-Sd galaxies has been classified as type Ia indicating a very low frequency. Only two (1921C and 1964F) of those classified as SNI are intrinsically very bright (M_B = -20) and could therefore possibly be of type Ia while the rest of the SNeI and the unclassified SNe all are observed to be fainter than -16.6. Thus adopting two SNeIa in the sample of nearby face-on Sbc-Sd galaxies we derive a frequency of 0.08 SNu (= $0.3h^2$ SNu) very similar to the value of $0.2h^2$ SNu given by Evans et al. (1989) for Sbc-Sd galaxies of all inclinations in their sample. h is the Hubble constant in units of 100 km/sec/Mpc. Assuming that the last three SNeI are of type Ib together with one of the unclassified SNe we get a total of around 7 SNeIb corresponding to a frequency of 0.29 SNu. In total we find ν(SNI)=0.37 SNu (= $1.5h^2$ SNu), which is a factor of 3 more than found by Evans et al. (1989). For the frequency of SNeII we derive 1.07 SNu (= $4.3h^2$SNu) again up by a factor of 3 compared to Evans et al. (1989). Tammann (1982) gives nearly equal rates of SNeI and SNeII in Sbc-Sd galaxies in his Table 2, while we derive a rather different ratio of 1:2.9. However, Tammann notes that the ratio is 1:2.0 in his sub-sample of face-on Sbc-Sd galaxies.

Traditionally correction factors have been applied to correct for inclination effects of the observed numbers of SNe. Tammann (1977) found that the frequency of SNe in inclined Sc galaxies ($i \geq 30°$) is lower by a factor of 5.5±1.7. In Table 1 we also give updated information on somewhat inclined

Sbc-Sd galaxies with $30° \le i \le 45°$ ($0.07 \le \log R_{25} \le 0.15$). It is a remarkable fact that not a single SNII is seen in this subsample while the SNeIa and SNeIb seem present at roughly the expected numbers adopting a control time of 29.2 years. The SNeII could obviously be so strongly obscured that they fall below the detection limit.

1.2 The Survey by Evans

Evans, van den Bergh and McClure (1987, 1989), van den Bergh (1989) and van den Bergh and McClure (1989) have discussed SN rates based on observations of 855 Shapley-Ames galaxies made during the period 1980 November 1 - 1988 October 31. They find for the average Shapley-Ames galaxy the following rates: SNIa $(0.28 \pm 0.10)h^2$SNu, SNIb $(0.27 \pm 0.15)h^2$ SNu and SNII $(1.04 \pm 0.30)h^2$SNu. The total number of SNe is 24 which is broken down on SN types and types of galaxies. The rates for the different types of galaxies are of course very uncertain being based on few SNe in each galaxy type. Only SNeII have been detected in statistically significant numbers in Sab+Sb galaxies (6 SNeII) and Sbc-Sd (5 SNeII). Increased control times are necessary to improve the situation. To check for completeness and to compare with other estimates, especially with those by Tammann (1982) and the rates derived above in this review, we will again discuss SNe in Sbc-Sd galaxies.

As already pointed out this review has led to a SNIa rate (Table 1) in Sbc-Sd galaxies for the Tammann sample ($v_0 \le 1200$ km/sec) similar to the Evans et al. (1989) rate. However, the Evans et al. rates of SNeIb ($0.4h^2$SNu) and SNeII ($1.3h^2$SNu) are down by a factor of 3 compared to the estimates in this review ($1.2h^2$ and $4.3h^2$SNu). Such a difference of a factor of 3 has existed for some years and is not easily explained, see e.g. van den Bergh et al. (1987). In particular Evans should have seen 29 SNI in the sub-group of Sbc-Sd galaxies using Tammann (1982) rates, while he saw only 3 as discussed by van den Bergh et al. (1987). This review has to some extent solved part of the problem, since it is now clear from Table 1 that only a very small fraction of the SNe are of type I. As already pointed out there seems no longer to be any serious discrepancy between the rates of SNIa for the sample of nearby face-on Sbc-Sd galaxies and the sample by Evans. There is no reason to believe that Evans is missing any SNeIa.

The problem with a factor of 3 has therefore switched to the frequencies of SNIb and SNII. There is no obvious solution but Evans et al. (1989) have pointed to a number of factors which could cause a low detection rate in their survey: (a) effective magnitude limit vary during the night and from night to night due to changes in transparency, seeing and zenith distance, (b) various forms of human errors and low probability of detection within 1 magnitude of the observational threshold, (c) for visual observers it is often difficult to know the total extent of a galaxy and searches may not be complete, (d) SNe superimposed on a bright galactic nucleus or in emission nebulae may be particularly difficult to see.

From Table 1 it is clear that the number of detected SNeII drops strongly for inclinations i≳30° in agreement with large correction factors for inclined Sbc-Sd galaxies. The inclination factors may in fact go up to any arbitrarily large value if obscuration brings the SNeII below the detection limit. Evans et al. (1989) assumes that each SN suffers the same absorption as the statistically determined absorption for the total stellar component that Sandage and Tammann (1981) adopted for the galaxy in which it occurred. Computing control times from mean light curves this dimming is included by Evans et al. (1989). Although a reasonable procedure it is certainly not easy to justify it. SNeII may well be situated in regions with much more absorption than average.

Fig. 2 in van den Bergh et al. (1987) shows the distribution of apparent distance moduli for their total sample of Shapley-Ames galaxies. The peak in the distribution occurs at apparent $(m-M)_V=32.5$. SNeII with a maximum brightness of $M_V=-18$ and a limiting magnitude of 14.5 can therefore be seen to a distance of $(m-M)_V=32.5$. If the limiting magnitude is somewhat brighter than the assumed value of 14.5 there is zero control time for a large number of galaxies. From their Table 5 in the same paper it is clear that the total surveillance (in units of years$\times 10^{10} L_{B\odot}$) decreases from 778 to 362 for Sbc-Sd galaxies assuming a limiting magnitude of 14.0 instead of 14.5. The rates of SNeII in Sbc-Sd galaxies is therefore critically dependent on limiting magnitude going up by more than a factor of 2 decreasing the limit by only 0.5 magnitude.

Similarly the control time decreases if the SNeII are absolutely fainter at maximum than the adopted value of $M_V=-18.0$. If the SNeII are fainter by 0.5 mag then the effect on surveillance corresponds again to the reduction from 778 to 362 (year$\times 10^{10} L_{B\odot}$) as mentioned above and the derived rates would go up by more than a factor of 2. van den Bergh et al. (1987) used the value at maximum of -18.0 from Kowal (1968) but recalculated control times using $M(max)=-17.8+0.65(M^{ot}_{B_T}+20.9)$ found by Evans et al. (1989). The existence of such a relation is not easily understood but suggested by the authors to be a metallicity effect. However, it is not quite clear if the relation results from selection effects discriminating against both faint galaxies and faint SNe at large distances. Anyway the mean of M(max) is close to -18.

van den Bergh and McClure (1989) have particularly looked into the intrinsic frequency of faint supernovae resembling SN1987A ($M_V(max)=-16.0$). The resulting rate is 0.0-0.4SNu for their sample of Shapley-Ames galaxies depending on the two detections of SN1982F and SN1987A. Comparing to the rates given by Evans et al. (1989) it is concluded that the non-detection of faint supernovae, of luminosity comparable to that of SN1987A, has probably not resulted in underestimation of the intrinsic frequency of SNeII by more than a factor of about 2.

Young and Branch (1989) have studied absolute light curves of type II SNe giving the absolute magnitudes in their Figure 6 for H_o=75 km/sec/Mpc. The median value turns out to be M_B(max)=-17.6 using H_o=50 km/sec/Mpc for SNe in galaxies with $v_o \leq 1200$km/s. However, no correction was made for extinction of the SNe in the parent galaxies and we are therefore back to a value around -18 correcting for that extinction very similar to the value used by Evans et al. (1989). It is interesting to notice that the maximum luminosity of the SNe increases with distance in their Figure 6 discriminating against low luminosity SNeII at large distance. Also their figure indicates a small but separate group of very bright SNeII beyond v_o=1200km/s.

Young and Branch (1989) also demonstrates the well known large scatter in M(max) for SNeII over an interval of 6 magnitudes. Using a mean light curve with such a large scatter in M(max) to calculate control times may well lead to unreliable rates although it is not straight forward to tell how large the effect really is. Ideally control times should be calculated for small intervals of M(max) and weighted with the distribution of M(max).

Since the sample of galaxies by Evans et al. (1989) is different from the sample of nearby face-on Sbc-Sd by Tammann (1982) it is not obvious that the rates should be the same. It will later in this review be evident that e.g. SNeII are mainly produced in prolific Sbc-Sd galaxies. Therefore we shall discuss the nearby subsample with $v_o \leq 1200$ km/s and $i \leq 30°$ from the list by Evans et al. (1989) and also compare to galaxies with $30° \leq i \leq 45°$.

Table 2. Evans face-on and nearly face-on galaxies with $v_o \leq 1200$km/s. Surveillance $\sum L_B t$ is given where L_B is in units of $10^{10} L_{B\odot}$ and t in years.

n_{gal}	L_B	SNIa $\sum L_B t$	N	SNIb $\sum L_B t$	N	SNII $\sum L_B t$	N	i
28	83.7	450	0	344	1	293	1	$\leq 30°$
35	66.8	337	0	273	0	179	0	$30°-45°$

Out of the 28 nearby face-on Sbc-Sd galaxies in Table 2 21 are common with the Tammann sample. The total surveillance $\sum$(luminosity × control time) is determined using the survey times given by Evans et al. (1989). Only one SNIb (1983N) and one SNII (1986I) are seen in the nearby face-on galaxies and none in nearly face-on galaxies. With ν(SNIb)=0.29SNu and ν(SNII)=1.07SNu as estimated above we would expect 1.0 SNIb and 3.1 SNII in the Evans et al. (1989) sample of nearby face-on Sbc-Sd galaxies. Although the numbers are small there seems to be a deficiency of SNe detections by Evans unless the detection rate of 12 SNe per 10 years (1960-69) in the Tammann sample is a gross chance overestimate. Also Evans did not see any SNe in nearby Sbc-Sd galaxies with small inclinations ($30°$-$45°$) although he should have seen 3 using the control times as given by Evans et al. (1989), which includes dimming from inclination.

1.3 Rates from the Asiago Supernova Search

Capellaro and Turatto (1988) used a sample consisting of all galaxies from RC2 included in the 70 fields of the Asiago Supernova Search. The sample is therefore not based on the RSA Catalogue of Bright Galaxies. The SNe and their types were taken from the updated Asiago Supernova Catalogue by Barbon et al. (1988). The SNe had to satisfy the requirements: (a) appearance in a galaxy of the sample and (b) appearance on an Asiago survey plate. The final sample consists of 736 galaxies and 51 SNe (SNI: 26, SNII: 14, SN(unclassified): 11). Surprisingly none of the SNeI belong to the SNIb subgroup but on the other hand it is not clear from Capellaro and Turatto (1988) how many of the SNeI are genuine SNIa and how many could be SNIb.

Cappellaro and Turatto have corrected their control times for lost SNe in overexposed central regions assuming that up to 50% are lost in the most distant galaxies. In particular they have discussed inclination effects and find a factor of 3 to correct the number of SNe in spirals with $i \geq 30°$ to the expected face-on value except for Sc galaxies whose correction factor was adopted to be 5. These values are intermediate to the values found by Tammann (1977) (≈ 6) and Maza and van den Bergh (1976) (≈ 2) for Sc galaxies.

To compare with rates mentioned above we derive ν(SNI)=0.54 SNu, ν(SNII)=0.63 SNu and a total rate ν=1.17 SNu for Sbc-Sd galaxies from Table 4 of Capellaro and Turatto (1988). For the nearby sample of face-on Sbc-Sd galaxies we find rates in this review as mentioned above for SNI: 0.37 SNu and SNII: 1.07 SNu. For the SNeI there is reasonable agreement between Capellaro and Turatto and this review, while our rate of SNeII is somewhat higher than theirs. However, their total rate of 1.17 SNu is not very different from ours of 1.44 SNu considering the limited numbers of SNe.

1.4 Multiple SNe Events

Richter and Rosa (1989) have discussed statistics of multiple SNe events in galaxies. They find that normal galaxies belong to two distinct classes as far as the formation of SNII progenitors is concerned. Most galaxies are rather inactive while only a small fraction of galaxies are fast producers in a burst phase. The rate of SNeII in the prolific galaxies is up by a factor of roughly 70 compared to the inert galaxies. The recurrence time scale of such bursts is several hundred times larger than their duration of around 10^7 years. Richter and Rosa find that 6% of the galaxies known to have produced SNe have indeed produced more than one SNe. The average time interval between two successive SNeII events is 15±3 years, while it is 26±10 years for SNeI events.

In this review we calculate the total frequency of SNe in the sample of nearby face-on Sbc-Sd galaxies from the subgroup of galaxies with more than one SN assuming that all SNe were caught in these galaxies by

$$\langle \nu^{-1} \rangle = N^{-1} \sum \tau L_B / 10^{12} L_{B\odot}$$
$$\nu = \langle \nu^{-1} \rangle^{-1}$$

where N=22 is the total number of time intervals τ between subsequent SNe. The total number of SNe is 31 found in 9 prolific galaxies. These 31 SNe perform a subset of the 35 SNe mentioned in Table 1. We derive ν=1.1 SNu this way. These 9 prolific galaxies radiate $42.8 \times 10^{10} L_{B\odot}$ or 52% of the light in the total sample of nearby face-on Sbc-Sd galaxies. The other 25 galaxies in the sample have each produced none or just one SN (in total only 4 SNe) although they contain as much light as the 9 prolific galaxies.

However, the 31 SNe in the prolific galaxies show a distribution on time of detection very similar to Figure 1 and all the SNe in prolific galaxies have obviously not been seen. The frequency is therefore higher than the above mentioned value of 1.1 SNu. The control time is again around 29.2 years and we find a total frequency ν=2.5SNu of SNe in prolific nearby face-on Sbc-Sd galaxies. The 4 SNe in non-prolific galaxies give a rate of only 0.3SNu (=1.4h^2SNu). Although this method may exaggerate differences between prolific and non-prolific galaxies it is obvious that the observed SNe mainly occur in prolific galaxies.

Table 3. Nearby face-on Sbc-Sd galaxies divided into a high and a low luminosity group. L_B is in units of $10^{10} L_{B\odot}$.

	n_{gal}	L_B	SNIa	SNIb	SNI	SNII	SN uncl.
high lumin.	6	42.7	-	1	3	11	3
low lumin.	28	40.4	-	2	2	12	1

Surprisingly, there is a number of high luminosity nearby face-on Sbc-Sd galaxies with no SNe detections e.g. NGC628 ($7.8 \times 10^{10} L_{B\odot}$). We have therefore further subdivided the nearby face-on Sbc-Sd galaxies into two groups: a high-luminosity and a low-luminosity group each containing an equal amount of luminosity. See Table 3. The two luminosity groups contain roughly equal numbers of SNe and give equal frequencies since prolific galaxies occur in both luminosity groups.

2 SNe AND FAR-INFRARED LUMINOSITIES

The results discussed above indicate that the blue luminosity is not a very good indicator of the formation rate of massive stars in an individual galaxy. The SNeII are mainly produced in prolific Sbc-Sd galaxies and very few in inert galaxies although they contain as much blue luminosity. E.g. de Jong (1986) has demonstrated that the FIR luminosity as measured by IRAS in the 60μ and 100μ bands is a much better indicator of star formation activity than the blue light. This conclusion has been strengthened by the work of Sage and Solomon (1989) and Trinchieri et al. (1989). We use the FIR luminosities from (1) Rice et al. (1988) for the large optical galaxies, (2) the IRAS Small Scale Structure Catalog (1986) and (3) the IRAS Point Source Catalog (1985). Introducing the unit SNuIR as 1 SN per $10^{10} L_\odot$ FIR luminosity per 100 years we derive the frequency $\nu=2.8$SNuIR of SNe in the prolific nearby face-on Sbc-Sd galaxies assuming that all SNe in these galaxies were caught and using the time intervals between subsequent events as described above. However, adopting a control time of 29.2 years for the prolific galaxies this number increases to $\nu=6.5$SNuIR. For the whole sample of nearby face-on Sbc-Sd galaxies we find $\nu=4.8$SNuIR (=$19.2h^2$SNuIR) adopting the control time of 29.2 years. We consider this last value of ν to be representative when scaling numbers with FIR luminosities.

Using this frequency also for an Sb galaxy we derive the rate of SNe in the Andromeda galaxy to be only 0.43 SNe per 100 years or $\tau = 230$ years, since the FIR luminosity ($9 \times 10^8 L_\odot$) of M31 as given by Rice et al. (1988) is extremely low. This is a much smaller rate of SNe in M31 than usually quoted, e.g. by Tammann (1982), who gives $\tau = 21$ years based on the blue luminosity. Our very low rate in M31 is in a way reasonable since only one supernova has ever been detected in M31 (1885A = S And, type Ia) although the SNeII are expected to be brighter than magnitude 8 at maximum and therefore easily seen in small amateur telescopes. We prefer to explain the missing SNeII in M31 by its very low massive star formation rate rather than by a too high estimated frequency of SNe for the nearby galaxies.

The prolific galaxies have in most cases $L_{FIR} \gtrsim 1 \times 10^{10} L_\odot$ but although the FIR luminosity correlates well with the star formation rate some Sbc-Sd galaxies with that luminosity have no detected SNe. E.g. NGC628 has a $L_{FIR}=1.4 \times 10^{10} L_\odot$ and is therefore expected to have $\tau=15$ years but no SNe are detected in this galaxy as mentioned above.

3 THE RATE OF SNe IN DISTANT CLUSTERS OF GALAXIES

The rates of SNe in very distant galaxies is important in several aspects, e.g. in relation to the star formation rate at earlier epochs and the chemical evolution of galaxies. Recent spectroscopic and photometric studies of such galaxies suggest a spectacular increase in star formation activity with look-back time for some subset of galaxy population (Butcher and Oemler (1978), Dressler and Gunn (1983), Couch and Sharpless (1987), MacLaren et al. (1988)). It is interesting to determine whether this activity has produced any increase in the rates of the various types of SNe. However, the SNe are so faint even at maximum that they are hard to find. At a redshift of $z=0.3$ we expect a magnitude for SNeIa at maximum of $V=21.7$ and for SNeII around 2 magnitudes fainter. The look-back time for $z=0.3$ is around 5 Gyrs. Only the survey with the Danish 1.5m telescope on La Silla by Hansen, Nørgaard-Nielsen, Jørgensen, Ellis and Couch as described by Hansen et al. (1989) and Nørgaard-Nielsen et al. (1989) has had any success in finding very distant SNe. Using clusters of galaxies as targets they are concerned with the rate in dense environments populated by early type galaxies and they expect therefore primarily SNe of type Ia to be detected.

The total sample consists of 65 rich clusters. 50% of the observations are on dense regular Abell clusters with redshifts $0.2 \leq z \leq 0.4$. The mean richness is estimated from the Bautz-Morgan classes to be similar to that of the Coma cluster. About 30% of the observations are on clusters from the deep AAO survey by Couch et al. (1988) estimated to have a mean richness of about 70% of Coma. The remaining 20% of the observations are on distant clusters found by various techniques (Butcher and Oemler (1978), Koo (1981), West and Frandsen (1981)) and classified as rich or richer than Coma. The mean richness of the total sample is thus comparable to that of the Coma cluster whose B luminosity within the Danish 1.5m telescope CCD frame at $z=0.3$ would be around $5 \times 10^{11} L_{B\odot}$. The galaxy population sampled will thus be dominated by E/S0 galaxies whose SNIa frequency has been estimated to be within the range $\nu=(0.2-0.8)h^2$SNu by Tammann (1982), Evans et al. (1989) and Cappellaro and Turatto (1988). Hansen et al. (1989) predict a number of detections of 2-9 SNeIa above the threshold of $V=24$ based on the quoted frequency. Having seen only one SN in their survey until April 1988 is not in serious disagreement with the lower limit of the frequency range as given by Evans et al. (1989). A new reduction and a more careful search for SNe in the frames from the Danish 1.5m telescope search have been initiated.

Acknowledgements. Suggestions and critical comments on the final version of the manuscript by L. Hansen and H. U. Nørgaard-Nielsen are greatfully acknowledged.

References

Barbon, R. Cappellaro, E. and Turatto, M.: 1988, The Asiago Supernova Catalogue. Version July 26, 1988. Preprint
Butcher, H., Oemler, A.: 1978, Astrophys. J. **219**, 18
Cappellaro, E. and Turatto, M.: 1988, Astron. Astrophys. **190**, 10
Couch, W.J., Sharpless, R.M.: 1987, Monthly Notices Roy. Astron. Soc. **229**, 423
de Jong, T.: 1986. In Proceedings of the Erice Workshop on 'The Spectral Evolution of Galaxies'. Eds. C. Chiosi and A. Renzini, Reidel Publ. Co., Dordrecht, p. 111
Dressler, A., Gunn, J.E.: 1983, Astrophys. J. **270**, 7
Evans, R., van den Bergh, S., and McClure, R.D.: 1989, Astrophys. J. **345**, 752
Hansen, L., Jørgensen, H.E., Nørgaard-Nielsen, H.U., Ellis, R.S., Couch, W.J.: 1989, Astron. Astrophys. **211**, L9
IRAS Point Source Catalog. 1985, Joint IRAS Science Working Group (Washington, DC: GPO)(PSC)
IRAS Small Scale Structure Catalog. 1986, prepared by G. Helon and D. Walker (Washington, DC: GPO) (SSSC)
Koo, D.C.: 1981, Astrophys. J., **251**, L75
Kowal, C.T.: 1968, Astron. J. **73**, 1021
Mac Laren, I., Ellis, R.S., and Couch, W.J.: 1988, Monthly Notices Roy. Astron. Soc. **230**, 249
Maza, J. and van den Bergh, S.: 1976, Astrophys. J. **204**, 519
Nørgaard-Nielsen, H.U., Hansen, L., Jørgensen, H.E., Salamanca, A.A., Ellis, R.S., Couch, W.J.: 1989, Nature **339**, 523
Rice, W. et al.: 1988, Astrophys. J. Suppl. Ser. **68**, 91
Richter, O.-G. and Rosa, M: 1988, Astron. Astrophys. **206**, 219
Sage, L. and Solomon, P.M.: 1989, Astrophys. J. **342**, L15
Sandage, A. and Tammann, G.A.: 1981, A 'Revised Shapley-Ames Catalogue of Bright Galaxies', Washington: Carnegie Institution
Tammann, G.A.: 1974, 'Supernovae and Supernova Remnants', ed. C.B. Cosmovici, Astrophysics and Space Science Library, Vol. 45, p. 155. Reidel, Dordrecht
Tammann, G.A.: 1977, 'Supernovae', ed. D.N. Schramm, Astrophysics and Space Science Library, Vol. 66, p. 95. Reidel, Dordrecht
Tammann, G.A.: 1982, 'Supernovae: A Survey of Current Research', eds. M.J. Rees and R.J. Stoneham. Reidel, Dordrecht
Trinchieri, G., Fabbiano, G., and Bandiera, R.: 1989, Astrophys. J. **342**, 759
Vaucouleurs, G. de, Vaucouleurs, A. de, and Corwin, H.G.: 1976, 'Second Reference Catalogue of Bright Galaxies', Austin: Univ. of Texas Press
van den Bergh, S., McClure, R.D., and Evans, R.: 1987, Astrophys. J. **323**, 44
van den Bergh, S.: 1989, Preprint 'Galactic and Extragalactic Supernova Rates'. Santa Cruz Workshop on Supernovae to be published by Springer Verlag (Berlin). Ed. S.E. Woosley

van den Bergh, S. and McClure, R.D.: 1989, 'The Intrinsic Frequency of Faint Supernovae Resembling SN1987A'. Preprint from Dominion Astrophysical Observatory
West, R.M. and Frandsen, S.: 1981, Astron. Astrophys. Suppl. Ser. **44**, 329
Young, T.R. and Branch, D.: 1989, Astrophys. J. **342**, L79

ASTRONOMICAL INSTRUMENTATION

Scientific Organizers: B. Fort, J. P. Picat
Local Organizer: P. Alvarez

E.S.O site evaluation for the V.L.T

MARC SARAZIN

European Southern Observatory
Karl-Schwarzschild Str. 2, D-8046 Garching bei Munchen, F.R.G.

Abstract
The methodology and instrumentation of the site evaluation campaign for the V.L.T innovate in several ways. In particular, seeing monitors based on the measurement of differential image motion have been developed for that purpose. The availability of such instruments opens a new era in the understanding of site related parameters, and permits to plan a more flexible use of modern telescopes in the future.

1 Methodology for a site evaluation campaign

1.1 Large scale geographic survey

Some site evaluation campaigns have constraints which limit already from the start the number of options, or which even reduce the task to merely confirming a choice made in advance.

The campaign for the choice of the V.L.T site started under quite different assumptions. Despite considerations on logistics and on the good quality of the science made at La Silla, it was decided not to set geographical limits to the initial site survey.

Many parameters enter in the composition of the site study but, assuming that if one can expect from modern technology new ways of compensating atmospheric distortions it is not probable that man will soon be able to wash out clouds! For an optical astronomical observatory, cloud cover had to be chosen as the main driver of a large scale survey. Another parameter related to sky clarity is the precipitable water vapor, a major source of noise in the infrared.

Starting in 1982, cloudiness and water vapor large scale distribution patterns were analyzed from satellite charts and meteorological statistics [Woltjer 83]. This led to select Northern Chile as the main area of study to be compared with La Silla and its surroundings.

1.2 Suitability of the terrain

The survey of Northern Chile between $-20°$ and $-25°$ of geographical latitude lasted 4 years during which many summits were visited and studied with respect to accessibility, shape, microclimate and dryness. It ended with a selection of 8 sites ranging from 2300m to 6300m in altitude, distributed along three main chains between the pacific coast and the cordillera [Ardeberg 86]. Fortunately, the more accessible coastal sites were also the most promising ones for their photometric quality.

At that time, the size of the summit itself was not considered as a driving parameter. The V.L.T linear shape had been proposed initially as the best arrangement considering the marked prevailing direction of the wind at La Silla. Because most potential sites are rather small in one or both dimensions, it is now clear that the final configuration of the V.L.T array will be determined by the topology as much as by wind patterns and considerations of (u,v) coverage in the interferometric mode.

Cerro Paranal, a 2664m high coastal peak, 800km North of La Silla observatory, was chosen as the best probable competitor against La Silla (and its immediate surroundings) and permanent monitoring of meteorological parameters started there as early as 1983.

1.3 Identification of the optical parameters of the atmosphere

Does the optical quality of the atmosphere above a site really matter for the future use of a telescope? The answer to this question would not have been so clear some decades, or even years, ago when observations were mainly limited by the quality of instruments and telescopes. These were designed according to the experience gained at traditional observatories often located in poor but convenient sites where 1 arcsec was considered as an excellent seeing. Nowadays, it suffices to consider the impressive amount of money which is spent in space astronomy, mainly to get rid of atmospheric effects.

In September 1984, the first V.L.T Site Selection Working Group [1] was constituted to advise on the instrumentation to be developed for the site study and to investigate the parameters to be monitored. This working group was composed of physicists of the atmosphere and of astronomers with a long experience in observing atmospheric related parameters: it contributed actively to the site study until the beginning of 1988 when most of the instrumentation was already in operation.

Quantitatively, the dependence of site quality on atmospheric parameters is a function of the type of observation (imaging, spectroscopy, photometry or inter-

---

[1] Constitution of the W.G : H. van der Laan (chairman), A. Ardeberg, J. Vernin, G. Weigelt, H. Wohl (permanent members), M. Sarazin (secretary), D. Cadet, F. Roddier (permanent consultants), D. Enard, J. P. Swings (permanently invited)

ferometry) and of the wavelength range. The relations for direct imaging of faint objects are simple when the sky photon noise dominates over the source and detector noises, the limiting detectable flux varying like the image size. If the focal ratio of the telescope is not made variable with seeing conditions, the exposure time necessary to image a star of a given magnitude will increase as the square of the image size.

For infrared observations, a nearly perfect correction of atmospheric effects will be possible only when the conditions are already fairly good. In the adaptive optics system foreseen for the VLT [Merkle 86], the number of elementary cells to be used for a perfect correction varies as the square of the initial image size. The efficiency of such a system is not only dependent on the total seeing, but also on the altitude profile of the turbulence which determines the isoplanatic angle, and on its velocity profile which gives the temporal frequency spectrum of the corrections.

In interferometric imaging limited by thermal background , the signal to noise ratio for an image corrected by adaptive optics is still dependent on the $17/12$ power of the seeing because of the spectral bandwidth. The dependence increases to the $35/12$ power for the incoherent addition of signals coming from an array of detectors [Vlt 86].

In spectroscopy, the exposure time may decrease also as the square of the image size , but only down to the limit for which the instrument was designed.

The seeing came out as the main parameter to be monitored and it was clear that an accurate instrument was to be designed. Because of the non linear dependence of efficiency of most observing modes with image size, a small difference in seeing could be enough to elect or discard a summit under study. Secondly, it was considered of interest to quantify the contribution of the various atmospheric layers to the total seeing.

1.4 Monitoring station

Only one system allows to remotely determine the altitude of the turbulent layers, it is the Scidar (Scintillation Detection and Ranging) [Azouit 80]. It is based on the spatio-temporal analysis of shadow patterns produced by double stars in the pupil plane of a large telescope. Though a transportable version with a plastic refractor has been successfully tested, it remains a rather heavy instrument due to the large aperture (80cm) required.

Speckle analysis is the only optical method which can provide information on most parameters [Weigelt 86] but only with large apertures (2m) and at the cost of much computing time.

After a proposal by F. Roddier, it was decided to build a differential image motion monitor (DIMM) as the main tool for the measurements of seeing quality. This technique, described in details in [Sarazin 89], uses 35cm diameter movable cassegrain telescopes built through a collaboration between E.S.O and Amos S.A (Liege, Belgium).

The first DIMM started routine measurements on Cerro Paranal in April 1987, followed by a second system in September 1988 on Cerro Vizcachas, and a third one in April 1989 on Cerro La Montura, 2500m, 4km from Cerro Paranal.

Besides the DIMM, a complete monitoring station includes a scintillometer, a set of microthermal sensors, and a meteorological station. All data are displayed on line every minute, automatically archived and may be statistically processed on the spot if required.

For more detailed studies of a particular site, the DIMM software includes provision for monitoring tilt lifetime (correlation of two times series with 64ms delay) and the tilt isoplanicity (differential motion of double stars as a function of their angular separation). One also uses an acoustic sounder for the analysis of the turbulence in the boundary layer, and a dust meter which measures the number of particles (0.3 to 10 μm in diameter) per unit volume of air.

The sky emissivity is monitored twice an hour with a KPNO infrared radiometer in manual mode. In the same time, the observer notes down the status of the cloud cover.

1.5 Decision making process

According to the VLT project schedule, the final decision on the site has to be taken by Council on proposal from the E.S.O Director General in December 1990, ie: eight years after the beginning of the site evaluation campaign, and three and a half year after the first regular seeing monitoring station was installed. The Council will receive the input of the Scientific Technical Committee and of the second V.L.T Site Selection Working Group [2].

This working group was set up in December 1988 and is composed of astronomers chosen *"to cover the main areas of visible and infrared observing as well as interferometric imaging"*. Its current assignment is *"to advise the Director General on the analysis and interpretation of V.L.T site data"*. The group has been visiting the areas under study and is also working on the determination of criteria of merit relative to each type of observing. The following assignment stage is *"to analyse the astronomical, operational and financial pros and cons of the several site options"*; and *"to prepare and submit to ESO management a recommendation on the best possible choice for the V.L.T site"*.

---

[2] Constitution of the W.G : J. P. Swings (chairman), I. Appenzeller, A. Ardeberg, P. Charvin, G. Lelievre, C. Perrier, H.E. Schuster, P. Shaver (permanent members), M. Sarazin (secretary), J. Beckers, D. Enard, M. Tarenghi (permanently invited)

2 Understanding the atmospheric physics

2.1 General theory

The atmosphere interacts with light through the spatial and temporal variations of air temperature which locally affect the instantaneous index of refraction. In a well developed turbulence - such as usually exists a few meters above ground - the thermal structure of the air is characterized by its isotropic temperature structure parameter [Clifford 78]:

$$C_T^2(\mathbf{r}) = \frac{<(T(\mathbf{r}) - T(\mathbf{r} + \Delta \mathbf{r}))^2>}{\Delta r^{2/3}} \quad °C^2 m^{-2/3}. \tag{1}$$

The index of refraction n of the atmosphere, for a wavelength λ, pressure P, absolute temperature T, and water vapor pressure e is:

$$n - 1 = 77.6 \, 10^{-6}(1 + 7.52 \, 10^{-3}\lambda^{-2})(P + 4810 \, e/T)/T. \tag{2}$$

The local variations of the index n are responsible for the perturbations induced on the optical wavefront. Most authors agree to neglect pressure and humidity fluctuations for optical wavelength at a fixed height over land. In this case, the same statistical law as for temperature also applies and one may define an atmospheric structure parameter for the index of refraction at $\lambda = 0.5 \mu m$:

$$C_n^2 = \left(80 \cdot 10^{-6} \frac{P}{T^2}\right)^2 C_T^2 \quad m^{-2/3}. \tag{3}$$

2.2 Long exposure image size

In the presence of atmospheric turbulence, one notices that when a telescope has a diameter D larger than a value r_0, it is not anymore limited by diffraction.

The resolution θ of a large telescope limited by the atmospheric turbulence may be defined according to the Strehl criterion:

$$\theta = (4/\pi)\lambda/r_0. \tag{4}$$

r_0 is also called the Fried parameter and the effect of the atmosphere on images by means of the long exposure optical transfer function [Fried 66], as a function of the spatial frequency f in radians $^{-1}$:

$$T(f) = \exp -3.44(\lambda f/r_0)^{5/3}, \tag{5}$$

which corresponds to images slightly different from a gaussian profile. The atmospheric index structure function is then related to the Fried parameter r_0 along [Roddier F. 81]:

$$r_0 = \left[0.423 k^2 (\cos\gamma)^{-1} \int C_n^2(H)\,dH\right]^{-3/5} \text{ m}, \qquad (6)$$

where $k = 2\pi/\lambda$ and γ is the zenith angle.

An equivalent long exposure image size FWHM (Full Width at Half Maximum) may then be defined, since it is the parameter directly accessible to the astronomer. A numerical computation gives [Dierickx 88]:

$$\text{FWHM} = 0.98 \lambda / r_0 \text{ radians}. \qquad (7)$$

The FWHM is the expression of the well known **Seeing** which, contrary to r_0, is not much dependent on wavelength (FWHM $\propto \lambda^{-1/5}$). It is related to the resolution of Strehl along:

$$\theta = 1.26 \text{ FWHM}. \qquad (8)$$

A commonly used parameter to characterize turbulence strength is the FWHM in arcsec and at $\lambda = 0.5\mu m$ for turbulence occurring between heights H_0 and H_1 above ground:

$$\text{FWHM} = 2.0\,10^7 \left[(\cos\gamma)^{-1} \int_{H_0}^{H_1} C_n^2(z)\,dz\right]^{3/5} \text{ arcsec}. \qquad (9)$$

2.3 Altitude of the turbulence layers

With the development of new techniques to correct the optical wavefront disturbances created by the atmosphere, it appeared that site quality was not only determined by the path integral of the turbulence in the troposphere but also by the altitude at which the turbulence occurs and by the speed of its motion. Scintillation, non-isoplanicity and short speckle lifetime are the main observable consequences.

In addition to the standard Fried parameter r_0, scale height parameters were introduced for interferometry, speckle interferometry, and adaptive optics. All refer to the wavefront temporal (speckle lifetime) and spatial (isoplanicity) coherence.

Describing the effect of isoplanicity on the performances of adaptive optics systems, [Fried 82] used the parameter:

$$h_F = \left[\frac{\int_0^\infty C_n^2(z) z^{5/3}\,dz}{\int_0^\infty C_n^2(z)\,dz}\right]^{3/5} \text{ m}. \qquad (10)$$

A slightly different form was given by [Roddier F. 82b] for stellar speckle interferometry with large apertures:

$$\overline{h} = \left(\left[\frac{\int z^2 C_n^2(z)\,dz}{\int C_n^2(z)\,dz}\right] - \left[\frac{\int z C_n^2(z)\,dz}{\int C_n^2(z)\,dz}\right]^2\right)^{1/2} \text{ m}. \qquad (11)$$

It is then possible to determine an approximate isoplanatic patch size:

$$\theta = 0.36 \frac{r_0}{h} \text{ radian.} \tag{12}$$

The same formulae apply also to speckle life time, replacing altitude by wind speed as a weighing parameter [Roddier F. 82a]. For the speckle boiling time, [Aime 86] found the relation:

$$\theta = 0.47 \frac{r_0}{\Delta v} \text{ s}, \tag{13}$$

where Δv is the dispersion of the horizontal wind speed as a function of altitude.

Finally, the index of scintillation defined as the variance of relative irradiance fluctuations as seen through a finite aperture was defined by [Roddier F. 81] as:

$$\sigma_I^2/I^2 = 19.12 \lambda^{-7/6}(\cos\gamma)^{-11/6} \int_0^\infty C_n^2(z) z^{5/6} \, dz, \tag{14}$$

which holds for small apertures only (a few centimeters). For a large telescope of diameter D, one uses a slightly different relation:

$$\sigma_I^2/I^2 \propto D^{-7/3}(\cos\gamma)^{-3} \int_0^\infty C_n^2(z) z^2 \, dz. \tag{15}$$

Both relations are not valid close to the horizon where saturation occurs.

2.4 Image motion

Short exposures taken at the focus of a large telescope limited by the atmospheric turbulence have a speckle structure. The speckles are moving in time and one calls image motion the motion of their center of gravity. It is a consequence of the average of phase fluctuations over the whole pupil of the telescope.

The Fried parameter r_0 or the resulting image size FWHM are defined for long exposures where the integration time is long enough to include all motions of the individual speckles in the image. For a large telescope, a minimum equivalent exposure time of 10s is required.

On such a time base, the fluctuations of the angle of arrival of the wavefront obey Gaussian statistics and their variance is isotropic. It is related to the Fried parameter by [Roddier F. 81]:

$$\sigma_x^2 = \sigma_y^2 = \sigma_{xy}^2/2 \simeq (3.44/2\pi^2)\lambda^2 D^{-1/3} r_0^{-5/3} \simeq 0.18 \lambda^2 D^{-1/3} r_0^{-5/3} \tag{16}$$

The standard deviation of image motion is independent of the wavelength and decreases when the averaging aperture increases. It is thus possible to evaluate r_0 and consequently the image quality of a large telescope by measuring image motion through any aperture of known diameter. The smaller aperture leads to the highest accuracy at equivalent signal to noise ratios, but also requires shorter exposure times to freeze the motion properly.

Direct image motion monitoring was until recently the most commonly used method for site studies worldwide. It is more reliable with fixed telescopes (star trails) since mechanical spurious motions can hardly be distinguished from atmospheric effects.

In that respect, monitoring differential image motion is simpler since the telescope spurious motions are subtracted out. Moreover it does not need a good optical quality, nor a perfect tracking contrary to direct image size measurements.

The problem of measuring differential angle of arrival through two relatively small apertures of variable separation has been considered by [Fried 75] and its analytical solution given by [Roddier F. 81] and in a more detailed manner in [Sarazin 89].

The variance of the differential motion between two pupils of diameter D at the distance d from eachother is equal to the sum of the two motions (uncorrelated statistics) minus the sum of the common motions (given by the covariance function):

$$\sigma_{\parallel}^2 \simeq 2(0.18 D^{-1/3} - 0.097 d^{-1/3})\lambda^2 r_0^{-5/3} \qquad (17)$$

$$\sigma_{\perp}^2 \simeq 2(0.18 D^{-1/3} - 0.145 d^{-1/3})\lambda^2 r_0^{-5/3}. \qquad (18)$$

We recall that, in general, if the subscript $*$ represents either type of above mentioned motions ($\parallel$ or $\perp$):

$$\sigma_*^2 = K_* \lambda^2 r_0^{-5/3}. \qquad (19)$$

In other words, for motion measurements made with a star at zenith angle γ, the integrated turbulence along the zenital path is:

$$\int C_n^2(H)\,dH = 0.06 \cos\gamma \; \frac{\sigma_*^2}{K_*} \; \mathrm{m}^{-2/3}. \qquad (20)$$

It is important to note again that the wavelength is not present in this equation, ie: for a given amount of fluctuations of the atmospheric index of refraction, the apparent motion of the image does not depend on the detector wavelength sensitivity.

3 Analysis of results

3.1 Calibration test

The seeing monitor normally operates in the open air on a 5 m high tower. The only way to determine whether this instrument was producing data representative of long exposures on large telescopes was to have it operating in rigorously similar conditions. Two nights were allocated at the La Silla Observatory on the 2.2 m telescope for the comparison of the differential motion seeing with large telescope seeing, from May 26 to 28, 1988 [Pedersen 88]. The seeing monitor was attached with its 35 cm diameter optics on the side of the 2.2 m primary mirror cell. Both instruments were thus inside the dome, observing the same stars close to zenith position.

Full widths at half maximum (FWHM) of stellar images were measured on 50 s exposures in direct CCD imaging mode on the 2.2 m. They compare precisely with seeing monitor data, even during events of a few minutes temporal scale. The comparison of DIMM results and 2.2 m data gives a correlation coefficient of 0.97 for the best linear fit.

3.2 Short time behaviour

It was often believed that seeing variations were slow, of the order of a few hours of even of a night. This is indeed true when the image size increase is mainly produced by dome seeing or even mirror seeing (a thermally active layer may exist at the surface of large primary mirrors which are not adequately ventilated).

On the other hand, seeing monitor data have shown that the atmospheric seeing may change by a factor of two or more in a matter of a few minutes. This is due either to local turbulence usually linked to changes in wind direction or wind speed close to ground level, or to activity in higher layers triggered by temperature variations due to pressure/altitude changes (gravity waves). Turbulence is also believed to lay at the interfaces of air streams of different velocity or direction (wind shear) [Coulman 88].

Such rapid changes have many consequences in astronomical work. For instance it makes it really hard to accurately focus a telescope in the conventional way, which consists in measuring the diameter of several consecutive short exposures. The smallest image size may correspond as well to the best focus or to the best seeing.

A site should of course present seeing gradients as low as possible, keeping in mind that this condition is not sufficient. The site must principally provide the astronomer with as much good seeing as possible.

It is thus useful to define a new parameter taking both effects into account: the night average seeing of a site, weighed by the integral of the normalized temporal autocorrelation function of the instantaneous seeing θ_i ($\theta_i, i = 1, n$ representing seeing samples recorded at constant intervals in the course of the night). Such a study is underway.

3.3 Long term statistics

One of the main question marks of a site survey is the length of the observing period necessary to be able to take a decision.

The 5 year long cloud survey made at Paranal and La Silla showed that the seasonal trends were well reproduced from year to year and that, even if there were good and bad years which could be linked to large scale climatic changes, they would never have modified the conclusions on the respective merits of the sites.

Seeing statistics of Cerro Vizcachas are shown on Fig. 1. The seasonal variations of the monthly median may exceed 20% which, elevated at some exponent cited in 1.3, is far from negligible. It is even larger than the difference between sites. A duration of one year is thus a minimum for site assessment. A shorter period may be enough on subsites presenting a good correlation with a main and well studied summit.

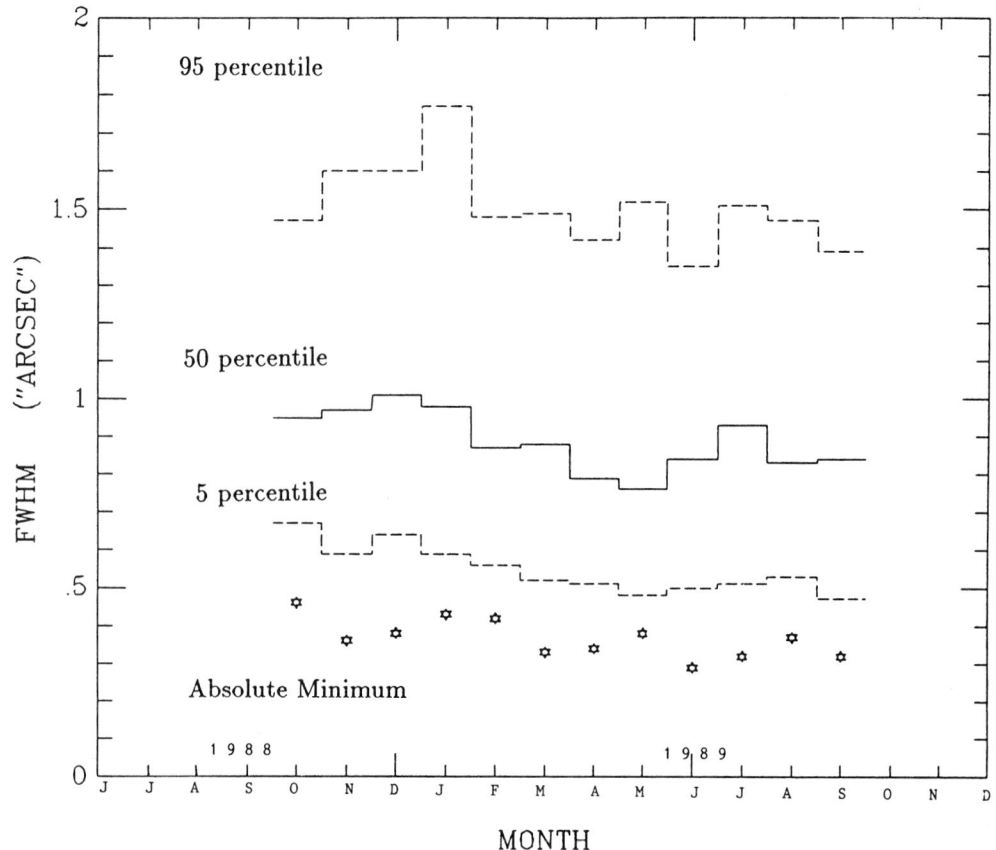

Figure 1: Monthly statistics of the seeing at Cerro Vizcachas obtained with the differential motion monitor. A comparison of this method with others is underway.

3.4 Comparison with existing observatories

From the start, the La Silla observatory was designated out as a calibrator for all potential alternatives. The first Working Group on site evaluation invested much effort in first assessing the intrinsic quality of La Silla as a site independently of telescope characteristics. This culminated with the LASSCA campaign [Vlt 87], 10 nights in February 1986 during which three large telescopes were solely dedicated to measuring seeing and related parameters. Current statistics obtained on Cerro Vizcachas, a V.L.T potential site close to La Silla, are in good agreement with the LASSCA conclusions.

A similar campaign was conducted at Mauna Kea during 1988-1989 and should also deliver useful terms of reference for what is believed one of the best sites worldwide. A seeing monitoring campaign was also conducted there for the N.N.T.T site survey in 1984-1986 which gave excellent results, yet not directly comparable to other methods.

Another new seeing monitoring method is currently used by the Carnegie Institution on several potential locations at Las Campanas Observatory. A joint program is planned with E.S.O to compare Las Campanas with Cerro Vizcachas, using identical differential motion monitors on each site for a few weeks.

It is otherwise rather difficult to use past litterature to assess site quality seeing-wise. Official statistics of conventional telescopes are always pessimistic because of the dome seeing and optical aberrations, and seeing monitor data are scarce and not always properly calibrated . The most complete review of that subject was written by [Walker 83].

4 Preservation and use of site quality

4.1 Telescope enclosure and optics

As was stated before, data obtained from a seeing monitor operated in open air a few meters above ground, are invariably smaller or equal to what is obtained at the focus of a conventional telescope.

The most detrimental cause is turbulence in the dome which may sometimes contribute even more than the whole atmospheric path to the total seeing. The second problem, which is noticeable on large telescopes during good seeing conditions is the poor control of optical aberrations.

It is possible, by an appropriate design of the telescope enclosure to minimize dome contribution. It is also possible, with the implementation of active control of the primary mirror shape, to permanently maintain the status of the telescope at its intrinsic best limits [Tarenghi 89].

It is thus now possible for the astronomer working with a modern and well designed telescope to take advantage of the moments during which most of the atmosphere is at rest. Such events may last tens of minutes providing the well known *lucky observer conditions* defined in [Barletti 76], ie: seeing remaining inside the 0.2 to 0.3 arcsec. range.

4.2 Proximity of buildings

The V.L.T is unique in many senses, and in particular because nobody ever dared to locate several large telescopes so close to each other. It is of prime importance to make sure that the array configuration be such that the seeing of one V.L.T unit is minimally disturbed by the neighbouring telescope.

Figure 2: Measurements of the seeing in the wake of a the NTT building, the arrow shows the location of the seeing monitor

By lack of previous experience, it is hard to determine how detrimental this effect could be. To answer this question, an E.S.O seeing monitor has been located during a few nights 40 meters south of the NTT building, so that by northern wind, it was in the wake of the enclosure (Fig. 2).

The seeing measured out of 1 minute exposures at the NTT has been compared to the seeing monitor data reduced at similar airmass and wavelength. The differential equivalent seeing was computed in terms of turbulence (C_n^2) which is linearly additive and then reconverted to arcseconds.

It was seen that, by northerly wind, the seeing downwind was larger than the seeing upwind and that this additional seeing, or "wake seeing" was stronger for low wind velocity ($\approx$ 1 arcsec at 2m/s windspeed) than for average wind velocity ($\approx$ 0.3 arcsec at 7m/s windspeed), and was not noticeable for strong wind (15m/s).

4.3 Telescope operation

After the choice of the V.L.T location, the site evaluation equipment will be used as part of a larger scheme intending to get the best use of observing time in accordance with observing conditions.

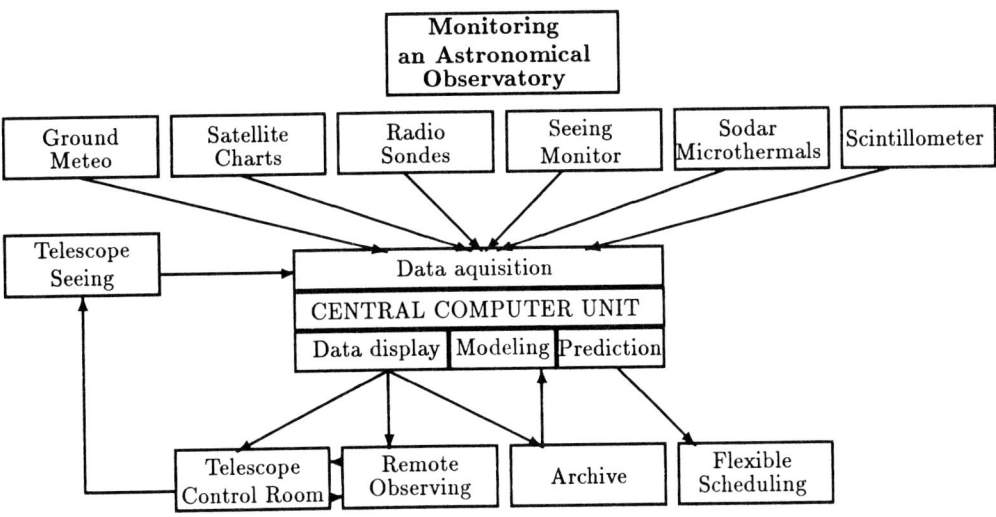

Figure 3: Organization diagram of a site monitoring unit

The general organization of an astronomical site monitoring unit is shown on Fig. 3. The most ambitious aim is of course to develop a model so as to be able to predict the observing conditions some time in advance. This must be done with enough accuracy to be of some use for decisions on the instrument configuration and for flexibly scheduling observing time.

Bibliography

[Aime 86] C. Aime, J. Borgnino, F. Martin, R. Petrov, G. Ricort, S. Kadiri; *Contribution to the space-time study of stellar speckle patterns*; JOSA A, Vol.3, p1001, July 86.

[Ardeberg 86] A. Ardeberg; *ESO-VLT site evaluation I*; Nov. 86, Proc. Second Workshop on ESO's Very Large Telescope, Venice, 29 Sep.-2 Oct. 1986.

[Azouit 80] M. Azouit, J. Vernin; *Remote investigation of tropospheric turbulence by two-dimensional analysis of stellar scintillation*; J. Atmos. Sci., **37**, 1550 (1980).

[Barletti 76] R. Barletti et al.; *Mean vertical profile of atmospheric turbulence relevant for astronomical seeing*;JOSA, Vol.66, No.12. 1976.

[Clifford 78] S.F. Clifford, *The classical theory of wave propagation in a turbulent medium*; Topics in Applied Physics. Vol. 25

[Coulman 88] C.E. Coulman, J. Vernin, Y.Coqueugniot and J.L. Caccia, *Outer scale of turbulence appropriate to modeling refractive-index structure profiles*; App. Opt. Vol.27, n^o1, January 88.

[Dierickx 88] P. Dierickx, *IMAQ 0.7, Diffraction Analysis Software Package*; E.S.O , Nov. 1988.

[Fried 66] D.L. Fried; *Optical Resolution through a Randomly Inhomogeneous Medium for Very Long and Very Short exposures*; J.O.S.A , **56** No:10, October 1966.

[Fried 75] D.L. Fried; *Differential angle of arrival: Theory, evaluation and measurement feasibility*; Radio Science, **10** No:1, 71-76, January 1975.

[Fried 82] D.L. Fried; *Anisoplanatism in Adaptive Optics*; J.O.S.A , **72**, No:1, January 1982.

[Merkle 86] F. Merkle, *Adaptive Optics for ESO's Very Large Telescope(VLT) Project*; VLT Report $n_o.47$, May 1986.

[Pedersen 88] H. Pedersen, F. Rigaut, M. Sarazin *Seeing measurements with a Differential Image Monitor*; The Messenger n^o53, 8-9 , 1988.

[Roddier F. 81] F. Roddier; *The effect of atmospheric turbulence in optical astronomy*; E.Wolf ed., Progress in Optics Vol. XIX, 1981.

[Roddier F. 82a] F. Roddier, J. M. Gilli, J. Vernin; *On the isoplanatic patch size in stellar speckle interferomety*; J. Optics, **13**, 2, 1982.

[Roddier F. 82b] F. Roddier, J. M. Gilli, G. Lund; *On the origin of speckle boiling ant its effects in stellar speckle interferomety*; J. Optics, **13**, 5, 1982.

[Sarazin 89] M. Sarazin, F. Roddier; *The E.S.O Differential Image Motion Monitor* ; to be published in A.& A. journ. Eso techn. preprint n^o5.

[Tarenghi 89] M. Tarenghi, R.N. Wilson; *The ESO NTT(New Technology Tele-*

	scope): The first active optics telescope; symposia on "Aerospace sensing, Orlando, 27-31 March 1989, SPIE 1114.
[Vlt 86]	*Interferometric Imaging with the Very Large Telescope*; June 1986, VLT Report n°49.
[Vlt 87]	Lassca; *La Silla Seeing Campaign, Data analysis Part I, Seeing*; December 1987, VLT Report n°55.
[Walker 83]	M.F. Walker, *High quality astronomical sites around the world*; Proc. Eso Workshop on "Site testing for Future Large Telescopes"; La Silla, 4-6 Oct. 1983.
[Weigelt 86]	G. Weigelt et Al.; *Speckle masking, speckle spectroscopy and optical aperture synthesis* Proc. Second Workshop on ESO's Very Large Telescope, Venice, 29 Sep.-2 Oct. 1986.
[Woltjer 83]	L. Woltjer, *Site testing for the V.L.T in Northern Chile*; Proc. Eso Workshop on "Site testing for Future Large Telescopes"; La Silla, 4-6 Oct. 1983.

THE LEST PROJECT

ODDBJØRN ENGVOLD
Institute of Theoretical Astrophysics
University of Oslo
P.O.Box 1029, Blindern
N-0315 Oslo 3, Norway

ABSTRACT.
LEST is a most ambitious ground based solar telescope program. The multi-national, non-profit organization behind the project, the LEST Foundation, presently counts 9 member countries. The construction of LEST could start in 1992 and the telescope may be ready for first light in 1995.

The current design is based on a 2.4 m and the diffraction limit of the modified Gregorian system is 0.05 $arcsec$ at $\lambda 5,000$ Å. Evacuation becomes impractical for such a large telescope since it will require a rather thick entrance window. It is instead suggested to fill the internal light path with helium gas at nearly ambient air pressure to make possible to use a thin entrance window.

The LEST will be placed either on *La Palma*, Canary Islands (2 360 m a.s.l.) or on the cinder cone *Pu'u Poli'ahu* at Mauna Kea, Hawaii (4 150 m a.s.l.). Besides utilizing a very good site for daytime observations LEST will seek to achieve near diffraction limited resolution within a small field of view by means of image sharpening techniques mentioned above.

1 Introduction

The future of solar physics lies in understanding the basic astrophysical processes that can be observed on the Sun. This includes understanding the magnetohydrodynamical processes that occur, and an important, even crucial, tool for understanding these processes is polarimetry, combined with spectroscopy, of small spatial domains over extended periods of time.

The objective of LEST is to establish, in international collaboration, such a "next-generation" type solar observing facility, which should be far superior in its scientific capabilities to existing solar telescopes. To allow high spatial resolution ($\leq$ 0.1 arcseconds) in combination with high spectral resolution and polarimetric accuracy, LEST will be located at the most optimum site, have a large aperture (2.4 m), a "polarization-free" design, and make use of new technologies to enhance the performance.

2 The LEST Organization

2.1 General

The LEST project is run by the LEST Foundation which has its seat at the Royal Swedish Academy of Sciences in Stockholm. Being initially a European project the LEST grew to include also non-European members such as Australia and USA. The other member countries are F.R.G., Israel, Italy, Norway, Spain, Sweden, and Switzerland. Five countries are presently represented in the LEST Council, and the everyday affairs are run by the Project Director and Executive Secretary. Further administrative details can be followed in an overview of the project by Wyller (1986), by Stenflo (1985), and in various LEST Annual Reports (Hauge 1983-87).

LEST is assisted in its fund raising efforts by the LEST International Advisory Council (IAC). The IAC is mainly concerned with solicitation of funding from private donors. The coordination of efforts to raise financial support from national funding agencies is the responsibility of the Executive Secretary.

2.2 Management and Execution

The technical concept of LEST is separated in (i) a "basic configuration" telescope, which will be designed and constructed as one unit, and (ii) systems that may be subcontracted as "work packages" to groups that possesse the specialized know-how and competence for designing and building them (Engvold and Hillerud 1988).

In 1988 the LEST Foundation concluded a contract with the Nordic Optical Telescope Association, under which a consultant group at Risø National Laboratory in Denmark, shall perform the design of the "basic configuration" telescope. This group consists presently of 9-10 people with expertise in the fields of telescope mechanics, electronics and optics. The Risø-group will present its report on the conceptual design by July 1990. This design will be reviewed by the LEST Foundation and the detailed design shall be completed by end of 1992. The design work is supervised by the LEST Director in consultation with the Scientific and Technical Advisory Committee (STAC) of LEST (Drs J.M. Beckers, O. von der Lühe (chairman), G. Scharmer, and A. Title).

For the execution of "work packages" it is crucial to the result that they are performed by specialists within the specific technical and scientific diciplines. These activities is expected to be performed by scientific-technical groups within the member countries. Supervision of the execution of the "work packages" will be done by the LEST Director and the STAC.

3 The LEST Design

3.1 General

The current design for the LEST telescope is as described in the LEST Technical Report No. 7 by Andersen *et al.* (1985), and it is aiming to meet the specifications for

- High spatial resolution

- High system throughput (time resolution)

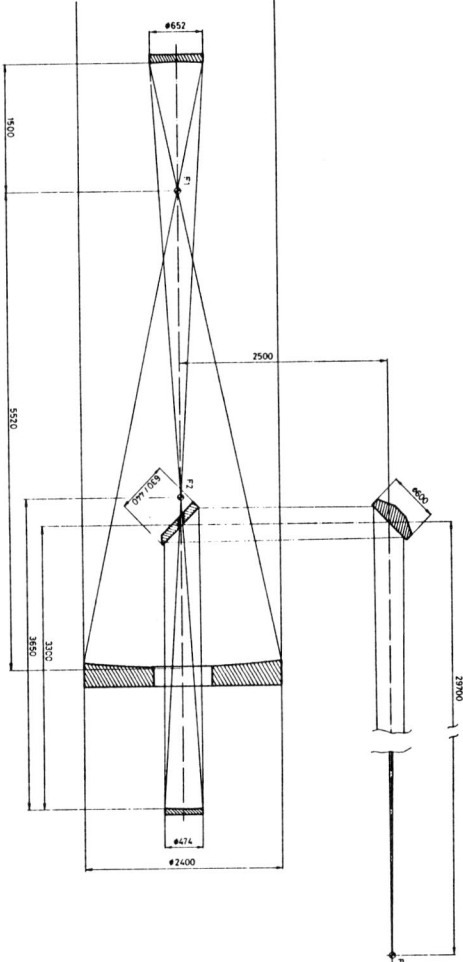

Figure 1: The optical system for LEST

- Low instrumental polarization

Additional important constraints are

- Low straylight level
- Near IR capabilities
- Good capabilities for data management
- Efficiency in telescope usage

Figures 1-3 shows the optical and mechanical layouts of the current design.

3.2 The LEST Aperture

The 2.4 m diameter aperture of LEST is necessary to secure high spatial and time resolution (high throughput) in a telescope for Stokes polarimetry. The following example serves to show that this is the case.

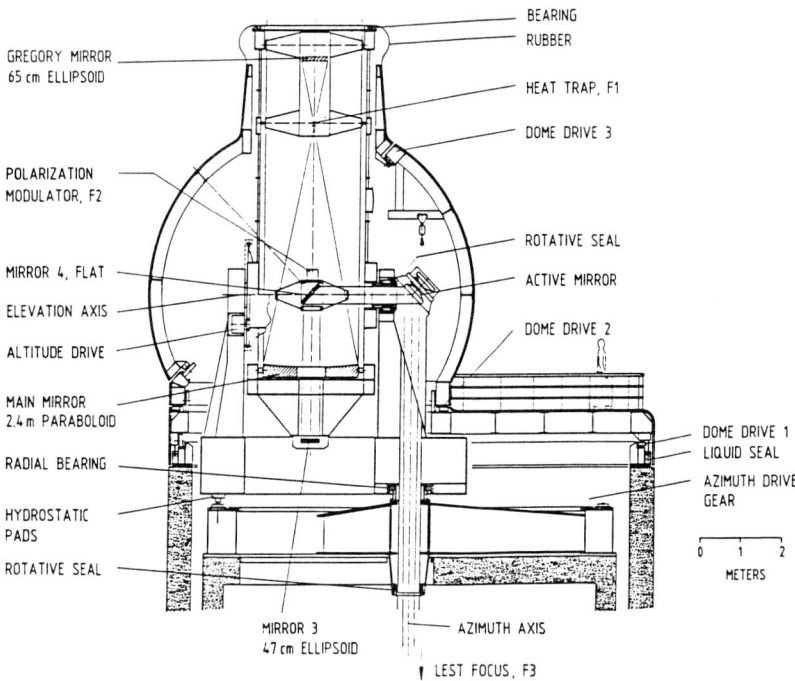

Figure 2: The current tube, mounting, and dome structure of LEST

The number of available photons in the detector plane are generally abundant in solar observations at medium resolution (1-2 arcsec) and over medium to broad spectral bands ($\Delta\lambda > 1$ Å). On the other hand, in the case of high resolution, Stokes polarimetry one needs a large aperture solar telescope with high throughput in order to attain the required signal to noise ratio. Besides increasing the collecting aperture area of the telescope one also have to use good detectors. Highly efficient charge-couple devices ("CCD") detector arrays offer quantum efficiencies up to 50% and more in the visible and near IR, compared to 2 - 25% for photomultipliers, and $\sim$1% for photographic emulsions.

The number of photons per unit time at the detector plane of a telescope with aperture diameter D and focal length f can be expressed by:

$$N = 0.5 P(\lambda)\ t(\lambda)\ a\ \Delta\lambda (\frac{D}{f})^2 \qquad (1)$$

$P(\lambda)$ is the flux number of solar photons (the factor 0.5 is atmospheric transmission), $t(\lambda)$ the overall system efficiency, a the detector element area, and $\Delta\lambda$ the spectral element. The system efficiency is the total light loss in reflecting surfaces, transmission optics, beam splitters, monochromators, gratings, and the detectors quantum efficiency. The net through-put is the product of the contribution from the telescope and from the post focus instrument, i.e. $t(\lambda) = t_t(\lambda)\ t_{pf}(\lambda)$.

The number of reflecting surfaces in modern vacuum telescopes are 4 - 5 (the Swedish telescope in La Palma has only three mirrors), and two transmission windows. Assuming that each reflecting surface is freshly coated with aluminum which gives mirror reflectivity will be $\simeq$ 0.88 at $\lambda \simeq$5,500Å, and window transmission 0.92 we get $t_t(5,500$Å$) = 0.45$ to 0.51. The major light loss usually takes place in the focal plane instruments. For example,

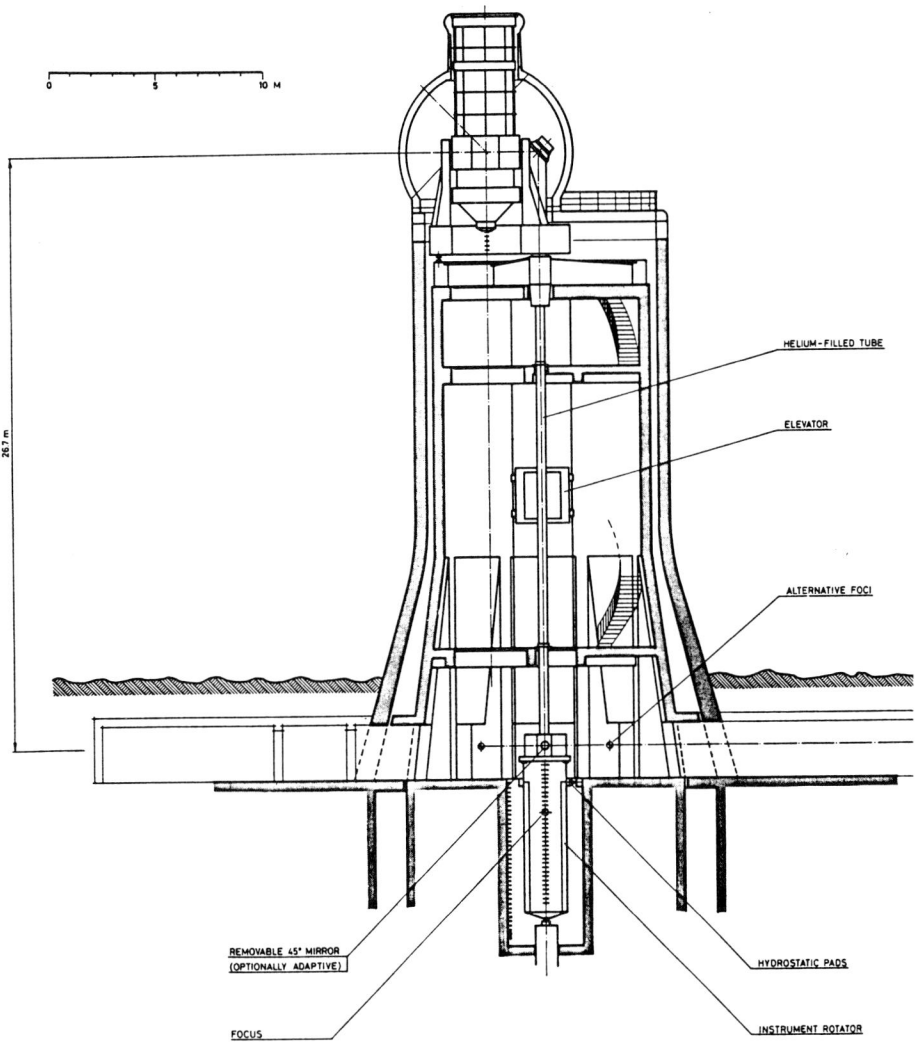

Figure 3: Cross section of the current double tower and the telescope

the estimates of Lites (1987a) for a LEST polarimeter give $t_{pf}(5{,}500\text{Å}) \leq 0.016$. The net efficiency will then be $t(5{,}500\text{Å}) \sim 0.008$. One may note that a major gain in the overall through-put of a solar telescope is likely to be achieved by clever design of focal plane instruments.

Assuming that photon noise dominates the signal to noise is given by $S = \sqrt{N}$. Let us assume that the angular subtense of the detector pixel elements $\Delta\Theta$ is about one-half of the angular resolution of the system and we get $\Delta\Theta \simeq 2\sqrt{a}/f$. We may then express the relation between the telescope diameter D and S as (cf. MacQueen 1987):

$$D \simeq \frac{2S}{\Delta\Theta} \frac{1}{\sqrt{\pi 0.5 P(\lambda) t(\lambda) \Delta\lambda}} \quad (2)$$

A signal to noise ratio $S \simeq 300$ will be appropriate for studies of spectral line profiles with $\Delta\lambda = 0.010$ Å. We take $P(5{,}500\text{Å}) = 5.5\ 10^{13}\ [cm^{-2}A^{-1}s^{-1}]$ according to Lites (1987a), $\Delta\Theta = 0.1$ arcsec and $t(\lambda) \simeq 0.008$ and find a requisite telescope diameter $D \sim 2.6\ m$ for an integration time of 1s. We may conclude that solar (Stokes) polarimetric work is generally photon starved unless the telescope aperture exceeds 1-2 m.

3.3 Residual Polarization

Observations and interpretations of spectral line profiles recorded in polarized light are invariably complicated by telescopic and instrumental polarization and the effects of yet unknown spatial averaging.

The stringent requirements on instrumental polarization in solar telescopes arises from their inability to provide high spatial resolution, - the polarimetry signal is blurred and therefore substantially weakened. Vector polarimetry will be more practical and simpler when the inherent spatial resolution is ≤ 0.3 arcsec.

The aims of LEST, as well as of the concurrent French THEMIS project (Mein and Rayrole 1988), are to reduce the parasitic polarization by using pointed telescopes with rotationally symmetric optics, at least before the position of the polarization analyzer. A recent study by McGuire and Chipman (1988) showed that even rotational symmetric optical systems give a net polarization effect. The effect depends largely on the f-ratio and the reflective coatings of the primary mirrors. In the case of LEST the upper bound of this polarization is .68%. However, since the effect is constant for a given telescope it can in essence be calculated and corrected (see Stenflo 1988; in McGuire and Chipman).

3.4 Technical Features of LEST

The LEST design includes the use of a thin (1-3 cm) entrance window. The techniques for manufacture (support, polishing and testing) of such a large optical window is being investigated in China (Nanjing Astronomical Instrument Factory) and in USA (Dunn 1984).

Most modern solar telescopes have evacuated optical light paths in order to eliminate internal seeing (Dunn 1969, 1972; Mayfield et al. 1969; Zirin 1969; Livingston et al. 1976; Nakai and Hattori 1985; Wyller and Scharmer 1985; Soltau 1989). The major drawback of such telescopes is that thermal and mechanical stresses in the entrance windows give rise to optical aberration and polarization (Dunn 1984).

In the case a 2.4 m aperture vacuum telescope the window thickness would have to be $\geq 15 cm$. Notable interest is therefore invested in the possible use of helium gas in the telescope light path which may allow the use of a thin entrance window. The idea of filling a telescope with helium was first put forward by B. Lyot and J. Rösch (Rösch 1965) and later tested by filling the Kitt Peak vacuum telescope with helium with promising result (Engvold et al. 1983). The advantages in using helium stem from its low refractive index, high thermal conductivity and relatively high viscosity. Detailed studies are presently carried out for the LEST project using a full scale mock-up steel tank of the telescope (Engvold et al. 1990). The microthermal fluctuations are apparently quenched by filtering and circulating the gas.

There will be intensified efforts within the LESTproject with the aim of designing an appropriate adaptive optics system for LEST. The proceedings of a LEST workshop on adaptive optics in Freiburg (Merkele et al. 1987) provides further details on this topic.

Smithson (1988) has recently demonstrated on a limited scale an adaptive optics systrem that improves the solar image in the vacuum tower telescope of NSO/Sac Peak. This experiment has verified the validity of the theoretical models used to describe the atmospheric effects and the imaging with an adaptive optics system. A partly corrected optical wavefront will lead to a point spread function that is the sum of a diffraction pattern of the telescope aperture in question and a broad "halo" whos width is given by the seeing (r_o) (Smithson et al. 1987).

Particular issues related to the technical development, including such as, post focus instrumentation, wind tunnel studies of the LEST structure, its polarimetry system, calculations of its optical and mechanical system, manufacture and testing of LEST optics, telescope control system, and the prospects of using *adaptive optics* in solar observations, are being published in the Technical Report Series of the LEST Foundation.

4 The Siting of LEST

The effective spatial resolution attainable with a given telescope system is a function of the optical quality of the system itself and the seeing. The main contribution to the seeing is from the atmosphere, even if a non-negligible share comes from the local telescope environment. The careful choice of site is equally important to the optical and mechanical quality of the telescope system. On the other hand, a good site may easily be spoiled if the telescope and tower structure generate strong air turbulence.

4.1 The LEST Site Survey

The LEST Foundation has run a fairly ambitious site survey program in its search for a LEST site. The program is designed and executed by the Site Investigation Team (SIT) of LEST (Drs. P.N. Brandt, D.A. Erasmus, U. Kusoffsky, U., A. Righini, A. Rodriguez, and recently, D. Sime). The results of the meteorological phase of this campaign have been reported in the LEST Technical Reports Nos. 38 and 39 (Brandt et al. 1989a and b). As a result thereof, the LEST Foundation has selected one candidate site on La Palma, Canary Islands, and one on Mauna Kea, Hawaii.

The candidate site finally chosen on La Palma is most likely going to be a location close to the Swedish solar telescope site which also served as a reference site during the campaign is shown in Figure 4. The SIT will be making further measurements in the fall of 1989 to

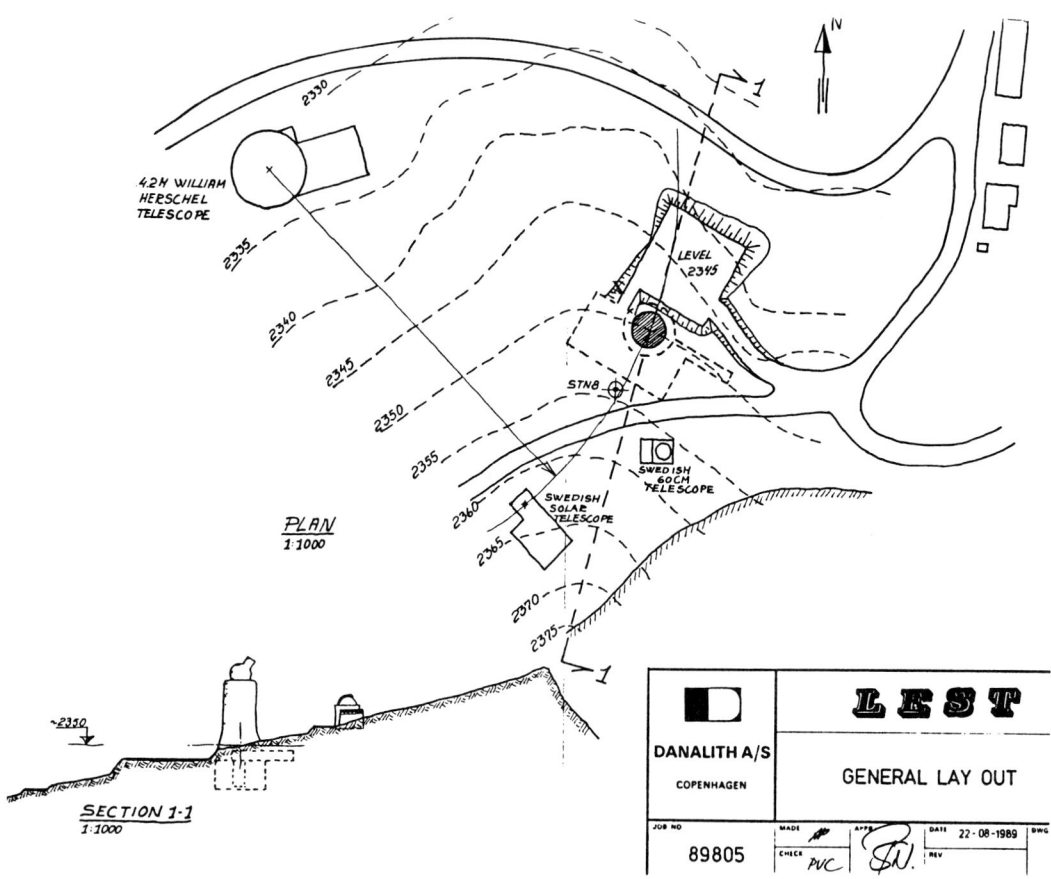

Figure 4: The candidate LEST site on the slope of the northern rim of the the Caldera. The locations of the Swedish solar telescope and the British William Herschel telescope are shown.

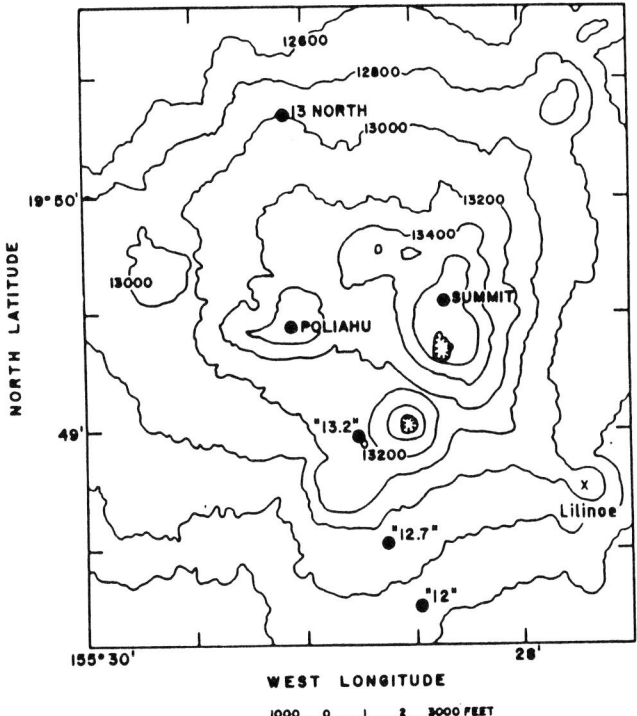

Figure 5: Topography of the summit region at Mauna Kea with the LEST site candidate Puu Poliahu

settle the exact location of this site (Rodriguez et al. 1989). Figure 5 is a map of the Puu Poliahu site on Mauna Kea.

4.2 General remarks on the candidate sites for LEST

The following general remarks on the background and basis for the LEST site survey may be made. In the past decades both test measurements and actual astronomical observations have shown that high level island and coastal sites in certain latitude belts around the earth show superior performance to inland sites as far as night-time seeing is concerned (Walker 1984). The extensive JOSO testing campaigns performed in the Mediterranean, at the Western coast of the Atlantic ocean and on the Canary Islands (Brandt and Wöhl 1982; Brandt and Righini 1985a) have added some evidence to the validity of this general statement also for solar observations. The location of such good sites is closely connected to the large-scale global circulation pattern, with ascending motion near the equator and descending air masses in subtropical latitudes, forming the so-called trade wind system. High level sites in these latitudes are generally located in semi-permanent high pressure systems above an inversion layer, and are immersed in subsiding, dry and stable air masses (McInnes et al. 1974; Erasmus 1988).

Both the Hawaiian and the Canary Island archipelagos fulfill these conditions and their excellent suitability for astronomical observations is demonstrated by the fact that about half a dozen telescopes had been built and are being operated successfully on each of them. However, their actual performance, especially during daytime, depends critically on the microthermal conditions in the boundary layer (0 to 300 m above the sites), which are strongly influenced by the topography of the sites, heating of the ground, slope winds etc..

The *Canary Islands* are situated at a latitude of approx. 28 N, longitude 17 W in the Azores high pressure system between 350 and 450 km from the African main land. The general climatological situation is well comparable with the one at Hawaii, i.e. a trade wind system with an inversion layer at heights between 1,200 and 1,600 m and subsiding stable air masses above these heights (Brandt and Righini 1985a,b).

With basis in in-situ measurements of the temperature fluctuations with radio-sondes and optical measurements (Barletti *et al.* 1977; Brandt and Wöhl 1982; Kusoffsky 1988; Brandt *et al.* 1988; Scharmer 1989a; Title *et al.* 1989; Soltau 1989) one believes that the conditions at these sites may often be good for seeing better than $\frac{1}{4}$ $arcsec$.

References

Andersen, T.E., Dunn, R.B., and Engvold, O.: 1985 LEST Technical Report No. 7

Barletti, R., Ceppatelli, G., Paternò, L., Righini, A., and Speroni, N.: 1977 Astron. Astrophys. **54**, 649

Brandt, P.N. and Wöhl, H.: 1982, Astron. Astrophys. **109**, 77

Brandt, P.N. and Righini, A.: 1985a, LEST Technical Report No. 11

Brandt, P.N. and Righini, A.: 1985b, Vistas in Astronomy **28**, 437

Brandt, P.N. Scharmer, G.B., Ferguson, S., Shine, R.A., Tarbell, T.D. and Title, A.M.: 1988, Nature **335**, 238

Brandt, P.N. Erasmus, D.A., Kusoffsky, U., Righini, A., Rodriguez, A. and Engvold, O.: 1989a, LEST Technical Report No 38.

Brandt, P.N. Erasmus, D.A., Kusoffsky, U., Righini, A., Rodriguez, A. and Engvold, O.: 1989b, LEST Technical Report No 39.

Dunn, R.B.: 1969 Sky & Telescope **38**, 368

Dunn, R.B.: 1972 Space Research **XII**, 1657

Dunn, R.B.: 1984 LEST Technical Report No. 3

Dunn, R.B.: 1985 Solar Physics **100**, 1

Dunn, R.B.: 1987 LEST Technical Report No. 28, p.243

Engvold, O., Dunn, R.B., Livingston, W.C., and Smartt, R.: 1983 Applied Optics **22**, 10

Engvold, O., Hillerud K.-I. : 1988 LEST Technical Report No. 29

Engvold, O. *et al.* : 1990 (To be published)

Erasmus, D.A.: 1988, LEST Technical Report No. 31

Hauge, Ø. (ed) : LEST Foundation Annual Reports 1983-1988

Kusoffsky, U.: 1988 (unpublished)

Lites, B.W.: 1987, LEST Technical Report No. 22

Livingston, W.C., Harvey, J., Pierce, A.K., Schrage, D., Gillespie, B., Simmons, J., and Slaughter, C.: 1976 Applied Optics **15**, 33

MacQueen, R.M.: 1987 LEST Technical Report No. 24

Mayfield, E., Vrabec, D., Rogers, E., Janssens, T., and Becker, R.: 1969 Sky & Telescope **37**, 208

McGuire, J.P. and Chipman, R.A.: 1988 LEST Technical Report No. 36

McInnes, B., Hartley, M. and Gough, T.T.: 1974, Observatory **94**, 14

Mein, P. and Rayrole, J.: 1988 *High Spatial Resolution Solar observations* The Tenth Sac Peak Summer Workshop, Agust 22-26 (Ed.:O. von der Lühe)

Nakai, Y. and Hattori, A.: 1985 Mem. Fac. Sci., Kyoto University, Ser. Physics, Astrophysics, Geophysics and Chemistry **36**, No. 3, 385

Rösch, J.: 1965 Applied Optics 4, 1672

Smithson, R.C., Peri, M.L., Benson, R.S.: 1987 LEST Technical Report No. 28, p.179

Smithson, R.C.: 1988 *High Spatial Resolution Solar Observations*, Proc. Tenth Sacramento Peak Summer Workshop, August 22-26 (Ed.: O. von der Lühe)

Soltau, D.: 1989 "Solar and Stellar Granulation", Eds.: R.J. Rutten and G. Severino, NATO ASI Series, Kluwer Academic Publishers, p.17

Stenflo, J.O.: 1985 Vistas in Astronomy **28**, 571

Title, A.M.: 1989 "Solar and Stellar Granulation", Eds.: R.J. Rutten and G. Severino, NATO ASI Series, Kluwer Academic Publishers, p.29

Title, A.M. et al.: 1989, publication in preparation

Walker, M.F.: 1984, in "Site Testing for Future Large Telescopes", ESO Conf. and Workshop Proc. No. 18, ed. Ardeberg and Woltjer, 3

Wyller, A.A.: 1986 *LEST Large Earth-based Solar Telescope - An Overview.* LEST Foundation, Royal Swedish Academy of Sciences, Stockholm

Wyller, A.A., Scharmer, G.B.: 1985 Vistas in Astronomy **28**, 467

Zirin, H.: 1969 Sky & Telescope **51**, 215

The Instrumentation Plan for the Very Large Telescope of the European Southern Observatory

Sandro D'ODORICO

European Southern Observatory

ABSTRACT

In December 1987, the Council of the European Southern Observatory approved the Very Large Telescope (VLT) project, an array of four 8-m telescopes to be erected in Chile within this century. The four telescopes can be operated independently one from the other or the optical beams can be combined incoherently to achieve the collecting power of a 16-m telescope or coherently to work in an interferometric mode.

The European Southern Observatory retains the full responsibility for the design and the construction of the telescopes and of a set of common user instruments. It was, however, decided early in the project that the instrumentation effort would be undertaken in collaboration with institutes in the ESO member countries to be able to cope with the ambitious task without too large an increase of the staff at ESO. As a first step in the implementation of this policy ESO distributed in June 1989 a document with the title "ESO VLT Instrumentation Plan: Preliminary Proposal and Call for Responses" where the main principles of the ESO policy in this field are spelled out and a global, preliminary instrumentation plan is presented to the community. Each of the 8-m telescopes is equipped with two Nasmyth foci, a Cassegrain and potentially a Coudé focus. One central laboratory for incoherent combination of the light from the 4 telescopes is also foreseen. The instrumentation plan describes a number of instruments, their distribution among the different foci and an implementation schedule. For some of the instruments a detailed optical design is presented and the expected performance at the 8-m telescope is discussed.

Figure 1 shows a view of the instruments which are included in the plan and of their distribution at the telescopes. No instruments are assigned to the 4th telescope and to some of the foci of the 3rd telescope. These are reserved for replicas of the instruments which are built for the first telescopes or for new devices which could be proposed at a later stage of the project. Copies of the ESO document on the VLT instrumentation can be obtained from the author.

In November 1989, the ESO community has commented on the plan and suggested a number of modifications. A revised version will be ready by the beginning of 1990 and a Call for Proposals for the construction of some of the instruments for unit telescope one will be distributed in April 1990.

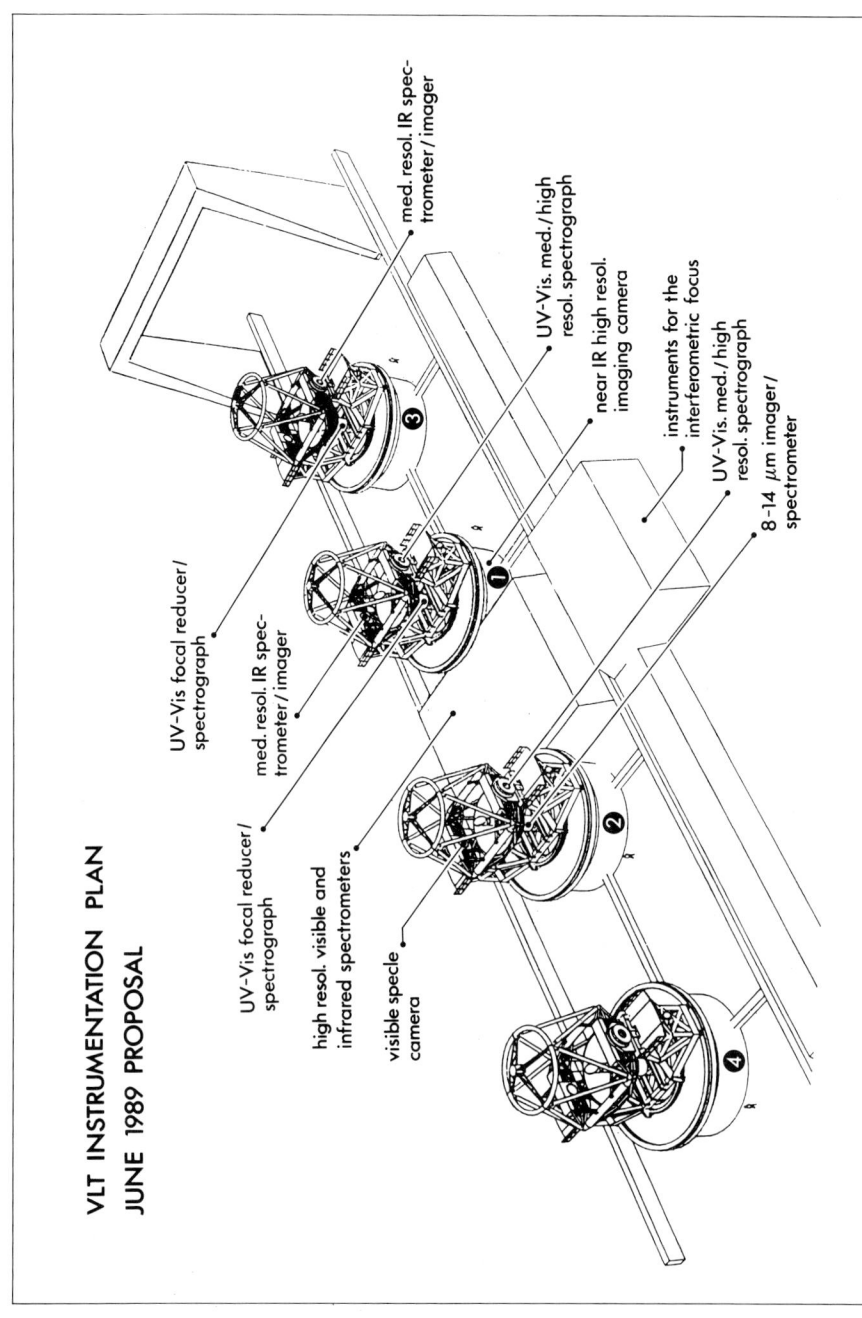

Figure 1: In this artist's view of a possible configuration of the four 8-telescopes, the distribution of the instruments for the first three telescopes is indicated. All telescopes are equipped with two Nasmyth and one Cassegrain foci. Telescope One has also an instrument at the Coudé focus. High resolution spectrometers are foreseen for the central laboratory.

CCDs for the 1990s

Paul R Jorden
Royal Greenwich Observatory
Madingley Road
Cambridge
CB3 0HA

1. INTRODUCTION

The Charge-coupled-device was invented in 1969 and although conceived for other purposes its advantages as an optical light sensor were soon recognised. By 1980 a few manufacturers had made and sold such silicon devices to astronomers as visible imaging detectors. Over the next decade, up to the present time, there has been a proliferation of these devices and now all major observatories use them as powerful tools for astronomy.

In the following sections, I shall review the developments of the past decade and make extrapolations into the future. This is rather a risky procedure in an area of still rapid change and I shall limit myself to guesses up to the year 1995! The extrapolations, based on "average" device performance, only give one view of future directions and so I shall complement them by making a few more specific guesses about realistic future achievements.

This review would not be complete without presenting a "shopping list" of currently-known detectors. Following this tabulation of present devices will be a discussion of prospects for improved devices, of various sorts, in the near future. Lower readout noise and larger area sensors are the main thrusts of development.

In order to retain clarity and keep to a reasonable length I shall not discuss non-CCD arrays (of which a few exist). Similarly, uses of CCDs for far UV, X-ray or particle detection are important, but not covered here. This paper will mainly deal with visible-wavelength applications of scientific CCDs.

A useful description of CCD-development and usage is to be found in the book by McLean (1989).

2. A 10-YEAR REVIEW

For the purposes of this exercise I have taken 3 time-slices; 1980, 1985 and 1989. In each year the available CCD devices have been considered - and these are tabulated in Table 1.

Table 1. "Useful" CCDs considered in survey/review

Year	Manufacturer	Pixel Format	Peak QE %	Read-noise e$^-$	Cost($)	Pixel Size μm
1980	RCA	320x512	70	80	5K	30x30
	GEC	385x578	40	~20	2K	22x22
	FAIRCHILD	380x488	12	30	?	18x30
1985	RCA	320x512	70	60	7K	30x30
	RCA	640x512	70	50	10K	15x15
	EEV (GEC)	385x578	50	10	2K	22x22
	Th-CSF	385x576	40	10	2K	23x23
	TI	800x800	70	5-10	?	15x15
	FAIRCHILD	380x480	12	30	?	18x32
1990	EEV	385x578	55	<5-10	1.5K	22x22
	EEV	298x1152	55	3-10	5K	22.5x22.5
	EEV	770x1152	55	3-10	8K	22.5x22.5
	Th-CSF	385x576	40	5-10	1.5K	23x23
	Th-CSF	1024x1024	40	5-10	10K	19x19
	FORD	516x516	50	~5	?	20x20
	RETICON	400x1200	40	~5	25K	27x27
	TEKTRONIX	512x512	50	~10	7K	27x27

The selection was made on the following basis:

1. Devices must be commercially available, ie purchased on the open market without special conditions. The Texas Instruments devices do not quite satisfy this requirement, but are sufficiently widespread within the USA that I felt they ought to be included (in 1985).

2. Devices must be of good scientific quality, and should generally have been actually used on a telescope.

3. Experimental or prototype devices, even with exceptional quality, are not included if only one device has been made. This sort of qualification has the effect of biasing the results towards "average" performance. I recognise this; but it seems appropriate when we are discussing general trends. I also feel that it is important to recognise that a device is only really useful to any user if more than one is available.

4. The performance parameters, especially readout noise, are quoted as typical values for those devices - rather than the more optimistic minimum figures that can be obtained sometimes.

The list of devices, and parameters, is not perfect and one may argue about its completeness, but it does represent fairly the number of devices and their typical characteristics in each selected year. The following sections discuss specific parameters in detail.

2.1 CCD availability

Figure 1 shows how the number of commercial CCD manufacturers has been increasing. The most important curve (a), despite having an uncertainty of about ±1 shows the trend of increasing numbers. The number of companies making products under special contract, curve (b), and the total numbers (c) including consumer devices are clearly increasing.

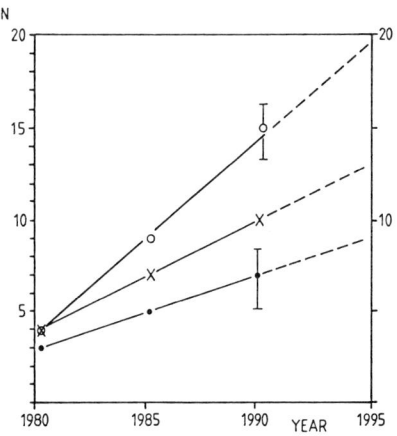

Figure 1. Number of CCD manufacturers, with "uncertainty indicated.
    .   Scientific, commercially available devices.
    x   Scientific, standard and special devices.
    o   Total, of all types, including consumer products.

Of greater interest is the number of devices available for scientific use, as shown in Figure 2. The numbers of "standard" devices and the numbers of other experimental ones are growing rapidly.

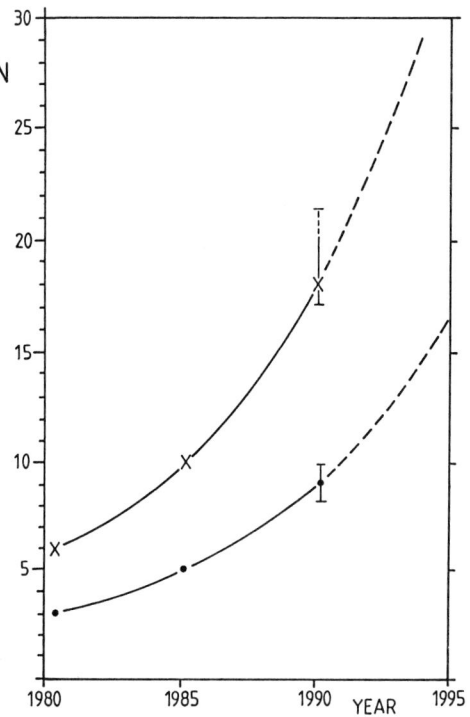

Figure 2. Number of types of scientific CCD.
.   Genuinely available and useful devices.
x   Total, of all types, including consumer products.

The market may saturate eventually, when the demand for such devices is fully met - but there are no signs of this in the near future. Indeed, the home-video market (which uses the compact low-power CCD) and the forthcoming high-definition TV development help to stimulate CCD production in industry.

Another item, associated with availability, is the actual cost of each device. The cost/pixel has dropped slightly since the "early" days of 1980, but is now fairly constant at about $14 per 1000 pixels. Of course, the cost in real terms, has effectively dropped - and performance has also improved.

2.2 Quantum efficiency

Firstly, for reference, Figure 3 shows the wavelength sensitivity of a variety of CCD types. It can be seen that performance varies, particularly at the UV end of the spectrum. However, even in 1980, a peak QE of ~80% was achieved and this cannot be improved upon significantly.

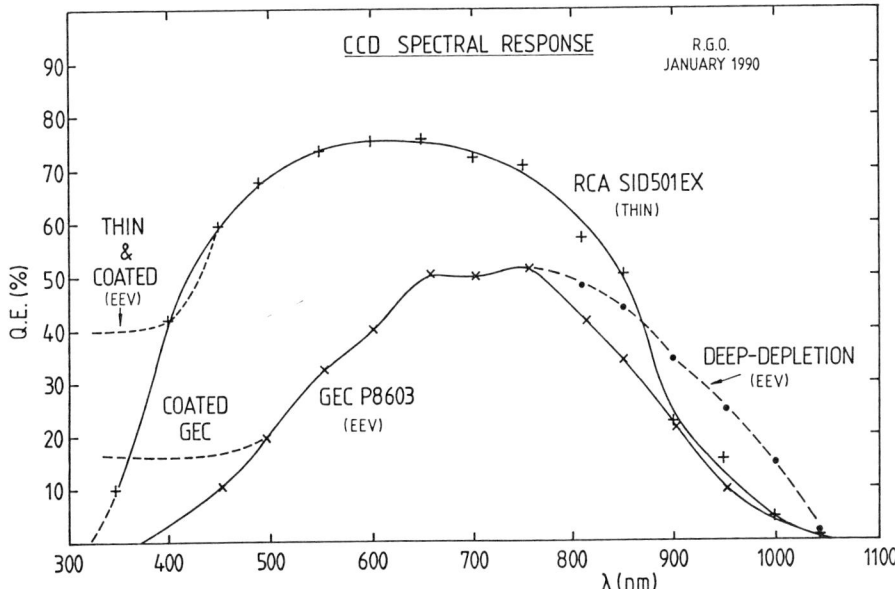

Figure 3. Spectral response of various CCD types.

Figure 4 shows the "review-graph" and although the peak has not increased significantly over the years, we now have less devices of poor efficiency. Most devices have 40-50% peak efficiency since the less efficient interline-transfer CCD is hardly manufactured now. Improvements are being made to the short wavelength response - see section 4.1 later.

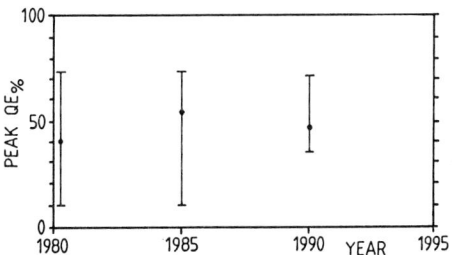

Figure 4. "Average" Quantum Efficiency at three time-slices. Maxiumum and minimum known values are shown.

2.3 Readout noise

Two main parameters contribute to the very high efficiency of the CCD as a light sensor. We have already seen that the quantum efficiency can be very high (peak ~80%) and the combination with a low readout noise allows detection of very low light levels. The readout noise is the random variation associated with the measurement of a signal from each pixel. In well-designed electronic systems the dominant component is normally the intrinsic noise from the on-chip output transistor.

The evolution of readout noise is shown in Figure 5 with an indication of the range of values within each time-slot. It can be seen that of all CCD parameters, the readout noise shows the clearest trend - that of a reduction of the average value by about a factor of four over the last 10 years, with minimum values even lower. This has been brought about by improvements in design, more attention to manufacturing technology and improved silicon quality.

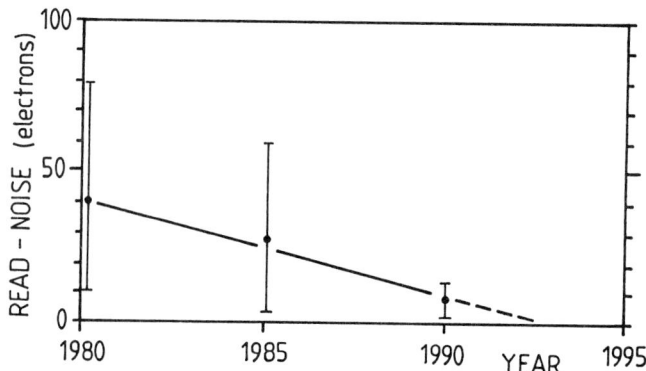

Figure 5. Readout noise (o) evolution with time. Average values are plotted, with maximum and minimum indicated.

Devices of high readout noise (>20 e$^-$ rms) are hardly sold for scientific use now, and sub 5 e$^-$ arrays are common. It is interesting to note that the fundamental causes of noise (in the buried channel MOS-FET transistor of the CCD) are still not understood. However, empirical lessons have allowed it to be characterised and reduced steadily.

2.4 Detector size

The principle of plotting a graph of average detector size versus time does not work very well since one large-format CCD can change the result significantly. Nevertheless, Figure 6 indicates the average trend of commonly-available devices in curve (a). This "pessimistic" curve leads to an extrapolation of a rather modest size increase even over the next 5 years. Curve (b) includes experimental devices and shows that at least some large devices are made and even larger ones are due. Past experience does show that it is very difficult to increase CCD size because the device yield, during manufacture, is a strong function of area.

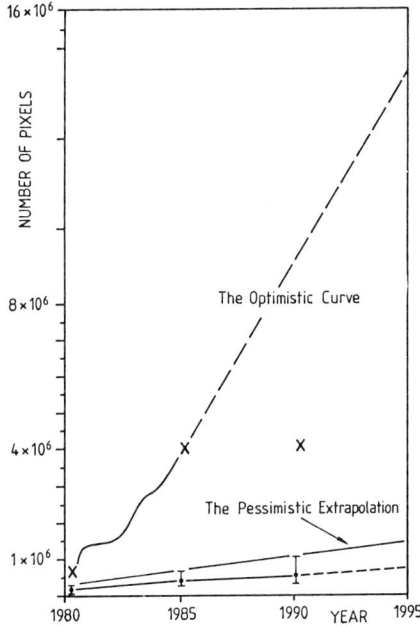

Figure 6. Detector size evolution.
    . (a) Commonly available devices.
    x (b) All scientific devices (minor variants are not included.

The pixel size is also important. It is clear that the average pixel size (of about 22μm) has not changed much, although a few devices have been made with small pixels. Nevertheless, pixel size is likely to decrease as more large-format devices are made. (Devices of small total area have a higher manufacturing yield).

Finally, I have calculated detector storage capacity and this average curve is given in Figure 7. The device capacity is calculated by multiplying pixel-number by pixel data capacity. (In 1980 the dynamic-range per pixel was ~ 14 bits, whereas in 1995 it will be ~ 18 bits.) The curve obtained increases steadily, but rather modestly. Of course, the sudden commercial availability of several large CCDs (2048 x 2048) would change the result - but previous history implies that we should not expect this too suddenly.

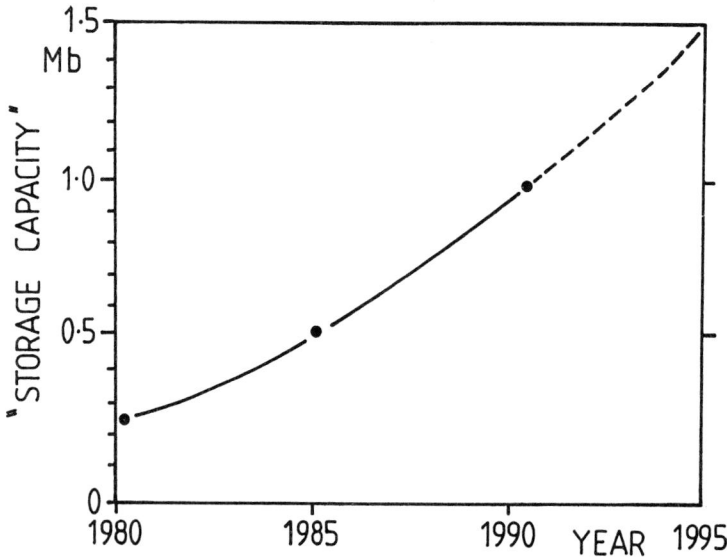

Figure 7. CCD storage capacity (assuming "average" device size).

2.5 Detector Systems

Along with the development of the detector arrays, there has been an evolution of their associated electronic systems. As an example I have taken 3 generations of RGO CCD-controllers and their parameters are illustrated below. Figure 8 (curve (a)) shows that the number of circuit boards required to operate a CCD drops rapidly as a result of increasing compactness of modern components; similarly, the power dissipation of the system is reduced (curve (b)) and this is important if we wish to minimise dome-seeing problems.

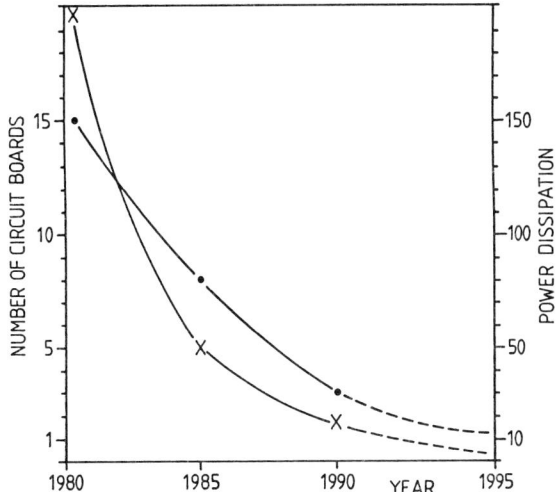

Figure 8. CCD controllers - system performance evolution.
. (a) Number of circuit boards needed.
x (b) Power dissipation.

The latest RGO CCD controller is an example of current developments. A prototype transputer-based controller has been constructed for use in a precision slow-scan CCD system. Very few components are needed, and the unit has been built using 2-3 circuit boards. Full details are to be found in a paper by Waltham, van Breda and Newton.

The paper by J P Picat in this session also emphasises how much more data we expect to have to cope with - yet modern computer power is increasing at a faster rate than detector capacity and so we do not need to worry too much.

3. THE-CURRENT-STATE-OF-THE-ART

As a means of illustrating the present position, Table 2 lists most CCD detectors that are in current use, or are "known of" as astronomical sensors. I include some widely used, but obsolete, detectors as well as others that are not yet made. The total of nearly 30 detectors includes about 15 that are certain to be in use, worldwide, on telescopes this year (1990). As before, only detectors classified as "scientific" CCDs are included; Philips devices and many Japanese and other CCDs are excluded.

Table 2 - Scientific CCD Selection Chart 1989

MANUFACTURER	NAME	FORMAT	PIXEL SIZE	PEAK QE %	σ_r	Notes
EEV	P8603	385x578	22x22μm	~ 50	5e⁻	STD
	Thin-8603	"	"		"	NEW
	P88100	298x1152	22.5x22.5	~ 50	"	NEW
	P88200	770x1152	"	"	"	NEW
	P88300	1242x1152	"	"	"	NEW
	P88500	2186x1152	"	"	"	PROTO.
RCA	SID-501EX	320x512	30x30	75	60	OBSOLETE
	SID-503	640x1024	15x15	"	"	"
THOMSON	TH7882	384x576	23x23	40	5	STD
	THX31159	512x512	23x23	"	"	"
	BUTTED -	400x479	23x23		"	NEW
	THX31156	1024x1024	19x19	"		NEW
RETICON		400x1200	27x27	30	5	NEW
		800x2400	13.5x13.5	30?	"	PLANNED
	Thin	400x1200	27x27	60	"	PROTO.
TEKTRONIX	TK512	512x512	"	30-40	10	STD
	512-THIN	512x512	"	~ 50 -	"	NEW/PROTO
	TK2048	2048x2048	"	~ 30	"	NEW
	2048-THIN	"	"	~ 50	"	PROTO.
	TK1024	1024x1024	"	-	-	PROTO.
FORD	PM512	516x516	20x20	~ 50	5	NEW
	SA1C-1024	1024x1024	18x18	30-50	6.5	NEW
	2048	2048x2048	15x15	-	-	PROTO
	4096	4096x4096	7.5x7.5	45	5	PROTO
T.I.		800x800	15x15	60		"STD"/OBS.
KODAK	KAF1400	1035x1320	7x7	~ 35?	15	NEW
		2048x2048	9x9	-	-	PROTO.
		4096x4096	?SMALL(4.5)	-	-	PLANNED
MIT/LINCOLN		~ 450x500	~ 20?	~ 35?	5	PROTO.

STD.   = Routinely available
NEW   = New Device, should be available
PROTO. = Prototypes made; not readily available yet

As a benchmark for later discussion, I consider that the current "standard" CCD that most represents those in use at all observatories is as given below in Table 3.

Table 3. The "1990 Standard CCD"

Format = 512 x 512 or 385 x 580
Pixel size ~ 20 x 20 μm
Peak QE ~ 40% (unthinned), 70% (thinned)
Readout noise ~ 5 e⁻ rms
Operating temperature ~ 150-160K
Cosmetic quality: No column defects, a few image-area traps (≤ 10)
                  1-2% pixel-pixel variation.
Cost: $1-2K

Astronomers who use such devices at many observatories will be aware of the level of performance that has been reached. The main features have been discussed already, however, there are may secondary characteristics that are of interest. For example, there is increasing commercial interest in "ambient" temperature CCDs, and these require the lowest possible dark current. This has resulted in design and operational techniques being applied in order to minimise dark current - this can allow operation at higher temperature which often has the benefit of improved charge transfer efficiency.

Good charge transfer efficiency is essential for the precise measurement of low-level signals - especially in zero background situations. This parameter can be measured, but it's effects are non-linear and therefore its quantitive nature is less easily characterised. There is no doubt that design "defects" which contribute to low CTE are being eliminated and that improved silicon quality has also helped significantly. The now universal use of the Epitaxial layer also contributes to the improved quality that is currently seen. Similarly, microelectronic techniques now allow greater precision at sub-micron levels and this results in more uniform and reliable devices.

4. EXPECTATIONS FOR THE FUTURE

In section 2 we saw that an analysis of the last decade pointed to certain trends for the future. Here I shall elaborate on these and give some specific indications of expected future developments.

4.1 Quantum efficiency

The basic CCD sensitivity function, and the factors that determine it, are well known [see Janesick, J., 1989 and references therein]. A lot of work has been done into improving the short wavelength response, and various methods are used. These are described below and illustrated in Figure 9.

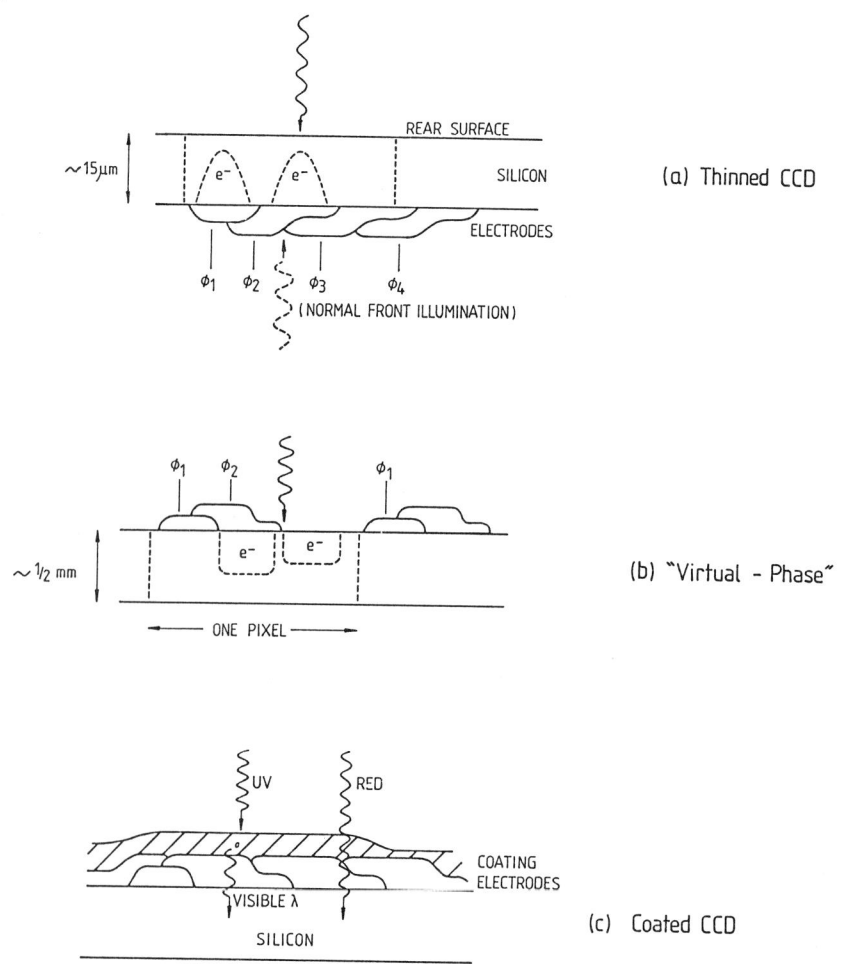

Figure 9. Techniques for improving quantum efficiency. Edge views of the CCD are shown. (See text.)

(a) Thinned, backside illumination (Fig 9a)

A normal "thick" device suffers particular attenuation of short-wave photons as they pass through the semi-transparent electrodes. The technique of thinning and rear illumination - as demonstrated well by RCA and others - gives a very good peak response and improvements in the UV. (Front and rear illumination are indicated in the figure.) However, the manufacturing difficulties are severe, and even after more than 10 years of development, the technique is not in widespread use. More manufacturers are refining the process, but it remains a specialised technique that is hard to adapt to mass-production.

(b) The "virtual-phase" structure (Fig 9b)

A useful alternative to full-scale thinning was pioneered by TI and other manufacturers have used variants of the technique. The CCD is manufactured with part of each pixel bare so that the maximum response is achieved over at least part of the area - thus increasing the peak sensitivity and giving a useful UV response. The necessary electrical potential over the bare area is obtained by suitable semiconductor processing. This approach is favoured because it avoids the difficult mechanical thinning process, although care is still needed if good UV response is to be obtained. More devices of this type will probably appear in the near future - it seems to offer a practical solution to achieving good blue sensitivity for colour TV cameras.

The wavelength response of silicon is always less at shorter wavelengths due to surface absorption characteristics. Anti-reflection coating can only partly compensate for this. The techniques described next allow wide-range UV sensitivity with a fairly flat response.

(c) Dye-coating (Fig 9c)

A third technique for achieving UV sensitivity consists of the use of fluorescent dyes. Originally demonstrated in the USA [Viehmann W., 1979], and then refined by ESO [Deiries S., 1987]; this process is now in common use. The dyes are deposited on the surface of a normal CCD and allow the chip to respond to UV - with efficiencies of ~20%, and a good flat wavelength response. An even better result is obtained if the dye-coating is added to a thinned CCD; in this case we achieve a high peak and good, flat UV response. The coating causes no degradation of peak response.

(d) Phosphor coatings

An alternative coating technique is also available. A "Lumogen" phosphor is evaporated onto a normal CCD. This material also achieves a useful UV response and has been used on a variety of CCDs. It is offered commercially by some manufacturers and appears to offer a more robust coating than the laser-dyes

(e) Deep depletion CCDs

It is possible to enhance the long-wavelength response ($\lambda > 800$ nm) by a modification to the CCD design. The use of a higher resistivity silicon allows deeper active regions and gives greater absorption of long wavelengths. This technique has been developed primarily to enhance X-ray sensitivity. In the case of optical CCDs it has the disadvantage of degrading the spatial resolution and therefore has a limited usefulness.

4.2 Readout noise

The trend of reducing readout noise is very clearly seen. It is very likely that the 1 e^- readout noise will be demonstrated in a fully working array. Techniques capable of breaching the 1 electron barrier have been described by Janesick et al, 1989. Even if this one particular technique is insufficient, there is no fundamental reason why such low levels should not be achieved. Sub-electron readout noise allows photon counting even at the lowest signal levels.

It is interesting to consider that already the photon counting system is being displaced by the CCD. A (direct) CCD, with say 4 e^- readout noise and high quantum efficiency can compete with a traditional (intensified) photon counting system. This latter situation applies on a 4m telescope with intermediate dispersion spectrograph, since the sky background noise soon exceeds the read-noise.

4.3 Array size

This characteristic is particularly difficult to predict. At the time of writing, by far the most common device is one of 385 x 576 format. However, there are also devices of 1024 x 1024 or similar size in use on telescopes. In addition, small mosaic's have been constructed and also some much larger arrays are in prototype form.

I think only time will show how quickly these large arrays come into use. I feel that the "workhorse" array within the next 5 years will be the 1K x 1K detector, but major observations will have arrays of 2K x 2K, with a few experimental 4K x 4K sensors.

The size of a silicon wafer must always impose a limit on the maximum size of a single device (this has tended to increase slowly from 3 to 4 to 5 inches in the last decade). Even with a large diameter wafer, the problems of precise manufacture and high quality across the whole area will limit the growth rate. A reduction in pixel size is anticipated. At present ~ 20 micron is common, with a few devices having sizes down to 7 micron. Probably pixels will have sizes of 10-15 microns over the next 5 years.

The commercial interest in "High-Definition" TV means that arrays of size ~ 1K x 1K will be actively developed (and probably with smaller pixels). Arrays of larger size remain as specialised scientific devices and will therefore have a cost premium.

There will remain a need for mosaics in order to achieve large focal plane areas.

4.4 Other Improvements - what else can we expect?

It is possible to conceive of an array of size 10,000 x 10,000 elements - this could be made with $10 \mu m$ pixels on a 10cm square of silicon. Nobody has proposed making such a large device yet but if excellent charge-transfer can be achieved then it could be possible. The readout-time would be a problem. Pixel times of $10 \mu s$ would lead to a 15 minute readout period; multiple outputs with some novel architectures would seem essential. Multiple serial registers, and multiple outputs are possible and we should expect further developments of "multi-channel" CCDs.

It should also be possible to tailor the wavelength response across the chip surface. Manufacturers already make colour-stripe filters on to CCDs for TV cameras. One can imagine a wavelength-gradient of anti-reflection coatings or other enhancements to optimise the array for spectroscopy. There is plenty of scientific interest in broad-wavelength sensitivity, including X-ray or UV.

The commercial interest in "machine-vision" cameras should lead to improvements in readout rate and developments of multiplexed outputs, eg for 1000 frames/sec images. These techniques do not give low noise performance, but we should benefit from developments in array structure.

4.5 Summary of specific predictions

Table 4 presents a summary of the trends discussed here, and shows my general extrapolations for the middle of the next decade. The most remarkable expectation is that of one electron readout noise. In contrast, quantum efficiency can only be expected to improve by a limited margin. The situation with regard to other parameters looks quite healthy - especially if one does not expect very large arrays too soon!

Table 4. Summary of General Extrapolations for 1995

No of manufacturers:	Scientific, commercial ~8 + others (R+D) ~5 + consumer only ~8 (including HDTV)
No of chip types:	20 min, ~30 max
Q.E.	Peak ~80% (unchanged) - better UV
Readout noise:	$\sigma_r \leq 1$ e^- rms
No of pixels:	~ 10^6 TYP > 2K x 2K "Special"
Cost/pixel:	$0.014
Pixel size:	~ 22μm still common + smaller on big arrays ($\geq 2K^2$)
Chip storage capacity:	> 1.5 Mb
No of system cards:	1 - 2
System dissipation:	~5W

This leads to my conception of what a "standard" CCD array will look like in 1999 - and this is shown in Table 5. This may be compared with the "1990 standard CCD" of Table 3. The "standard" CCD, as discussed here, refers to a typical device that any observatory would be expected to use - at several instrument focii. This is a device that can be purchased readily from several manufacturers.

Table 5. The "1995 Standard CCD"

Format	=	1024 x 1024
Pixel size	~	15 x 15 μm
Peak QE	~	80% (thinned)
Readout noise	~	2 e⁻ rms
Operating temp	~	180K - 200K
Cosmetic quality:		No column defects, very few image area traps (<2)
		<1% pixel-pixel variation
Cost	~	$ 3K

5. ACKNOWLEDGEMENTS

I should like to thank J Janesick, M Cullum and D Burt for helpful discussions prior to preparing this paper. In addition, G Newton, D Thorne, I van Breda and N Waltham contributed to many of the CCD developments at RGO. Finally, I must thank the IAC for their hospitality during the IAU colloquium.

6. REFERENCES

Deiries S, 1987, Proc ESO/OHP Workshop (1986) **25**, 73

Janesick J, Elliott T, Arzhan D, Bredthauer R, Chandler C, Wesphal J, Gunn, J, 1989, CCDs in Astronomy, ed G H Jacoby, in press.

Janesick J, 1989, Proc SPIE **1159**, (in press) Optical & Optoelectronic Applied Science & Engineering, "Open Pinned-Phase CCD Technology".

McLean I S, 1989, Electronic and Computer-aided Astronomy, Ellis Horwood.

Viehman W, 1979, Proc SPIE **196**, 90

Waltham N R, van Breda I G, Newton G M, 1990, Proc SPIE 1235, (in press) Instrumentation in Astronomy VII, Tucson.
"A simple transputer-based CCD camera controller"

NEW TRENDS IN GROUND-BASED ASTRONOMY: FAST REAL TIME PROCESSING NEEDED

Picat Jean Pierre
URA 1281
Observatoire Midi Pyrénées
14 rue Ed. Belin
31400 Toulouse

ABSTRACT

The next decade will be of major importance for ground-based astronomy development with the coming of very efficient detectors, of Very Large New Technology Telescopes and of new observing techniques.

The major result will be a tremendous increase in the information to be processed which calls for developing new real time processing facilities.

This need is shown and discussed through some examples already implemented on present telescopes.

INTRODUCTION

Things have evolved very fast during the past decade in optical ground-based astronomy.

Several 4 meter telescopes are in operation and Very Large Telescope projects will be in operation before the end of the century.

Progress in detectors has been very impressive and the technology is not very far from being perfect at least in the visible range. Photocounting systems have progressed a lot but still need development for fast, reliable read out in large fields. CCD's are already in operation on all telescopes, in a very efficient way. The format is increasing, the noise decreasing and the operating techniques more and more sophisticated and efficient. The infrared is improving very fast and in the near future, the performance will not be very far from that in the optical.

The instrumentation has also become more and more efficient with the introduction of new material (glass, optical fiber, electronic components, microcomputers, etc) and new techniques (Scanning Fabry Perot, Infra red Imagery and Spectroscopy, Multi aperture spectroscopy...).

As a consequence more and more data are coming from telescopes than in the past: the calibrations must be more accurate, the exposure times are often shorter, the detectors are better filled.

The amount of data is increasing so fast that real time and fast processing is necessary to prevent bottleneck problems in data reduction with too much data to be recorded and to check the results at the telescope with enough accuracy.

Real time observing support software is also badly needed for optimizing the observations.

Very Large Telescopes of the 8 meter class will certainly be a plateau for some years and as the improvements of detectors will not drastically change the landscape (except perhaps in the infrared domain) it seems that in the future a way of greatly improving instrumental efficiency will be better use of telescope time. This should be done not only by improving optical transmission and adaptation, but in developing support software for preparing and reducing the observations, optimizing telescope scheduling, developing new observing techniques.

All these topics call for improvement of the computing systems in use in ground astronomy in the sense of fast, real time processing.

Some work is already oriented towards this goal and this evolution will be illustrated through some new developments.

1 WHY THIS QUESTION SHOULD BE ADDRESSED TODAY

Several topics call for fast real time processing, already on 4 meter telescopes.

1.1 Detector Evolution

Up to this decade, most two dimensional detectors used in astronomy have been based on photographic or photoelectric processes. The dynamic range was not very high and the quantum efficiency no higher than about 10 to 15% which gives few images taken during an observing night.

Photography is a special case as very large field images are concerned and has led the development of very fast digitisation machines with automatic software for finding objects, making photometry and classification (recent examples can be found in Heydon-Dumbleton et al 1989 or Maddox et al 1988).

Data reductions were made off line and, except for photocounting or TV like systems (which were generally one dimensional), checking the results was not possible in real time at the telescope.

Things changed with CCD's because of larger dynamics, lower noise and higher quantum efficiency. Furthermore, the photometric and geometric stabilities of these devices instigated the implementation of new observing techniques allowing very accurate photometry or spectroscopy in two dimensional fields.

Several groups are developing techniques for mosaicing CCD's in order to cover large fields with enough resolution. (Bell Laboratories, ESO and Toulouse Observatory, University College of London...). These developments are very important in the context of New Technology Telescopes as they will provide very high resolution in a large field. For example, covering the field of view of the Cassegrain Focus of the ESO VLT (15 arcmin) with a scale 0.2"/px would need a 5000×5000 CCD.

As a result, the amount of data to be reduced will grow very fast because of increasing size, reduced elementary exposure times, higher complexity of calibration procedures and better filling of detectors.

As the noise is going down very fast, higher signal to noise and higher dynamics can be requested. Most of the CCD's work with a 16 bit digitization which is not enough (if the ADU is only some electrons) to cover all the dynamic of CCD's, several hundred thousands of electrons.

It is no longer reasonable and feasible to record all the raw data for an a posteriori image cleaning and processing. A cleaning step in real time is necessary.

There are related problems of data storage, archiving and retrievial on such a large amount of information. Even if the data are recorded in clean form, a strong compression will have to be done.

Because of technical needs, photocounting and IR detectors have in common (for different reasons) very short exposures which already implied the development of real time processing systems.

On the contrary, it seems that development of fast real time processing is always in its infancy with CCD's. Perhaps a reason is that CCD's are new detectors but in the line of photography, electronography or TV imagery and, most of the CCD software has been adapted from existing algorithms.

The important question is knowing if astronomers on the ground are ready to lose some of the calibration images and to be quite confident in automatic preprocessing, just keeping the clean images.

1.2 New Observational Techniques

New observational techniques are now implemented on most Large Telescopes and two particular examples will be shown in section 3: multiaperture spectroscopy which leads to very efficient use of two dimensional detectors and very faint photometry which calls for a tremendous number of individual pictures. These techniques have been pushed by the specific properties of CCD (low noise, high quantum efficiency, linearity, photometric and geometric stabilities).

With multiaperture spectroscopy, either by fiber optics, or by masking techniques, several spectra are recorded on the same exposure, so that the information per exposure is multiplied by a factor of 20 to 100 compared to classical long slit spectroscopy.

With very deep photometry one typical exposure on the telescope needs in fact about 30 images and 50 to 100 individual exposures for calibration. This application in a large field could be -if not the most important -a very efficient way of using a VLT for cosmological applications.

1.3 New Observing Philosophy on the Ground

Astronomers on the ground now try to optimize their observations by using flexible scheduling on multimode instruments to match observing conditions to the program needs.

This kind of flexible scheduling will be combined with remote observation. This implies transfer of data to Europe from often very distant countries, and the constitution of a very efficient network to support the remote control of a large number of parameters without collisions. A few experiments are now working and it seems that one of the most elaborate is the system in use at the 2.2 m Telescope at ESO from La Silla to Garching (Raffi, 88). An attempt has been made from Pic du Midi to Toulouse Observatory using a public facility called Transcom and working at 64 Kb/sec. Other new systems seem more promising for a European network: for example, Numeris or Transfix which will be public facilities in France in the near future.

1.4 New Facilities

There are many reasons for real and fast processing. The idea is not very new but the facilities now exist on a commercial basis: digitized detectors, computers, fast processors, image processing workstations, communication satellites, public communication facilities, image quality monitoring etc.

2 AN EXAMPLE OF MOSAIC DEVELOPMENT: THE ESO / INSU CAMERA

Making a mosaic of CCD's is certainly the only way to providing large detectors with pixel size of about 20 to 30μ which seems to be the optimum on Very Large Telescopes (Fort, 88). The ESO/INSU mosaic is one of the first to be implemented on a Telescope.

2.1 The Thomson CCD'Products

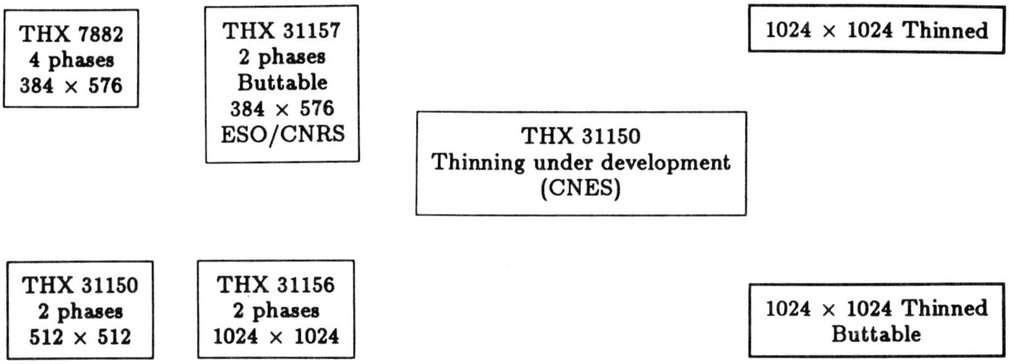

THX 7882	THX 31157		1024 × 1024 Thinned
4 phases	2 phases		
384 × 576	Buttable		
	384 × 576	THX 31150	
	ESO/CNRS	Thinning under development	
		(CNES)	

THX 31150	THX 31156		1024 × 1024 Thinned
2 phases	2 phases		Buttable
512 × 512	1024 × 1024		

Figure 1: Status of the Thomson products used in astronomy. Apart from the existing products (thin line boxes) we would like to push for development of thinned 1024*1024 CCD's in a buttable technology (thick line boxes).

Thomson has already announced a 512*512 thin CCD to be on the market in the beginning of 1990. Some tests made in the Toulouse laboratory and at CEA (L. Vigroux) on preliminary models show that with some cosmetic improvements the technique will lead to very good CCD's. At the time this paper is written, this new product is under tests at CEA and Toulouse.

2.2 The Buttable Thomson CCD

Figure 2: Picture of a mosaic of 4 CCD's in place in an ESO type cryostat head.

A specially designed Thomson device (THX 31157) has been developed with ESO and INSU to build large mosaics. The concept allows butting two rows of CCD's giving mosaic sizes of 1150 pixels by 384 times the number of CCD's in a row (4 CCD per row leads to 1536).

A mosaic of 4 CCD (1150*718) is now working at ESO and Toulouse Observatory. The dead zones between the CCD's is 400 μm (or 20 pixels) between the rows and 200 μm between the CCD's.

A special machine has been constructed at Toulouse Observatory, the Mosaic Mounting and Alignment Machine (MAM) to butt the CCD on a saphir substract with target accuracy of 15 μm in coplanarity and 2 μm in alignment. Fig 2 shows the mosaic of 4 CCD's in an ESO type cryostat head.

2.3 The ESO/INSU Controller

A newly designed controller has been built to monitor 16 CCD's at a time. Reading is made in parallel allowing a short reading time. The image is reassembled in real time during read out. Results obtained on the mosaic are quite good: the noise is about 4 electron rms, at a correlated double sampling time 2*40 μsec., with typical quantum efficiency for thick CCD's (40 % at 7000 Å). The transfer efficiency is good but remanence could remain for several hours if these CCD's are used at a level of more than about 100000 electrons (which roughly corresponds to the upper limit fixed by the ADC, with a gain of 1.7 e/ADU).

There are two possible ways of linking that controller to the host computer-either by parallel link or by serial fiber link.

A detailed description of the ESO/INSU controller and CCD Mosaic is given by Reiss (1989).

2.4 Preprocessing System

Here is a philosophical choice: the data can be sent either to a local host computer on which every step of the processing is done, from image cleaning to astrophysics, or to a dedicated station where preprocessing can be done, giving accurate clean data to be recorded and used elsewhere for astrophysical reduction.

In Toulouse, we chose the latter solution because we feel it is the best way to optimize the use of a detector and to be free of the environment.

We developed a preprocessing station around a 68020 CPU with 8 Mbytes of memory, graphic boards ELTEC OPAC 1280*1024 with LUT. This station is intended to preprocess the images for quicklook at the telescope and accurate image cleaning, in almost real time whenever possible. Ways to accelerate these procedures will be studied in the future, involving fast processors like arrays and/or transputers.

The clean data will be recorded on a standard compatible medium - optical disk or high capacity tape like Exabyte system - to be sent to another computer.

3 NEW TECHNIQUES

3.1 Two Dimensional Spectroscopy

General references are given in "Instrumentation for Ground Based Optical Astronomy", the Ninth Santa Cruz Summer Workshop in Astronomy and Astrophysics, july 13-24 1987, edited by L.B. Robinson.

This is a concept which is now commonly used on telescopes and which presents different approaches. Depending on several parameters like aperture size, crowdiness of the field, technique employed, it is possible to take between 20 and several hundred spectra per exposure, which improves observing efficiency tremendously.

3.1 a) Multi-object Spectroscopy

One way is to put a mask at the focus of the telescope with holes at the place of the object to be analyzed, depending on observing requirements. This method allows several spectra (typically 10 to 50) to be taken in medium size fields (10') at the same time, with the efficiency of the spectrograph. The detector is not used with the maximum coverage except that all unused columns can be filled with sky references.

Another way is to use optical fibers, put at the telescope focus on the objects of interest and then rearranged on the slit of the spectrograph. This method allows a very large number of spectra (typically one to several hundreds) to be recorded in large fields (1 to 2°). Computer control for object-fiber matching and fiber anti collision has already been done (Felenbok et al, 1988). The coverage of the detector can be perfect. Small efficiency losses, depending on the wavelength, can be due to the fiber transmission or to optical adaptation.

3.1 b) Integral Field Spectroscopy

Integral field systems allow one to get spectra on different points of a two dimensional image in a small field (10 to 20").

One system (Vanderriest et al 1988) consists in using a bulk of fibers rearranged on the slit of the spectrograph (typically 400). This leads to a very good filling of the detector, with a usable wavelength range which depends only on the dispersion and the size of the detectors. A loss in efficiency is due to the fibers (transmission and dead space).

A second system (Courtes et al 1988) consists in using lens arrays, each lens giving an image of the telescope pupil used as a single slit. This is a very efficient mounting which allows very high spatial resolution on extended objects but the field, spatial resolution and wavelength range are not quite independent.

Let's mention Fabry Perot which is a very efficient way of measuring radial velocities in large fields but in very narrow spectral bands at a time. Scanning Fabry Perot has led to extensive developments of real time fast processing systems like for example Taurus (Atherton et al 1982) or Cigale (Laval et al, 1987).

3.1 c) An example of Multi-object Spectroscopy: masking technique developed at Toulouse Observatory

In the case of mask technique, which is mainly devoted to very faint object spectroscopy, the first step is to image the field for photometry, in order to search for the objects and select them, given different criteria. At this level, support software must help the astronomer optimize the observation by taking into account the science, the nature of the objects and their dispersion in the sky. This whole procedure should preferably by made in real time. It is the way multiaperture spectroscopy has been implemented on the CFH Telescope.(Picat 1988). This step is very important in saving telescope time (removal of false objects, objects at the same signal to noise in the same exposure time...)

Fast real time processing allowing the spectroscopic observation just after the imagery is done, the mask can be matched to seeing and refraction conditions which would not be the case if the mask was prepared off line (for the following night for example).

That kind of software must also help the astronomer control in real time the robot preparing all the setup for the observation (make the mask or put the fiber arms in good position), and control the mask alignment during the exposure.

Different techniques have been employed in making the mask. One is to drill or punch metallic plates but this has the disadvantage of having little versatility on aperture shape and size. In fact, the best quality is certainly obtained by the photographic process but it is not easy to implement for general use on a telescope..

Another approach is using slitlets which allows in situ robot and very accurate slits. A disadvantage is the minimum size of the slitlets which prevents centering the aperture on the object and can limit the number of objects to be observed in the field. This technique can probably be applied better on a VLT because of the larger scale at the telescope focus.

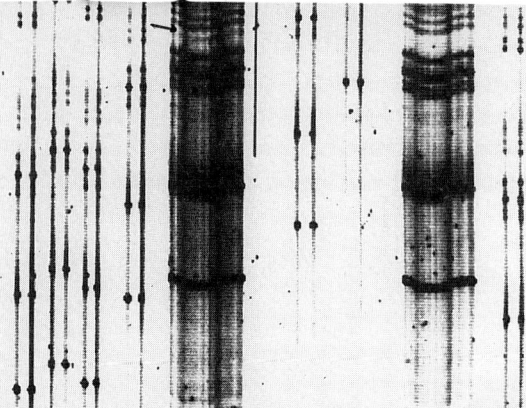

Figure 3: Example of use of a slit matched to the shape of the arc in the galaxy cluster A370. Spectrum has been taken at ESO with the system EFOSC/PUMA2. Right spectrum is the sky reference. Note on the left spectrum the emission line $[OII]\lambda 3727$ redshifted at $z = 0.724$.

A method developed for CFHT is using a YAG laser cutting technique on thin metallic sheets. Tests (Lemaitre 1989) are very encouraging and open a very versatile way of working, as the aperture shape can be optimized on each object. Example of an application is given on fig. 3 showing the spectrum of the arc A370 made by Soucail et al (1988) with a punching machine. In this way, a very careful analysis of the field, the seeing and the refraction must be done prior to optimizing the robot work. It seems that using Excimer laser on plastic film (Znotins et al, 1987) would lead to much higher aperture quality, but the difficulty is insuring the mechanical support of such a film with enough accuracy.

The second step is the reduction of data which becomes very complex because in this multiaperture mode the detector is fully used in two dimensions. Spectra must be considered as images with additional problems coming from the spectral dispersion varying from one spectrum to the other and of the different spectral sensitivity of individual pixels. In the case of mask application, spectra must be calibrated each time, as the aperture can be anywhere in the field. In the case of fiber approach, the spectra are always made on the same part of the slit which can facilitate the calibration procedure.

On very faint object spectroscopy, one crucial step is the removal of the sky that can be ten to fifty times brighter than the object. At this level, all defects must be carefully corrected and calibrated, implying complex and long reduction procedures. Interest is to have an automatic routine, on line in order to check the quality of the exposure and be able to react on time on the observation.

This facility will become more and more efficient as the readout detectors noise will decrease, allowing very short time segmentation of the total exposure time. In this respect, a fast real time processing would allow to work, like counting techniques where it is possible to look at the evolution of the signal to noise in real time and stop the exposure at the optimum.

A reduction software has been developed in Toulouse Observatory which enables an almost total automatic procedure from the image cleaning to the flux calibration. This package will be put under Midas very soon.

3.2 Deep Photometric Survey

This is an other very promising technique already in use on 4 meter telescopes. Time segmented shifted exposures are added to get the high signal to noise necessary to obtain an accuracy of a few thousands on the sky level. As the read out noise of CCD's becomes lower and lower, this mode can be very efficiently put in use on a VLT.

Several groups are now working in this field but most of the developments have been pionered by J.A. Tyson (see J.A. Tyson and P. Seitzer 1988 for references).

Most of the work on CCD photometry has shown that limitation was caused by systematic errors: -charge skimming, bias or interference fringes, Q.E. variations and transmission variations, rather than photon noise, so that the high dynamic of CCD's could not be used at its best.

The systematic errors are corrected by obtaining very high signal to noise flat fields and darks and by time segmenting the exposures on different part of the CCD - with telescope offset, larger than the mean size of the objects - to average the sensitivity response from pixel to pixel. The more images in different places, the better the average.

Approximately 100 exposures are needed for one image. As the scientific impact of very deep photometry is stronger on very large fields, this represents a tremendous amount of information to be handled.

Another point is that the deeper the photometry, the larger the filling of the CCD, which leads to more complex sky substraction procedures and a to a higher number of useful pixels.

As an illustration, with a 1000*1000 CCD which will seem to be common very soon on all telescopes, very deep photometry as proposed by Tyson leads to about 30 offset exposures per images (60 Mbytes) plus 60 to 100 darks, which is 2 to 300 Mbytes of information, a superflat being made from exposures.

With 5K*5K CCD which can be guessed from mosaic technique in a few years, the information will be more than 1Gbyte per night.

3.3 Scheduling Optimization

Efficient optimization of observations will come from better scheduled use of the telescope, depending on the observing conditions: seeing, photometric quality, availability of equipment, etc. The concept of flexible scheduling is proposed for the ESO VLT (Johnston 1988) and a project calleds Spike is on the way for the H.S.T.

A lesson can be received from space but with a fundamental difference on the ground where things are evolving relatively fast with a time constant much smaller than the life of a telescope. This is the case for example, for detectors and computer facilities. Things must be kept open on the ground to follow technical advances.

Very efficient software must be developed to help decisions in real time by simulating the observation which can lead to long calculations, with integration of a large number of parameters (meteorological conditions, seeing, instrument status, detectors, astrophysics...) including comparison to data base. Some developments are already underway, for example at ESO (Rosa et al 1986, Prieur 1989).

The decision software must be accompanied by real time ways for controlling all the equipment and checking the result quality.

Such an "expert" system is very time consuming and certainly needs fast real time processing (as parameters are evolving in real time) probably on parallel processors (Fosbury et al 1988)

4 AN APPROACH FOR REAL TIME IMAGE PROCESSING

It is interesting to discuss the first experiment in real time image processing which has been developed for very deep photometry (Tyson and Lee 1988).

The purpose was to adapt an existing image processing system (FOCAS under IRAF) to very deep photometry on large CCD fields.

This development is intended to be used on images from a CCD mosaic of 4 2048*2048 Tektronix CCD's. With the technique of very deep photometry discussed in paragraph 3.2 something like 1 Gbyte of data will be moved from disk to memory several times during the reduction process of one image.

The concept is to save only the corrected images and to develop a very powerful computing system which has been based on a 68020 CPU with 24 slot VME bus, a 1.2 Gbyte Winchester, 72 Mbyte bulk memory and arrays processors (up to 5 boards) doing the arithmetics. The result is a 100 Mflops processor which allows 30 5K*5K images to be processed in 100 sec.

5 CONCLUDING REMARKS

In the 1980's (keeping in mind the 2K*2K Tektronix CCD) people worried about the data reduction process. At that time, the computers on telescopes (HP 1000, Vax 750) where not powerful enough to handle such a large amount of information.

In fact, computing facilities have grown much faster than the size of the CCD's. This simple example shows how difficult it is to project a situation in the future in the case of ground astronomy where things can evolve very fast.

Everything is possible with new processors and pioneering Tyson development could very soon be done with commercial hardware.

Developing real time fast processing in the near future must be considered a high priority.

Many arguments converge towards this conclusion. The number of data bytes is growing very fast, because of the size of CCD's and above all, because new observational techniques lead to a tremendous increase in the number of exposures per night and to more complex reduction procedures. This behaviour will be reinforced by flexible scheduling and observation optimization.

With high quality CCD's already available, very large collectors coming, and high angular resolution, it is the only way for drastically improving the efficiency of ground based observations in the next decade.

But can we decide today? This is very difficult in the computing domain. Table I shows some results of trials that have been made at Toulouse by Dang Duc Hung (1989) on different machines, on procedures used in image processing. Row 1 is a fast fourier transform written in C, row 2 is a procedure used to prepare a flat field (which implies very numerous QIO), row 3 is a model for gravitational effects (essentially CPU). This table shows how different are the machines, depending the job made.

machine	vax 11/750	vs 3200	ds 3100	sun 4/110	sun 4/330	cdc 910/437	cdc 920/252	appollo DN	ardent titan	mips rs2030
1	1500s	393s	32s	160s	61s	46s	22s	28s	62s	30s
2	363s	92s	107s	234s	110s	183s	82s	127s	35s	113s
3	786s	227s	86s	334s	99s	108s	52s	27s	209s	85s
power 1	0.7	2.7	14	7	16	10	40	15/30	10	12
power 2			1.6		2.6	0.9	7.5	5.8	6	1.8
processor	cisc vax	cisc vax	risc mips	risc sparc	risc sparc	risc mips	risc mips	risc prism	risc + asic	risc mips

Table 1: power 1 in Mpis, power 2 in Mflops

New efficient techniques, such as the Risc processors have multiplied the computing power and other new processors like the Intel 860 will give a Cray I on the desk (Delemarre, 1989).

It is then possible to choose either a host computer with a big software system like Midas or Iraf or a dedicated workstation with more specialized procedures.

I am personally in favor of the second solution which is supported by the results shown in Table I. For real time processing it seems easier to optimize procedures on dedicated workstations. This calls for the definition of standards in order that people have a common langage from one equipment to another.

How to do that is not clear now because it is possible to use commercial processors or develop special systems involving Array processors for computing or Transputer for fast peripheric control or even analogical cicuitry for repetitive procedures.

All these developments will be needed very soon but keeping in mind that we work for the future. Even if we do not know the hardware we can imagine some constraints.

The first one is to work on the Unix system using software easy to optimize on vectorial and parallel processors which will become very common and cheap.

The second one is to assess that the instruments must be driven by dedicated and very open systems that can be optimized and support easy evolution: that is the case for 680-0 with VME bus.

The prereduction and quicklook procedures must be done on that equipment which can be upgraded with arrays or anything else. The handling must be easy with very well determined procedures.

More sophisticated data reduction must be done on workstations or a host computer (which will be a powerful workstation in the very near future) with a very powerful system allowing easy implementation of general procedures, like Midas or Iraf.

The greatest efficiency of ground based telescopes will come from this fast real time processing development as it is the only way of providing true optimization of telescope use.

Who could say today what is the exact useful amount of time on a telescope? We cannot take the risk that a non optimized 8 meter telescope would be not more efficient than an optimized 4 meter telescope.

Acknowledgements:

I am very grateful to M. Dang Duc Hung for providing the results of long tests on several micro and mini computers and to J.P. Dupin for providing information on the preprocessing system developed in Toulouse Observatory.

References

ATHERTON P.D., TAYLOR K., PIKE C.D., HARMER C.F.W., PARKER N.M., HOOK R.N., 1982, Mont. Not. R. Astr. Soc. **201**, 661

COURTES G., GORGELIN Y., BACON R., MONNET G., BOULESTEIX J., 1988 in Instrumentation for Ground-Based Optical Astronomy ed. L.B. Robinson, Springer-Verlag p. 266

DANG DUC HUNG, 1989, Private communication

DELEMARRE H., 1989, Minis & Micros, **317**, 39

FOSBURY R.A.E., ADORF H.M., JOHNSTON M.D., 1988 in Very Large Telescopes and their instrumentation, ESO Conference, Garching 21-24 March Ed. M.H. Ulrich, p. 1283

FELENBOK P., GUERIN J., FERNANDEZ A., TOURNASSOUD P., VAILLANT R., 1988 in Very Large Telescopes and their instrumentation, ESO Conference, Garching 21-24 March, Ed. M.H. Ulrich, p. 1207

FORT B., 1988 in Very Large Telescopes and their instrumentation, ESO Conference, Garching 21-24 March Ed. M.H. Ulrich, p. 929

HEYDON-DUMBLETON N.H., COLLINS C.A., MAC GILLIVRAY H.T., 1989, Mont. Not. of the R. Astr.Soc.

JOHNSTON M.P., 1988 in Very Large Telescopes and their instrumentation, ESO Conference, Garching 21-24 March Ed. M.H. Ulrich, p. 1273

LAVAL A., BOULESTEIX J., GEORGELIN Y.M., GEORGELIN Y.P., MARCELIN M., 1987, Astron. and Astrophys., **175**, 199

LEMAITRE G., 1989 Private communication

MADDOX S.J., EFSTATHIOU G., LAVEDAY J., 1988 in Large scale structure of the Universe, UAI symposium n° 130, ed. J. Audouze, Kluwer Academic Publishers p. 151

PICAT J.P., 1988 in Instrumentation for Ground-Based Optical Astronomy ed. L.B. Robinson, Springer-Verlag p. 209

PRIEUR J.L., 1989 Private communication

RAFFI G., 1988 in Very Large Telescopes and their instrumentation, ESO Conference, Garching 21-24 March Ed. M.H. Ulrich, p. 1061

REISS R., BAUER H., DEIRIES S., D'ODORICO S., LONGINATTI A., 1989, in SPIE Proceedings Volume 1130 of ECO 2 Conference, session "New Technology for Astronomy"

ROSA M., BAADE D., 1986, The Messenger, **45**, 22

SOUCAIL G., MELLIER Y., FORT B., MATHEZ G., CAILLOUX M., 1988, Astron. Astrophys. **191** L19

TYSON J.A. and SEITZER P., 1988 Astrophys. Journal **335**, 552

TYSON J.A. and LEE R.W., 1988 in Very Large Telescopes and their instrumentation, ESO Conference, Garching 21-24 March Ed. M.H. Ulrich, p. 1051

VIGROUX L., private communication

VANDERRIEST C., LEMONNIER J.P., 1988 in Instrumentation for Ground-Based Optical Astronomy ed. L.B. Robinson, Springer-Verlag p. 304

ZNOTINS T.A., PAULIN D., REED J., 1987 Laser/Focus Electro-Optics May 87

The ISO Instrumentation and Expected Performances

Th. de Graauw

Department of Space Research
P.O. Box 800
9700 AV Groningen
The Netherlands

1 INTRODUCTION

With the instrumentation of the Infrared Space Observatory, in an advanced development stage, to be launched mid 1993, it is now possible to make a more reliable estimate of the scientific performance of these instruments. In this contribution I will give a brief outline of the instruments and their expected sensitivities.

2 THE INFRARED SPACE OBSERVATORY

The ISO spacecraft will contain a.o. a payload module which is essentially a large cryostat tank with a 60 cm telescope and its four instruments suspended in the center of a toroidally shaped helium tank with about 2300 liters of superfluid helium. The 20' total unvignetted field of view of the telescope is distributed to the four instruments by a pyramid mirror and each instrument receives a 3' unvignetted field. Each instrument occupies a 80° sector in a cylindrical volume behind the primary and the restriced space has put severe constraints on the design of the instruments. The system and instrumental requirements are given in table 1. For comparison those of IRAS and SIRTF, a mission to be flown in the late nineties, are given as well.

Table 1: SYSTEM AND INSTRUMENT REQUIREMENTS FOR IRAS, ISO and SIRTF

Requirements	IRAS	ISO	SIRTF
Mirror Diameter	60 cm	60 cm	95 cm
Wavelength	8-120 μm	2.5-200 μm	2-700 μm
Diffr. Limit of Telescope	$\geq$12 μm	$\geq$5 μm	$\geq$4 μm
Angular Resolution	60 arcsec	1.5-60 arcsec	1 arcsec
Pointing Stability	10 arcsec	3 arcsec	.15 arcsec
Modulation	Scanning	Scanning Focalplane Chopper	Chopper
Field of View	30 arcmin	20 arcmin 3 arcmin/instr	7 arcmin
Sensitivity			
10 μm	20,000 μJy	60 μJy	6 μJy
60 μm	80,000 μJy	600 μJy	150 μJy
Number of Detectors	60	2000	20,000
Spectral Resolving Power	20	50-20,000	>2000
Lifetime	10 months	22 months 18 guaranteed	5 years

It shows that the ISO telescope will be made diffraction limited for wavelengths of five micron or longer. Another important characteristic determining the capabilities of an observatory is the pointing accuracy. Here ISO is not so competetive as one would like. The pointing specifications are given in table 2. Two sigma values are given and one calibration per orbit is assumed.

Table 2: ISO POINTING SPECIFICATIONS

ERROR TYPE	Specifications (arcsec)	Current Estimated Performance (arcsec)
Absolute Pointing Error (APE)	11.7	7.32
Absolute Pointing Drift (APD) 1 hour	2.8	0.64
Relative Pointing Error (RPE) 30 secs	2.7	2.52

The APE is generated by the thermal drift of the structures that connect the startrackers to the payload tank. The RPE arises from the jitter in the startracker - gyro - reactionwheel control loops. The first error can be circumvented by sufficient frequent in-orbit calibration between a quadrant star sensor located in the helium tank, and the startrackers. The jitter cannot be eliminated and the design of the instruments had to take these inaccuracies into account.

3 THE ISO INSTRUMENTATION

The ISO scientific payload will consist of four instruments: a photo polarimeter (ISOPHOT), a camera system (ISOCAM), and two spectrometers (SWS and LWS). An overview of the characteristics is given in table 3.

Table 3: CHARACTERISTICS OF SCIENTIFIC INSTRUMENTS

Instrument and Principal Investigator	Main Function	Description Wavelength (λ) (microns)	Spatial Resolution	Spectral Resolution
ISOCAM (C. Cesarsky, CEN-Saclay, F)	Camera and Polarimetry	Two 32 x 32 detector arrays λ: 2.5-5.7 λ: 4-18	Pixel sizes: 1".5, 3", 6", 12"	Broad-band Narrow-band CVF (R=50)
ISOPHOT (D. Lemke, MPI für Astronomie Heidelberg, D)	Imaging Photo-polarimeter	Four subsystems: i) Photo-Polarimeter P_1 (λ: 3-18) P_2 (λ: 15-30) P_3 (λ: 40-120) ii) FIR camera C_{50} (λ: 31-56) C_{100} (λ: 61-113) C_{200} (λ: 122-200) iii) Mapping array λ: 8-31.3 iv) Spectrophotometer λ: 2.5-5 λ: 6-12	i) P_1: 5", 7".6, 13".8, 18", 23" P_2: 52" P_3: 99", 180" ii) C_{50}: 30".5 x 30".5 C_{100}: 43".5 x 43".5 C_{200}: 89".4 x 89".4 iii) 19" x 19" iv) 24".	Broad-band and Narrow-band Filters See figure 1 and 2 Near IR Grating Spectrometer with R = 90
SWS (Th. de Graauw, Lab. for Space Research Groningen, NL)	Short-wavelength Spectrometer	Two gratings and two Fabry-Perot Interferometers λ: 2.5-45	7.5" x 20" and 12" x 20"	1000 across range and 2×10^4 from 15 - 40 μm
LWS (P. Clegg, Queen Mary College, London, GB)	Long-wavelength Spectrometer	Grating and two Fabry-Perot Interferometers λ: 45-180	100"	200 and 10^4 across wavelength range

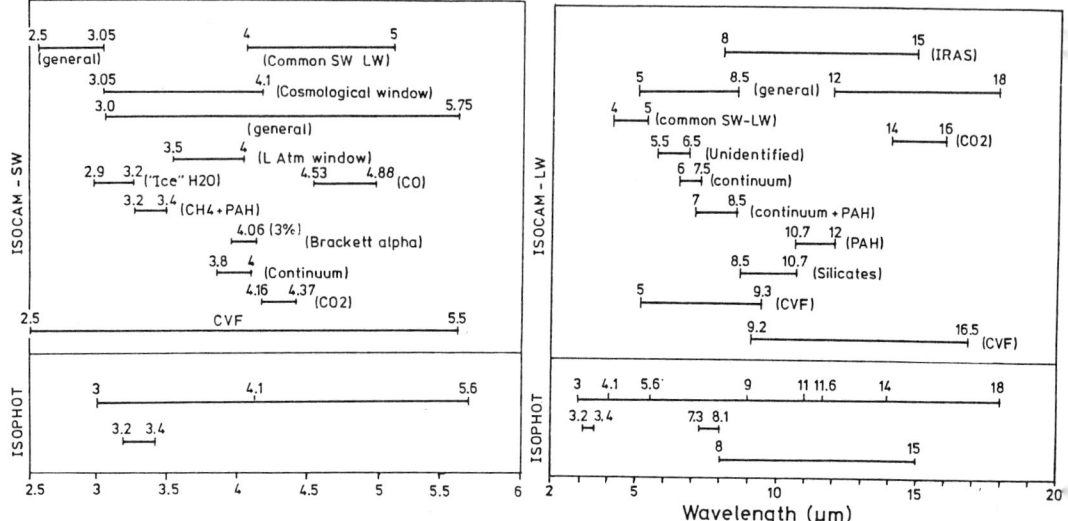

Figure 1: ISOCAM filters of the SW and LW section as a function of wavelength.

ISOPHOT is composed of 4 subsystems, which can be used one at a time:

- ISOPHOT-P, a multiband, multiaperture photopolarimeter for the range 3–120μm,
- ISOPHOT-C, a photometric camera for the range 30–200μm,
- ISOPHOT-A, a two-dimensional array for photometric mapping in the range 8–28μm,
- ISOPHOT-S, two grating spectrometers, operated simultaneously, for the range 2.5–12μm (excluding 5–6μm).

ISOCAM consists of two optical channels (one used at a time) each of which contains an array of 32 × 32 elements. It is designed to map selected areas of the sky in the wavelength range from 2.5–17μm at various spatial resolutions. Polarisation mapping is also possible.
An overview of the ISOCAM filters used with the short wavelength (SW) array and the long wavelength array (LW) are given in figure 1. The ISOPHOT-P filters for this wavelength are given as well.

The SWS instrument consists of two nearly independent grating spectrometers together covering the wavelength region 2.4–45μm with an overall spectral resolution

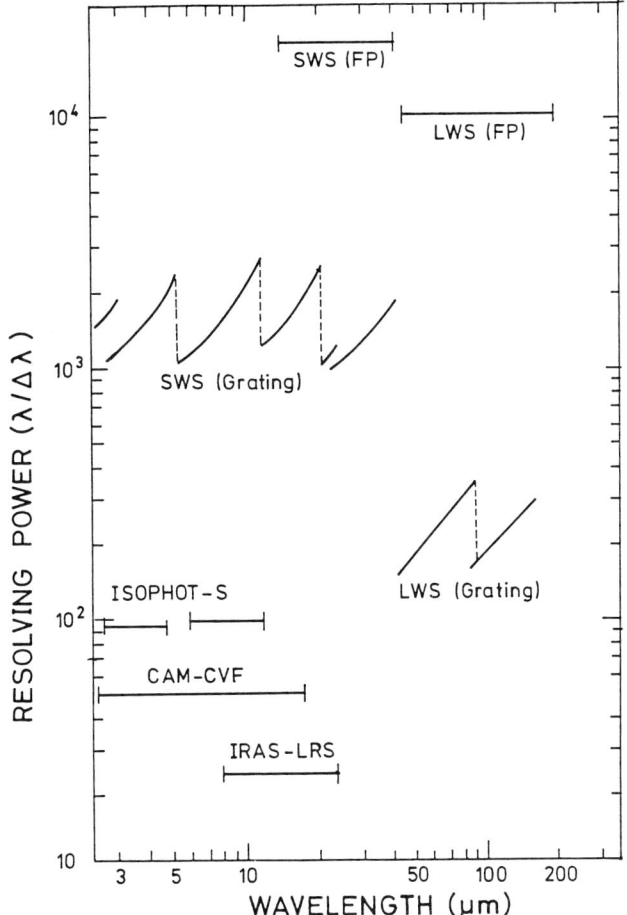

Figure 2: ISO spectral resolving power capabilities.

of ~1000 (i.e. 300 km/s). By inserting Fabry-Pérot (F–P) filters, one for the region 15–25µm and the other for the region 25–35µm (with capability to 40µm at reduced sensitivity), the resolution can be increased to $\sim 2\times 10^4$ (i.e. 15 km/s).

The LWS is a grating spectrometer providing spectral resolving power of ~200 from 45–200µm. Fabry-Pérot interferometers can also boost the resolution to 10,000 across this entire wavelength range.

In figure 2 the overall spectral resolving power capabilities are given as a function of wavelength.

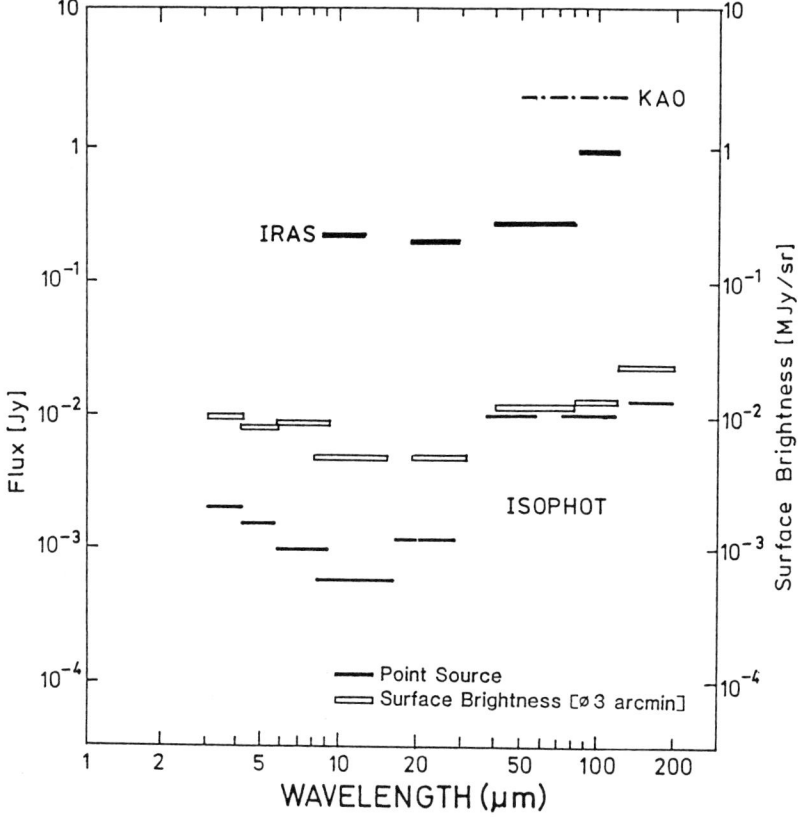

Figure 3: Calculated ISOPHOT sensitivities.

4 THE EXPECTED PERFORMANCE

The expected performances for all instruments are based on measured sensitivities for the detectors. As all instruments are in the assembly phase the optical transmission used in the calculations are calculated using data from measurements at component level (filters, mirrors, etc.). In all cases shown the limiting flux levels have been calculated for observing times of 1000 secs and the signal-to-noise ratios required were 10.

The ISOPHOT sensitivities for point sources and for extended sources (3 arcmin diameter) have been calculated using the detectors specifications of the industrial contract. The aimed goals are usually a factor two better. Results taken from the calculations of the ISOPHOT consortium are shown in figure 3.

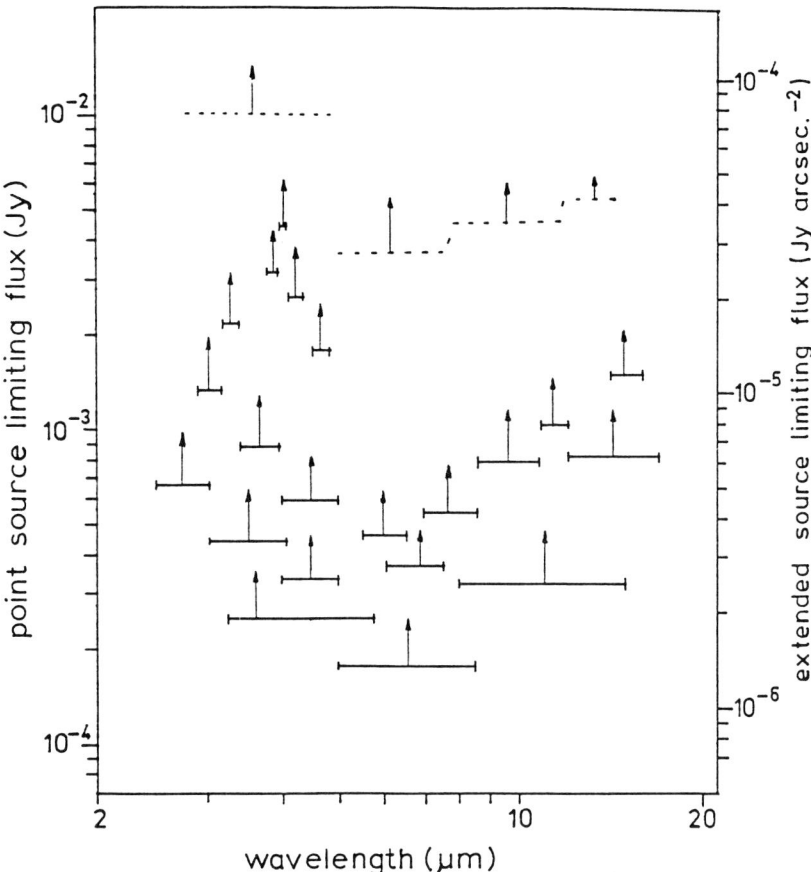

Figure 4: Calculated ISOCAM sensitivities for point sources and extended sources. The 6 arcsec pixel field of view has been used to achieve SNR=10 in 1000 seconds.

The expected performance of the two ISOCAM channels is given in figure 4. Sensitivities for point sources and extended sources, calculated from input of the ISOCAM consortium are given for the 6 arcsec pixel field of view. Also here the detector sensitivities used in the calculation are based on current laboratory measurements. With the wideband IRAS filter and 6 arcsec field of view pixel the LW array detector noise and zodiacal background shot noise are already comparable after 10 seconds of integration.

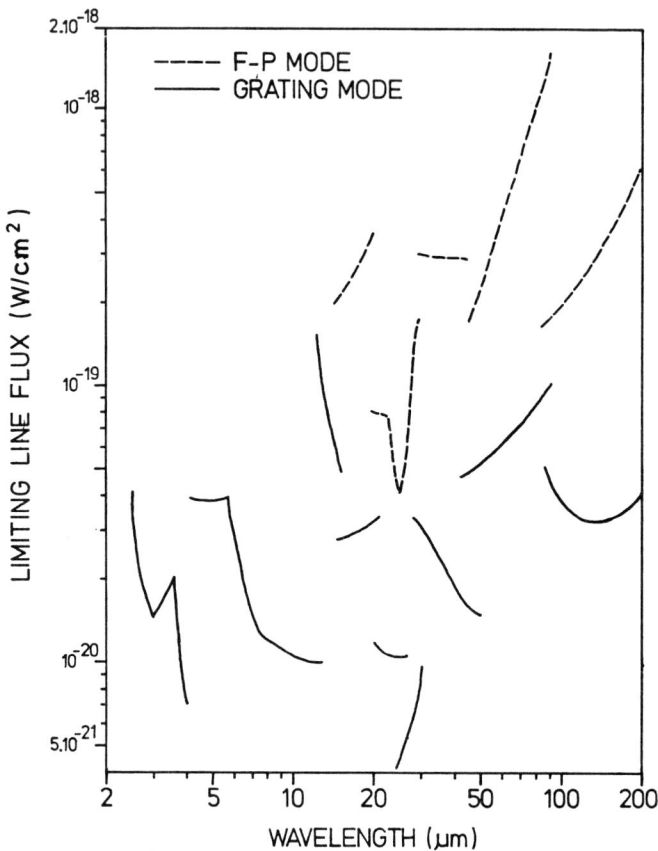

Figure 5: Limiting line flux of the ISO spectrometers.

The expected performance of the spectrometer is given in figure 5. Although laboratory measurement indicated NEP values of the order 2×10^{-18} WHz$^{-1/2}$, I took, for all detectors, a NEP value of 1×10^{-17} WHz$^{-1/2}$. It is highly uncertain whether the excellent detector NEP values can be made effective. Responsivity changes due to cosmic particle hits are likely to occur. The large variation in the sensitivity is due to the variation in the transmission filters, grating efficiencies and detector response as function of wavelength. The dotted curves are for the spectrometers in the Fabry-Pérot mode. Here I have assumed that five spectral points across a line were observed twice for 1000 seconds. A similar assumption was made for the LWS in the grating mode. As the SWS uses 1×12 element detector arrays, covering a complete line instantaneously, I took twice 1000 seconds of observing time.

5 CONCLUSION

With the present predicted performances of the ISO instrumentation an improvement of two orders of magnitude or more in sensitivity and spectral resolution can be expected and will provide the European astronomical community a new class of tools to study interesting astrophysical problems at infrared wavelengths.

REFERENCES

The second draft of the document: "Scientific Capabilities of ISO Payload" has been used as input for the calculations. This document will eventually become available as a userguide for ISO.

INTERFEROMETRY WITH LARGE OPTICAL TELESCOPES

PIERRE J. LENA
Université Paris VII and Observatoire de Paris
92195 Meudon Cedex, France

1. The Development of Optical Interferometry

The development of optical interferometry from ground-based observatories, at infrared or visible wavelengths, is proceeding fast. A recent Summer School (Alloin & Mariotti 1989) gives a thorough overview of the various experimental programs, the construction of instruments and the fields open for scientific investigation in a number of astrophysical subjects.

The current operating optical interferometers are characterized by a moderate telescope diameter (at most 1.5 meter, more frequently a few decimeters) and a number of cophased telescopes limited to two. Most of the planned instruments aim for similar diameters but for a larger number of unit telescopes, with the specific goal to restore not only the amplitude, but also the phase of the visibility of the object, hence to reconstruct actual images. The European Very Large Telescope is the first program where it is planned to coherently combine 8-meter class telescopes, in order to obtain the inherent sensitivity of large collecting areas.

Although this concept appeared early in the project (Léna 1983), it was only proposed after a thorough analysis (ESO/VLT Working Group on Interferometry 1986) and included in the final VLT Proposal (European Very Large telescope Proposal 1987). The initial concept of the VLT interferometric mode may be found in Léna (1987) and Merkle (1988a); it is now deeply reviewed by a dedicated Panel, who performs detailed studies for the implementation of this mode and will report conclusions at the beginning of the year 1990.[1]

The coherent combination of large optical telescopes requires to analyze a number of parameters, design constraints, site impacts, cost considerations, etc. As there exists yet no practical experience on the feasibility of such a goal in an astronomical context, great care should be exercized to properly establish a phased, step-by-step program where no option is made which prevents future long term developments, while, on the other hand, the strategy is comforted by results obtained

---

[1] The VLT Interferometry Panel is chaired by J.M.Beckers and advises the Scientific and Technical Committee of the European Southern Observatory.

at each step [2]. Moreover, a number of specific and critical points should be examined early, as they will condition the feasibility of the whole project. We like to discuss here several of these, namely the *sub-array concept*, the *limiting sensitivity* of large interferometric optical telescopes, including the recent perspectives offered by the development of adaptive optics and finally the *mechanical stability* of large structures at interferometric accuracies. There are indeed many other aspects, some quite difficult, but probably not as critical as the three ones just mentioned.

The interferometric mode of the VLT was a logical consequence of the choice of the array configuration, made for the overall design of the program. In the future, plans for new large telescopes should be carefully analysed in terms of possible coherent combination with other partners or even with existing telescopes of smaller diameter, as the efficiency of multi-telescope arrays is rapidly increasing with the number of aperture [3].

2. The Sub-Array Concept

Several considerations led to plan the 8-meter array with additional, satellite, smaller diameter telescopes :

- provide a flexible development tool where technical solutions for delay lines, telescope control, beam combining optics and detectors be elaborated without loading the precious observing time of the large mirrors;

- capitalize scientific results during the development period;

- cover with movable telescopes the u-v spatial frequency plane, which is inherently covered in an incomplete manner with the fixed large telescopes;

- reduce the pressure on the large telescopes, as the combination of uneven apertures leads to an equivalent diameter as the geometric mean : i.e. combining 2-m and 8-m leads in most of the pplanned spectral range to the sensitivity of an interferometer made of 4-meter telescopes;

These considerations led to include in the VLT program at least $N = 2$ additional 2 m class telescopes (the *Sub-Array*), movable along and perpendicular to the main array line and able to feed the common interferometric beam combination area. With $N = 4$, the Sub-Array alone would provide 6 independent baselines, a number increased to

---

[2] The construction phase of the four main telescopes of the VLT will lead to completion of the project before the year 2000. The implementation of the interferometric mode should begin on the site by approximately the year 1996 with the Sub-Array as described below, and progressively combine coherently the large telescopes. It is likely that the four 8-meter telescopes be combined together simultaneously at the beginning of the next century.

[3] The NOAO 8-meter Telescopes Proposal (1989) suggests such a long-term option for building a coherent array on the summit of Mauna Kea (Hawai).

10 when combiit is combined to a single 8-meter, and to 15 when combined to two of these.

This flexibility could be further exploited, either by adding extra telescopes on existing stations and/or tracks, or, if site allows, by providing a fixed station at a kilometric distance from the main array for higher resolution devoted to objects otherwise unresolved. Preliminary studies have been carried for a dedicated type of movable interferometric telescope (Plathner 1988) and are fairly conclusive on its feasibility and low cost.

3. Sensitivity Perspectives

The sensitivity of a ground-based interferometer is a many parameters problem. It involves at least the analysis of the following ones :

1. The *wavelength range*, divided between 0.3-1.4 μm (signal photon noise), 1.4-2.5 μm (detector noise) and 2.5-20 μm (background thermal noise).

2. The *atmospheric turbulence*, characterized by the coherence diameter $r_o(\lambda)$ and the coherence time $\tau_o(\lambda)$, both quantities being heavily wavelength dependent.

3. The availability (or not) of *adaptive optics*, a system able either to fully correct for atmospheric phase distorsions over each telescope area, or only to achieve partial correction. This result is again wavelength- and telescope diameter-dependent.

4. The possibility (or not) to correct for *phase fluctuations* between two telescopes (cophasing). This correction provides the only possibility to increase arbitrarily the integration times.

5. The number N of telescopes being *simultaneously* used in the interferometric array, and their relative configuration. A fully redundant configuration gives less data points in the u-v plane than the $\frac{N(N-1)}{2}$ points obtained in a fully non-redundant one, for the same total integration time.

6. Finally, the astronomer thinks in terms of *image*, while the interferometer provides data in the u-v plane. The final image quality is not only depending on the signal-to-noise ratio Λ of these data, but also on the dynamic range, the total spatial frequency coverage, the efficiency of the image restoration methods.

A first analysis was given by Roddier & Léna (1984). It showed that the full gain of large apertures could only be obtained by full or at least partial phasing of these apertures, stressing therefore the vital importance of *adaptive optics* for the development of interferometry on large telescopes in the presence of atmospheric phase distorsions. The negative conclusions reached by Dyck & Kibblewhite (1986) are not valid for phased apertures. Phasing the VLT in the wavelength range 2 to

12 µm appears feasible with current technologies (Merkle 1988b) and a vigorous development program has therefore been undertaken (Kern *et al* 1988). This spectral range is therefore designated as the prime target of the VLT interferometric operation, but this does not exclude the operation at visible wavelengths of the Sub-Array or of the 8-meters, whenever feasible.

We shall not give here a full discussion on the complex issue of interferometric sensitivity but only give some partial results and indicate some trends [4].

We restrict the analysis to a pair of apertures, observing during a total time T and giving *one data point* in the u-v plane. The final value of the signal-to-noise ratio Λ in the final image heavily depends of the total number of baselines and of the beam combination strategy. To achieve actively phased apertures or cophased telescopes, a reference phase must be determined and fed to the servo-control system. When this reference is the source under study itself, the technique is called *source referenced*. The source must be sufficiently bright at a certain wavelength λ_{ref} to define a relative phase of the wavefront limited in area (of the order of r_o^2) and in integration time (of the order of the atmospheric coherence time τ_o) by the atmospheric perturbations. Such a source, called a *bright* source in the interferometrists jargon, may be studied by the interferometer in a straightforward way and its magnitude is independent of the telescope diameter.

Detector- or background-noise limitation (1.4 to 20 µm).

We give in Table Ia and Ib the magnitudes and the fluxes of objects for which an acceptable value of Λ is obtained in the snapshot mode (*bright* objects) for a visibility unity (unresolved sources) and a system efficiency of 0.3. The value of $\Lambda = 50$ is selected in Table Ia as a conservative estimate of what is required to make an active control system with an adequate band-pass and a more prospective value of $\Lambda = 10$ is adopted in Table Ib.

- for the detector noise-limited case (photometric bands H and K), sensitivities are computed in Table Ia with a detector read-out noise of 300 electrons, a value consistent with existing detectors. In Table Ib, a more prospective value of 30 electrons is adopted, representing most likely the state of the art in 5 years.

- for the thermal background noise limited case (photometric bands L, M, N, Q), assuming the photon count to be Λ times the background noise computed for a temperature of 273 K, an emissivity of 1.0 and a diffraction-limited cold stop appropriate to the coherence area.

---

[4] This discussion is part of a complete analysis to be published (Ridgway & Léna 1990) and has received helpful comments from Dr.J.Beckers. It is part of a work carried for the VLT Interferometric Mode and also in the preparation of the NOAO 8-m telescope proposal.

Table Ia. Source-referenced Mode Sensitivity
Phasing & Cophasing. *Conservative Case*
Telescope Diameter 2-m

Seeing	H	K	L	M	N	Q
0.5"	6.9	8.1	7.2	5.4	2.6	-0.3
	1700	370	400	1 200	4 000	14 000
1.0"	4.7	5.8	5.3	3.5	2.3	0.7
	14 000	2 900	2 600	6 600	5 300	5 500

Units: upper lines are magnitudes ; lower are milliJy
Phasing : $V = 1$ or $V \neq 1$. Cophasing: $V = 1$.
1.0" seeing $\Leftrightarrow r_o(0.5\mu\text{m}) = 10$ cm.

Table Ia. Source-referenced Mode Sensitivity
Phasing & Cophasing. *Conservative Case*
Telescope Diameter 8-m

Seeing	H	K	L	M	N	Q
0.5"	6.9	8.1	7.2	5.4	4.4	2.7
	1700	370	400	1 200	760	880
1.0"	4.7	5.8	5.3	3.5	2.6	1.5
	14 000	2 900	2 600	6 600	4 000	2 600

Units: upper lines are magnitudes ; lower are milliJy
Phasing : $V = 1$ or $V \neq 1$. Cophasing: $V = 1$.
1.0" seeing $\Leftrightarrow r_o(0.5\mu\text{m}) = 10$ cm.

Table Ib. Source-referenced Mode Sensitivity
Phasing & Cophasing. *Long-Term Case*
Telescope Diameter 2-m

Seeing	H	K	L	M	N	Q
0.5"	11	12	8.9	7.1	4.4	1.4
	34	7	90	230	750	2 700
1.0"	8.9	10	7	5.3	4	1
	270	60	510	1 300	1 100	3 800

Units: upper lines are magnitudes ; lower are milliJy
Phasing: $V = 1$ or $V \neq 1$. Cophasing: $V = 1$.
1.0" seeing $\Leftrightarrow r_o(0.5\mu\text{m}) = 10$ cm.

Table Ib. Source-referenced Mode Sensitivity
Phasing & Cophasing. *Long-Term Case*
Telescope Diameter 8-m

Seeing	H	K	L	M	N	Q
0.5"	11	12	8.9	7.1	6.2	4.4
	34	7	90	230	140	170
1.0"	8.9	10	7	5.3	4.3	3.2
	270	60	510	1 300	800	510

Units: upper lines are magnitudes ; lower are milliJy
Phasing: $V = 1$ or $V \neq 1$. Cophasing: $V = 1$.
1.0" seeing $\Leftrightarrow r_o(0.5\mu\text{m}) = 10$ cm.

When phasing or cophasing can be achieved with a reference distinct from the source under study, the sensitivity is greatly increased. Techniques such as *laser referencing* aim to create an artificial reference star within the isoplanatic angle and use it for phasing and cophasing (Foy & Tallon 1989). This allows to perform long time integration, since the referencing is now independent of the source under study. The limits become independent of seeing and are given in Table II for a time integration $T = 10$ minutes.

Table II. Off-source Referenced Mode Sensitivity
Integration Time $\Delta T = 10$ minutes. *Long Term case*

Telescopes	H	K	L	M	N	Q
2-m + 2-m	23.6	23.3	13.2	11.2	8.0	4.6
	-	-	1.6	5.1	27	150
8-m + 8-m	26.5	26.4	15.2	12.8	10.6	7.6
	-	-	0.25	1.2	2.9	9

Units: upper lines are magnitudes ; lower are milliJy

The values given in Table II may appear surprisingly low, but this is a direct consequence of the *source independent* phasing and cophasing process. They therefore reach the sensitivity of a large telescope with good detectors. This points to the immense scientific field open to the interferometric operation of large telescopes.

Signal-noise limitation (0.4 to 1.4 μm).

This domain slightly extends in the infrared, covering the I and J bands as long as detectors are signal noise limited. In this domain, there are two difficulties for aperture phasing : the technology required to build an adaptive mirror with a large number of actuators ($D/r_o \approx 1500$), and the small value of the wavefront coherence area, drastically limiting the number of available reference sources for phasing.

In the most immediate configuration without adaptive optics, the complex visibility must be deduced from the analysis of a multi-r_o (or multi-speckle) image by triple correlation (*speckle masking*). Detailed simulations (Weigelt 1989) indicate that a magnitude

$$m_V = 17.7$$

is accessible to the interferometer with $\Lambda \approx 5$ and $\Delta T \approx 6$ mn. Even if complete laser-referenced phasing is not possible in a near future, *partial phasing* in the visible, resulting from a complete adaptive correction in the near infrared, leads to significant gains.

4. Mechanical Stability

The mechanical stability of an interferometer is a fundamental requirement. The stability of the optical path, within a fraction of the operating wavelength, must be maintained between the entrance pupil of each telescope, namely the primary mirror, and the combined interferometric focus. The main sources of phase noise are indeed vibrations excited by wind, drives or other mechanical sources (pumps, fans, etc.). It is likely that large structures with a great inertia may be more stable than small interferometric telescopes used to date. The tolerable level of vibrations is set by the phase noise amplitude of the atmosphere itself. First experiments are carried (Bourlon & Léna 1988) to measure at the required interferometric accuracy the vibration behaviour of a number of interferometric instruments as well as large conventional telescopes (the ESO 3.6 meter) or open-air structures (the SEST sub-millimetric antenna in Chile). It is interesting to note that the observed levels are not significantly above the interferometric requirements at or above a wavelength of 2 μm. These measurements are continued and will provide design inputs for the VLT mechanical concept.

5. Conclusion

Optical interferometry with large telescopes of the emerging 8-meter class is a new concept which will be gradually explored by the European Very Large Telescope. This concept is heavily dependent of the developement of adaptive optics, but recent and spectacular results demonstrate its feasibility in astronomy (Rousset *et al* 1990).

The concept of a Sub-Array creates in the vicinity of the large telescopes an expandable set of auxiliary, smaller size, movable interferometric telescopes, with the possibility of flexible combination between themselves or with the large ones. This approach could be considered for the new projects of large telescopes, whether they are of the array type or not. It indeed imposes more severe site requirements and also a larger degree of coordination with other instruments possibly available on the site. The extraction of coherent beams from conventional telescopes may be, in the future, eased by the use of rapidly developing single mode optical fibers.

The sensitivity of large telescopes in the interferometric mode becomes exceptional, as soon as cophasing with adaptive optics is achieved. This opens an entirely new domain in astronomy.

REFERENCES

Alloin, D. and Mariotti, J.M., Ed., 1989, *Diffraction-limited Imaging with Large Optical Telescopes*, NATO-ASI Cargèse Summer School, Kluwer.

Bourlon, P. and Léna, P., in *High resolution Imaging by Interferometry*, F.Merkle ed., (European Southern Observatory, Garching).

Dyck, M. and Kibblewhite, E., 1986, P.A.S.P., **98**, 260.

European Southern Observatory Interferometry Working Group, 1986, VLT Report 49, Garching.

The VLT Proposal, 1987, (European Southern Observatory, Garching).

Foy, R. and Tallon, M., 1989, *Astron.Astrophys.*, submitted.

Kern, P., Léna, P., Rousset, G., Fontanella, J.C., Merkle, F., Gaffard, J.P., 1988, in *Large Telescopes and their Instrumentation*, (European Southern Observatory, Garching).

Léna, P., 1983, in *ESO's Very Large Telescope*, Garching.

Léna, P., 1987, *Messenger*, **50**, 53.

Merkle, F., 1988a, *J. Opt. Soc. Am.*, **A5**, 904.

Merkle, F., 1988b, in *Large telescopes and their Instrumentation*, Garching.

The NOAO 8-meter Telescopes, 1989, *Proposal to the National Science Foundation*, (NOAO, Tucson).

Plathner, D., 1988, in *High Resolution Imaging by Interferometry*, Merkle, F., ed., (European Southern Observatory, Garching).

Ridgway, S.R. and Léna, P., 1990, in preparation.

Roddier, F. and Léna, P., 1984, *J. Optics*, **15**, 171 and 363.

Weigelt G., 1989, Private Communication.

New Developments, Concepts, and Dreams

Summary Talk of the Instrumentation Meeeting

I. Appenzeller

Landessternwarte Heidelberg-Königstuhl, FRG

Scientific progress in astronomy has always been closely related to advances in observational techniques and methods. Therefore, it was reassuring to learn at this meeting about many powerful new instruments which have been completed recently or are being planned for the near future. Some of the described devices are already producing data, others are under development, for some there exist only concepts, and some of these exciting new concepts may remain dreams. As all the papers presented will be included in the two volumes of the conference proceedings, this written version of my summary will *not* deal explicitly with the individual contributions. Instead, I will use this opportunity to add a few general remarks and reflections on the scientific results of this meeting and its likely impact on the future of our field.

The 25 talks and 7 poster papers presented at this meeting covered three main topics: (1) the design, ancillary instrumentation, detectors, and optimum site selection for large groundbased telescopes, with particular emphasis on new projects, such as the ESO-VLT, Columbus, POST, LEST, etc.; (2) space telescopes (such as ISO, ORFEUS, SANTA MARIA, etc.) and their instrumentation and observing stategies; and (3) new concepts for obtaining very-high resolution images of astronomical objects by means of interferometric methods.

Among the highlights of this meeting were the progress reports on the giant new ground-based telescopes now under construction or in advanced planning stages. The most obvious property of this new telecope generation is a greatly increased light collecting power. However, for many astrophysical applications their improved imaging performance, which has become feasible as a result of active optical systems, will be an at least as valuable advantage. Thus, these new telescopes will not only be larger but also better instruments, and the latter aspect must be regarded of equal importance as the light collecting power.

On the other hand the high optical quality of the new telescopes also makes the selection of a suitable site much more important than in the past. Some of the usual site criteria (cloud cover, wind, atmospheric water vapor content, etc.) can be measured using straightforward methods. But, as was also pointed out at this meeting, measuring the seeing properties of a new site with small test instruments is not a trivial problem and the atmospheric physical processes affecting astronomical seeing are still not fully understood. Nevertheless, the reports presented at this meeting illustrate that data obtained with suitably designed and calibrated small test telescopes can indeed be extrapolated to derive the full seeing properties of a site. From the work carried out at ESO, at La Palma, and in the US it seems clear that the new methods allow a least on a differential basis qualitative measurements of all atmospheric contributions to seeing. This meeting convinced me for the first time that the quality of a site can be tested objectively, if an adequate effort is made.

A major fraction of this meeting was devoted to new spectrometers, cameras, and detector systems for existing and future optical telescopes. The reported or predicted performance data of some of these instruments were impresssive. However, in the case of some of the instruments sugested for the future telescopes I am quite worried about the overall efficiency. Although the very large telescopes now under construction or on the drawing board will give us many more photons than their predecessors, these photons will be expensive and we cannot afford wasting any of them. Moreover, with the introduction of the CCDs the efficincy of the optical detectors has reached almost unity, and a significant further increase of telescope sizes will probably have to wait for the next century. Therefore, our only hope of increasing the S/N of optical observations with the new telescopes rests with the improvement the ancillary instrumentation efficiency, which (e.g. in the case of spectrographs) so far often falls well below unity.

In other words, in order to achieve an unobstructed view through the large "new windows to the universe" represented by the new generation optical telescopes, we have to match their impressive light gathering performance by highly efficient back-end instruments.

Among the spacebased instrumnets discussed at this meeting were one very large observatory-type project (ISO) and several smaller specialized missions (ROZHEN, ORFEUS, SANTA MARIA). While the importance of spacebased observatories (such as HST and ISO) is quite evident, the discussions at this meeting also showed that for many scientific objectives specialized smaller and simpler space missions often are not only scientifically superior but may also produce comparable scientific results at considerably lower costs.

The formation of high resolution images by means of long-baseline interferometry has long been an astronomer's dream. During the past decade for the radio astronomers this "dream" became a routine procedure. Best known examples are the beautiful high-resolution images of various types of cosmic radio sources produced at the VLA and other modern radio interferometers. These images became possible by the development of efficient phase recording or phase reconstruction methods. At shorter wavelengts phase reconstructoin is, of course, much more difficult. Nevertheless, as we also learned at this meeting, progress is being made and there seems to be realistic hope to achieve interferometric imaging at least in the infrared. As a particularly exciting future prospect we learned at this meeting about the possibility of using the ESO VLT for high-sensitivity interferometric imaging.

After having been discussing astronomical instrumentation for three full days, one may ask how this exercise will possibly influence the future development of our field. I believe that there will be two valuable effects. Firstly, such meetings are the most efficient means for exchanging information and arranging new cooperations between different groups. This aspect appears particularly important in view of the growing role of large international observatories and space experiments. Secondly, such meetings are an excellent opportunity to focus attention on common problems.

As an example lets look back one generation to the instrumentation meeting of the IAU General Assembly of 1958 at Moscow. If the proceedings of this meeting (published in the IAU Transactions Vol. X, 1960) are a correct measure, more than two thirds of the 1958 session were devoted to the single topic of optical detectors. Although many different (by now mostly obsolete) types of detectors were described, there was a clear feeling that all these devices were inadequate and that it was important to invest a major amount of effort in improving the situation. As we all know, this effort (greatly helped by the developments in microelectronics) led to fundamental changes in the detector field, and much of the scientific progress in optical astronomy during the past decades can be directly traced to these advances in detector performance.

I am confident that the problems which we focused on during this meeting will develop as auspiciously and that this meeting will not only be remembered for its excellent organization, an exceptionally pleasant location, and and the great hospitality of our hosts, but that it will also have a lasting effect on the future development of astronomical instrumentation.

GENERAL LECTURES

Ground-based European Astronomical Projects

A. BOKSENBERG

Royal Greenwich Observatory
Madingley Road, Cambridge, UK

Electromagnetic radiation carries virtually all the information on which our understanding of the remote parts of the Universe is built. The production of electromagnetic radiation is intimately related to the physical and chemical conditions prevailing in the source object, and in its propagation to the receiver is modified by the conditions along its path. The optical region happens to contain more information about the Universe than any other region of the electromagnetic spectrum observationally explored in astronomy and most of the fundamental astrophysical advances to date have come from the use of optical telescopes. Practically, the 'optical region' covers the range from near-ultraviolet wavelengths reaching close to 3000A at the atmospheric cutoff, to near-infrared wavelengths which nowadays some consider encompass the region to about 2 microns. This observational range includes the highly redshifted information originating from the astrophysically-related far-ultraviolet region manifest from distant objects in the Universe. And because of the richness of the optical region, astronomers investigating in the radio, far-infrared, X-ray and other regions need to complement their observations with optical measurements to discover physical characteristics or even to recognize what class of object they have revealed.

There has been progressive improvement in telescope light gathering capability over the centuries. Galileo's simple telescope gave a thirty-fold sensitivity improvement over the unaided eye; William Herschel's largest telescope exceeded 10^4 times; the Mount Wilson 2.5m telescope bettered 10^5 times; and the Mount Palomar 5m telescope then the Soviet 6m telescope have reached nearly 10^6 times improvement. With the advent of photographic recording near the beginning of this century, followed by the introduction of electronic detectors which now almost entirely have taken over from photography, the progressive gain in sensitivity over the eye is very much greater still, and this is further enhanced by increases in telescope angular field of view.

An important factor crucially affecting the effectiveness of a telescope is the quality of the observing site: the progress of observational astronomy has required the establishment of remote mountain observatories with clear sky, transmission and seeing qualities far exceeding what was available or even possible from the traditional local sites. But it is only in the last few years that the full potential of the best sites has been realized, from appreciation of the potentially destructive effect of 'dome seeing' and its avoidance in the design of modern telescope facilities.

Europe is very well endowed with a wide range of superbly instrumented optical telescopes placed on the best available astronomical sites. The continued vitality of this observational field is demonstrated by the several very large telescopes which are now in design or development stages to be completed within the coming decade. Here I will confine my attention to the major optical telescope facilities which nations in Western Europe are operating now or planning for the near future.

Observational techniques in astronomy are developing ever more rapidly. Astrophysics above all is built on the achievements of spectroscopy and great efforts have gone into developing techniques of spectral analysis which are efficient as well as achieving the required resolving power. Pre-eminent among these is the application of multi-fibre-optic feeds which lead the light from many individual objects in a large dispersed field to alignment along the entrance slit of a spectrograph for simultaneous analysis. Effective gains in efficiency over single object spectroscopy exceeding fifty times have been achieved in this way. Supplementing such advances are developments in detectors both of the image photon counting system (IPCS) and charge-coupled device (CCD) varieties. These detectors are of course also used for direct imaging applications and have almost totally supplanted the use of photography except in applications such as the wide field Schmidt telescope surveys. New direct imaging techniques exploiting speckle interferometry, related techniques employing apertures in the field of one telescope or arrays of telescopes, and real-time adaptive optical systems, now allow imaging resolution from well within the conventional seeing disc down to the diffraction-limited angular discrimination of the system. Finally, there has been an explosive growth in data reduction and processing methods coupled with widespread and almost unlimited access to computer systems both on and off the telescopes, which has transformed the process of astronomical investigation out of all recognition of what it was only two decades ago. One hopes that this apparently progressive technological process has not left the necessary thought process in the practitioners wanting.

There are more than 150 professional optical observatories belonging to or associated with Western European nations, many having a variety of telescope facilities. Included amongst these are seven of the world's twelve largest existing optical/infrared telescopes, listed in the Table: two of these, the European Southern Observatory 3.6m telescope and 3.5m New Technology Telescope, are operated at La Silla, Chile; two more, the 3.6m Canada-France-Hawaii Telescope and the 3.8m United Kingdom Infrared Telescope are operated on the Mauna Kea site at Hawaii; the 3.9m Anglo-Australian Telescope is located at Siding Spring in Australia; the Max-Planck-Institute for Astronomy 3.5m telescope is operated at the

4M CLASS TELESCOPES

1	6.0M	1976	BOL'SHOI TELESKOP AZIMUTAL'NYI SPECIAL ASTROPHYSICAL OBSERVATORY ZELENCHUKSKAIA MT PASTUKHOV (2100M)
2	5.08M	1948	GEORGE ELLERY HALE TELESCOPE PALOMAR OBSERVATORY PALOMAR MOUNTAIN (1700M)
3	4.5M	1979	MULTIPLE MIRROR TELESCOPE WHIPPLE OBSERVATORY MT HOPKINS (2606M)
4	4.2m	1987	WILLIAM HERSCHEL TELESCOPE ROYAL GREENWICH OBSERVATORY OBSERVATORIO DEL ROQUE DE LOS MUCHACHOS (2332M)
5	4.0M	1976	CERRO TOLOLO INTER-AMERICAN OBSERVATORY CERRO TOLOLO (2160M)
6	3.9M	1974	ANGLO-AUSTRALIAN TELESCOPE ANGLO-AUSTRALIAN OBSERVATORY SIDING SPRING MOUNTAIN (1165M)
7	3.8M	1973	NICHOLAS U MAYALL TELESCOPE KITT PEAK NATIONAL OBSERVATORY KITT PEAK (2100M)
8	3.8M	1979	UNITED KINGDOM INFRARED TELESCOPE ROYAL OBSERVATORY EDINBURGH MAUNA KEA (4200M)
9	3.6M	1979	CANADA-FRANCE-HAWAII TELESCOPE MAUNA KEA (4180M)
10	3.6M	1976	EUROPEAN SOUTHERN OBSERVATORY CERRO LA SILLA (2400M)
11	3.5M	1985	GERMAN-SPANISH ASTRONOMICAL CENTRE CALAR ALTO (2160M)
12	3.5M	1989	NEW TECHNOLOGY TELESCOPE EUROPEAN SOUTHERN OBSERVATORY CERRO LA SILLA (2400M)

Table

The world's twelve largest optical/infrared telescopes.

German-Spanish Astronomical Centre at Calar Alto, Spain; and the 4.2m William Herschel Telescope is operated by the Royal Greenwich Observatory on behalf of the UK, the Netherlands and Spain at the Observatorio del Roque de los Muchachos on La Palma, Canary Islands, belonging to the Instituto de Astrofisica de Canarias.

Several new telescopes are in firm planning or early construction phases. The most ambitious of these, to be the most powerful telescope facility in the world, is the European Southern Observatory Very Large Telescope (VLT) consisting of four separately mounted 8m telescopes. Italy with institutes in the USA is planning the Columbus Telescope: two 8m telescopes on a common mount. Italy also is to start construction on a 3.5m telescope closely similar in design to the ESO New Technology Telescope. Finally, the United Kingdom is planning to join in a partnership with Spain and other nations to build an 8m telescope; a design for this intended for the Observatorio del Roque de los Muchachos has been taken to an advanced stage by the Instituto de Astrofisica de Canarias and the Royal Greenwich Observatory, based on the William Herschel Telescope.

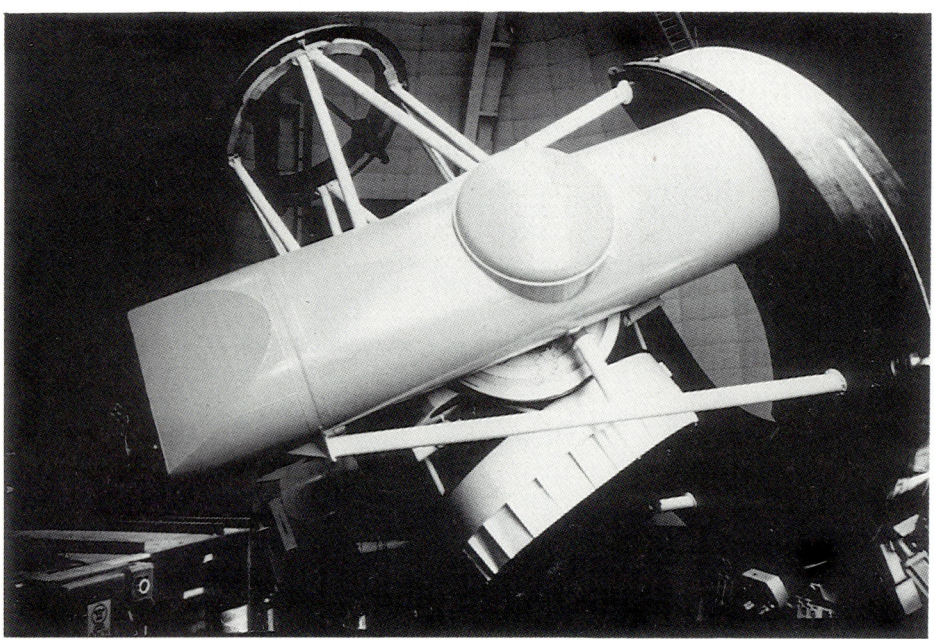

The equatorially-mounted 3.6m Canada-France-Hawaii Telescope at Mauna Kea, Hawaii.

The Max-Planck-Institute for Astronomy 3.5m equatorially-mounted telescope at Calar Alto, Spain.

The equatorially-mounted 3.8m United Kingdom Infrared Telescope operated at Mauna Kea, Hawaii, by the Royal Observatory Edinburgh.

The equatorially-mounted 3.9m Anglo-Australian Telescope at Siding Spring Observatory, Australia.

Several telescopes at the Observatorio del Roque de los Muchachos site of the Instituto de Astrofísica de Canarias, on La Palma, Canary Islands (upper). Star trails showing little attenuation down to the horizon (lower).

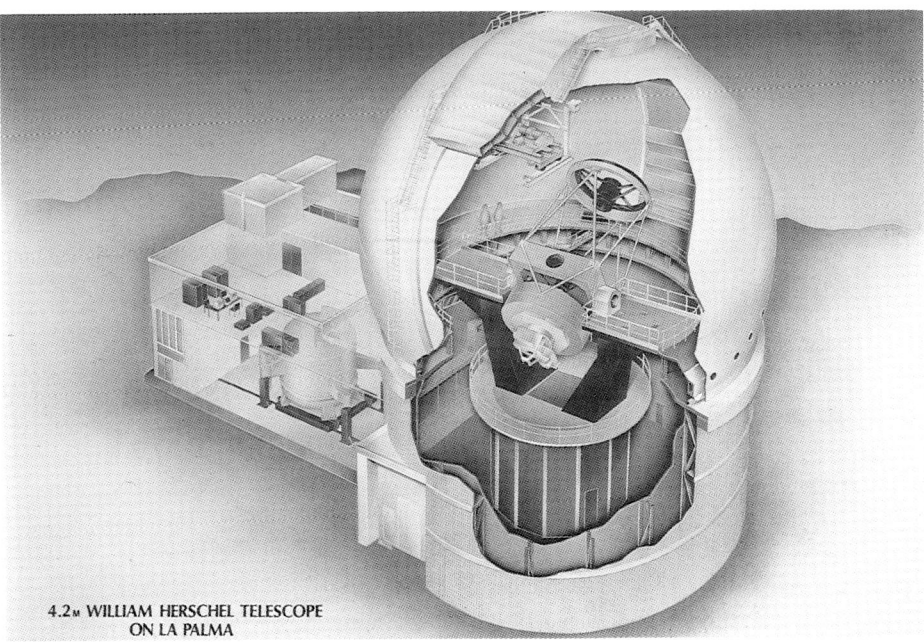

The compact dome and building of the 4.2m William Herschel Telescope operated at the Observatorio del Roque de los Muchachos, La Palma, Canary Islands, by the Royal Greenwich Observatory (upper). Cutaway section showing the alt-azimuthally-mounted telescope in a clear dome space without local heat inputs (lower).

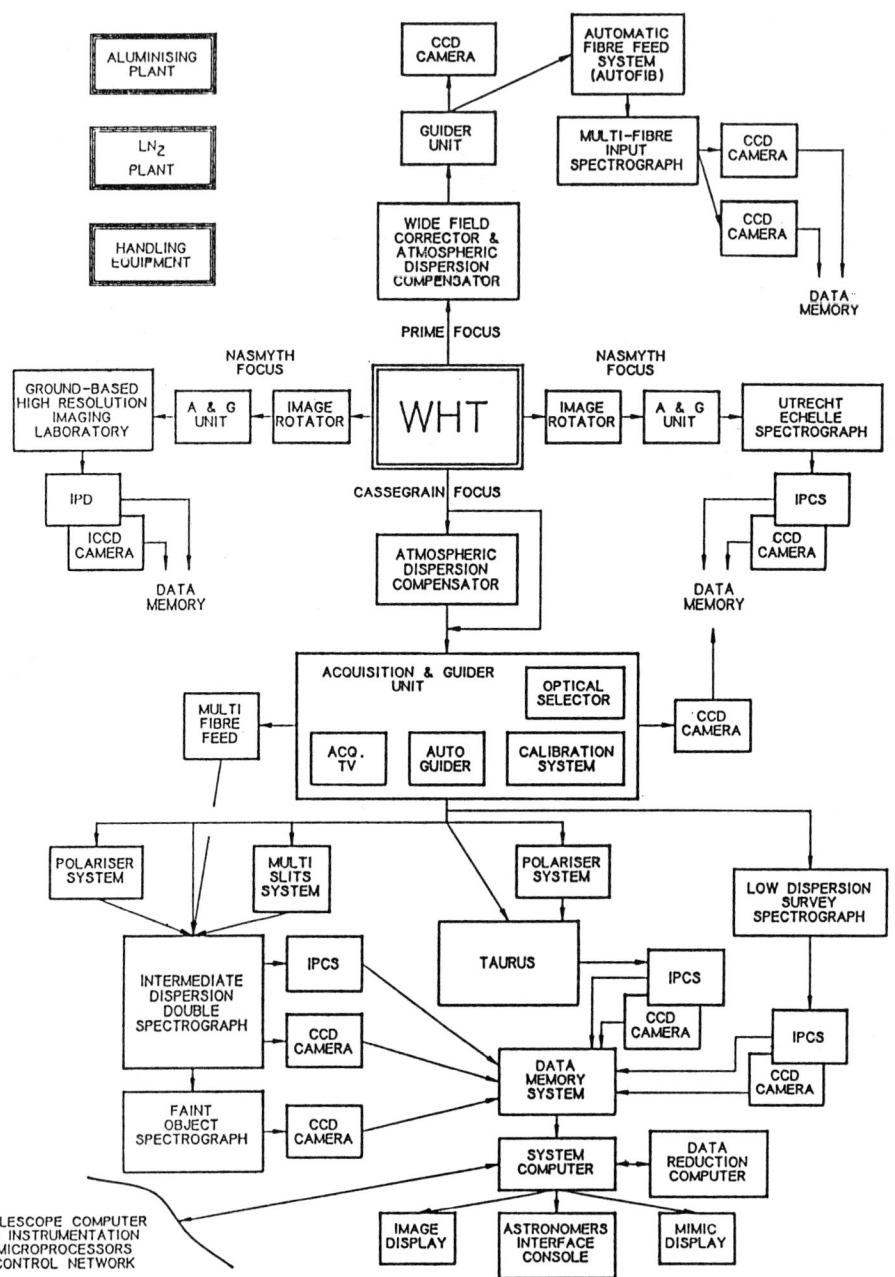

Block diagram of the auxiliary instrumentation and facilities of the 4.2m William Herschel Telescope, indicating the technical sophistication of a modern telescope. At one of its Nasmyth foci the WHT contains a unique facility – the Ground-based High Resolution Imaging Laboratory (GHRIL) – built jointly by institutes in the United Kingdom, the Netherlands and Spain to provide well-supported access for experiments on image improvement through the application of adaptive optics and other techniques.

The European Southern Observatory's 3.6m telescope. The telescope is connected with the 1.4m Coudé Auxiliary Telescope located in an attached building.

The European Southern Observatory's alt-azimuthally-mounted 3.5m new Technology Telescope. This is a revolutionary new telescope with a thin primary mirror and incorporating an 'active optics' system. The building has been designed to achieve the smallest possible dome-induced air turbulence around the telescope. In several respects this telescope serves as a prototype for the ESO Very Large Telescope.

Possible form of the European Southern Observatory Very Large Telescope consisting of four thin-mirror alt-azimuthally-mounted 8m telescopes. By appropriately combining the light of the individual fixed telescopes and movable auxiliary telescopes the VLT also will function as an interferometer.

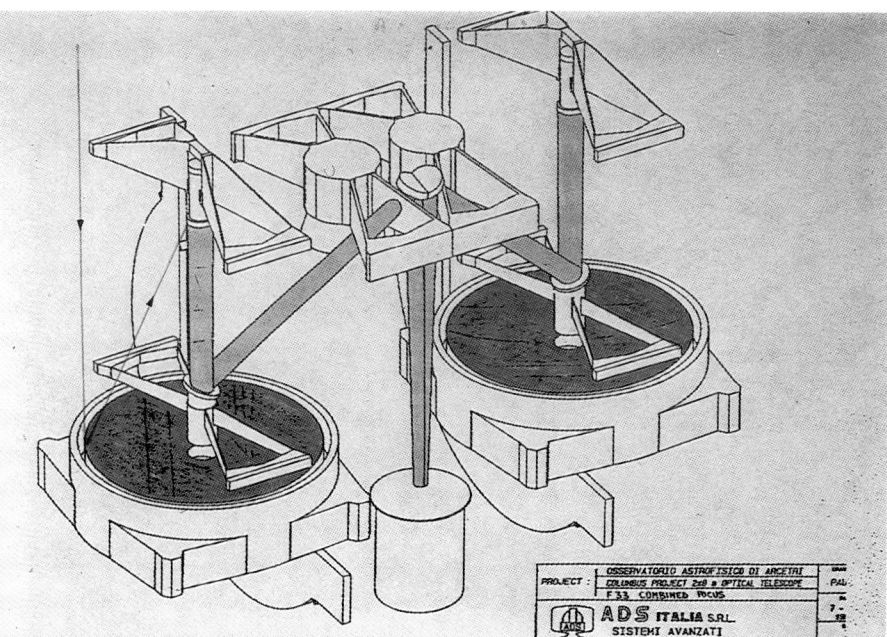

Schematic diagram of the configuration of the proposed Italy-USA two-8m Columbus Telescope. A single mount will carry the two optical trains and instruments will be operated at the combined focus.

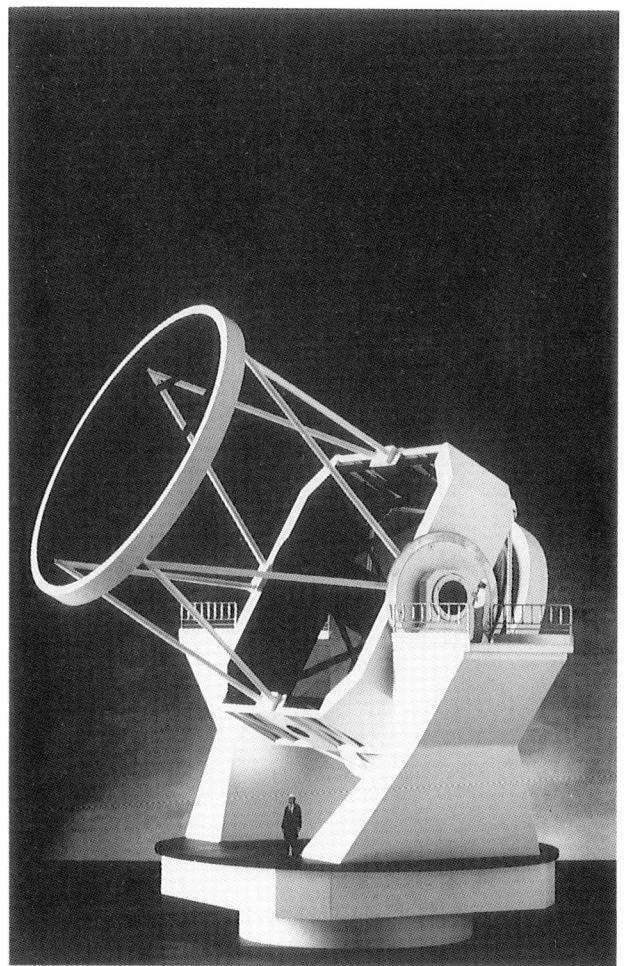

Model of the proposed 8m optical/infrared telescope designed by the Royal Greenwich Observatory and the Instituto de Astrofísica de Canarias for construction at the Observatorio del Roque de los Muchachos, La Palma, Canary Islands. The design is based on the William Herschel Telescope and features a thin meniscus primary mirror, of comparable weight to the thicker WHT mirror, actively supported on more than two hundred indivdually controllable support pads.

ESA ASTRONOMICAL PROJECTS

Martin C.E. Huber *
European Space Agency
Space Science Department
ESTEC, Postbus 299,
NL-2200 AG Noordwijk

1. INTRODUCTION

When we look at the past missions flown by the Scientific Directorate of the European Space Agency, ESA (or by the European Space Research Organisation, ESRO, before the founding of ESA in 1975), we find that only 5 of the 15 free-flyers were astronomy missions. This is a result of ESA's traditional distinction between solar system and astronomy projects.

The past, current and future ESA free-flyers and two multi-disciplinary payloads are listed in Table 1. A closer look at this Table shows that the GIOTTO probe which flew by the nucleus of comet Halley is listed as a solar-system (S) rather than an astronomy (A) mission. Thus the distinction is rather artificial. It is practical, because the techniques employed in solar-system exploration - mostly in-situ probing - are different from those used in 'astronomy' missions - remote sensing exclusively. Yet most of the objects investigated are astronomical objects or are at least examples (like the terrestrial magnetosphere) of phenomena observed around astronomical objects (e.g. planets or neutron stars). Furthermore, many 'astronomy' mission do observe solar system objects.

---

*   On leave of absence from the Swiss Federal Institute of
    Technology Zurich (ETHZ), Institute of Astronomy

Also progress in astronomy is based more and more on inter-disciplinary collaboration. (Multi-wavelength observations are adopted as the standard database already.) We therefore interpret our topic 'Astronomical Projects' in the broad sense, and select the missions to be discussed here accordingly.

All the current and future missions carried out by ESA's Scientific Programme are regularly described in ESA reports (Taylor 1989, Appourchaux 1990). In the following we can therefore restrict ourselves to a brief tour d'horizon of some of these missions.

We will start with the highlights of four past and current missions and then give an outlook towards four future missions. Subsequently we will summarise the aims, features and consequences of ESA's long-term scientific programme 'Space Science - Horizon 2000' and, in concluding, will also try to look beyond Horizon 2000, where the Moon might become a natural space station.

2. PAST AND CURRENT ASTRONOMICAL MISSIONS

2.1 IUE, the International Ultraviolet Explorer

The International Ultraviolet Explorer (IUE) is a collaborative mission between ESA, the United Kingdom Science and Engineering Research Council (SERC) and NASA. This satellite carries a 45-cm spectroscopic telescope, which covers a wavelength range extending from the Lyman-alpha line of hydrogen to 300 nm with either high or low dispersion, affording spectral resolutions of, respectively, 10^4 or 300.

Launched into a geostationary orbit in January 1978, IUE has operated continuously. For 8 hours every day the spacecraft and telescope are operated from the ESA groundstation at Villafranca (Spain) and for the remaining 16 hours control is exercised by the NASA ground station at Goddard Space Flight Center (GSFC). The

European share of the observing time has regularly been oversubscribed by a factor of three.

The objects observed by IUE range from solar-system objects to quasars. In fact, observations of the quasi-stellar object (QSO) HS1700 + 6416 at a redshift z = 2.72 enabled IUE to get a glimpse of radiation that - in the rest frame of the QSO - is emitted around 50 nm, i.e. beyond the Lyman continuum (Reimers et al. 1989).

The low-dispersion spectra taken by IUE are made available to a wide scientific community through the Uniform Low Dispersion Archive (ULDA) which has been installed in nine host institutes that serve 11 of the 13 ESA member states. The fact that more than 10 000 spectra have been retrieved within less than a year after the archive had been distributed, underlines the need for this easily accessible database.

2.2 EXOSAT

The EXOSAT X-ray observatory was operational during three years, namely from May 1983 until April 1986. EXOSAT's highly eccentric orbit whose period was 90 hours made it possible to monitor variable X-ray objects for uninterrupted intervals, which largely exceeded the then usual one-hour observing window typical of low earth orbit.

The energy region accessible to the EXOSAT instruments extended from 0.05 to 50 keV, and the spatial resolution ranged from ca. 15 arc sec to 45 arc min. EXOSAT could observe a wide range of X-ray objects: from stellar coronae, cataclysmic and other X-ray binaries and supernova remnants over active galactic nuclei to clusters of galaxies. Quasi-periodic oscillations - implying rapidly rotating neutron stars (van der Klis et al. 1987) - are one of the discoveries made with EXOSAT.

A survey of recent results on compact galactic and extragalactic X-ray sources - an area with important contributions from EXOSAT observations - is given in White (1990).

As in the case of IUE, EXOSAT data are made available to the astronomical community in the form of an archive. The EXOSAT database is probably the most advanced astronomical archive today. It is currently accessible through computer networks and will be available on optical discs in future.

2.3 GIOTTO's Flyby near the Nucleus of Comet Halley

The first interplanetary probe launched within ESA's programmes obtained, in March 1986, images of the nucleus of Comet Halley, as well as in-situ probings of the environment outside and inside the coma. Among the six spacecraft sent to encounter Halley's Comet, GIOTTO made the closest approach (ca. 600 km) and obtained the most detailed data about the environment of a comet.

In particular, the GIOTTO probe penetrated beyond the cometary ionopause into the field-free cylindrical cavity of ca. 4000 km radius that existed around the nucleus. The pictures taken by the camera aboard GIOTTO revealed an elongated cometary nucleus - roughly an ellipsoidal of $16 \times 8 \times 8$ km^3 - which had a relatively low density (ca. 0.3 g cm^{-3}) and an albedo of only four percent. The surface appears hilly with a coarse roughness at least down to a scale of 0.5 km, which may be an inherent feature and possibly be characteristic of 'building blocks'.

If GIOTTO can be reawakened during its approach to Earth in 1990, and if the health of spacecraft and experiments is judged to be sufficient, the probe may be retargetted towards Comet Grigg-Skjellerup for an encounter in July 1992 - thus leading to a GIOTTO Extended Mission (GEM).

2.4 Hipparcos

The astrometry satellite Hipparcos, which was launched in August 1989, did not reach its foreseen geostationary orbit, because its apogee boost motor could not be ignited. It is possible nevertheless to carry out observations from the current orbit having an apogee and perigee of 35 000 km and ca. 500 km, respectively. By use of three ground stations (instead of the one foreseen for the nominal geo-stationary orbit) it will be possible to operate with an efficiency of about 75 percent (versus 90 percent nominally).

Although the regular crossings of the van Allen radiation belts lead to some degradation of the solar cells, Hipparcos may have a lifetime long enough to achieve the originally specified astrometric performance: the measurement of positions, parallaxes and annual proper motions with an accuracy of 2 milli-arc-seconds for 120 000 stars brighter than B = 13 mag. Hipparcos would thus establish a uniform and dense whole-sky reference frame and a very large number of stellar distances of unprecedented accuracy. The infrastructure for the data analysis for Hipparcos has been provided by two consortia, who will independently establish an output catalogue. This permits cross-checking the results. The final Hipparcos catalogue will be published jointly by the two consortia - if necessary, after discrepancies between their results have been eliminated.

A third consortium will derive two-colour photometry for all stars down to B = 10 to 11 mag. This will result in the so-called Tycho catalogue of ca. 500 000 stars.

A detailed description of the satellite, the input catalogue and the data reduction procedures has been published by Perryman et al. (1989 a, b and c).

3. PROJECTS UNDER DEVELOPMENT: OUTLOOK TO FUTURE MISSIONS

3.1 The Hubble Space Telescope

The Hubble Space Telescope (HST), currently scheduled for a launch into low-Earth orbit in April 1990, is a collaborative mission of ESA and NASA. ESA's contribution - nominally a fifteen-percent share - is the provision of a contingent of staff members in the Space Telescope Science Institute (STScI) in Baltimore, one of the major four focal-plane instruments, namely the Faint Object Camera (FOC) and the solar arrays of HST.

In addition, ESA is operating a European Coordinating Facility for Space Telescope (ST-ECF) jointly with the European Southern Observatory (ESO). This facility, located at ESO's headquarters in Garching near Munich, is providing (i) support for proposers from ESA member nations, (ii) maintains data reduction software and (iii) will make available to European scientists an archive of all the HST observations, as soon as they are in the public domain.

3.2 ISO, the Infrared Space Observatory

The cryogenically cooled Infrared Space Observatory (ISO), to be launched into an eccentric 24-hour orbit in 1993, will provide access to the wavelength range 2.5 to 200 μm for a duration of (nominally) 18 months.

The 60-cm telescope will have a complement of focal-plane instruments comprising a camera (Isocam), a photo- and polarimeter (Isophot) and two spectrometers for 2.5 - 45 μm and 45 - 180 μm (SWS and LWS, respectively). Various spectral resolutions between 200 and 2×10^4 will be available.

ISO is a European project - carried out by ESA in collaboration with the institutes providing the focal plane instruments. Nevertheless, the contribution of a second ground station by another space agency - for better orbital data coverage - is currently under discussion. Observing time will be made available to the wide scientific community upon application and after evaluation of the proposals by an observing programmes committee. As usual, part of the observing time can also be allocated to astronomers from countries not belonging to ESA.

An outlook towards the astrophysical problems to be investigated by ISO is given in Glasse, Kessler and Gonzales Riestra (1990).

3.3 STSP, the Solar Terrestrial Science Programme

Five spacecraft will be flown in the Solar Terrestrial Science Programme (STSP): the Solar and Heliospheric Observatory (SOHO), which will be launched into a halo orbit around the metastable Lagrange point L1 on the Earth-Sun line, and the four-spacecraft Cluster, which will fly in an (adjustable) formation in an eccentric near-polar Earth-orbit. Launch of both missions, SOHO and Cluster, is foreseen for 1995.

SOHO will "look into the inside of the Sun": it will permit the study of structure and dynamics of the solar interior by the methods of helioseismology. The required measurements will be made by instruments aboard SOHO that will observe the subtle oscillations of the solar photosphere in the velocity, radiance and irradiance domains.

SOHO will also investigate the physical structure and the dynamics of the outer atmosphere of the Sun as well as its expansion into the solar wind. For this purpose the spacecraft will carry a complement of extreme and far ultraviolet (EUV and FUV) spectrometers, a soft-X-ray and EUV imaging telescope and two coronagraphs with spectroscopic capabilities (one observing the coronal emission and scattering in the ultraviolet, the other in the visible), as well as a set of in-situ solar-wind and energetic-particle analysers.

Cluster, in turn, will investigate the plasma structures and processes in the terrestrial magnetosphere by in-situ measurements that - owing to the simultaneous four-point measurements made by the four spacecraft - will yield information in three dimensions. (Note that four spacecraft in a non-planar arrangement establish the origin and the three axes of a coordinate system. Measurements at the three points along these axes and at the origin permit the derivation of differential operators of the temperatures, densities and velocities of the particles and also of the fields encountered in magnetospheric structures.)

Among the important physical processes to be studied by Cluster are the acceleration of particles in plasmas and plasma heating by field-line reconnection. As these processes are thought to occur both in the Earth's magnetosphere as well as in the solar atmosphere (albeit with somewhat different plasma parameters), Cluster and SOHO have in fact very similar goals: the investigation of physical structures and their dynamics in various contexts within the solar-terrestrial environment. SOHO carries out these studies by remote sensing as well as by in-situ measurements, Cluster will make in-situ measurements exclusively, and while SOHO observations might usually aim at investigating both the overall context and the details of the plasma structures, Cluster will focus on the detailed study of such structures. STSP may therefore be considered to be a 'research system'. Indeed, it is to be hoped that the analysis of STSP observations will go beyond traditional single-technique studies of a given phenomenon in one regime, and will lead to a deeper and perhaps more general understanding of cosmic plasmas (cf. Haerendel, Huber and Haskell 1985).

STSP, which is a collaborative programme (in the ratio 2:1) between ESA and NASA, will be supplemented by additional spacecraft enhancing Cluster that will be supplied by Interkosmos (IKI). Further, STSP will be coordinated with the numerous solar-terrestrial physics missions that will fly in 1995 and beyond. The coordination is stimulated by the Interagency Consultative Group (IACG), an informal association of the space organisations of Europe, Japan, the U.S. and the USSR, i.e., respectively, ESA, ISAS (The Institute of Space and Astronautical Science), NASA and IKI. The research goals and payloads of the two STSP missions SOHO and Cluster, are described by Domingo and Poland (1989) and Goldstein and Schmidt (1988).

3.4 The High-throughput X-ray Spectroscopy Mission, XMM

XMM (the X-ray Multi-Mirror mission) is to be launched into a highly eccentric 24-hour orbit in 1998. The spacecraft will be the largest ever flown by ESA. Its length is governed by the 7.5-m focal length that is necessary to afford the glancing incidence reflections on the nested mirror shells. XMM will carry three mirror modules, each consisting of 58 shells of concentric light-weight Wolter type-I mirrors, with a total effective collection area of ca. 5×10^3 cm^2 in the energy range 0.1 to about 10 keV.

The instrumentation associated with these telescope modules will consist of three CCD cameras covering a field of view of 30 x 30 (arc min)^2 with a spatial resolution of 30 arc sec, and having an energy discrimination that results in a spectral resolving power of 10 and 50 at 1 and 7 keV, respectively. Reflection-grating modules, which will be fitted to two of the telescope modules will feed separate detectors and afford a spectral resolving power of 200 to 400 in the 0.4-to-5 nm range.

An optical monitor will yield simultaneous spectro-photometric images (in the waveband 200 - 600 nm and with limiting magnitude 24.5) of the field observed by the co-aligned X-ray telescopes.

The XMM mission and its scientific aims have been described in Peacock (1988).

4. SPACE SCIENCE - HORIZON 2000

The long-term scientific programme 'Space Science - Horizon 2000' (or, for short, Horizon 2000) has been elaborated in response to the aims of ESA as stipulated in its Convention. Horizon 2000 contains four major missions, the so-called 'Cornerstones', which together cover the broad interests of the space scientists served by the Scientific Programme Directorate of ESA. (It may be worthwhile mentioning here that other scientific space activities, namely Earth observations and microgravity fall within the purview of another ESA Directorate - that of Earth Observations and Microgravity.)

The first two of the four Cornerstones - STSP and XMM (with launches in 1995 and 1998) - are described above. The other two Cornerstones - Rosetta, a Sample Return from a Comet Nucleus and FIRST, a Submillimetre Spectroscopy Mission - are currently scheduled for launch ca. 2002 and 2006, whereby the sequence of the two missions is to be determined in 1992.

In addition to the large Cornerstones with pre-determined scientific aims, Horizon 2000 contains also four medium-sized missions, whose subjects are not pre-determined. These projects thus provide flexible elements within Horizon 2000. They are selected after three-year study cycles. A cycle starts with a call for mission proposals that is addressed to the wide scientific community, is followed by Assessment Studies of the best submitted ideas, continues with Phase A studies after a further selection and ends with the selection of a new project after the results of the Phase A studies have been presented to the wide community.

The first medium-size mission selected within the framework of Horizon 2000 is the Titan probe 'Huygens'. It will be launched in 1996 on NASA's Saturn orbiter 'Cassini', will arrive at Saturn in the year 2002, and will then make a descent through Titan's atmosphere.

The cycle for the second medium-size project has just been initiated and has resulted in 22 proposals for new missions - with topics covering not only solar-system exploration and astronomy, but also basic physics. Six of these mission proposals will be subjected to Assessment Studies and one year later, about four of these will be recommended for Phase A studies.

A tangible result of the design of Horizon 2000 - and indeed a condition for its implementation - was an increase of the budget of ESA's Scientific Programme by five percent per year leading to an eventual annual budget ca. 50 percent larger than that available before Horizon 2000 was started.

4.1 What about the Moon?
The impact of Space Station on the scientific long-term programme Horizon 2000 is still somewhat limited. A possible 10-to-15 percent share of the payload on ESA's Polar Platform may become available for experiments in solar system exploration and astronomy, but other benefits from this space infrastructure have not been identified <u>and</u> ascertained at the time of writing.

Astronomers and solar system explorers may, in the meantime, dream about another, natural space station, namely the Moon. There are many advantages for astronomical observations that are performed from the Earth's satellite, rather than from the ground or even from Earth orbit - provided, of course, pollution of the lunar environment by excessive commercial exploitation of lunar resources can be avoided (Bonnet 1989). The Moon has no atmosphere, no magnetic field to speak of and is outside the Earth's magnetosphere most of the time; thus the measurement of electromagnetic radiation over the entire spectrum and of charged particles is possible. Radio observations from the Moon's far side are an obvious additional example of the advantages that a lunar site would give to astronomy. The stability of the lunar surface also provides an excellent base for interferometers, and - in view of the continuous degradation of terrestrial observatory sites (Garstang 1989) - it may in the long term become necessary to locate large telescopes outside the Earth's atmosphere anyway. The best and most convenient location may indeed be the Moon.

It may be useful to recall that the first scientific experiments carried out on the Moon led to the first estimate of the average baryon density in the Universe (Geiss and Reeves 1972). If science were to return to the Moon - by the use of robotic systems or even with a small number of human beings - we might look forward to new avenues being opened through international collaboration in the exploration of the Universe.

REFERENCES

Appourchaux T 1990, ed., ESA Report to the 28th COSPAR Meeting, ESA SP (Noordwijk: ESA Publications Division, ESTEC), in preparation

Bonnet RM 1989, La Recherche 20, 1566

Garstang RH 1989, Ann. Rev. Astron. Astrophys. 27, 19

Geiss J and Reeves H 1972, Astron. Astrophys. 18, 126

Glasse ACH, Kessler MF and Gonzalez Riestra R 1989, eds, Infrared Spectroscopy in Astronomy, ESA SP-290 (Noordwijk: ESA Publications Division, ESTEC)

Haerendel G, Huber MCE, Haskell G 1985, eds, Future Missions in Solar, Heliospheric and Space Plasma Physics, Garmisch-Partenkirchen Workshop, ESA SP-235 (Noordwijk: ESA Publications Division, ESTEC)

van der Klis M, Jansen F, van Paradijs J, Lewin WHG, Sztajno M and Trümper J 1987, Astrophys. J. (Lett.) 313, L19

Peacock A 1988, ed., The High-throughput X-ray Spectroscopy Mission, Mission Science Report, ESA SP-1097 (Noordwijk: ESA Publications Division, ESTEC)

Perryman MAC and Hassan H 1989a, eds, The Hipparcos Mission, Vol. I, The Hipparcos Satellite, ESA SP-1111 (Noordwijk: ESA Publications Division, ESTEC)

Perryman MAC and Turon C 1989b, eds, The Hipparcos Mission, Vol. II, The Input Catalogue, ESA SP-1111 (Noordwijk: ESA Publications Division, ESTEC)

Perryman MAC, Lindegren L, Murray CA, Hog E and Kovalevsky J 1989c, eds, The Hipparcos Mission, Vol. III, The Data Reduction, ESA SP-1111 (Noordwijk: ESA Publications Division, ESTEC)

Reimers D, Clavel J, Groote D, Engels D, Hagen HJ, Toussaint F, Naylor T, Wamsteker W and Hopp U 1989, Astron. Astrophys. 218, 71

Taylor BG 1989, ed., Report on ESA's Scientific Satellites, ESA SP-1110 (Noordwijk: ESA Publications Division, ESTEC)

White NE 1990, ed., Two Topics in X-ray Astronomy, I. X-ray Binaries, II. AGN and the X-ray Background, SP-296 (Noordwijk: ESA Publications Division, ESTEC), in press.

General lectures

Table 1. ESRO/ESA Scientific Spacecraft

	Launch Date	End of Operational Life	Mission
Launched			
ESRO II	17 May 1968	9 May 1971	Cosmic rays, solar X-rays (S)
ESRO-IA	3 October 1968	26 June 1970	Auroral and polar-cap phenomena, ionosphere (S)
HEOS-1	5 December 1968	28 October 1975	Interplanetary medium, bow shock (S)
ESRO-IB	1 October 1969	23 November 1969	As ESRO-IA (S)
HEOS-2	31 January 1972	2 August 1974	Polar magnetosphere, interplanetary medium (S)
TD-1	12 March 1972	4 May 1974	Astronomy (UV, X- and gamma-ray) (A)
ESRO-IV	26 November 1972	15 April 1974	Neutral atmosphere, ionosphere, auroral particles (S)
Cos-B	9 August 1975	25 April 1982	Gamma-ray astronomy (A)
Geos-1	20 April 1977	23 June 1978	Dynamics of the magnetosphere (S)
ISEE-2	22 October 1977	26 September 1987	Sun/Earth relations and magnetosphere (S)
IUE	26 January 1978		Ultraviolet astronomy (A)
Geos-2	24 July 1978	October 1985	Magnetospheric fields, waves and particles (S)
Exosat	26 May 1983	9 April 1986	X-ray astronomy (A)
FSLP	28 November 1983	8 December 1983	Multi-disciplinary
Giotto	2 July 1985		Comet Halley encounter (S)
Hipparcos	9 August 1989		Astrometry (A)
Planned Launches			
Space Telescope	April 1990		UV/optical astronomy (A)
Ulysses	October 1990		Out-of-ecliptic, solar polar (S)
Eureca	May 1991		Multi-disciplinary
ISO	May 1993		Infrared astronomy (A)
Soho	March 1995		Solar-terrestrial science (S)
Cluster	December 1995		
Huygens/Cassini	1996		Titan probe (S)
XMM	March 1998		X-ray astronomy (A)

Prospects of the Development of Ground-Based and Space Astronomy in the USSR

A. A. BOYARCHUK

Astronomical Council of the Academy of Sciences of the USSR, Moscow

1 INTRODUCTION

The development of astronomy in the Soviet Union began long ago. In Armenia, archeologists found the remains of a very ancient observatory whose foundation goes back to the year 2000 B.C. (Parsamian, 1988). World-wide known in the Middle Ages was Ulughbek Observatory in the town of Samarkand. The history of modern astronomy starts at the beginning of last century when observatories in Tartu, Kharkov and Pulkovo were established. A lot of effective work in astrometry was carried out at these observatories. Although some excellent work was perforemed, the geographic location and the lack of large telescopes impeded effective astronomical observations.

After World War II, a rapid development of astronomical observatories took place in the areas of Crimea and the Caucassus. Both in the number of clear nights and in seeing, these observatories excel much of those located in the Northern part of Russia.

As a result of the development, Soviet astronomy possesses nowadays several large telescopes.

Table 1 shows summarized data about these observatories. Besides that, there are about ten 1 meter telescopes used mainly for polarimetric and photometric observations.

Despite the fact that the conditions for astronomical observations in the areas of Crimea and the Caucassus are considered the best for the European part of this country (50% of nights are clear and 10% of nights have seeing around 1 arcsec)

they are still far from the best in the world and do not allow the operation of larger telescopes most effectively.

Table 1
SOVIETIC TELESCOPES WITH MAIN MIRROR LARGER THAN 1.5 m

Telescope	Observatory	Altitude, location	Types of observation
6 m BTA	Special Astrophysical Observatory	2040 m North Caucasus	Direct photography, photometry, spectroscopy (1-200 Å/mm), measurements of magnetic fields
2.6 m ZTSh	Crimean Astrophysical Observatory	570 m Crimea	Spectroscopy (1.5-200 Å/mm), photometry, polarimetry, measurements of magnetic fields
2.6 ZTA	Bucaran Astrophysical Observatory	1450 m Armenia	Direct photography, photometry, spectroscopy (30-200 Å/mm)
2.0 m Zeiss	Shemakha Astrophysical Observatory	1435 m Azerbaijan	Spectroscopy (1.5-200 Å/mm), photometry, direct photography
1.5 m AZT-22	Tartu Astronomical Observatory	63 m Estonia	Spectroscopy (10-200 Å/mm), photometry

During the last twenty years, the astronomical climate of Central Asia has been investigated thoroughly. There, on the mountains of Maydanak (Uzbekistan) and

Sanglok (Tadjikistan) several telescopes of up to 1 meter have been in operation for several years. The analyses performed have proved the astronomical conditions in the region to be similar to those in the Canary Islands in the number of clear nights, seeing and even dust days.

Central Asia was very long ago believed to have good conditions to carry out astronomical observations. But poor communications and retarded economy as well as the absence of investigations on the astronomical climate in that region, did not allow Central Asia to be regarded as fit for mounting large telescopes there.

But now the situation has changed drastically. The region of Middle Asia is considered at present as the most favourable site in the Soviet Union for building large optical telescopes. Several projects for the emplacement of optical and radio telescopes are currently in various stages of realization.

2 THE UNITED ASTRONOMICAL OBSERVATORY

A project aimed at building a United Astronomical Observatory of socialist countries is currently being elaborated. Table 2 gives a list of telescopes proposed for emplacement at this Observatory. The use of new technology, i.e. thin adaptive mirrors which will compensate for the wavefront distortions in the atmosphere (see Table 2), is planned. Several possible versions of its configuration are discussed now for the largest telescope whose main mirror diameter is 25 meters. The main part of these versions are (a) one mirror of 25 meters in diameter (Steshenko, 1981), and (b) several telescopes with diameters of order 12 meters, whose images are optically concentrated in one focal plane. A final decision concerning the configuration of this giant telescope has not yet been taken.

We hope that the Observatory will be built before 2005. Mounting of telescopes belonging to individual institutions will be considered favourably.

Of the medium size telescopes which are planned within the near future, I should mention a 2 meter Zeis-Jena telescope of the Main Astronomical Observatory of the Ukranian Academy of Sciences. It is planned that this telescope, located in the Northern Caucassus, will come into operation with its first observations by next year.

Table 2
TELESCOPES PLANNED FOR THE UNITED OBSERVATORY
OF SOCIALIST COUNTRIES

Telescope	Types of observations
Two 2-m telescopes	Interferometry, photometry, spectroscopy, direct photography
4-m telescope (monolitic)	Spectroscopy, direct photography, photometry
25-m telescope (segmentic) Version I: One dish	Spectroscopy, photometry, measuremens of magnetic fields
Version II: Four 12-m telescopes	Interferometry, spectroscopy, photometry, measurements of magnetic fields
2-m solar telescope	Spectroscopy, measurements of magnetic fields, monochromatic images
Automatic astrometric telescope	Precise measurements of the position of astronomical objects

3 RADIOTELESCOPE PT - 70

A radiotelescope with a parabolic antenna of 70 meters is currently being built on the Sufa Plateau of Uzbekistan. The telescope is intended for measuring fluxes of radiation from various radio sources as well as in the system of space interferometer in combination with the 10 meter radiotelescope mounted on board the high apogee satellite. The telescope is planned to begin observations in the early 90s in the range from 1 cm to 1 m. After setting the system of mirror adaption, it will be possible to carry out observations effectively in the mm range.

4 RADIOINTERFEROMETER WITH A VERY LONG BASE

Achievements in some radiointerferometric observations with intercontinental bases and the long and fruitful operation of the radio-complex VLA, have proved the necessity of creating an interferometer with the base of several tens of thousands kms. Such an interferometer will solve many astrophysical problems concerning high spatial resolution and astrometric problems connected with the determination of coordinates of point sources. A radiointerferometer which would cover the entire country is being built at present in the USSR. Information relevant to this is given in Table 3. Observational data will be transmitted to the Institute of Applied Astronomy of Leningrad with their further processing there. Beginning of this work is planned for the early 90s.

Table 3
RADIOINTERFEROMETER WITH A VERY LONG BASE

Number of radiotelescope	6
Maximum distance	15000 km
Diameter of main mirror	32 m
Diameter of secondary mirror	4 m
Pointing	±10 arc sec
Wavelengths (cm)	1.35; 2.6; 3.5; 6.0; 13; 21.
Efficiency	0.65
Accuracy of coordinate system	0.1 marc sec.

5 SPACE ASTRONOMY

There is a rather expanded programme of astronomical space observations.

During the past years, several spacecrafts specially designed for astrophysical observations have been launched into space. Among them are "Astron" with the 80 cm UV telescope and X-ray spectrometers, which operated during 1983-1989 (Kovtunenko, 1984), Modulus "Kvant" on board the station "Mir" for X-ray and UV observations, both having been operated successfully since 1987 (Sunyaev, 1987).

At present a new spacecraft of the series "Spektr" has been worked out. It will orbit a set of scientific equipment having the weight of up to 1500 kg. At least three types of experiments, i.e. UV, X-ray-γ and the radiotelescopes, are planned

5.1 UV Telescope

The main part of this telescope is T-170 one with the main mirror diameter of 1.7 m, built on basis of the Ritchey-Cretien scheme. Table 4 gives its main parameters. The telescope is equipped with a field camera and a set of spectrometers allowing to investigate spectra of different resolution. The main spectrum range to be investigated is 3500-1100 Å. However, about one third of the mirrors' surface will consist of a special coating for observations in the 500-1500 Å spectral range. The light from these parts of the mirror, after passing the slit of the spectrometer, will enter a special spectrometer built on basis of Rowland circle. Besides, four small telescopes (D~20 cm) with multi-layer coatings will be mounted on this spacecraft. They will allow the registration of radiation coming from the stellar objects, in four bands with $\Delta\lambda \approx 40$ Å in the spectral range 100-500 Å. Thus, we shall be able to investigate astronomical objects in a wide spectral range - from the extreme UV range to a visible region.

The main problem to investigate in the process of the UV experiment is non-stationary objects in the Universe. The main objects for observation are non-stationary stars, including neutron stars and other relativistic objects, galaxies' nuclei, quasars, etc. The station must be put into a high apogee orbit ($\sim$ 2000000 kms apogee, 40000 kms perigee). In this case, the spacecraft will always be beyond the radiation zone and during 90within the zone of direct visibility with communication stations located in the territory of the Soviet Union. The experiment is made on basis of international cooperation. Besides the Soviet Union, its active participants are Italy, GDR and Czechoslovakia.

5.2 Radioastronomy

To obtain high resolution images is one of the main problems of modern astrophysics. Methods of interferometry using ground-based telescopes will help obtain the angular resolution of order 0.001 arcsec. But such a resolution is, however, not sufficient to solve such important problems as, for example, the structure of quasars. So the launch of radiotelescopes into space becomes very essential.

In 1975, on board the station "Salyut-6" in the Earth's orbit, there was a radiotelescope with an antenna of 10 meters in diameter. The main purpose of that experiment was to improve the design of the antenna and to measure the radiation of space objects.

At present projects are being developed in the USSR to use the ground-based 70 meter telescope and the space telescope located in a high apogee orbit. The main parameters of the telescopes are being worked out and are given in Table 5.

Table 4

SPACE ULTRAVIOLET TELESCOPE T-170

Optical scheme	Ritchey-Chrétien
Main mirror	
diameter	170cm
thickness	10cm
coating	Al+MgF$_2$ 2/3 of surface
	Re 1/3 of surface
D/F	1/3
Secondary mirror	
diameter	50 cm
thickness	5 cm
coating	the same as for main mirror
Spectrometer A	
spectral range	3500-1900 Å
resolution	0.10-0.15, 3 Å
Spectrometer B	
spectral range	1900-1100 Å
resolution	0.08-0.10; 3 Å
Spectrometer C	
spectral range	1200-500 Å
resolution	6 Å
Spectrometer D	
spectral range	5000-1100 Å
resolution	30 Å
The field camera U	
field	2 x 2 arcmin
pixel size	0.12 arcsec
spectral resolution	about ten filters with $\Delta\lambda = 100\text{-}300$ Å in the spectral range 1200-3200 Å
The field camera V	
field	5 x 5 arcmin
pixel size	0.3 arcsec
spectral range	5000-9000 Å

Table 5
THE PROJECT "SPEKTR R"

	Radioastron-cm	Radioastron-mm
Frequency range (GHz)	0.327; 1.67; 4.83; 22.2	22.2; 45; 87; 115; 230
Radiotelescope diameter (m)	10	10
Maximum spacecraft distance from the Earth's center (km)	$8\ 10^4/7\ 10^5$	$8\ 10^4$
Interferometer beam width at a limiting frequency (μarcsec)	30/3*	3
Sensitivity (rms value for 24 hours interaction at 22.26 Hz) (10^{-29} W/m² Hz)	2	2
Mass (tons)	1.5	2.0
Power (W)	500	1000
Pointing accuracy (arc sec)	60	5
Revolution period (1 day)	1/27*	1

5.3 High energy astronomy
Three large projects are under development in the Soviet Union.

The space X-ray observatory "Granat" was designed in order to obtain the images of different astronomical objects in the energetic range 3-2000 Kev. The spectroscopy with low energetic resolution $\Delta E/E \sim 0.2$ and time resolution 100 mc will be possible also. Scientists from France, Denmark and Bulgaria have taken part in this project. The observatory "Granat" should commence operations by the end of November 1989. Table 6 contains some details about this observatory.

Table 6
THE SPACE X-RAY OBSERVATORY "GRANAT"

Coding aperture telescope ART-P	
Energetic range	3-100 kev
Angular resolution	5 arcmin
Field size	1,9 x 1.8°
Coding aperture telescope "Sigma"	
Energetic range	0.005-2 Mev
Angular resolution	16 arcmin
Field size	4.2 x 4.25°
Telescope ART-S	
Energetic range	1-130 kev
Field size	2 x 2°
Time resolution	100 mc
Detectors for gamma-bursts	
Total weight of scientific equipment	2300 kg

The astronomical modul "Gamma" has scientific equipment which offer the possibility of observing in X-ray (2-25 Kev) and gamma (0.1-5 Mev) ranges simultaneously. Table 7 gives some details of modul "Gamma". French and Polish scientists participated in this project. Launching is scheduled for Spring 1990.

Table 7
THE SPACE MODUL "GAMMA"

Telescope "Gamma-1"
    Energetic range   50-5000 Mev
    Angular resolution   1.°5
    Field size   20°
    Effective surface   200 cm^2

Telescope "Disk"
    Energetic range   0.1-5 Mev

Telescope "Pulsar X-2"
    Energetic range   2-25 kev
    Field size   10 x 10°
    Sensitivity for bursts   0,3 Crab/c
    Accuracy of location   10 arcmin

The X-ray observatory "Spektr-Rentgen-Gamma" is one of the most important projects in high energy astronomy developed in the USSR during recent years. Its two incidence telescopes have high sensitivity and rather good angular resolution. They offer the possibility of investigating many scientific problems in different fields of astronomy. The Bregg-spectrometer has very high energetic resolution. It enables the study of spectral lines in the X-ray range. Scientific payload contains the equipment of X-ray and gamma-ray burst all sky monitor. Table 8 shows the most important characteristics of scientific payload. Scientists from the United Kingdom, USA, Denmark, West Germany, Italy, Finland, Switzerland, Bulgaria, East Germany, Poland and the European Space Agency have taken part in the project. Launching is planned for mid 1990.

5.4 Project "Coronas"
This project is designed for the investigation of solar activity. In the process of realization of the project it is planned that several satellites capable of carrying 100 kg of scientific devices on board, will be launched.

The satellites will be equipped with the following devices: (1) Devices for registration of γ-quants of 0.1-100 Mev energies; of neutrons with up to 30 Mev; of fluxes and nuclei with energies of up to 1 Mev/nuclon; of electrons with energies of up to $>$ 0.05 Mev/nuclon. (2) An X-ray telescope, a coronograph, a heliometer, a polarimeter and a photometer, investigating a region of radiation of 0.1-30 Kev. (3) Photometers in the optical range for investigation of the Sun's pulsations with various periods. (4) A radiospectrometer

The penetrators are designed for a detailed study of the soil's properties and for obtaining images of the planet's surface in the landing region.

5.6 A permanent space astronomical observatory

As a perspective for 2010-2020 a question is being considered now as to the possibility of creating a permanent astronomical observatory in space by means of the carrier "Energy" which would be equipped with large telescopes.

The main problems are studying the fundamental laws of space, time, matter, search for life in the Universe, search and analysis of new forms of matter through a wide international cooperation.

The observatory consists of three stations (1) A 400 m radiotelescope (1mm range - 30 cm difractionally, 20 mkm - 1mm bolometrically, orientation 0.1 arcsec). (2) A 10 m optical telescope (0.09 - 20 mkm range, orientation 0.1 arcsec) (3) γ and X-ray telescopes with an aperture of 100 m2 (orientation 0.1 arcsec).

Each station weights approximately 100 tons. The main working orbit - antisolar liberation point (1.5 mln km from the Earth).

REFERENCES

Parsamian, E.S.: 1988, "Istoriko-Astronomicheskie Issledovania" p.136, Moscow, Nauka.
Steshenko, N.V.: 1981, "Izvestia Crim. Astroph. Obs" 64, 161.
Kovtunenko, V.M.: 1984, "Vestnik Akademii Nauk SSSR", 8, 12.
Sunyaev, R.A.: 1987, "Pisma v Astronom. Zhurnal", 13, 1027.

Table 9
THE PROJECT "MARS-94"

SPACECRAFT

Orbit	300 km
Mass	200-250 kg
Mapping of surface	(TV, IR, Radop)
Chemical Composition of surface	(Gamma-and-IR-Spectrometer)
Distribution of t°	(IR-radiometer)

Chemical composition and structure of the atmosphere.
Magnetic field

BALOONS

Duration	up to 15 days
Altitude	up to 5 km
Mass	6 kg

Detailed mapping, chemical composition

SMALL STATION

Number	3-6
Mass	2-3 kg

Monitoring of t°, pressure, humidity

PENETRATORS

Number	3-4
Mass	3-4 kg

Investigation of soil and magnetic field

After entering Mars's orbit, the station delivers onto the planet aerostats, 2 or 3 penetrators, and 3 to 6 small stations, while it itself remains in the orbit as Mars's satellite. The satellite will carry out detailed cartography of the planet's surface, investigating its permafrost, monitoring the processes of erosion on the surface, studying its atmosphere, the magnetic field and its interconnection with the solar wind.

The aerostats are designed mainly for investigation of Mars's atmosphere. Temperature, pressure, humidity and optical density of the atmosphere of the planet will be measured. A large-scale survey of the planet's surface and distance measurements of the composition of rocks will also be made.

The small stations placed in various parts of the surface of Mars will be monitoring temperature, pressure, humidity and optical thickness of the atmosphere, making seismic measurements as well as direct measurements of the soil's composition.

for registration of the corona in 0.30-30 MHz range.

The satellite's orbit is quasisynchronous. It will allow repeating periods (~20 days duration), when the satellite does not enter the Earth's shadow. It will enable continuous long-period observations of the Sun.

5.5 Project "Mars"

Unlike in previous years, when the Soviet space programme primarily envisaged investigations of Venus, in the coming decades, Mars will be the main object of space investigations. This programme began in 1988 with the launch of the 2 space-stations "Phobos". Its scientific equipment was created by scientists of several countries.

The main aim of investigating Mars is a detailed study of its surface, its geological and climatic history, search for any traces of life in the present epoch or evidence that life might have existed on the planet in the past.

At present a rather vast programme is being established, which will include detailed cartography of Mars and the investigation of its surface and atmosphere including automatic sampling of soil as well as the launching of cosmonauts to the planet approximately in 2015.

The programme "Mars-94" is established in more detail. Its main parameters are given in Table 9.

The following articles were also presented at the conference:

R. Sancisi: HI Distribution in Galaxies
G. Efstathiou: The Epoch of galaxy Formation
R. D. Joseph: IR Observations of Interacting galaxies
G. Tenorio-Tagle: Large Scale Propagating Star Formation and the Physics of Self-Regulation
S. Lizano: Present Status of the Theory of Star Formation
A. Boksenberg: Galactic and Intergalactic Gas at High Redshifts from Observations of QSO Absorption Lines
P. Salinari: The Columbus Project

Table 8
X-RAY OBSERVATORY SPEKTR-RENTGEN-GAMMA

Thin foil high throughput telescope SODART

Energetic range	0.2-20 keV
Field	1° x 1°
Sensitivity (0.5-10 kev)	$3 \cdot 10^{-14}$ erg cm^{-2} c^{-1} for the exposure time 2000 sec.
Angular resolution	2 arcmin
Energetic resolution	$\Delta E/E \sim 500$
Effective surface 1 kev	4000 cm^2
6 kev	3000 cm^2

Grazing incidence telescope JET-X

Energetic range	0.2-10 keV
Field	40 x 40 arcmin
Sensitivity (0.5-10 kev)	$3 \cdot 10^{-15}$ erg cm^{-2} c^{-1} for one day exposure time
Angular resolution	10-30 arcsec
Energetic resolution	$\Delta E/E \sim 350$
Effective surface 1kev	350 cm^2

Coding aperture telescope MAPT

Energetic range	4-100 keV
Field	5° x 5°
Sensitivity (4-10 kev)	$3 \cdot 10^{-12}$ erg cm^{-2} c^{-1} for one day exposure time
Angular resolution	7 arcmin
Effective surface	650 cm^2

Coding aperture telescope ART-SP

Energetic range	2.5-35 keV
Field	7.3° x 7.3°
Sensitivity (3-10 kev)	$5 \cdot 10^{-12}$ erg cm^{-2} c^{-1} for one day exposure time
Angular resolution	7 arcmin

All sky X-ray monitor

Energetic range	3-12 keV
Angular resolution	1° x 1°

Gamma-ray burst monitor

Energetic range	3 keV - 100 MeV

Scientific payload	2.5 tons
Orbit	20000 km